Experimental Methods in the Physical Sciences

VOLUME 34
CUMULATIVE SUBJECT INDEX
VOLUMES 1–32

EXPERIMENTAL METHODS IN THE PHYSICAL SCIENCES

Robert Celotta and Thomas Lucatorto, *Editors in Chief*

Founding Editors

L. MARTON
C. MARTON

Volume 34

Cumulative Subject Index

Volumes 1–32

ACADEMIC PRESS

San Diego London Boston New York Sydney Tokyo Toronto

This book is printed on acid-free paper. ♾

Academic Press
a division of Harcourt Brace & Company
525 B Street, Suite 1900, San Diego, CA 92101-4495, USA
http://www.apnet.com

Academic Press
24–28 Oval Road, London NW1 7DX, UK
http://www.hbuk.co.uk/ap/

International Standard Serial Number: 1079-4042/99

International Standard Book Number: 0-12-475981-5

PRINTED IN THE UNITED STATES OF AMERICA
98 99 00 01 02 IP 9 8 7 6 5 4 3 2 1

CONTENTS

PREFACE

The first volume of the series *Methods of Experimental Physics*, the forerunner of the current series, was published in 1959 under the thoughtful guidance of Series Editor Dr. Ladislaus Marton. Dr. Marton envisioned the series as providing a helpful guide for the experimental physicist. The original statement of editorial policy was very clear on the purpose of the enterprise:

> The experimental physicist—the person for whom this book should be written—is normally a specialist, working in a relatively narrow domain. He is presumed to know his own specialty very well. In the course of his work, however, he is often confronted with the necessity of using methods borrowed from neighboring fields, with which he is less familiar. In such cases he may have to make a literature search or ask advice from a specialist in the neighboring field. The existence of a concise presentation of the most important methods used in experimental physics would considerably simplify the task.
>
> The methods used may be purely experimental, but very often they are partly theoretical or computational. They may be qualitative, or may require the ultimate in accuracy. In an unfamiliar field, the experimenter will need a guide to ways and means best adapted to his own investigation.
>
> Thus the book should be a concise, well-illustrated presentation of the most important methods, or general principles, needed by the experimenter, complete with basic references for further reading. Indications of the limitations of both applicability and accuracy is an important part of the presentation. Information about the interpretation of experiments, about the evaluation of errors, and about the

validity of approximations should also be given. The book should not be merely a description of laboratory techniques, nor should it be a catalog of instruments.

These volumes should be written so as to be of value to all research workers who use physical methods. Finally, the volumes should be organized in such a way that they will provide essential tools for graduate students in physics.

In the years that followed, Dr. Marton, assisted by his wife Claire Marton, fulfilled this promise with great success. He worked tirelessly to be sure that every important field was represented, that the most knowledgeable editors and authors wrote for the series, and that the spirit of the series, as stated above, was always maintained. Sometimes a topic would require multiple volumes to provide complete coverage. In other cases, the field was moving so rapidly that it was necessary to update the topic in a subsequent volume. The Martons delighted at traveling to key scientific meetings to seek out the most knowledgeable researchers and to convince them of the importance to the field that they edit a volume or contribute a chapter.

In 1983, Robert Celotta and Judah Levine became the Editors-in-Chief for the series. Since 1993, Robert Celotta and Thomas Lucatorto have served in that capacity. Changes have come to the scientific enterprise and the series as well. In fact, science, as practiced today, has made the series all the more valuable. It was always important to be able to understand and utilize the accumulated knowledge of the neighboring discipline, but today, when many of the best scientific opportunities lie at the boundaries between disciplines, the value of guides, such as this series provides, is greatly enhanced. We also live in a time when more emphasis is placed on multidisciplinary teams working together to solve important, but very difficult, problems. Guides to the knowledge base and language of other disciplines are more useful than ever. Perhaps most important is the trend toward scientific agility, that is, the requirement that scientists, over their professional life, be capable of learning the methods of, and making contributions to, fields different from their original training. Science is making rapid gains, and consequently the areas seen to be most fertile, or perhaps most worthy of support, are apt to change more rapidly than in the past. A book designed to provide ready access to the important experimental methods of a new field is exactly what is needed to help researchers make the transition.

The increasingly rapid pace of technological development over the past forty years has greatly reduced the migration time for measurement technology to leave the research laboratory and arrive in the industrial laboratory or the factory floor where it is extended and put to practical use. Descriptions of the methods used in industry, along the lines set out for this series, can be very useful to a wide variety of readers. Of course, it is just what is needed by the new scientist or engineer beginning a career in an industrial setting. It is also important to the university professor eager to be able to contribute to the solution of a major technological problem. Finally, the technology devised to make a commercial microelectronic device may be the best, or even the only, way to produce the physical sample needed to answer the most academic of questions for a basic researcher. Because of the trends mentioned above, in 1996, the series title was changed to Experimental Methods in the Physical Sciences. It was our intention to recognize that the information presented was not just of interest to physicists but instead to a broad range of scientists and technologists.

As we head into the next millennium, we can reflect on the series created by Dr. Marton to provide the *essential tools* for graduate students to learn the important methods of related disciplines. Now extended to a include broader range of readers and technologies, it is our hope and desire that it continue to provide the technical community with an important tool for achieving scientific and technological progress.

Robert J. Celotta and Thomas Lucatorto, Washington, D.C., 1998

SUBJECT INDEX

A

α particle energy spectrum, **24A,** 115–116
α, *n* neutron sources, **5B,** 560
  γ-radiation from, **5B,** 561–562
α-cyano-f-hydroxycinnamic acid, **30,** 418
α-particles, emission, **28,** 190–191
  americium-241, **28,** 214–215
A. C. bridge methods, **6B,** 43–45, 59–60
A/D (analog-to-digital) converters, **2A,** 416–419
ABAP, *see* Accelerator-based atomic physics
Abbe comparator, **1,** 54
  refractometer, **1,** 345
Abbe prism, **25,** 66
Abbe refractometer, **25,** 187
Abbe sine condition, **25,** 24; **4A,** 87
Abbe theory of the microscope, **26,** 86
Abel inversion, **7A,** 122; **9A,** 426; **18,** 394, 742
  process, **7B,** 315–319
Aberrated wavefront, **25,** 44
Aberration
  definition, **31,** 151
  nonsymmetrical systems, **31,** 163–166
  primary, **31,** 154
  reduction, **31,** 172–173
  representation, **31,** 172–173
  secondary, **31,** 158
  spherical, **31,** 170
  third order, **31,** 361–364
  zone plate, **31,** 296–298
Aberration correction, **31,** 30
  circular mirror, **31,** 170–172. *See also* Coma correction; Astigmatism correction
Aberration(s), **1,** 331ff, 337; **12C,** 280; **26,** 83
  correction by holography, **26,** 197
  in holography, **26,** 172
  polynomial, **25,** 40, 44
  in spectrometers, **12A,** 477–479
  spherical, **5A,** 353, 354
  in the electron microscope, image, **16B,** 199–205
Ablation
  definition of, **30,** 15
  mechanisms, **30,** 78–106, 485–486
  mid-infrared, **30,** 102–104
  near-infrared, **30,** 102
  off-axis, in PLD, **30,** 39
  threshold, **30,** 186, 428, 584
  ultraviolet, **30,** 90–91, 94–95, 103
  visible, **30,** 102
Abrasion, effect of on desorption, **30,** 151, 156
  absolute charged particle, measurement by counting, **5B,** 494–495
  atomic, deflection due to magnetic field, **5B,** 69
  colliding, **5B,** 686–689
  high momentum, using counters as detectors, **5B,** 752–755
  monitoring based on induction electrodes, **5B,** 495–498
  neutron, from reactors, **5B,** 572–580
  parallel, of charged particles, **5B,** 485ff
  polarized, **5B,** 216–223
  separators, **5B,** 747–760
    degrader type, **5B,** 747–748
    electromagnetic, **5B,** 748–750
    nuclear interaction, **5B,** 751–752
    radiofrequency, **5B,** 750–751
  size, finite, correction for, **5B,** 419
  transport systems, **5B,** 691–760
    examples, **5B,** 752–760
Absolute counting, *see* Counting

Absolute determination of neutral-neutral total cross sections, **7A,** 289–292
Absolute flux, laser plasma, **31,** 85–86
Absolute sensitivity, in impurity determinations, **6A,** 30
Absolute temperature scale, **1,** 32
Absolute value circuit, **2A,** 395
Absolute zero, **1,** 325
Absorber-shift method, **5B,** 336–337
Absorbing films for radiant energy, **4A,** 443–445
Absorptance, **1,** 334
Absorption
 absorbers, high Z, **21,** 7
 air, **10,** 91–93
 atmospheric, **12B,** 149
  gas, **12B,** 142–176
  of infrared, **3,** 89
 in biological tissues, **19,** 13
 bridge, **3,** 392–393
 cadmium, of thermal neutrons, **5A,** 518
 in a cavity containing a paramagnetic sample, **5B,** 88
 cells
  cavity as, **3,** 27–28
  gas, **5B,** 24–28
  Stark effect, **3,** 662–663
 by chemically reacting systems, **19,** 183, 188
 classical, nonadiabatic sound propagation, **3,** 685–686
 coefficients, **3,** 8–9; **3A,** 111, 114; **6B,** 250, 263; **10,** 94, 342, 344, 345; **13A,** 22, 110; **13B,** 46; **19,** 18, 20–21, 24–26; **21,** 5, 6, 8; **22,** 176, 179; **23C,** 92; *see also specific gases*
  atomic, **3,** 303
  gamma ray, of NaI, **5A,** 621–623
  for ideal gas, **3,** 244
  linear, **3,** 303
  of a medium, **1,** 233, 374, 394
  photons, **7A,** 144
  due to single relaxation process, **19,** 183, 188
  sonic waves, **3,** 681
  of various materials, **19,** 10–13
  in waveguide, **3,** 25
  X-ray mass, **1,** 416
 counter, total, for photons and electrons, **5A,** 189–191
 cross section, **5A,** 504
  photons, **7A,** 143, 145
 definition of, **19,** 20, 107
 depth, **30,** 335–339, 583–590
 edge(s), **22,** 175
  X-ray, **1,** 416
  X-ray critical, **5A,** 675–676
 enhanced, **13B,** 336
 factor, neutron, **3,** 355
 free electrons in cyclotron motion, **10,** 189, 190, 230–238, 259–263
 in a gas-filled waveguide, **5B,** 87–88
 general expression for, **12B,** 145
 index, **13B,** 46, 135
 infrared spectroscopy, **22,** 26
 intensity per molecule, **3,** 242
 ionospheric, **12B,** 125–126
 introductory theory, **19,** 20–27
 laser studies, **13B,** 325
 laws, **3,** 8–10
 longslit attenuation, **21,** 109
 magnetic resonance, **5B,** 106–111
 measurement of gamma ray energy, **5A,** 671–676
 microwave in dielectrics, **3,** 650
 molecular, of sound waves in gas, **3,** 689–693
 relaxation, **3,** 691
 resonance, trans-dichlorethylene, **3,** 501
 self-absorption, **21,** 6, 8, 66
  with β-ray source, **5B,** 290
 soft-tissue, **30,** 461–462
 of sound, **1,** 213–215, 223–224, 232–234
  in gases, **3,** 693–700
  in liquids, **3,** 700–708
 spectra, **13A,** 22, 248
  picosecond continuum monitoring of, **15B,** 203
 spectroscopy, sources, **3,** 161–162
 spectrum
  of scintillator, **5A,** 126
  total, large crystal, **5A,** 629
 state, **14,** 196
 superconductors, **10,** 433–441
 theory, **13A,** 87
 transitions, weak, red shift of, **3,** 234
 two-photon, **13B,** 32
 of vacuum ultraviolet, **1,** 404–405
 water, **10,** 354–355
 water vapor, **10,** 91–93, 321
 of X-rays, **1,** 415–417; **3,** 302–306
Absorption coefficient, calculation, **31,** 258

Absorption coefficient, *see* Compton total linear absorption coefficient.
Absorption cross sections, **8,** 239, 245
Absorption of radiation
atomic attenuation coefficient, **18,** 406
for chemical composition, **18,** 612–634
for density measurement, **18,** 405–408
linear attenuation coefficient, **18,** 406
mass attenuation coefficient, **18,** 406
Absorption of ultrasonic waves
definition, **16C,** 70
in various polymers, **16C,** 73
Absorption-emission event, in Raman spectroscopy, **16A,** 77
Absorptivity, **1,** 302
spectral, **18,** 465, 472–475
Abundance, **22,** 408
Ac measurements, **2B,** 63–70
average or peak, **2B,** 67–70
common-mode rejection ratio in, **2B,** 74
common-mode voltage in, **2B,** 74
current comparators in, **2B,** 70
digital value in, **2B,** 76
"guarding" in, **2B,** 74
harmonics errors in, **2B,** 69–70
Johnson noise in, **2B,** 72
noise factors in, **2B,** 71–76
operational rectifiers in, **2B,** 68–69
oscilloscopes in, **2B,** 71
phase-lock and signal averaging in, **2B,** 75–76
rectifier instruments in, **2B,** 68
shielding and guarding in, **2B,** 73–74
square-law characteristics of, **2B,** 67
thermoelements and transfer standards in, **2B,** 63–65
Ac method, for energy distribution curves, **11,** 94
Ac motors, for servomechanisms, **2B,** 439–440
Ac Stark effect, phase shifts generated by, **29C,** 302
Ac Stark shifts, **15B,** 242–243
Ac voltmeter, **2B,** 64
AC, *see* Alternating current
AC-cut quartz, **19,** 52
Accelerating lens system, of parallel-plate analyzer, **17,** 440, 442, 450
Accelerating-decelerating system, **17,** 183
Acceleration, **1,** 78–81
angular, **1,** 86
centrifugal, **1,** 63
of gravity, **1,** 35–36, 66
Accelerator, characteristics, **31,** 2
Accelerator(s), **5B,** 623ff; **17,** 4–21; **23A,** 107; *see also* specific accelerators
beam-foil experiments, **17,** 530, 533–535
for charge state fraction measurements, **17,** 83–84
d.c., **5B,** 580ff
fixed-field alternating-gradient, **5B,** 678–686
radial sector, **5B,** 678–680
spiral sector, **5B,** 680–681
linear, **5B,** 634–645
electron, **5B,** 642–645
proton, **5B,** 635–642
mass spectroscopy, sensitivity for isotopic selection, **29A,** 301
particle trajectory, **1,** 521pulsed, as neutron source, **5B,** 392–393
systems, channeling experiments, **21,** 297
van de Graaff, increased pulsed neutron yield, **5A,** 548
Accelerator-based atomic physics
accelerators, *see also* Electrodynamic accelerator; Electrostatic accelerator
hybrid facilities, **29A,** 318–319
ion sources, **29A,** 301
tuning, **29A,** 300–301
types, **29A,** 306–307
applications, **29A,** 299–306
collision energy for shell electron ionization, **29A,** 301–302
electron capture, **29A,** 305–306
facilities, **29A,** 299–300
ion penetration in solids, **29A,** 302–303
Accelerator-decelerator stripping technique, **17,** 281–282
Accelerograph, strong motion, **24B,** 11
Accelerometer, **1,** 92, 218, 219
cantilever beam, **1,** 80
force-balanced, **24B,** 7, 65–73
manometer, **1,** 78, 79
mechanical, **1,** 81
optical, **1,** 80–81
pendulum, **1,** 78, 79
piezoelectric, **1,** 78, 80
Acceptance
ion beam, **29A,** 73–74
simulation, **28,** 451–452
Acceptors per unit volume, **5A,** 270

Accommodation
coefficient, **1,** 160, 298, 299; **14,** 7, 52; **7B,** 269
thermal, **1,** 161
Accuracy, **1,** 2–3; **2,** 2; **3,** 710
instrument, **1,** 448
of mass spectroscope mass values, **5B,** 7–9
of measurement, **5B,** 808
ACE transformation method, **28,** 1, 273–274
Acetic acid, **19,** 193, 202
Acetone, **14,** 499; **19,** 189, 190, 195
in leak detection, **14,** 512
Acetylene, rotational Raman spectrum, **3,** 142, 145
Achromatic devices 1 and 2, **26,** 133
Achromatic lens(es), **1,** 336; **29C,** 351
Achromatic objectives, **25,** 164
Achromatic wave plates, **26,** 144
Acid deposition, **28,** 376–378
Acid pickle, **14,** 443, 444
Acid-string saw, **11,** 10
Acoustic anemometer, **18,** 315–318
Acoustic branch, **16A,** 101–102
modification by crystal field, **16A,** 117–118
Acoustic calorimetry, **19, 192**
Acoustic cavitation, *see* Cavitation
Acoustic cyclotron resonance, **11,** 430
Acoustic delay line, **1,** 112
Acoustic detector, **1,** 225
Acoustic devices, **1,** 230–234
acoustooptic, characterization of, **19,** 477, 480, 486–492
Acoustic dispersion, **19,** 41, 183, 187
Acoustic Doppler velocimeter, **18,** 317
Acoustic drying, **19,** 342–345
Acoustic emissions, **30,** 191
focal mechanism, **24A,** 161–164
location, **24A,** 157
moment tensor, **24A,** 167
power spectrum, **24A,** 167
transducer, **24A,** 155, 164–165
waveform, **24A,** 161
Acoustic flowmeter, **18,** 337–340
Acoustic gain, **19,** 192, 217
Acoustic holography, *see* Holography
Acoustic impedance, **1,** 201, 216, 232; *see* Impedance
Acoustic interferometer, **1,** 222, 223, 279, 280
Acoustic oscillation, **19,** 192
Acoustic samples, **1,** 228
Acoustic streaming, *see* Streaming
method, attenuation measurement, **19,** 132–133
Acoustic waves, *see* Seismic waves
speed, **7B,** 179
wavelength and damping length of, **9A,** 15
Acoustical densitometer, **16C,** 100
Acoustical holography, **26,** 197
Acoustics, **1,** 200ff
aerial, **1,** 201
communications, **1,** 231
room, **1,** 229
Acousto-optic effect, **26,** 150
Acousto-optical measurements, **24A,** 23
Acoustooptic devices, lithium niobate, **29C,** 390
Acoustooptic figures of merit, **19,** 463–464
Acoustooptic isolation, **29C,** 359
Acoustooptic measurement of sound wave
amplitude and intensity, **19,** 476, 478–481, 484–492
attenuation, **19,** 474–481, 487
beam diffraction, **19,** 475–477, 484–492
beam propagation, **19,** 475–477, 484–492
phase, **19,** 484–492
power, **19,** 476–492
presence of, **19,** 458, 471–473, 482
propagation, **19,** 475–477, 484–492
reflection, **19,** 475–477, 484–487, 490
standing wave ratio, **19,** 480, 484–487, 490
velocity, **19,** 471–473, 487
Acoustooptic modulator (AOM)
AOM/EOM for dye lasers, **29C,** 129
free spectral range using, **29C,** 109
frequency stabilization with, **29C,** 300–301, 359
principles, **29C,** 123–124
tuning with, **29C,** 113–114
Acoustooptic phenomena, **19,** 302, 326, 455–493
Acoustooptic techniques, **19,** 467–492
classification of, **19,** 467–469
dc detection of light, **19,** 467–469
diffraction orders, observation of, **19,** 470–471
Fabry-Perot interferometry, **19,** 478–481
heterodyne, **19,** 469, 481–492
knife edge, **19,** 484–487, 490–492
photoelectric detection, **19,** 474–478
Schlieren studies, **19,** 472–474
Acrylics, **14,** 414
Actinides, **23C,** 217

Action function, **4A,** 109–110
Activation cross-section, neutron, **5B,** 407–411
Activation energy, **14,** 232; **19,** 221
  determination, **21,** 443, 448, 477
  for diffusion, **16C,** 324
  for methyl side-group motion, **16A,** 311
  for molecular motion, various polymers, **16A,** 305
Activation energy, for crystallization, **16B,** 349
  from melt, **16B,** 363
  from solution, **16B,** 364
Activation methods of impurity detection, **6A,** 31
Activation of cathodes, **4A,** 4, 19–20, 26
Activators, introduction by neutron bombardment, **6B,** 297
Active filters, **2A,** 28, 333–336
  defined, **2A,** 333
  design considerations for, **2A,** 333–336
Active modelocking, **29C,** 175–180
Active plasma, **3A,** 332
Activity, **19,** 201
  coefficient, **19,** 201–207
  optical, **3,** 679–680
Acylation of polystyrene, **16B,** 255
ADA, *see* Ammonium dihydrogen arsenate crystals
Adamantane, **23A,** 507
ADC, *see* Analog-digital converters
ADC, *see* Analog-to-digital converter
Added variable plot, **28,** 270
Additive coloration, **6B,** 326–328
Additive detectors for atoms, **4A,** 409–410
Additive-error model, **28,** 246
Additives, **16A,** 9
Adenosine, **19,** 203
Adenosine–5'-phosphate, **19,** 202
Adhesion, role of electrostatic charge, **16C,** 433
Adhesives
  microstain tensile test for, **16C,** 128
  molecular vibrational modes, **16A,** 167–168
    infrared studies, **16A,** 127
    small-angle neutron scattering studies, **16A,** 505–506
  shear testing of, **16C,** 128
  testing in joints, **16C,** 126–128
    effect of bond-line thickness, **16C,** 133
Adiabatic approximation, **23A,** 371
Adiabatic calorimeter, **6A,** 378
Adiabatic condition, **21,** 459, 464
Adiabatic demagnetization, **1,** 542
  in rotating field, *see* NMR, ultraslow motion techniques
  in the rotating frame, *see* ADRF
Adiabatic fast passage, in atomic beam, **4B,** 246
Adiabatic isolation, **1,** 306
Adiabatic passage, **3,** 389–390
Adiabatic potential energy surface (APES), **30,** 49–56
Adiabatic process, **1,** 307, 310, 322
Adiabatic screen, **1,** 268
Adiabatic shield, **1,** 264; **6A,** 380
Adiabatic slow passage, **5B,** 108–110
Adiabatic wall temperature, **18,** 458–459, 665
Adjacent reentry model of folding, **16B,** 229–230, 238, 342, 344
Adjustment distance, **25,** 165
Admittance, **1,** 461, 467; **2,** 21–22; **2A,** 6
  bridges and null devices, **6B,** 8–15
  comparator, **1,** 459
  driving-point, **2,** 25; **2A,** 9
  instruments for measuring, **6B,** 8–23
  transfer, **2,** 26; **2A,** 9
  voltmeter, **1,** 447
  wave, **1,** 202
ADP, *see* Ammonium dihydrogen phosphate
ADP, *see* Ammonium dihydrogenate phosphate
ADRF, **23C,** 247
Adsite symmetry, **22,** 57
Adsorbate diffusion, **22,** 363
Adsorbates, in desorption, **30,** 76
Adsorbates, metal surfaces
  scanning tunneling microscopy imaging theory, **27,** 14
Adsorbed atoms, *see* Channeling applications
Adsorbed monolayers, **23B,** 1
  experimental, **23B,** 6
  inelastic scattering, **23B,** 53
Adsorption, **14,** 25, 29, 196, 231; **22,** 3, 356, 359
  covalent, **22,** 4
  dipole layer, **22,** 3
  of gases on glass, **14,** 323
  heat of, **22,** 366, 456
  in hermetically sealed devices, **14,** 182
  ionic, **22,** 4
  isosteric heat of, **22,** 438, 440

isotherm(s), **14,** 238; **23B,** 5
molecular, relaxation, **3,** 429–430
molecular sieves, **14,** 139, 140
porous structure, **14,** 353
of pumping fluids, **14,** 167–169
rate, **22,** 457
ratio, **14,** 196
on reduced metal clusters, **22,** 103
states, **22,** 152, 154
statistical thermodynamics of, **22,** 457
van der Waals, **22,** 4
Adsorption, selective
amount, **4B,** 320
characteristics, **4B,** 321
materials, **4B,** 321–322
Advancement of chemical reaction, **19,** 184
Advantages and disadvantages of NMR diffusion technique, **21,** 439–441
Aerodynamic force
coefficient, **1,** 152
principle, **18,** 254–256
on vane anemometer, **18,** 254–258
on whirling arm anemometer, **18,** 256–259
Aerodynamic noise, study in wind tunnel, **18,** 779
Aerosol OT, in magnetic domain studies, **11,** 688–692
Aerostatic balance, **1,** 120
AES, *see* Auger electron spectroscopy
Affinities
electron, **3,** 552–558
of halogen atoms, **3,** 557
Affinity of chemical reaction, **19,** 184
Afocal systems, **25,** 132, 138
Afterglow, **1,** 171
nitrogen, **3,** 168
Afterglows, atomic collision studies, **7B,** 113–124
electron-ion recombination, **7B,** 113–118
mechanism, **7B,** 118–123
cavity methods for measurements, **7B,** 87–96
definition of, **7B,** 78–79
excitation methods for, **7B,** 82–85
flowing, chemical aspects of, **7B,** 133–139
mass spectrometry of, **7B,** 131–133
free space microwave propagation measurements, **7B,** 96–99
guided wave propagation measurements, **7B,** 99–100
mass spectrometric techniques for, **7B,** 111–113
microwave radiation measurements, **7B,** 100–102
observational methods for, **7B,** 85–113
particle continuity equations, **7B,** 79–82
reflected microwave signal measurements, **7B,** 93
stationary, mass spectrometry of, **7B,** 124–131
apparatus, **7B,** 124–127
data analysis methods, **7B,** 128–131
studies using, **7B,** 78–79
Afternoon "bulge," whistlers and, **9B,** 110
Ag on Au(111)
epitaxy, scanning tunneling microscopy of, **27,** 296
Ag on Si(111)
scanning tunneling microscopy of, **27,** 199
Ag-O-Cs, **12A,** 24
Agammatism, **8,** 184
AGC, *see* Automatic gain control
AgCl transmission, **10,** 95
Age determinations, **5B,** 298–302
$AgGaS_2$, as OPO material, **29C,** 257, 258
$AgGaSe_2$, as OPO material, **29C,** 257–259
Agglomeration, **6A,** 134–139; **19,** 340–342
forming techniques, **6A,** 137–139
processing variables, **6A,** 134–137
Aggregation number, **23B,** 493
Agitation energy, **5A,** 92, 93
Aharanov-Casher effect, **29B,** 341
Aharonov-Bohm effect, **23A,** 309
Aiming error, **22,** 414
Air, physical properties, **14,** 421
Air, rare gas composition, **14,** 419
Air dispersion, **3,** 172–173
Air exposure, **14,** 446
Air resistance, **8,** 1
Air showers, **12A,** 317, 356
distinction between proton and gamma-ray induction, **12A,** 356
Air standard, **3,** 172
Air stripping
in rotary oil sealed pumps, **14,** 124
Air-dielectric line, **8,** 115
Air-insulated pressure tank Van de Graaff accelerator, **17,** 7
Air-spaced interferometer, **29C,** 350
Airy disc, **25,** 73, 168; **26,** 67, 80
Airy disk, diameter, **29A,** 53

Airy formula, **5B,** 52
Airy function, **12A,** 532
Alcohol, **14,** 499
Aldebo, single scattering of rain, **12B,** 182–183
Alfvén Mach number, **9A,** 217
  critical, **9A,** 237
Alfvén velocity, of magnetosonic waves, **9A,** 226
Alfvén waves, microwave scattering from, **9A,** 51–52
AlGaAs laser
  coatings, **29C,** 95
  properties, **29C,** 80, 81
  tuning, **29C,** 89
AlGaInP laser, **29C,** 80
Algebraic reconstruction technique, **19,** 575
Aliasing, **29B,** 365
Alignment
  electric hyperfine structure, **5B,** 183–184
  of magnetic hyperfine structure, **5B,** 176–183
  nuclear, **5B,** 163, 174
Alkali beams, **4B,** 167–168
Alkali elements, **7A,** 91
Alkali halide(s), **30,** 72–75
  color centers in, **15B,** 1–2
Alkali metal-antimony photocathodes, spectral response curves, **6B,** 388
Alkali metals, **22,** 403
Allan variance, **12C,** 201
Allotropic effect, on stopping cross sections, **17,** 70
Allowed radiative transtion, **17,** 579–583
Alloy junction, **6B,** 87
Alloyed contacts, **6B,** 51
Alloyed junctions, **6A,** 64–66
Alloys, *see* specific materials
  applications of positrons to, **21,** 94
  polycrystalline, attenuation in, **19,** 267–290
  polymeric, phase morphology, **16C,** 287–299
Alpha cutoff frequency, **2A,** 279
Alpha particle, **8,** 273
Alpha particle(s)
  calibrated sources, **21,** 65
  counters, **5A,** 107
    proportional, **5B,** 286–288
  decay, emission, energy, **21,** 2
  detection, photographic film, **21,** 44–48
  detection
    proportional counters, **21,** 50, 60
    scintillators, **21,** 50
    silicon surface barrier, **21,** 53
  differential scattering cross section in gold, **5A,** 291
  image in emulsion, **5A,** 237
  radioactivity, **5A,** 313–314
  range, **21,** 3
  response
    of CsI(Tl), **5A,** 415, 416
    of NaI(Tl), **5A,** 412
    of semiconductor detector, **5A,** 274–276
  sample preparation, **21,** 60
  scattering, **5A,** 309–313
  sources
    neutron, **5B,** 556–562
    radioactive, **5B,** 567–572
Alpha-particle projectile, **17,** 402–403
Alphatron, **14,** 157
  gage, **1,** 168–169, 190
Alternate decay models, **13A,** 118
Alternating beams, **7A,** 294
Alternating conditional expectation algorithm, **28,** 273–274
Alternating copolymer, **16A,** 3
Alternating current, **1,** 459–488
  instruments, indicating, **2,** 577–580
    instantaneous recording, **2,** 579–580
    peak, rms and average, **2,** 577–578
  measurements
    boxcar averaging, **29A,** 446
    digital signal processing, **29A,** 449–450
    gated integration, **29A,** 446
    lock-in amplifier, **29A,** 448–449
    null-detection methods, **2,** 580–583
    transient digitizer, **29A,** 447–448
Alternating current and voltage, regulation of, **2A,** 242–243
Alternating Josephson current, **11,** 235–238
Alumina, **30,** 76–77, 100, 232, 598–600; *see also* Sapphire; Ceramics
Aluminum (Al), **22,** 155; **30,** 113–114, 404
  boiling point, **14,** 350
  electrical conductivity, **14,** 363
  hydrogen solubility, **14,** 354
  ion laser, **15A,** 164–166
  melting point, **14,** 347, 350
  as mirror coating, **29C,** 347
  oxide, **14,** 363
  physical properties, **14,** 362–364
  thermal conductivity, **14,** 356
  unsuitable for ultrahigh vacuum, **14,** 480

vapor pressure, **14,** 347, 350
Young's modulus, **14,** 360
Aluminum nitride (AlN), **30,** 245–258, 284
Aluminum oxide, reflectance spectra, **31,** 222
Aluminum, reflectance spectra, **31,** 206–207
Aluminum-aluminum oxide junctions, **11,** 173
Alundum, *see* Sapphire; Ceramics
Am Sulf, **14,** 481
Am Tel, **14,** 481
Am Zirc, **14,** 480
Amarada temperature recorder, **24B,** 192, 194
Ambient gas, **30,** 245–247, 383–387
Ambiguity noise, **18,** 104, 126, 136, 160, 162, 166
effect on LDV signal processing, **18,** 162
Ambipolar diffusion, **7A,** 378
Ambipolar diffusion coefficient, in afterglow studies, **7B,** 80
Ambipolar mobility, **6B,** 93–94
Americium–241, α-particle emissions, **28,** 214–215
Amici lens, **25,** 129
Amici prism, **25,** 66, 187, 210
Amidal, **5A,** 217, 218
Aminobenzoic acids, **19,** 225–229
Aminolysis of polyesters, **16B,** 275–276
Ammeter
absolute, **11,** 284
analog-to-digital converters
integrating converter, **29A,** 431–432
sampling converter, **29A,** 431–432
voltage-to-frequency converter, **29A,** 431–432
components, **29A,** 421
current integrator, **29A,** 430–431
feedback integrator, **29A,** 430–431
noise model, **29A,** 431
principle, **29A,** 423, 429
reset switch, **29A,** 430
schematic, **29A,** 423, 429
shunt integrator, **29A,** 430
electrodynamic, **2B,** 64
ideal properties, **29A,** 421
impedance, **1,** 447
logarithmic ammeter
output voltage, **29A,** 428–429
principle, **29A,** 422
schematic, **29A,** 423, 428
transdiode amplifier, **29A,** 428
maximum input voltage, **29A,** 434
performance-affecting factors
air ionization, **29A,** 433
dielectric absorption, **29A,** 434
ground loops, **29A,** 434
humidity, **29A,** 433
light, **29A,** 433
room temperature, **29A,** 432–433
triboelectric charge generation, **29A,** 433
resistive shunt ammeter
noise model, **29A,** 425
performance-limiting effects, **29A,** 423–424
principle, **29A,** 422
schematic, **29A,** 423–424
transresistance ammeter
noise model, **29A,** 427
operational amplifier, **29A,** 425, 427–428
principle, **29A,** 422
risetime, **29A,** 426
schematic, **29A,** 423, 426
voltage burden, **29A,** 425
Ammonia, **23A,** 493
3–3 microwave line, **3,** 9
absorption coefficient of, **12B,** 171–172
atmospheric concentration, **12B,** 170
inversion, **3A,** 78
spectrum, **3,** 15–16
laser, **10,** 49–50; **15B,** 76–77
maser, **3,** 37
molecule, **10,** 318–319
shape, **3,** 15
physical properties, **14,** 421
Raman spectra, **3,** 150
spectrum, **10,** 318–319
Ammonium beam maser, **15A,** 6–7
Ammonium dihydrogen arsenate crystals, in SFG and SHG, **15B,** 177–180
Ammonium dihydrogen phosphate (ADP), **12A,** 378
Ammonium dihydrogen phosphate crystals, in SFG and SFG, **15B,** 177–180
Amorphic diamond, **30,** 189
Amorphous alloys, **23B,** 255
metal-metal, **23B,** 260
metal-metalloid, **23B,** 255
Amorphous magnets, **23C,** 192
Amorphous material,
in cellulose, selective hydrolysis, **16B,** 269–273
in melt-crystallized polymers, removal by solvent-etching, **16B,** 239–241

in poly(ethylene terephthalate), selective hydrolysis, **16B,** 274–275
selective halogenation, **16B,** 245–251
selective oxidation, **16B,** 261–267
Amorphous peak, in X-ray diffraction pattern, **16B,** 119
Amotic force microscopy (AFM), **30,** 291, 292, 299–306
Ampere, **1,** 33
Ampere's law, **1,** 431
Amplidyne, **2B,** 442
Amplification factor
pentodes, **2,** 64
tetrodes, **2,** 64
triodes, **2,** 58
Amplified spontaneous emission, **29C,** 154–162
for short-pulse dye lasers, **15A,** 271–276
excitation intensity and, **15A,** 275
and gain of amplifying medium, **15A,** 274
geometry for, **15A,** 273
optical feedback and, **15A,** 274
for single-pass short-pulse amplifier, **15A,** 283
Amplifier(s), **2A,** 245–339; Amplifiers, *see specific types*; *see also* Transistor(s), Transistor amplifier(s)
admittance and transconductance values for, **2A,** 271
alpha cutoff frequency in, **2A,** 279
amplitude modulation of, **2A,** 247–248
audio, **2,** 221–233
automatic gain control in, **2A,** 337, 404
basic stages of, **2A,** 253–287
beta cutoff frequency in, **2A,** 278
boxcar averaging, **29A,** 446
broadband, **2A,** 305–309
chopper, **2A,** 300–305
clipping, **2,** 249–250
collector-to-emitter voltage in, **2A,** 260
comparator, **2,** 445–446
computer-aided design in, **2A,** 285–286
current vs. resistance in, **2A,** 276
in dc measurement, **2B,** 50
defined, **2A,** 245
detector, phase sensitive, **3,** 438–439, 482–484
difference, **2,** 254–255; **2A,** 313; **2B,** 449
differential, **1,** 462; **2A,** 292–298
direct current, **1,** 450, 453
digital signal processing, **29A,** 449–450
direct-coupled, **2,** 233–240; **2A,** 287–305, *see also* Direct-coupled amplifier
distributed, **2,** 232–233
drift, **1,** 450
electrometer, **2,** 253–254;**2A,** 298–300
electron tube, **1,** 459
electronic current, **14,** 82
for electron spin resonance, **3,** 480–481
equivalent circuits and, **2A,** 267–272
flat-staggered single tuned, **2,** 244
frequency modulation of, **2A,** 248–249
frequency range of, **2A,** 247
gain
negative feedback effect, **2,** 503–505
in pulse amplitude measurement, **2B,** 81
stabilization, **2,** 503–506
gated integration, **29A,** 446
ionization chamber, **5A,** 105–107
input resistance vs. load resistance in, **2A,** 277
integrated, **2B,** 485
integrated-circuit, **2A,** 287–288
klystron, **2,** 720–721
two-cavity, **2,** 718–720
linear, **2A,** 245
lock-in, **2A,** 331; **2B,** 75
amplifier, **29A,** 448–449
logarithmic, **1,** 219
low-frequency response in, **2A,** 287
low-noise, **2A,** 325–333
magnetic, **2,** 785–794; *see also* Magnetic amplifiers
feedback behavior, **2,** 792–793
multicore circuits, **2,** 789–792
single-core circuits, **2,** 787–789
transient behavior, **2,** 792–793
manufacturer's power specification for, **2A,** 258–259
maximum power from, **2A,** 275
modulated, **2,** 238–240
noise, **2,** 808–818
figure, **2A,** 250–252, 326
specification for, **2A,** 250–252
temperature and, **2A,** 252
noise in, **2B,** 482–486
noninverting, **2A,** 313
nonlinear, **2A,** 245
with nonlinear transfer characteristics, **2,** 249–252
nonoverloading pulse, **2,** 252–253

operational, **2,** 639–641; **2A,** 309–315; **2B,** 446–452
for transresistance ammeter, **29A,** 425, 427–428
parametric, **2,** 817–818; **2A,** 332
perfectly balanced, **2A,** 294
performance
factors in, **2A,** 275–276
parameters for, **2A,** 310
phase modulation of, **2A,** 249–250
post detector, **10,** 225–226
power, **2,** 255–263; **2A,** 320–324
gain in, **2A,** 275
gain vs. load resistance in, **2A,** 276
neutralization, **2,** 261–263
tuned, **2,** 256–261
untuned, **2,** 255–256
pulse, **2,** 221–233
clipping time, **2,** 814
noise, **2,** 813–815
rise time, **2,** 813
signal specification for, **2A,** 247–250
signal-to-noise ratio in, **2A,** 77
single-stage, **2A,** 253–256
single-stage connection to multistage circuit, **2A,** 282–284
selection for noise minimization, **29A,** 443–444
semiconductor thin-film majority-carrier, **2,** 170–171
in servomechanisms, **2B,** 442
special purpose, **2,** 249–255
specifications for, **2A,** 246–253, 310
stage
gain accuracy and stability in, **2A,** 279
ideal, **2A,** 272–273
limitations in, **2A,** 272–281
switching bridge, **2A,** 323
transconductance amplifier
noise minimization, **29A,** 443–444
schematic, **29A,** 445
transdiode, for logarithmic ammeter, **29A,** 428
transfer functions for, **2A,** 254
transformer vs. capacitor coupling in, **2A,** 282–283
transient digitizer, **29A,** 447–448
transistor, *see* Transistor amplifier(s)
high frequency, **2,** 322–323
low frequency, **2,** 316–319
parameter values for, **2A,** 270
video, **2,** 319–321
triode, **2,** 213
common-cathode, **2A,** 262–263
tuned, **2,** 240–249; **2A,** 315–320; **2B,** 482
in feedback circuit, **2,** 247–249
in forward transmission circuits, **2,** 240–247
ultrasonic, **2,** 127
vacuum tube, **2,** 211–265
design notes, **2,** 218–221
elementary circuits, **2,** 213–218
specifications, **2,** 212–213
varactor bridge input, **2A,** 299
voltage amplification factor in, **2A,** 276
wideband, **2B,** 482
Amplitude gratings, **31,** 397
Amplitude(s), *see* acoustooptic measurement of sound wave, amplitude and intensity
channels, **2B,** 81
coherent, **5A,** 567
control
by grid-leak bias, **2,** 510
by nonlinear feedback element, **2,** 509
division, **1,** 349
factor, in de Haas-Van Alphen effect, **11,** 37–39
modulation, **2A,** 247–248
devices for, **2A,** 337–338
modelocking, **29C,** 176–179
response of seismic unit, **1,** 92
spectroscopy, **13B,** 73
stability, oscillator, **2,** 272, 273
stabilization of feedback oscillators, **2,** 508–509
AMS, *see* Accelerator mass spectroscopy
AMT, *see* Audiomagnetotellurics
Analog computer systems, components of, **2B,** 153–160
Analog computer(s), **2B,** 152–165
applications of, **2B,** 152
computing elements in, **2B,** 154
function generator in, **2B,** 156, 162
operational amplifier and, **2A,** 309
operational amplifier in, **2B,** 155
problem variables in, **2B,** 153
quarter-square multiplier in, **2B,** 159
Analog indicators, in dc measurements, **2B,** 49
Analog methods in Fourier analysis, **6A,** 236–237
Analog techniques, in simulation and signal processing, **2B,** 160–165

Analog-digital converters
  digital data acquisition in, **2B,** 121–122
  fast, **2B,** 120–121
  vs. multidiscriminator systems, **2B,** 119–120
  multi-user system in, **2B,** 129
  for pulse spectrometry, **2B,** 128–129
Analog-to-digital converters, **2A,** 416–419; **21,** 55, 56, 99–101, 110, 123–124
  intergating converter, **29A,** 431–432
  in pulse spectrometry, **2B,** 82
  sampling converter, **29A,** 431–432
  voltage-to-frequency converter, **29A,** 431–432
Analysis of spectra, **4B,** 90–96
  complex spectra, **4B,** 92
  Edlén formula, **4B,** 94
  effect of degree of ionization, **4B,** 91–92
  use of computer, **4B,** 94
Analyzer(s)
  for linearly polarized light, **12A,** 363
  magnetic, reaction product, **5B,** 431–440
  multichannel
    for high counting rates, **5A,** 557–559
    for low counting rates, **5A,** 555–557
  octave band, **1,** 227
  pulse
    1024 channel with magnetic core storage, **5A,** 561–562
    2000 channel, **5A,** 562–564
    five-hundred channel, **5A,** 559–560
    height, **1,** 474
    magnetic tape, **5A,** 564–566
  time-of-flight using digital storage techniques, **5A,** 559–566
Anamorphic magnification, **12A,** 464, 469
Ancillary components, beamline design, **31,** 35–36
AND/OR functions, **2A,** 409–410
Anderson-Darling test, **28,** 202
Andrade creep, **16C,** 6, 15
Anechoic chamber, **18,** 778
Anelastic strain recovery, **24B,** 404
Anelasticity, **6A,** 303–307
  definition, **6A,** 303
  dynamic methods, **6A,** 304–307
  elastic after effect method, **6A,** 304
  flexual oscillations method, **6A,** 305
  formula for measurement, **6A,** 303
  piezoelectric excitation, **6A,** 305–306
  quasistatic methods, **6A,** 304
  relaxation of stress at constant strain method, **6A,** 304
  torsional pendulum method, **6A,** 304–305
  ultrasonic method, **6A,** 306–307
Anemometer, hotwire, **1,** 157, 217
Aneroid gauge, **14,** 48–49, *see also* Manometer
Anger camera, **23A,** 210
Angle, **1,** 56–58
  of attack, **1,** 150, 152
  Brewster's, **1,** 378
  criterion, maximum, for outcoming particle identification, **5B,** 828–831
  critical, **1,** 329
  determined by three atoms, standard deviation, **3,** 301
  effective solid, **5A,** 346
  of ejection of delta rays, **5A,** 55, 56
  of emission of bremsstrahlung, **5A,** 60
  between two gamma rays, **5A,** 679
  of incidence, **1,** 329
  magnetic deviation, **5A,** 299
  mean absolute scattering, **5A,** 394, 396, 398, 400–401
  mean scattered, charged particle in emulsion, **5B,** 37–38
  measurements, **26,** 41
  phase, **1,** 473–474
  polarization, **5B,** 225
    time rate of change, **5B,** 225ff
  reflection, **1,** 329, 330
  refraction, **1,** 329, 330, 338
  scattering, **5A,** 73–76, 289
  screening for Fermi-Thomas potential, **5A,** 392
  solid, of charged particle beams, **5B,** 506–508
  spatial, mean of absolute value, **5A,** 390
  valve, **14,** 296
Angle-resolved energy distribution curve, **22,** 171
Angle-resolved photoemission, **22,** 133, 135, 159, 160, 161
  spectra, **22,** 170
Angle-resolved photon-stimulated desorption, **22,** 485, 491, 493
Angle-resolved spectrometers, **22,** 140
Angled physical contact connectors, *see* APC connectors
Angstrom, **4B,** 50

Angular correlation (of annihilation radiation), **21,** 79, 95, 96, 106–122
 angular correlations, 2-D, **21,** 79, 109–115
 data analysis, 2-D, **21,** 115–121
 deconvolution, 2-D, **21,** 119
 electronics
  1-D, **21,** 107, 111
  2-D, **21,** 110
 longslit attenuation, **21,** 109
 longslit, 1-D, **21,** 79, 107
 peakcounting, **21,** 86, 119
 position sensitive detectors, **21,** 98, 109–115
 position sources, **21,** 81, 121–122
 resolution, **21,** 94–97, 107, 108, 115
 scatter of annihilation radiation, **21,** 109
 scintillators, **21,** 97, 98, 107, 109–111
 stabilization, **21,** 115
 stopping profile of positron, **21,** 104, 109
 umklapp, 2-D, **21,** 116, 117
Angular correlations, **13A,** 134
Angular dispersion, **8,** 32; **25,** 200
Angular distribution, **22,** 333, 473
 in ablation, **30,** 481–482, 485–488
 ambient-gas effects, **30,** 488–491
 apparatus, **7A,** 29–35
 of atomic radiation, **7A,** 45, 59, 84, 106, 456
 of electrons from heavy particle ionizing collisions, **7A,** 449–454
 in ion-atom collisions, **7A,** 442–454
 measurements, **22,** 272, 279
 of neutrals, **22,** 497
 of photoelectrons, **22,** 164
 soft-size effects, **30,** 487–488
 of sputtered atoms, **21,** 262
Angular distribution studies, **8,** 261
Angular divergence, **8,** 167
 circular aperture, **29A,** 83
 dependence on perveance, **29A,** 85–86
 slit aperture, **29A,** 83
Angular energy distribution curves, **11,** 100–101; *see also* Energy distribution curves
Angular field of view, **25,** 131
Angular fluid motion, in superfluid, **11,** 469–470
Angular magnification, **25,** 133
Angular momenta, *see* Orbital angular momenta.
Angular momentum
 of e-m fields, **13A,** 40
 of molecules, **13A,** 106
Angular resolution
 in neutral-neutral collision experiments, **7A,** 279–292, 354–356
 of telescope, **26,** 81
Angular resolution, **8,** 156
Angular separation, **8,** 43
Angular separation, VLBI measurement of, **12C,** 268
Angular-profile measurements, **22,** 286
 LEED, **22,** 291
 RHEED, **22,** 275, 295
Anharmonicity, **23A,** 30
 of molecular vibrations, **3,** 348
Anion vacancy, in color center, **15B,** 3
 H ion trapped in, **15B,** 17
Anisotropic distortion, of electron microscope image, **16B,** 204–205
Anisotropic media, **1,** 490; **26,** 133
Anisotropic motion, **20,** 24, 25, 98–102
 time-dependent hyperfine interaction, **20,** 98
Anisotropic specimens, errors due to, **19,** 291–294
Anisotropic temperature tensor, **23A,** 446
Anisotropy, **17,** 191
 and beam spreading, **19,** 292
 calculation for martensite, **19,** 254–255
 constants
  cubic crystals, **6B,** 243–244
  hexagonal, **6B,** 244
  tetragonal crystals, **6B,** 244
 and elastic moduli, **19,** 293–294
 energy, **10,** 400, 411, 413
 field, **10,** 400, 411, 413
 macroscopic, in polycrystalline materials, **19,** 275–277, 292–294
 of pearlite, **19,** 267, 284–288
 of polarizability, **3,** 119
 preferred orientation of cause, **19,** 252–252, 275–277, 292–294
 in pyrolytic graphite, **19,** 292–293
 reduction by phase transformations, **19,** 253–256, 267, 280–290
 in scattering by grains, **19,** 241–244, 253–256, 267–290
 in tool steel, **19,** 293–294
 of velocity, **19,** 277
 of zinc bar, **19,** 275–277
Anisotype heterojunctions, **15A,** 217
Anitcrossing spectroscopy, **29B,** 335–339
Annealing, **6A,** 362–363; **16B,** 299–300

of polymer single crystals, **16B,** 230, 231
prevention of crystal thickening, **16B,** 254
simulated, **28,** 440–441
Annealing studies, positrons, **21,** 93
Annihilation
-in-flight
in polarized target, **5B,** 262–263
positron, **5B,** 238–239, 262, 485, 511–514
of positrons, **21,** 78
radiation, wavelength, **4B,** 14
of slow positrons, **5B,** 263–264
Annihilation quanta, **8,** 228
Anti-Compton arrangement, **8,** 215
Anti-Compton spectrometer, **8,** 213, 216, 218–220, 225
performances, discussion, **8,** 220
ANNNI model, **23C,** 143
Annular field-relief geometry, **2A,** 70
Annulus, **14,** 278
Anode
delay, **2,** 105
lens design for electron guns, **4A,** 104–107, 111–116
voltage drop, **2,** 111
Anodic oxidation in film preparation, **6A,** 133–134
Anomalous dispersion, **7A,** 127
Anomalous energy spread in electron beams, **4A,** 134–135
Anomalous magnetic moment
of electron, **4B,** 30, 35, 40–41
of positron, **4B,** 43
Anomalous scattering, **23C,** 342
Anomalous skin effect, **11,** 546
Anomalous transmission, **23A,** 318
Antenna arrays, *see* Array antennas
Antenna calibration, **12B,** 82–96
pointing, **12B,** 96
radio sources for, **12B,** 83, 86–93
radio sources, classes of, **12B,** 84
Antenna patterns
definition of, **12B,** 3
for paraboloidal reflectors, **12B,** 34
Antenna temperature, **12B,** 203; **12C,** 1
definition of, **12B,** 2
Antenna theorem, **18,** 126
Antennas, *see also specific types*; Reflectors
characteristics of, **12B,** 7
for electron wave transmission, **9A,** 19–20
for ion-wave transmission, **9A,** 13
microwave, **2,** 716
in microwave circuits, **2B,** 263
patterns and gain, **2,** 741–742
patterns and gain in, **2B,** 298–300
types of, **12B,** 7–28
for whistlers, **9B,** 90–91
Anthracene, **23A,** 445; **30,** 110–112
scintillation properties, **5A,** 421ff
Anti-Stokes Raman line, **18,** 422, 723
Anti-Stokes scattering, **10,** 70, 420, 421
Anti-Stokes spectrum, **16A,** 78
Antibodies, **20,** 114, 115
Anticoincidences, **8,** 218
Anticrossing, dispersion relation, **23A,** 469
Antiferroelectric crystal ($PbZrO_3$), **10,** 383
Antiferromagnetic domains, methods for, **11,** 740–743
Antiferromagnetic resonance, **6B,** 240; **10,** 407–415
frequency, **10,** 410, 411
Antiferromagnetic structures, **23C,** 119
nuclear, **23C,** 272
Antiferromagnets, **23A,** 70
Antigens, spin labels as, **20,** 110, 111
Antimony point, **1,** 257
Antimony-cesium cathode, **6B,** 385–388
formation, **6B,** 385–386
properties, **6B,** 386–387
Antineutrino
bombardment of protons, **5A,** 685ff
detector, **5A,** 687, 688
flux, **5A,** 686
reactions, **5A,** 682
Antiphase domains, **22,** 254
rotational, **22,** 254
translational, **22,** 254
Antiphase islands, **22,** 254
rotational, **22,** 254
translational, **22,** 254
Antiproton beam, **5B,** 752–755
Antiproton scattering, **5A,** 321
Antireflecting films, **10,** 96–98
Antireflection coatings, **26,** 33
Antireflective coatings, **29C,** 346, 348
Antoine's equation, **28,** 484–487
AOM, *see* Acoustooptic modulator
AOT-water-oil system, **23B,** 534
APC connectors, **29C,** 172
Apertometer, **25,** 166
Aperture
angular, **1,** 380

blocking, in paraboloidal reflectors, **12B,** 38–44
broadening, *see* Antenna theorem
effective, **5B,** 710–711
efficiency, *see* Efficiency, aperture
in electron microscope,
diffraction at, **16B,** 200–201
selected, **16B,** 212
finite, correction, **5B,** 247–248
limiting, **1,** 333–334
luminosity, **5A,** 346, 352, 357
numerical, **1,** 341
stop, in light-scattering optical systems, **9A,** 92
synthesis, **12B,** 24–28; **12C,** 149
earth rotation in, **12C,** 170
example of, **12C,** 173
one-dimensional, **12C,** 166–167
two-dimensional, **12C,** 170
VLBI observations for, **12C,** 252–256
Aperture stops, **8,** 31, 34, 36, **31,** 150–151
Apertures in molecular beam formation, **3,** 592
Apherical surfaces, **25,** 7, 53, 94
Apiezon
J, **14,** 414, 415
K, **14,** 414, 415
L, **14,** 415
M, **14,** 415
N, **14,** 415
Q, **14,** 416
T, **14,** 415
W, **14,** 415, 416
Aplanatic lens, **25,** 110
Aplanatic surface, **1,** 330
Apochromatic objectives, **25,** 164
Apodization, **10,** 142–144; **12A,** 404, 502, 508; **13B,** 62, 64; **25,** 226; **29B,** 390
Apparatus
for fluid dynamic research, **18,** 755–819
function, **12A,** 502
for optical pumping, **7B,** 13–18
Apparent barrier height
theory of, **27,** 20, 26
Apparent cross section, **7A,** 367
Apparent field, **25,** 61
Apparent resistivity, **24B,** 335–338, 504
Apparent size of adatom
theory of, **27,** 16
Appearance potential(s), **3,** 526, 527–542; **3B,** 579; **7A,** 40
negative ions, **3,** 536
formed in electron-induced dissociation, **3,** 552
Appearance-potential spectroscopy, **22,** 198
resolution, **22,** 205
Application
laser plasma XUV source, **31,** 88–89
wall stabilized arc, **31,** 46
x-ray laser, **31,** 107–109
Application of various techniques in work function measurements, **22,** 21
Approximation, **28,** 491
Aqua regia, **8,** 179
Aqueous solutions, **23B,** 471
Ar^+, resonant charge exchange by merging beams, **7A,** 480, 482
Araldite, **14,** 416, 417
Arc plasma discharge source, *see also* Duopigatron; Duoplasmatron; Penning discharge source
Calutron sources
Calutron/Berna/Nier conifiguration, **29A,** 113–114
design, **29A,** 112–113
lifetime, **29A,** 113–114
diffusion coefficient across the field, **29A,** 78
external field penetration, **29A,** 92–93
filament materials, **29A,** 100–101
Freeman source, **29A,** 111–112
geometry, **29A,** 75–76
hollow-cathode ion sources, **29A,** 105–106
hot cathode lifetime
estimation, **29A,** 98–101
mechanisms of wear, **29A,** 98
ion current density
atomic mass relationship, **29A,** 96–97
extraction voltage relationship, **29A,** 95
loss at column ends, **29A,** 78–79, 100
sheath edge, **29A,** 77, 99–100
ion extraction
design of system, **29A,** 93–94
intensity of ion beam, **29A,** 95–98
plasma
analytical approximation, **29A,** 81–84
computational simulation, **29A,** 85–91
solid emitters, analytical approximation, **29A,** 80–81
ionization, **29A,** 76–77
efficiency, **29A,** 96–98

ion optical effects in the plasma boundary, **29A,** 93–95
material utilization efficiency, **29A,** 98
minimum pressure requirements, **29A,** 79
multicusp, magnetic-field confinement
design, **29A,** 114–115
source types, **29A,** 115
perveance
concentric cylinder geometry, **29A,** 81
cylindrical emission boundary, **29A,** 81
dependence on angular divergence, **29A,** 85–86
optimum for minimum angular divergence, **29A,** 84–85
parallel plate geometry, **29A,** 80, 82
spherical emission boundary, **29A,** 81
plasma
potential, **29A,** 92
sheath formation, **29A,** 91–92
principle, **29A,** 74–75
stable discharge criteria, **29A,** 77
thermionic emission, **29A,** 75–76
three-electrode ion extraction system, **29A,** 84
voltage breakdown in a vacuum, **29A,** 98
Arc(s), **7A,** 120–123; **13A,** 267, 69
carbon, **4A,** 296–297, 299–300; **25,** 249
cold-cathode, **2,** 88–92
compact source, **25,** 247, 249
devices, **2A,** 172–186
hydrogen thyratrons as, **2A,** 176–180
ignitrons as, **2A,** 180–182
spark gaps and, **2A,** 183–184
discharge in mercury, **10,** 13–15
drop, **2,** 111
hot-cathode, **2,** 92–95
harmonic generator, **10,** 23–26
mercury, high pressure, **4A,** 301–305
sources, **1,** 373, 374
welding, **14,** 427
xenon, high pressure, **4A,** 305–308
zirconium, **4A,** 298–299
Arc-chamber box, **8,** 185
Arc-discharge source, **8,** 185
Arc-plasma tunnel, **18,** 462, 784
Archie's Law, **24B,** 284, 291, 462
Archimedes' principle, **1,** 66, 119
Archimedes spiral, **14,** 90
Area, **1,** 25, 58–59
ratio equation, **1,** 129
Arecibo Antenna, **12B,** 19
Argon, **14,** 419–420
in ari, **14,** 419
atmospheric concentration, **12B,** 170
double ionized, **15A,** 68
first spectrum of, **15A,** 60
fluoride lasers, **15B,** 125–156
instability, **14,** 222
in ionization chambers, **5A,** 92–93
ion
laser, **15A,** 61; **29C,** 68, 69
laser transitions of, **15A,** 65
liquid, **23B,** 466
purification, **4B,** 327
Argon II laser, **15A,** 92–95
Argon III laser, **15A,** 92–95
Argon-chlorine, **15A,** 116
Argon-oxygen laser, **15A,** 117
Argonne fast chopper, **5A,** 532
Arithmetic unit, in pulse amplitude measurement, **2B,** 122
Array antennas, **12B,** 98–118
building problems, **12B,** 104–111
cables and lines for, **12B,** 114–116
cost of, **12B,** 104, 117
design of, **12B,** 111
elements for, **12B,** 108–110
feeder networks for, **12B,** 104–108
frequency sensitivity of, **12B,** 105
gain measurements of, **12B,** 112–114
groundscreen for, **12B,** 110
materials, **12B,** 118
patterns, **12B,** 101–102
phasing techniques for, **12B,** 104–108
power line noise in, **12B,** 118
resistivity, and induced polarization
electromagetic coupling, **24B,** 304, 311–313
resolution, **24B,** 304, 311
selection, **24B,** 301, 304
types, **24B,** 303
site selection for, **12B,** 104
Arrested relaxation technique, in chemical lasers, **15B,** 98
Arrhenius equation, **19,** 172; **22,** 430, 451
for polymerization rate constants, **16B,** 332
plot, **22,** 395
Arrhenius law, of sorption and diffusion, **16C,** 324
Arrhenius plot, **23A,** 512
of ultrasonic data, **16C,** 89
Arrhenius relation, **24A,** 92

Arsenic ion laser, **15A,** 153–154
ART, *see* Algebraic reconstruction technique
As–1 × 1 adsorbed Ge(111), scanning
  tunneling microscopy of, **27,** 241, 243
ASD, *see* Auger-stimulated desorption
ASE, *see* Amplified spontaneous emission
ASLEP, **12C,** 273
Aspect ratio, **14,** 425
Aspherical gratings, **13A,** 229
Aspirator, **14,** 137, *see also* Ejector pump
  Gerdien, **1,** 504–505
Associated liquids, **3B,** 989
Association colloids, **23B,** 490
Associative analyzer, in pulse amplitude
  measurement, **2B,** 128
Associative ionization, **15B,** 227
Associative memories, in pulse spectrometer,
  2B, 127–128
Associativity of equilibrium, **1,** 309
Astable circuits, **2A,** 355–359
Astable multivibrator waveforms, **2A,** 358
Asterisms, **6A,** 238–240
Astigmatic resonator, for cw dye laser,
  15A, 343–344
Astigmatism correction
  beamline design, **31,** 28
  glancing incidence spherical mirror,
    31, 162–163
Astigmatism curvature, **31,** 367
Astigmatism, **1,** 332, 337; **13A,** 229; **25,** 42
  CW dye laser, **29C,** 58–59
  definition, **31,** 156
  in electron microscope image, **16B,** 205
  grating, **1,** 368
  prism, **1,** 365
  Rowland circle, **31,** 367
  singlehanded mirror, **31,** 160–161
ASTM test methods
  compression test, **16C,** 124
  electric breakdown, **16C,** 488
  flexural test, **16C,** 125
  lap-shear adhesive test, **16C,** 127
  permeability, **16C,** 363, 365
    cup method, **16C,** 366, 371
  tensile test, **16C,** 123
ASTRID storage ring
  beamline design, **31,** 30–32
  efficiency, **31,** 32–35
Astrometry, **12C,** 161–166
  VLBI applications in, **12C,** 261, 269
Astronomical applications, **13B,** 48, 100, 243
Astronomical radiometers, **26,** 249
Astrophysical studies, and beam-foil
  experiments, **17,** 599–605
Astrophysics, infrared spectroscopy in,
  3, 109–111
Asymmetric beam waist, **29C,** 196
Asymmetric collision, **17,** 384–385
  energy spread, **17,** 178
  inner-shell vacancy production, **17,** 485
  noncoincidence technique in study of,
    17, 157
  SCA calculations, **17,** 341
  x-ray production cross section, **17,** 285–288
Asymmetric interferograms, **13B,** 70
Asymmetric rotor, **3A,** 21, 39, 75
Asymmetric wiggler, **29C,** 32, 33
Asymmetrical optical system, **31,** 148–149
  aberration, **31,** 163–164
Asymmetrical top, **10,** 304, 320–321
Asymmetry, **29C,** 196
  function $S(\theta)$, **5B,** 250–252
  polarization, **4B,** 233
  ratio, polarimeter, **5B,** 141
Asymptotic period method, usage, *see* Rod
  worth determination.
Asymptotic positive period, **8,** 249
AT-cut quartz, **19,** 52, 54
Atactic chain configuration, **16B,** 19
Atmosphere, **12A,** 416; **14,** 2
  absorption of infrared by, **12A,** 416, 440,
    447–448, 453–454
  complete microwave spectrum of,
    12B, 172–176
  conductivity, **1,** 504–505
  currents, **1,** 504
  electric field in, **1,** 502–504
  emission of infrared radiation by, **12A,** 416,
    435, 438, 442, 444, 447–451
  gaseous constituents, **12B,** 170
  ionization of lower, **1,** 501
  neutral structure of, **12B,** 136–141
  potential gradient in, **1,** 501, 502–504
  propagation in, **12B,** 140–141
  transmission through infrared windows,
    12A, 447–448
Atmosphere research, *see* Upper atmosphere
  research.
Atmospheric effects, **12B,** 119–120
Atmospheric electricity, **1,** 501–506
Atmospheric fluctuations, compensation for,
  12C, 14

Atmospheric fluorescence, **12A,** 316–343
absorption of X-rays in, **12A,** 320–324
arrival direction of exciting radiation, **12A,** 329
background sources of light, **12A,** 338–343
decay time of, **12A,** 321
efficiency of, **12A,** 319
molecular transitions involved, **12A,** 318
sensitivity of, **12A,** 329
time duration of a pulse, **12A,** 327
Atmospheric gases, *see specific gases*
Atmospheric sounding techniques, **26,** 236
Atmospheric temperature fields, field statistics, **28,** 466–471
data, **28,** 466
field significance testing, **28,** 470–471
map statistical significance estimation, **28,** 467–471
preprocessing of data, **28,** 467
random time series generation, **28,** 468–470
Atmospheric transmission, **10,** 91–93
Atmospheric windows, **10,** 91–93
Atom(s)
buncher, **29B,** 179–180
compound, **3,** 2
exchange reactions, vibrational distributions and, **15B,** 98
excited, collisions of, **7B,** 78–124
expectation value for, **15A,** 20–21
in flight, detection of, **4A,** 414
interference, by separated-field excitation, **29C,** 295–299
interferometry
basic concepts, **29B,** 342–344
beam splitters, **29B,** 344–348
experimental realization, **29B,** 348–356
future developments, **29B,** 356–357
ionization in cloud chamber, **5A,** 444–445
in a magnetic field, **3,** 602–605
mu-mesic, formation of transient, **5A,** 329–330
positions
determination, **3,** 291–295
standard deviation, **3,** 300
Rutherford nuclear model, **5A,** 289
scattering
of electrons by, **3,** 335–338
of heavy particles by, **5A,** 55–56
of slow neutrons by, **3,** 349–352
state of, **15A,** 20–21
superposition model, **27,** 8
traps
dark spontaneous-force optical, **29B,** 140
dipole force, **29B,** 140
Ioffe, **29B,** 136, 141
magnetic quadrupole, **29B,** 135
magneto-optic, **29B,** 137–140
TOP, **29B,** 137
Atom-probe, **22,** 404, 407
10-cm, **22,** 414
imaging, **22,** 415
mass spectroscopy, **22,** 404
pulsed laser, **22,** 406
spectroscopy, **22,** 404
voltage-pulsed, **22,** 404
Atomic
beam light sources, **1,** 375
clocks, **1,** 29, 71–72
excitation potential, **1,** 47
heat, **1,** 261
polarization, **1,** 492
wavelength standards, **1,** 55
Atomic angular momentum, **4B,** 109–112
Atomic beam deflection, by laser, **29B,** 135
Atomic beam detector
electron ejection, **4B,** 133, 165–166
electron multiplier, **4B,** 168
photomultiplier, **4B,** 141, 143
Stern-Pirani, **4B,** 164
surface ionization, **4B,** 168
Atomic beam recoil experiments, **7A,** 89–95
Atomic beam resonance, **4B,** 154–170, 192–198
apparatus, **4B,** 154–170
hyperfine structure measurement
of H, **4B,** 163–166
of He, **4B,** 166–167
of other atoms, **4B,** 167–170
measurement of atomic *g* factor, **4B,** 192–198
Atomic beam source tabulation, **4A,** 192–195
Atomic beam sources, **7A,** 71, 74–75, 230–255, *see* Thermal beam sources
Atomic beam velocity selection, **7A,** 94
Atomic beams of radioisotopes, **4A,** 228–256
arsenic, **4A,** 228–299
astatine, **4A,** 229–230
bismuth, **4A,** 230–231
bromine, **4A,** 232–233
cesium, **4A,** 233–235
copper, **4A,** 236
gallium, **4A,** 236–238

gold, **4A,** 238–239
helium, **4A,** 239
indium, **4A,** 239–240
iodine, **4A,** 240–242
oxygen, **4A,** 242–243
potassium, **4A,** 243–244
rare earths, **4A,** 244–245
refractory elements (Ta, W, Rh, Ir), **4A,** 255–256
rubidium, **4A,** 245
silver, **4A,** 245–248
sodium, **4A,** 249
thallium, **4A,** 249–251
transuranic elements, **4A,** 251–254
tritium, **4A,** 254–255
Atomic beams, for optical pumping, **7B,** 13
Atomic binding, **8,** 268
Atomic clocks, relativistic effects, **8,** 163–164
exercises, **8,** 164
references, **8,** 164
Atomic clocks, types of, **12C,** 208
Atomic coherence, **15A,** 20
Atomic collisions, **8,** 35
of electrons, ions, and excited atoms, **7B,** 78–124
particle continuity equations, **7B,** 79–82
neutral atom laser excited by, **15A,** 116–120
noble-gas ion lasers excited by, **15A,** 121–124
Atomic emission spectroscopy (ICO-AES), **30,** 385
Atomic energy levels, **9A,** 164
Atomic force microscopy of charge density waves in
1T-TaS_2, **27,** 358, 359–372
1T-$TaSe_2$, **27,** 358, 374
charge density waves (CDW_, **27,** 350, 385
Atomic force microscopy, **27,** 86
force gradient microscopy, **27,** 89
low-temperature
instrument design, **27,** 355–357
thermal noise in, **27,** 89
van der Waal forces in, **27,** 87
Atomic frequency standards, comparison of, **12C,** 225–227
Atomic hydrogen maser, description of, **12C,** 214–219
Atomic hydrogen photoionization cross section, **7A,** 172
Atomic hydrogen-electron scattering, **7A,** 72, 73, 81, 83
Atomic lines, **8,** 32
Atomic mixing, **22,** 333
Atomic nitrogen, **7A,** 81, 88
Atomic number of projectile, and degree of target ionization, **17,** 379
Atomic oxygen, **7A,** 88, 91
Atomic partition function, **7A,** 120
Atomic potential, **22,** 311
Atomic processes in plasmas, **9A,** 115–168
atomic energy levels, data on, **9A,** 164–165
breakdown process, **9A,** 117
broadening of energy levels and spectral lines, **9A,** 115
charge transfer, **9A,** 159–161
charge exchange, **9A,** 159
cross sections, **9A,** 160
collision processes, **9A,** 118
deexcitation of atoms, **9A,** 115, 144–147
detailed balancing, **9A,** 144
emission (induced), **9A,** 147
forbidden transitions, **9A,** 146
multi-photon emission, **9A,** 146–147
oscillator strength, **9A,** 145, 147, 166
spontaneous radiative decay, **9A,** 145–147
transition probabilities, **9A,** 145
effective charge, **9A,** 117
excitation, **9A,** 115, 134–144
free-free transitions, **9A,** 161–162
Bremsstrahlung, **9A,** 161
continuum emission, **9A,** 162
free-bound emission, **9A,** 162
Gaunt factor, **9A,** 162
ionization potential, **9A,** 161
negative ions, **9A,** 162
recombination radiation, **9A,** 162
impurity ions, **9A,** 115
information centers for data on, **9A,** 167
ionization of atoms, **9A,** 115, 147–154
autoionization, **9A,** 153
configuration interaction, **9A,** 153
electron impact ionization, **9A,** 150
heavy particle ionization, **9A,** 151–152
innershell ionization, **9A,** 149
ionization coefficient, **9A,** 148
ionization rate, **9A,** 149
ionization rate coefficient, **9A,** 149
Lorentz ionization, **9A,** 154
photoionization, **9A,** 152

radiative recombination, **9A,** 153
three-body recombination, **9A,** 153
line broadening, **9A,** 162–164
Doppler effect, **9A,** 163
ion kinetic temperature, **9A,** 164
pressure broadening, **9A,** 164
radiative transfer, **9A,** 163
Stark effect, **9A,** 163
Voigt functions, **9A,** 163
multiplet tables, **9A,** 165
plasma parameters measurable by, **9A,** 116
recombination of electrons, **9A,** 115, 154–159
collision limit, **9A,** 157
collisional-radiative recombination, **9A,** 155
collisional recombination, **9A,** 155
dielectronic recombination, **9A,** 154, 157–159
rate coefficients, **9A,** 155
three-body recombination, **9A,** 156–157
velocity distribution function, **9A,** 115
Atomic scattering factor, **31,** 260
definition, **16B,** 29
of various atoms, **16B,** 30
Atomic spectra, **13A,** 6, 100ff, 148ff, 204ff, 253ff
analysis of, **13A,** 336
nuclear effects in, **13A,** 341
Atomic spectra in the infrared, **13B,** 2
Atomic structure
beam-foil measurements of, **17,** 596–599
synchroton radiation studies, **29C,** 40
Atomic time, **12C,** 263
definition of, **12C,** 182
Atomic trajectories, **4B,** 239–240
Atomic transition(s)
in solids, **17,** 139–144
laser frequency stabilization, **29C,** 120
probabilities, electric arc use in, **7B,** 347–348
Atomic-hydrogen-maser oscillator, **12C,** 208
Atomism, **3,** 1
Atomistic" description of solids, **7B,** 289–304
Atomization, **19,** 333–338
ATPase, mitochondrial, **20,** 245
ATR tube, in microwave circuits, **2B,** 285–286
Attachment
coefficient, measurement of, **7B,** 56
probability of electrons in gases, **5A,** 92
electron, **5A,** 100–101
Attenuated total reflectance (ATR), **13B,** 48
Attenuated total reflection, *see* Infrared spectroscopy
Attenuation, **22,** 244; *see* Absorption; *see also* Acoustooptic measurement of sound wave, attenuation; Internal friction
amplitude coefficient, **24A,** 52
atmosphere, **10,** 91–93
in biological tissues, **19,** 14
coefficient, **1,** 221, 222, 293; **22,** 242
photon total, **5B,** 518–520, 529–531
of sound, Kirchhoff's, **3,** 685–686
computation, models for rain, **12B,** 180
corrections, **23B,** 421
definition of, **19,** 27, 107
factors, **21,** 166
of gamma rays, **5A,** 671ff
light pipes, **10,** 124
logarithmic decrement, **24A,** 50
measurements of, **19,** 84, 107–111, 257–267, 474–481, 487
microwave, **2,** 739–740; **2B,** 295–296; *see also* Attenuation, measurements of
due to nonlinearity, **19,** 300, 326
oversize waveguides, **10,** 123
plasma, **10,** 280
Q, **24A,** 12–13, 33–34, 50
radiation, **1,** 396
scattering, **1,** 394
Snoek peak, **24A,** 36, 48
in solids
agreement theory and experiment, **19,** 268–275
in bainitic steel, **19,** 285–287
in carbon steel (low alloys), **19,** 284–290
changed by phase transformations, **19,** 279–290
changed by precipitates, **19,** 282–284
changes on tempering of martensite, **19,** 287–288
dependence on grain size, **19,** 242–247
dependence on prior autenite grain size in steel, **19,** 285–290
differences among grain size at constant microstructure, **19,** 288–290
differences among microstructure at constant grain size, **19,** 279–289
diffusion region, **19,** 246
experimental methods for, **19,** 257–267

frequency-dependence of, **19,** 242–247, 267, 272–275
graphs, log-log technique, **19,** 246, 272–274
intermediate region, **19,** 242–247, 267, 272–274
in iron-nickel alloy (polycrystalline), **19,** 268–274, 279–281
in martensitic steel, **19,** 279–281, 285–290
in nickel (polycrystalline), **19,** 268–275
in pearlitic steel, **19,** 267, 285–287
in polycrystalline alloys, **19,** 267–290
with preferred orientation, **19,** 275–277
Rayleigh region, **19,** 242–247, 267, 272–274
scattering as cause, **19,** 237–257, 267–291
scattering cross sections for grains, **19,** 245
scattering formulas for, **19,** 242–244
sources other than in polycrystalline materials, **19,** 246, 260–261, 267, 281, 286–289, 292
in steel, reduction due to transformations, **19,** 267, 285–289
of surface waves, **19,** 503, 506
in titanium alloy, **19,** 281–284
in zinc bar, **19,** 275–277
specific damping capacity, **24A,** 33
standard waveguides, **10,** 118, 120, 121, 123
superconducting films, **10,** 433–441
in superfluid helium, **19,** 437–438, 448, 449, 450
of surface waves, **19,** 503, 506
Attenuator, **1,** 484; **8,** 125, 152; *see also* Variable attenuator.
microwave, **6B,** 236–237
waveguide, **2,** 713–714; **3,** 26
Atwood machine, **1,** 75, 78
Au on Au(111)
surface diffusion, **27,** 292
Au on GaAs(110)
spectroscopy of, **27,** 268
Au on Ni(110)
epitaxy, scanning tunneling microscopy of, **27,** 295
Au(100)
surface reconstruction
scanning tunneling microscopy of, **27,** 291
surface reconstruction, missing row
scanning tunneling microscopy of, **27,** 286
spectroscopy of, **27,** 285
Au(110)
phase transition, scanning tunneling microscopy of, **27,** 287
roughening transition, scanning tunneling microscopy of, **27,** 293
Au(111)
scanning tunneling microscopy of, **27,** 292
spectroscopy of, **27,** 281
Au, *see* Gold
Au-adsorbed
Si(111)
scanning tunneling microscopy of, **27,** 200
Au/AlAs/GaAs
heterostructure interface
ballistic electron emission microscopy of, **27,** 337, 339
Au/GaAs
interface
ballistic electron emission microscopy of, **27,** 335
ballistic electron spectroscopy of, **27,** 341
ballistic hole spectroscopy, **27,** 340
carrier scattering spectroscopy of, **27,** 343
Au^{198} decay scheme, **5B,** 159
Audiomagnetotellurics, **24B,** 313, 315, *see also* Magnetotellurics
Auger analysis, **22,** 231
Auger decay, **17,** 524; **22,** 148
Auger deexcitation, **29B,** 194–195
Auger effect, **17,** 433–434
detector, **22,** 495
Auger electron emission, **7B,** 202; **17,** 479–482, 514–516; **17,** 191, 468–482, 561, 566–570
angular distribution, **17,** 479–482
anisotropies in, **17,** 514–516
in beam-foil research, **17,** 591
in coincidence experiment, **17,** 345
Auger electron, **21,** 4, 61, 251, 256, 272
appearance-potential spectroscopy, **22,** 202, 210
Coulomb ionization, **17,** 492–500
detector, **17,** 314
energy, **17,** 401
analysis, **17,** 544
measurement of, **17,** 88

coincidence measurements, **17,** 318–319
production and detection, **17,** 436–467
by electron promotion, **17,** 490–491
spectral identification, **17,** 347–350
Auger electron spectroscopy (AES), **22,** 33, 45, 154, 226, 324, 496, 500, 504, 507, 508; **30,** 291, 292, 318, 323–326; *see also* Ion-induced Auger electron spectroscopy
detector, **22,** 495
Auger emission, **7A,** 8, 85; **22,** 150, 227
Auger line, **5A,** 583–585; **17,** 468–469
of lithiumlike atoms, **17,** 506–514
shape, **22,** 230, 508
Auger linewidths, **17,** 524–527
Auger measurement, of Auger yield, **17,** 292–294, 393, 395
Auger microscopy, **22,** 506
Auger neutralization, **22,** 304; **29B,** 194–195
Auger peak, **22,** 179, 502
Auger process, **4B,** 133; **22,** 148, 174, 470
in Mössbauer spectroscopy, **11,** 109, 310
Auger production cross section, **17,** 479
Auger recombination, in germanium, **15B,** 208
Auger spectrometer, **22,** 487
Auger transition, **13A,** 178; **17,** 470, 472–477; **22,** 227
forbidden, **17,** 586–587
Auger yield, **22,** 190
Auger-stimulated desorption, **22,** 470
Austentite grain, scattering by, **19,** 267, 281, 285–290
Autocollimator, **1,** 56; **25,** 142
Autocorrelation spectrometers, **12B,** 280–298; **12C,** 22
amplitude discriminators in, **12B,** 290–291
constructional considerations for, **12B,** 293
examples of, **12B,** 297–298
measurement techniques with, **12C,** 46–57
prefilters for, **12B,** 287–288
quantizing considerations, **12B,** 281–284, 294–295
resolution in, **12C,** 55–57
samplers for, **12B,** 291–293
sidelobes, **12C,** 5
spectrum renormalization, **12B,** 285–287
standard deviation of spectral estimate, **12B,** 285
three-bit, **12B,** 283
two-bit, **12B,** 282
video converters for, **12B,** 287–288
weighting function for, **12B,** 286
weighting function in, **12C,** 55
zero-crossing in, **12B,** 290–291
Autocorrelation, **20,** 310–313; **29C,** 187–188; *see also* Spectrometers
frequency resolved optical gating, **29C,** 215, 216–220
function, in magnetic spin systems, **16A,** 247
higher-order autocorrelation, **29C,** 213–215
intensity autocorrelation, **29C,** 205, 209–215
interferometric autocorrelation, **29C,** 215–216
multishot autocorrelation, **29C,** 206
phase-sensitive autocorrelation, **29C,** 215–220
principles, **29C,** 203–208
second-order autocorrelation, **29C,** 205, 206, 208, 209–213
single-shot autocorrelation, **29C,** 208
Autocovariance
function, lag, **28,** 306
sequence, **28,** 314
Autoionization, **3B,** 858, 930; **7A,** 87, 151, 174, 449
Autoionizing electron, **17,** 191; *see also* Auger electron
Automatic gain control (AGC), **2A,** 337, 404
Autoradiography, **21,** 44–49
diffusion studies, **21,** 48
nuclear track materials, **21,** 48
photographic emulsion, **21,** 44–48
Autoregressive models, **28,** 326
Autoregressive process, first-order, **28,** 463
Autoregressive spectral estimate, **28,** 327
Avalanche
electron, definition, **7B,** 66–67
electron-initiated, **5A,** 118
Avalanche diode(s), **2A,** 240; **2B,** 275–277
applications of, **2B,** 280–282
power generation capabilities of, **2B,** 279
Avalanche multiplication noise, **2B,** 470
Avalanche photodiode, **2B,** 327–328
Avalanche rectifier, **2A,** 217
Average (mean) velocity, **29B,** 9
AVF cyclotron, **17,** 13–14
Avogadro's constant, **1,** 39–41
Avogadro's law, **3,** 3
Avogadro's number, **4B,** 3–4; **6A,** 284–285; **8,** 272

Avrami equation, **16B,** 336, 348
  applications of, **16B,** 366, 368, 371–372
Avrami integer, for various nucleation modes, **16B,** 348
Avrami theory of crystallization, **16B,** 347–348
Axes, reference, **3,** 269
Axial component, **8,** 183
Axial field, **8,** 83, 91, **8,** 100
  components, **8,** 74, 79
  coordinate, **8,** 98
  homogeneity, **8,** 102, 105
Axial symmetric systems, magnetic field calculation, **8,** 55–59
Axial symmetry, **8,** 67
Axial wire (in voltage clamp), **20,** 452–455
Axicon lens, for light-scattering optical system, **9A,** 95
Axis of symmetry, *see* Symmetry axis.
Axisymmetrical optical system aberration, **31,** 152–153
Ayrton-Mather shunt, **1,** 450
Azbel-Kaner cyclotron resonance, **10,** 249–251
Azbel-Kaner resonance, **13B,** 97

B

β-emitting radiotracers, **24A,** 110–111
β-Eucryptite
  ion beam emission, **29A,** 158
  synthesis, **29A,** 158
B-function, **1,** 8
B-J characteristic, **8,** 91
B^{10} energy level diagram, **5B,** 361
Ba^+ excitation, **7A,** 115
Babinet
  compensator, **1,** 382; **6B,** 259–261; **26,** 143
  principle, **26,** 68
Bacitracin, **19,** 189, 191
Back reflection method, **6A,** 216–217
Back scattering in Faraday cup, **5B,** 490–491
Back-scattered quanta, **8,** 205, 223
Back-scattering angle, **8,** 220
Back-scattering instruments, **23A,** 235
Backflow, forevacuum pump, **14,** 127
Background, **21,** 177, 218
  counts in tandem counters, **5A,** 467
  eliminator, time-of-flight, **5A,** 549–551
  epithermal and fast neutrons, **21,** 163
  fog, from multiple electron scattering, **16B,** 190
  from incoherent elastic scattering, **21,** 160, 163
  infrared, **12A,** 422–426, 429, 435–444, 447–449, 451
  in large shower detectors, **5A,** 666–667
  light-dependent, in image tubes, **12A,** 242
  in PIXE, **21,** 253–254
  rate in antineutrino measurements, **5A,** 691
  rejection, **28,** 441
  from sample and apparatus, **21,** 163
  signals, removal, in spectral analysis, **28,** 302–305
  sources, **12C,** 17
Background noise, **8,** 41, 53
  influence, **8,** 51–52
Backgrounds presence, counting rate measurements, **8,** 191–193
Backing material, *see* Mechanical loading
Backing pump, *see* Forevacuum pump, Diffusion pump, forevacuum pump
Backscattering, *see* Rutherford backscattering
  energy loss, **17,** 58–65
  spectrometer, **21,** 215
Backstreaming, **14,** 229, 234
  diffusion pumps, **14,** 159–159, 166–173
  forevacuum pump, **14,** 163–165, 517
    oil, **14,** 507
  mechanical pump, trapping, **14,** 482
  reduction, **14,** 159, 463
Backward diode, **2A,** 67, 367–368
Backward wave oscillator(s) (BWO), **10,** 36–38; **13B,** 107; **29B,** 276–277
$BaCo_2(AsO_4)_2$, **23C,** 134
Bacteriophages, **23C,** 389
Bacteriorhodopsin, **19,** 234
  photochemistry, **15B,** 206–207
Baffle(s), **14,** 234, 278, 299
  ring focus, **5A,** 367–368
Bakeout, **14,** 280, 443, 447, 448
  procedures, vacuum, **4B,** 289–290
Baking, **14,** 190
  cooling, **14,** 497
  cycled systems, **14,** 497
  of glass, **14,** 322–327
  heating methods, **14,** 497
  heating strips, **14,** 497
  oven, **14,** 497
  of photographic emulsions, **12A,** 233
  of pumps, **14,** 492
  temperature requirements, **14,** 492–494
  temperature uniformity, **14,** 493–495

thermal considerations, **14,** 495
tie-temperature relations, **14,** 494–495
typical cycle, **14,** 494–496
of ultrahigh vacuum systems, **14,** 491–498
valveless system, **14,** 463
Balance
aerostatic, **1,** 120
equal-arm, **1,** 61
hydrostatic, **1,** 119
Kelvin, **1,** 449
method of swings, **1,** 61
quartz torsion, **1,** 62
radiation pressure, **1,** 227
ring, **1,** 541
Schriever inertial, **1,** 63
spring, **1,** 62
torsion, **1,** 84, 541
Balanced diode linear gate, **2A,** 381
Ballast, *see* Rotary oil-sealed pump
Ballistic chronograph, **1,** 75
Ballistic electron emission microscopy, **27,** 307
basic principles, **27,** 309
carrier transport at interfaces
critical angle at
for electrons, **27,** 312
for holes, **27,** 321
quantum mechanical reflection, **27,** 317, 342
transverse momentum conservation at, **27,** 311, 331
collector current
expression for, **27,** 315
normalization by tunnel current, **27,** 316, 325, 327
for off-axis conduction band minima, **27,** 317
electron attenuation length
in Au, **27,** 331
experimental methods, **27,** 326
sample electronic properties, **27,** 327
sample preparation, **27,** 329
imaging, **27,** 334
of Au/AlAs/GaAs, **27,** 337, 339
of Au/GaAs, **27,** 335
of Au/Si, **27,** 335
interface resolution of, **27,** 313
low temperature, **27,** 327
spectroscopy, **27,** 316
carrier-carrier scattering, **27,** 322
electron, **27,** 316
of Au/GaAs, **27,** 332
of Au/Si, **27,** 330
theory, **27,** 311
threshold, **27,** 316, 325, 331
hole, **27,** 319
asymmetry compared with electron spectroscopy, **27,** 320
of Au/GaAs, **27,** 341
of Au/Si, **27,** 340
theory, **27,** 319
spectroscopy of
carrier-carrier scattering
of Au/GaAs, **27,** 344
of Au/Si, **27,** 343
multiple scattering, **27,** 324
theory, **27,** 322
threshold behavior, **27,** 325
tunneling, **27,** 314, 320
Ballistic pendulum, **1,** 75–76
Ballistic range, **18,** 779–781
Ballistic ring method, **6B,** 201
Balloons, use in antenna calibration, **12B,** 113
Balun impedance matching, **8,** 117–122
Balun, **12B,** 117
exercises, **8,** 121–122
Band assignment, in infrared and Rama spectroscopy, **16A,** 90
Band broadening, isotropic scattering, **3,** 146
Band gap states, scanning tunneling microscopy of, **27,** 255, 270, 273, 275
Band mode(s), **10,** 368; **21,** 186, 189
Band progressions, in infrared spectra, **16A,** 110–112
Band structure **23,** 23, 584, 589
of adsorbed overlayer, **22,** 168
mapping, **22,** 165
technique, **22,** 166
three-dimensional, **22,** 169
two-dimensional, **22,** 165, 168, 181
Band theory of solids, **6A,** 16
Band-gap energy, **12A,** 8
Bandpass, *see* Trapezoidal monochromator bandpass.
Band-pass filters, **10,** 105–109; **25,** 287; **26,** 33
Bandpass filter(s), **28,** 295–296; **29C,** 350
Bands, **13A,** 21
in alkali halides, **6B,** 325–326
parallel, **10,** 327
perpendicular, **10,** 327
pivotement, **3,** 152
Raman, intensities, **3,** 124–125

Bandwidth, **8,** 23, *see also* Half-width bandwidth; Resonance.
finite-, corrections, **2,** 661–662
resolution, **28,** 323
$BaNi_2(PO_4)_2$, **23C,** 153
Bar gauge pressure measurement, **18,** 593, 602
Bare projectile nucleus, **17,** 299
Barium, **22,** 374, 376
Barium titanate ($BaTiO_3$), **19,** 30, 54
pressure gage sensor, **18,** 542
Barkas effect, on stopping cross section, **17,** 71
Barkhausen oscillations, **14,** 62
Barnett equation, **1,** 160
Barns, *see* Nuclear cross section
Barometer, **14,** 42, *see also* Manometer
Barometric technique
for permeability, **16C,** 362
for sorption, **16C,** 352–354
Barrel distortion, **16B,** 204
Barrel shock, **29B,** 28
Barrier determination, **3A,** 89
Barrier diodes, **2A,** 57–60
Barrier emission processes, in normal-metal tunneling, **11,** 167–169
Barrier layer, **10,** 19, 20, 160
capacitance, **10,** 164, 165
diode, **2,** 135
rectifier(s), **2,** 132; rectifiers, **2A,** 59–60
Barrier(s)
anticreep, **14,** 278
cadmium or zinc sulfide types, **11,** 191
carbon, **11,** 192
oxidation procedure for, **11,** 189–190
preparation of, **11,** 188–189
in tunnel junctions, **11,** 161–167
Base
candelbra, **25,** 242
mogul, **25,** 242
prefocus, **25,** 242
spread, **5A,** 344
Basel convention, **4B,** 233; **5A,** 465
Baseline
correction, in DSC melting curve, **16B,** 318–324
curvature, spectra with, **12C,** 48
fitting, spectra with, **12C,** 51
interferometer determination of, **12C,** 139, 162–163
shift, in pulse amplitude measurements, **2B,** 97–98
VLBI measurement, **12C,** 229, 261–276
Basic laws of geometrical optics, **25,** 10
Basis reactions, **19,** 210
Basset-Boussinesq-Oseen (BBO) equation, **18,** 8
Bathobacteriorhodopsin, **15B,** 207
$BaTiO_3$, **10,** 373–375, 379, 381–383; *see also* Barium titanate
Bayard-Alpert gauge, **1,** 190; **14,** 61, 67–72; **22,** 444, *see also* Ionization gauge
Bayes's theorem, **28,** 406
continuous version, **28,** 407
gamma posterior distribution, **28,** 410
Bayes, Thomas, **28,** 403
Bayesian credibility interval, **28,** 413
Bayesian estimation, **28,** 413
Bayesian maximum a posteriori estimator, **28,** 408
Bayesian methods, **28,** 430–430
climate change, **28,** 418, 423–425
Gibbs sampler, **28,** 427, 429–430
hyperparameters, **28,** 426
hyperprior distribution, **28,** 429
maximum entropy method, **28,** 415–416
neutron scattering, **28,** 414–417
nuclear magnetic resonance, **28,** 417–422
prior distribution, **28,** 404, 407–413
conjugate priors, **28,** 410–411
improper priors, **28,** 409–410
maximum entropy priors, **28,** 411–413
noninformative priors, **28,** 408–409
prior choice consequences, **28,** 407–408
rocket photometry, **28,** 426–428
Bayesian spectrum analysis, **28,** 417–422
high signal-to-noise ratio, **28,** 418–420
low signal-to-noise ratio, **28,** 418, 421–422
Bayesian statistical interference, **28,** 403–407
posterior distribution, **28,** 405
prior distribution, **28,** 404
subjective probability, **28,** 405
BBO, as OPO material, **29C,** 257–259, 261, 262, 263
BCS theory, **11,** 131, 137, 144–145, 153, 550
Be (α,*n*) reaction, **5B,** 556–560
Be^{2+} ion, **19,** 202
Be^9 bombardment with deuterons, **5B,** 357–358
BEA, *see* Binary encounter approximation
Beam absorption densitometry, **18,** 405–408
Beam angle
aperture radius relationship, **29A,** 191–192

consideration in combining lenses, **29A,** 202–203
Beam benders, **23A,** 204
Beam deflection probes, **30,** 217, 218
Beam delivery, **30,** 191–201
Beam direction, **8,** 273
Beam divergence, monochromator, **31,** 183
Beam dump, in light-scattering optical systems, **9A,** 93
Beam energy fluctuation, **8,** 168
Beam energy, fluctuations in, **17,** 187
Beam expanders, **25,** 133
Beam expansion in holography, **26,** 200
Beam foil spectroscopy, **13B,** 213, 218
astronomical applications, **13B,** 243
high energy beams, **13B,** 232
sources, **13B,** 216
Beam impurities, **17,** 84, 535–536
Beam intensities, digital measurement, **7A,** 295
Beam lines, **22,** 144, 150, 179
Beam maser spectroscopy, **3B,** 839
Beam pattern, **12A,** 466
Beam profile, **30,** 200, 201
Beam pulsing technique, **7A,** 109–113
Beam rastering, **22,** 313
Beam shaping, **30,** 198–200
Beam splitter(s), **10,** 146, 147, 149; **12A,** 428–429; **18,** 379; **26,** 33; **29C,** 348–349
Beam spreading
acoustooptic measurement of, **19,** 475–477, 484–492
with buffer method, **19,** 151, 260–262, 267
corrections for attenuation, **19,** 71–76, 239, 260–262, 276, 292
dependence on symmetry, **19,** 292
preferred orientation influences, **19,** 276, 292
and texture, **19,** 292
Beam switching measurements, **12C,** 14
Beam transport system, **17,** 464–465
Beam transversal dimensions, **8,** 167
Beam tubes, **2,** 62–65; **23A,** 149
Beam waist, **29C,** 196
Beam, *see* Light source
Beam-beam instability, **9A,** 299
Beam-current problems, **22,** 320
Beam-foil decay curve, **17,** 572–578, 580–582
Beam-foil experiments, **17,** 371–373, 387, 529–606
allowed radiative transitions, **17,** 579–583
applications of, **17,** 595–606
Auger measurements
cylindrical-mirror analyzer, **17,** 458
lithiumlike atoms, **17,** 512–514
cascading, **17,** 572–575
energy-loss measurements, **17,** 47–51
experimental arrangements, **17,** 530–548
accelerators and ion sources, **17,** 533–535
beam analysis and source purity, **17,** 535–536
filters, spectrometers, spectrographs, detectors, **17,** 538–548
foil targets, **17,** 536–537
normalization procedures, **17,** 537–538
forbidden Auger transitions, **17,** 586–587
forbidden radiative transitions, **17,** 583–586
inner-shell vacancy measurements, **17,** 260, 519–524
Lamb shift in heavy ions, **17,** 587–590
lifetime measurements, **17,** 570–590
for measurement of excited states, **17,** 88
population changes, **17,** 572–575
postfoil
beam velocity, **17,** 575–578
charge distribution, **17,** 551–552
quantum beat phenomena, **17,** 590–595
source characteristics, **17,** 548–550
spectra, **17,** 550–570
Auger-emitting states, **17,** 566–570
charge state identifications, **17,** 552–554
Doppler broadening and shifts, **17,** 554–557
doubly excited states, **17,** 569–564
hydrogenic (Rydberg) states, **17,** 564–566
singly excited states, **17,** 557–560
Beam-foil spectroscopy, **7A,** 136
applications, **29A,** 303
for lifetime measurements, **29B,** 232–235
Beam-lead air-isolated transistors, **2A,** 113–114
Beam-lead bipolar integrated circuits, **2A,** 141
Beam-plasma interactions, **9A,** 130
Beamline characteristics, **31,** 27
Beamline design
ancillary components, **31,** 35–36
ASTRID storage ring, **31,** 30–32
efficiency, optimization, **31,** 35
process, **31,** 27–30
radiation protection, **31,** 36
Beamline development, **31,** 27
Beams, modulated secondary, **7A,** 295
Beams, molecular, **3,** 589–636

detection, **3,** 594–596
formation, **3,** 590–593
magnetic and electric resonance methods, **3,** 615–626
measurement of dipole moment, **3,** 669–670
optics, **3,** 607–615
sources, **3,** 593–594
velocity distribution, **3,** 593, 597–598
Bearing materials in vacuum, **4B,** 287–288
Becke line, **25,** 189
Beckmann thermometer, **1,** 244
Becquerel, *see* Radiation units
phosphoroscope, **1,** 397
BEEM, ballistic electron emission microscopy, **27,** 307
Beer's law, **3,** 302; **18,** 407, 475, 623–624, 629, 632
Beer-Lambert equation, **16A,** 77
Behren's test, **1,** 12
Bel, **1,** 200
Bell jar, **14,** 277
Bellow convolution, **14,** 298
Bellow corrugation, **14,** 298
Bellow seal, **14,** 277, 278
Bellow's gage for pressure measurement, **18,** 511
Bellow(s), **7A,** 29; **14,** 298, 300–306
Belt drive, **14,** 450
Bending angle, computation of, **12B,** 191
Bending magnet
brightness, **31,** 11–12
power, **31,** 10
radiation, **31,** 3–7
Bending magnet radiation, accelerating electrons, **29C,** 24–28
Bending, beam, **5B,** 692–747
Bending, molecular, **3,** 187
Bendix-type instrument, **8,** 169
Benioff short-period seismometer, **24B,** 51–52
Bennet-Koehler reflectometer, **26,** 274
Bennet-type instrument, **8,** 169
Bent crystal spectrometers, **31,** 408–411
Bent mirror, spectroscopic application, **31,** 179–180
Benzene, **22,** 421
vapor pressure, **28,** 485
Berlin electron storage ring (BESSY), **31,** 71
conditions for comparison standard, **31,** 130
primary source standard, **31,** 125
Bernal chart, **3,** 312–313
Bernoulli equation, **1,** 128ff, 144
Bernoulli formula, **18,** 243–245, 325, 333, 776
Bernoulli pressure, **18,** 503
Bernoulli principle, **1,** 118
Bernoulli trials, **28,** 16–17, 32
Berthelot's equation
for virial coefficient, **3,** 685
of state, **1,** 276, 278
Bertrand lends, **25,** 173
Beryllia, *see* Ceramics
Beryllium
copper alloy, **14,** 379–382
filter instruments, **23A,** 241
ion laser, **15A,** 156–157
neutron production in, **5A,** 525–526
neutron transmission, **5A,** 507–509
reflectance spectra, **31,** 208–209
transition radiation, **31,** 94–95
Bessel function(s), **11,** 44, 55
first order, **26,** 66
modified, **28,** 369
Bessel functions, **31,** 5–6, *see* Spherical Bessel functions.
Bessel response, **2A,** 334
Best image
field, **22,** 383
voltage, **22,** 383, 387
Best value, **6A,** 1
BET gas absorption, **24A,** 98
Beta
coincidence, double, spectrometer, **5A,** 363
cutoff frequency, **2A,** 278
decay
determination of neutron energy, **5A,** 683–685
momentum triangle, **5A,** 693
polarization of electrons from, 5A, 64
ray
range in matter, **5A,** 71–72
spectrometer, **5A,** 342–347
magnetic, **5A,** 588–590
Beta particles
calibrated sources, **21,** 65
detection
proportional counters, **21,** 60
pulse counters, **21,** 50, 60
detectors
liquid scintillators, **21,** 52, 60
solid state and plastic, **21,** 50
range, **21,** 5, 6
Beta-gamma correlations, **13A,** 178

Beta-gamma-radioactive isotope separator, **8,** 181
Beta-particle
back scattering, **5B,** 288
counters, **5B,** 80
sources
in nuclear orientation, **5B,** 203
radioactive, **5B,** 572–580
Beta-ray spectrometer, **8,** 188
Betatron, **5B,** 626–630
Bethe coupler, **1,** 484
Bethe formula, **7A,** 35
for electronic stopping, **17,** 34–37, 45
Bethe-Bloch stopping formula, **7B,** 235
Bethe-Bloch formula, **5A,** 4–7, 231
verification at relativistic energies, **5A,** 20–41
Bethe-Heitler theory, **8,** 227
Bethune cell, **29C,** 162
Betz-Grodzins model, **17,** 136, 519–520, 523
BF3 counters, filling pressure, **8,** 194–195
Bhabha scattering, **5B,** 242
Bhatia-Thornton formalism, **23B,** 251
Bi-orthogonal code, in telemetering, **2B,** 224
Bias, in particle tracking, **18,** 5, 174, 195
Bias circuits
transistor, **2A,** 256–261
for vacuum tubes and FETs, **2A,** 261–267
Biasing, for integrated circuit, **2A,** 284
Biaxial orientation, **16A,** 430
Biaxially stretched films, **16C,** 141–142
Biconcave lens, **29C,** 351
Biconvex lens, **29C,** 351
Bicritical points, **23C,** 38
BIG (bismuth-substituted iron garnet), **29C,** 360
Bimodal rotational distributions, in hydrogen halide reaction products, **15B,** 117
Bimolecular reaction rate, enhancement of, **15B,** 294
Bimolecular recombination, **6B,** 348
Bimolecular relaxation, **15B,** 280
Binary counter, **2B,** 3
Binary encounter, **17,** 159–173
Binary encounter approximation, **17,** 200–202, 229, 237, 323, 328, 383
Coulomb ionization, **17,** 486, 492
K-shell ionization, **17,** 326
L-shell ionization, **17,** 240–242, 244, 329–330
relativistic electron wavefunctions, **17,** 210
Binary hologram, **26,** 189
Binary optics, **29C,** 360–361
Binary scattering, **22,** 300, 302
Binding energy, **22,** 189, 466; **17,** 204–207, 236–237, 385
Auger spectroscopy, **17,** 470, 473–474
Coulomb ionization, **17,** 487
and equilibrium charge-state distribution, **17,** 121
lithiumlike atoms, **17,** 506, 508
nuclear, **5A,** 335–336
Binding states, **22,** 428, 429
Binocular microscope, **25,** 170
Binocular telescope, **25,** 59
Binodal, **16C,** 277
Binomial distribution, **28,** 17
Binomial trials, **28,** 166
Biochemistry, infrared spectroscopy in, **3,** 111
Bioeffects of ultrasound and cavitation, **19,** 408–411
cylindrical bubble studies, **19,** 409
illustrative calculation, **19,** 410–411
in plants, **19,** 409
Biological model, **22,** 416
deposited from aqueous solution, **22,** 416
Biological polymers
ENDOR studies, **16A,** 451
fluorescent probe studies, **16A,** 409, 411
inelastic electron tunneling spectra, **16A,** 164–166
NMR studies, **16A,** 365–368
Biological processes, picosecond pulses in study of, **15B,** 206–207
Biot number, **18,** 668–669
Biot-Savart Law, **8,** 55
usage, *see* Current loop, magnetic fields.
Biphenyl, **23A,** 499
Bipolar transistor(s), **2A,** 67–80, *see also* Transistor
noise in, **2B,** 471–479
Birch coefficients, **11,** 434
Birefringence, **1,** 379–382; **6B,** 250, 264–278; **16C,** 149, 161–167; **29C,** 353
in crystals, **26,** 133
defined, **11,** 715–716
form, **16C,** 165
induced, **26,** 146
intrinsic, of polymer chain, **16C,** 162
magnetic, **11,** 715–718
measurement of, **16C,** 163–165

method, in magnetic domain studies, **11,** 707–721
optical-field-induced, **15B,** 260–261
positive and negative, **16C,** 162
tuner, **15B,** 31
Birefringent crystals, **11,** 699
Birefringent filter, dye lasers, **29C,** 63
Birefringent interferometers, **12A,** 401–406, *see also* Fourier spectrometers
Bismuth-silver-oxygen-cesium cathode, **6B,** 390–391
Bistatic measurement of spectrum and autocorrelation function of radio waves, **9B,** 150–153
Bit, **2,** 687
Bit detection, in telemetering, **2B,** 233–235
Bit synchronization, in telemetering, **2B,** 229–231
Bitter's solenoid, **1,** 527, 528
Bitter-type solenoid, in magnetoresistance measurements, **11,** 20, 595
BIV, *see* Best image, voltage
Bivariate normal model method, **28,** 464
Bivariate time series, **28,** 340, 346
spectral estimation, **28,** 343–346
BK approximation, *see* Brinkman-Kramers approximation
BK7, **29C,** 343–345
Black body, **1,** 253, 301, 302, 375
Black lipid membranes, *see* Lipid model membranes
Black polyethlene, **10,** 99, 100, 101
Black-body radiation, **4A,** 289–290
Black-body radiators, **29C,** 6–7
Black-light lamp, **25,** 252
Blackbody, **8,** 27
modifications, *see* Nonblackbody modifications.
radiation, **10,** 13, 16, 288, 289; **12B,** 2; **18,** 465–466, 690, 699; **26,** 219
equivalent temperature, **12B,** 144
from plasmas, **10,** 288, 289, 291–295
temperature, **30,** 206
theory, **8,** 28–30
Blackbody radiators, **31,** 120–121
Blackbody, Blackening, **12A,** 421
photographic, **1,** 411
Blade geometry in turbopump, **14,** 252
Blank-off pressure, **1,** 172ff
Blast wave solution, in self-similarity, **18,** 835–842
Blatt fluctuation theory, **5A,** 263
Blaze, of diffraction gratings, **12A,** 472
Blaze angle, **3,** 156
in grating, **26,** 78
Blaze grating, **31,** 370
efficiency, **31,** 386–394
Blaze wavelength, **1,** 368
Blazed grating(s), **1,** 368; **10,** 134, 135, 138, 139; **13A,** 230, 288; **13B,** 16, 58; **25,** 214; **26,** 78, 158
Bleaching of color centers, **6B,** 332–333
Bleaching of holograms, **26,** 204
Bleeder resistance, **2A,** 204
Blending methods, **16C,** 276
Bleustein-Gulyaev wave, **19,** 500, 502
Blob
density, **5A,** 245–263
length distribution, **5A,** 258
Bloch
energy, **11,** 137
equation(s), **3,** 386–390; **3B,** 511; **20,** 6, 7, 72–75, 108
functions, **11,** 75
state of positron, **21,** 79, 80, 83
susceptibility components, **3,** 387
Block copolymer(s), **16B,** 340; **23C,** 311
crystalline, **16C,** 290, 298
definition, **16A,** 3
electron microscopy studies, **16B,** 234–235
ESR spin probe study of interphase, **16A,** 476
morphology, **16C,** 294–299
small-angle neutron scattering studeis, **16A,** 503
styrene-butadiene, **23C,** 311
thermodynamic theory, **16C,** 278–283
Block diagram, for feedback control systems, **2B,** 377
Block kriging, **28,** 397
Block theory of magnetic resonance absorption, **5B,** 106–111
Blockading, **7A,** 49
Blocking, **21,** 290, 350, 357
factor, **16C,** 329; **23B,** 149
oscillator, **2A,** 408
Bloembergen, Purcell, and Pound theory, **21,** 448–449, 456
Blood flow, **20,** 331–333
Blooming, **1,** 352
Blotzmann superposition

creep equation, **16C,** 9
definition, **16C,** 7
in dielectric theory, **16C,** 385
stress-relaxation equation, **16C,** 10
Blow-down wind tunnel, **18,** 758
Blue shift, *see* Gravitational blue shift.
Blue shifts, **3,** 235
Blue-green laser line, in Argon II lasers, **15A,** 93–94
Blue-green pulse xenon ion laser, performance of, **15A,** 100
Blume detection system, **11,** 403, 405
Bode diagrams, **2,** 468
Bode minimum-phase property, **1,** 231
Bode plot, **2B,** 391–394, 419–420, 424
Body waves, **24B,** 11–12, 581–584
Body-centered cubic lattice
definition, **16B,** 5
structure factor for, **16B,** 37
Bohr magneton, **4B,** 107
Bohr magnetron, **10,** 386
Bohr potential, **17,** 164
Bohr radius of impurity atoms in semiconductors, **10,** 177
Bohr-Lindhard model, **17,** 136, 519–520, 523
Boiler, *see* Diffusion pump, boiler
Boiling point(s); **1,** 243
of elements, **14,** 350–351
Boiling, **30,** 186, 229, 234, 281, 398
Bolometer beam detector, **7A,** 350–352, 398–399
Bolometer detector(s), **10,** 157, 168–172; **22,** 497
Bolometer(s), **1,** 276, 392, 393, 468; **3,** 84, 481–482, 493; **8,** 123, 124, 127; **13B,** 7, 8, 52, 113; **26,** 295; **29B,** 282
bridge, **1,** 468; **8,** 123–124
cryogenic, **4A,** 435–437
carbon resistor, **4A,** 436
single-crystal germanium, **4A,** 436
superconducting, **4A,** 435–436
detectivity, **4A,** 432–433
metal, **4A,** 433–434
in microwave circuits, **2B,** 284, 295
optimal design, **4A,** 431–433
thermistor, **4A,** 434–435
Bolt circle diameter, **14,** 279, 280
"Bolt" cathode, **4A,** 92
Boltzmann
constant, **1,** 26, 41; **14,** 6, 7, 232; **24B,** 286, 290's
distribution, **10,** 46; **18,** 464, 473–474, 480, 640–641; **29B,** 107; **30,** 33–35, 124–126, 209, 228, 241, 493, 600
of ejecta, **30,** 123, 125
equation, **4B,** 223; **10,** 266
factor, **3,** 376
formula, **9A,** 120
partition formula, **9A,** 121
statistical weights of, **9A,** 120
transport equation, **1,** 126
Boltzmann equation, **31,** 50
Boltzmann plot, **31,** 52
Boltzmann population ratio, XUV laser, **31,** 102
Boltzmann's aftereffect equation, **24A,** 46
Boltzmann's constant, **8,** 28; **10,** 2
Bond (chemical) density, **18,** 413, 419
Bond energies, concept of, **15B,** 273–274
Bond length accuracy, **3,** 300–301
Bond orbitals, **30,** 28–30
Bond strength, **14,** 231
Bond-dissociation energies, **3B,** 609
Bond-orbital calculations, **30,** 120
Bonding of transducers and thin slabs, **19,** 150, 161, 259–260
Bonferroni
intervals, **28,** 261
method, **28,** 175–176, 178
tests, **28,** 261
Bonn Telescope, **12B,** 13
Booster pump, *see also* Lobe pump
booster-diffusion pumps, **14,** 329
foreline, in diffusion pump systems, **14,** 163–165
Boosters, **23A,** 121
Bootstrap methods, **28,** 491–498
applications, **28,** 495–498
implementation, **28,** 494–495
methodology, **28,** 492–494
motivation, **28,** 491–492
nonparametric bootstrap, **28,** 493–494
parametric bootstrap, **28,** 493–494
Bootstrap sample, **28,** 493
Borates, BBO, **29C,** 257, 258
Bordè atom interferometer, **29C,** 295–298
Bordoni peaks, Peierls barriers and, **11,** 422
Borehole
breakouts, **24B,** 401–403
deformation gage, **24B,** 383–386
thermal equilibrium, **24B,** 192–193, 245–246

Borman effect, **4A,** 147
Born approximation, **3B,** 869; **6A,** 247; **7A,** 36–38, 73, 82; **8,** 227, 228; **17,** 89, 92, 110–112, 114; **21,** 147, 149, 150; *see also* Plane-wave Born approximation
  derivations from, **3B,** 920–926
  for electron energies, **9A,** 135
Born von K·rm·n formalism, **23A,** 377
Born, M., **23A,** 371
Born-Oppenheimer approximation, **15B,** 271
Boron
  disintegration reaction, **5A,** 500
  filtering of neutrons, **5A,** 518
  trifluoride chambers, **5A,** 108–109
Boron nitride, *see also* Ceramics
  ultimate pressure, **14,** 321
  ultraviolet radiation outgassing, **14,** 329
  water, outgassing, **14,** 326
Borrman effect, *see* Anomalous transmission
Bose condensation, **23B,** 331
Bose-Einstein condensation, **29B,** 141
Botha-Spalding burner, **7B,** 192–193
Bouguer correction, **24B,** 151
Bouguer plots, **12A,** 142
Bouguer's law, **1,** 335
Bound carrier effects in semiconductors, **6B,** 112–113
Bound electrons, **8,** 221
Boundaries
  eliminating effects of, **19,** 296–297
  of grains, **19,** 238, 240, 246, 269–270, 277, 279, 283, 290
  lateral, **19,** 294–297
  multimode guided wave propagation in presence of, **19,** 294–296
  of narrow specimens, **19,** 294–297
  reflections at, **19,** 262–264
  small angle, **6A,** 339–341
Boundary conditions, **14,** 276
Boundary effect in proportional counters, **5A,** 471–472
Boundary imperfections, **31,** 275
Boundary layer, **1,** 130, 133, 134
  recovery factor for temperature probe, **18,** 459
  study in wind tunnel, **18,** 766–768
Boundary value solutions, references for, **12B,** 62
Bourdon gauge, **14,** 47–48, *see also* Manometer
Bourdon gauge for pressure measurement, **18,** 511
Bourdon tube, **1,** 117; *see also* Amarada temperature recorder
Bovine serum albumin, **23B,** 532
Bow shock, definition of, **9B,** 50
Bow shock wave, **18,** 358
Box plots, **28,** 510–511
Box resonator, **8,** 130
Boxcar averages, **29B,** 225
Boxcar averaging, signal-to-noise improvement, **29A,** 446
Boxcar integrator, in microwave apparatus, **9A,** 59
BOXCARS (variant of CARS), **18,** 432
Boyle's Law, **14,** 39
Bragg cell acoustooptic modulation, **19,** 487–490
Bragg cell, for frequency shifting in LDV, **18,** 193–194
Bragg condition, **23A,** 9, 25
Bragg crystal reflection spectrometer, **17,** 361–365
Bragg diffraction, *see* surface acoustic wave applications
Bragg effect in holograms, **26,** 172
Bragg equation, **5A,** 566, 601; **16A,** 219; **16C,** 151, 178, 301
Bragg law
  crystal optics, **31,** 403–404
  multilayer gratings, **31,** 380
Bragg law, **3,** 280, 310; **6A,** 196–197; **9A,** 62
Bragg position, **6A,** 255
Bragg reflection, **8,** 239
  intensity, **23A,** 342
  width, **23A,** 325
Bragg regime, **26,** 151
  for light scattering, **19,** 82–84, 461, 470
Bragg relation, **1,** 421, 424; **23C,** 77
Bragg scattering, **21,** 147, 148, 158, 159, 176
  of neutrons, **16A,** 209–210
  vector, **20,** 303, 326
Bragg's law, **20,** 230–232
  of X-ray diffraction, **16B,** 32, 161
    using degraded samples, **16B,** 266–267
Bragg's rule, **21,** 230
  for stopping cross sections, **17,** 69
Bragg-Fresnel lens, **31,** 296
Bragg-intensity-versus-temperature measurement, **22,** 279
Brahma press, **1,** 67

Brake horsepower, **1,** 95
Branch, **2,** 19
  in network theory, **2A,** 7
Branching fraction, measurement, **29B,** 243–248
Branching in low-density polyethylene
  carbon–13 NMR determination, **16A,** 285–288
  long chain, GPC determination, **16A,** 61–63
Branching ratio(s), **5B,** 297–298; **7A,** 48
  for mesons and hyperons, **5B,** 846
Brass, **14,** 382, 481
  vapor pressure, zinc, **14,** 346, 347, 351
Brass, β-, **23C,** 27
Braun electrostatic voltmeter, **1,** 442
Bravais lattice(s), **3,** 270; **16B,** 3, 5; **23C,** 77
  magnetic, **23C,** 120
Brazing, **14,** 277, 426
Breakdown, **5A,** 112, 118; **6B,** 24–30; **22,** 368, 369; *see also* Electric breakdown
  dielectric, **30,** 463
  direct-current type, **7B,** 75–77
    near atmospheric pressure, **7B,** 77
  electrical, **14,** 307
  pressure, **24B,** 393
  voltage, **22,** 368
Breakdown, dc, **2,** 71–80
  high-frequency, **2,** 79–80
  non-uniform field, **2,** 77–78
  uniform field, **2,** 72–73
Breeding, **8,** 177
Brehmsstrahlung, **5B,** 235–237, 509–511; **13A,** 186; **18,** 407, 486, 706; **21,** 4, 5, 61; **22,** 199, 202, 207; **23A,** 128; **31,** 36
Breit-Rabi
  diagram for Cs^{132}, **5B,** 64
  formula, **3,** 450–451; **4B,** 115, 119, 196, 236
  corrections, **4B,** 119
Breit-Wigner
  formula, **23C,** 267
  scattering formula, **3,** 350
  theory for isolated resonances, **5A,** 538–541
  from electrons, **5A,** 56–61
    angular distribution, **5A,** 60
    energy distribution, **5A,** 57
  emission from plasmas, **10,** 287–295
  inverse type, **9A,** 89
  in microwave scattering, **9A,** 53
  in particle bombardment, **21,** 254
  polarization, **5A,** 60–61
  radiation, **17,** 458
Brewster, **26,** 116, 121, 122
  angle of incidence, **11,** 704
  angle polarizers, **10,** 104
  fringes, **1,** 350, 351, 359
  law, **1,** 378
Brewster's angle, **15B,** 26, 30, 90
  surface, in cw dye lasers, **15A,** 351
Bridge
  alternating current, **1,** 462ff
  balanced microwave, with superheterodyne detection, **5B,** 26–27
  capacitance, **1,** 67
  circuit for d.c. comparison of capacitances, **5B,** 488–490
  direct current, **1,** 456ff
  impedance, **1,** 486–487
  Kelvin double, **1,** 246
  measurements of impedance, **13B,** 142
  methods
    in conductivity measurements, **6B,** 40–45, 54–55
    in dielectric measurements, **6B,** 8–15; **16C,** 397–400
    evaluation, **6B,** 45
    use in nuclear resonance, **6B,** 230–232
  Mueller, **1,** 246
  mutual inductance, **1,** 542
  networks, microwave, **3,** 471–473
  rectifier, three-phase, **2A,** 234–238
  Schering, **1,** 464, 498
  transfer impedance, **1,** 498
  transmission-line, **1,** 487
  type samples, **6B,** 52
  Wheatstone, **1,** 246
Bridgeman internal Peltier effect, **1,** 508, 515
Bridges, in equipment testing, **2B,** 176–177
Bright-field image, in electron microscopy
  description, **16B,** 190–192
  mode of operation, **16B,** 211–212
Brightness, **1,** 389; **12A,** 193; **25,** 195
  ion beam, **29A,** 73
  temperature, **1,** 255; **18,** 465–466; **26,** 241
    definition of, **12B,** 2
Brightness ("Richtstrahlwert") of electron beam, **4A,** 87–92, 94, 124, 144–145
  available brightness, **4A,** 88
  brightness limit, **4A,** 88, 144
  dependance on beam diameter, **4A,** 89

effect on thermal energy, **4A,** 88
use of virtual demagnified cathode, **4A,** 88
Brightness calculation
bending magnet, **31,** 11–12
undulators, **31,** 17
wigglers, **31,** 14
Brightness of light sources, **18,** 689
of black-body, **4A,** 290, 291
photometric, **4A,** 286–288
temperature, **4A,** 292–293, 295
Brillouin components
intensity of, **3A,** 407
polarization of, **3A,** 409
Brillouin function, **11,** 502; **24A,** 314
Brillouin linewidth, **16A,** 178
Brillouin peak, intensity of, **16A,** 170
Brillouin scattering, **3A,** 397–403, 418; **15B,** 260; **18,** 415–417; *see also* Rayleigh-Brillouin scattering
stimulated, **3A,** 434–446
Brillouin spectroscopy
diamond anvil high-pressure cell, **24A,** 22
Fabry-Perot interferometer, **24A,** 15–19
goniometer, **24A,** 20
photon-phonon interaction, **24A,** 14
spectral intensity, **24A,** 21–22
Brillouin splitting, **16A,** 181, 190
Brillouin zone, **10,** 360; **11,** 3; **16A,** 101; **23A,** 375
zone peak, **21,** 154
Brinell hardness number, **6A,** 307
Brinkman-Kramers approximation, **17,** 489
British thermal unit, **1,** 93
Broad band impedance matching, **8,** 107–115
exercises, **8,** 114–115
Broad crested weir, **18,** 334–336
Broad source, **8,** 44–49
Broad-window-ratio method, **24B,** 570–573
Broadband amplifiers, **2A,** 305–309
defined, **2A,** 305–306
design considerations for, **2A,** 306–308
feedback in, **2A,** 307–308
Broadband polarizating beam splitter cubes, **29C,** 354
Broadband transducers, **19,** 61–65
flat passband, **19,** 62–64
impulse response (optimal), **19,** 61–62
tuning effects, **19,** 62–64
Broadening, **22,** 247
Broadening effects, **17,** 188; *see also* Doppler broadening; thermal broadening
inelastic energy-loss measurements, **17,** 168–173
Brockhouse measurement, **8,** 200
Brockhouse's formula, **8,** 197
Bromine
potential energy surfaces for, **15B,** 272
reaction of with halide, **15B,** 289–291
Bromine atom, addition of to olefins, **15B,** 288
Bromine cyanide laser emission, **10,** 50
Bromine-hydrogen-chlorine system
correlation diagram for, **15B,** 276
ground state potential surface for, **15B,** 274
Bronze, **14,** 382, 481
Brookhaven National Laboratory, **11,** 590
Brown and Rubens double yoke, **1,** 533, 554
Brown-Lien grazing-incidence monochromator, **22,** 144
Brownian motion, **18;** **20,** 308–310, 316
effect on tracer method, **18,** 39
rotational, **16A,** 412, 424, 432
Brugger elastic constants, **11,** 383, 385–386, 414, 438, 440, 434
Brumlein laser electrodes, **15A,** 119
Brushite ($CAHPO_4 \bullet 2H_2O$), **30,** 142
BSA, *see* Bovine serum albumin
Bubble chamber(s), **5A,** 203–207; *see also* Radiation-induced cavitation
energy measurements, **5A,** 436–438
ionization loss measurements, **5A,** 40–41
measurement of mass, **5B,** 35–37
momentum measurements, **5A,** 357–388
neutron energy measurements, **5A,** 478–481
range measurements, **5A,** 436–438
separated beams for, **5B,** 755–759
velocity measurements, **5A,** 451–453
Bubble collapse, *see* Collapse of transient cavity
Bubble count, **5A,** 40–41
Bubble fields, **19,** 380–383
dynamics using holographic methods, **19,** 382
mutual Bjerknes forces, **19,** 380–381
streaming, **19,** 381, 405, 408
Bubble model, of negative ion and positronium atom, **11,** 443–445
Bubble motion, **19,** 372–403
adiabatic versus isothermal gas bubbles, **19,** 378
asymmetric shape oscillations, **19,** 382–383
damping of gas bubbles, **19,** 376–378

equations, **19,** 373–374
nonlinear effects of gas bubbles, **19,** 378–380
rectified diffusion, **19,** 317, 375–376
resonances of vapor bubbles, **19,** 384
subharmonic generation, **19,** 379, 382, 394–395, 405
vapor filled, **19,** 383–385
Bubbles in tissues, **19,** 409–410
Bucking, microwave, **3,** 493
Buckling, **23A,** 165
Buerger precession goniometer, **3,** 318–319
Buerger precession method, **6A,** 213–214
Buffer gas, **4B,** 172, 182, 186
Buffer gases, in flowing afterglows, **7B,** 133–136
in optical pumping, **7B,** 13
Buffer rod method, for attenuation, **19,** 101–104, 110–111, 262–265
Buffon's needle, **28,** 437
Bulb lamp, **25,** 252
parabolic, **25,** 240
reflector, **25,** 240
tubular, **25,** 240
Bulk and surface parameters, separation, **6B,** 99
Bulk laser, **29C,** 172
Bulk lifetime, **6B,** 95–96
Bulk modulus, **1,** 101, 120, 224; **6A,** 297; *see also* Elastic moduli, bulk
definition, **16C,** 97
measurement of, **16C,** 112
Bulk optics, **29C,** 361
Bulk waves, light scattering from
in bulk media, **19,** 460–463, 471–473, 475, 484–492
at surfaces, **19,** 463, 484–492
Bulk waves, *see* Modes, longitudinal; Modes, shear
Bullard probe, **24B,** 231–232, 238
Buna N, **14,** 407
Bunching system, magnetic, **5A,** 548–549
Bunsen calorimeter, **1,** 266
Buoyancy, **1,** 60, 115
Burch interferometer, **26,** 20
Burch's law, **13B,** 28
Burg's algorithm, **28,** 326–328
Burgerscircuit, **6A,** 325–327
Burgersvector, **6A,** 325–327
Burning-on method of film preparation, **6A,** 132
Burnup, **8,** 180, 248
determination, *see* Nondestructive burnup determination.
history, determination, **8,** 203, 207–210
Burr-Brown Research Corp., **2A,** 301
Burris diode, **25,** 256
Burrows permeammeter, **1,** 552, 554
Burst noise, **2B,** 567–470
in amplifier, **2A,** 251
Bursts, in pulse amplitude measurements, **2B,** 94
Butt weld, **14,** 432
Butterworth response, **2A,** 333
BWO, *see* Backward wave oscillator

C

C, *see* Carbon
C field region, **3,** 616
C waves, **1,** 113
C-curve, **28,** 502
C_6H_6 on Rh(111), chemisorption, scanning tunneling microscopy of, **27,** 297
Cable protection, **14,** 453
Cable samples, for electric breakdown, **16C,** 461
Cable temperature jump, **19,** 194
Cable theory, **20,** 452–455
Cadmium
absorption of thermal neutrons, **5A,** 518
boiling point, **14,** 350
melting point, **14,** 347, 350
physical properties, **14,** 367
red line, **1,** 27, 55
sulfide, **12A,** 379
tin oxide (Cd_2SnO_4), **30,** 554
unacceptable in vacuum system, **14,** 481
vapor pressure, **14,** 346, 347, 350
Cadmium ion laser, **15A,** 128–143
cataphoresis in, **15A,** 134–137
excitation mechanisms in, **15A,** 131–134
noise in, **15A,** 137–138
spectroscopy of, **15A,** 128–131
Calcite, **12A,** 368, 374
Calcium carbonate ($CaCO_3$), **30,** 104–106, 146, 151, 160
Calcium fluoride (CaF_2), **14,** 395; **30,** 94–95, 98–99, 585–588, 601–606
Calcium-stabilized lasers, **29C,** 294–295
Calculation sheets for thin lenses, **25,** 29
Calibrated photodiode, **22,** 485

Calibration
electron-beam excitation, **31,** 78–79
extreme UV, **31,** 133
laser plasma source, **31,** 86
plasma source, **31,** 130
secondary source standard, **31,** 127
soft x-ray, **31,** 129–131
spectrometers, **31,** 71–72
ultraviolet, **31,** 138–143
vacuum UV, **31,** 138–143. *See also* Source standard, Spectroscopic system.
Calibration, **22,** 226; *see also* Antenna calibration; Pressure measurement
camera, in chronophotography, **18,** 84–86
electron beam fluorescence, **18,** system, **18,** 450–451, 453
flowmeter, **18,** 322
heat transfer gauge, **18,** 677, 680, 682, 684
hot-wire anemometer, **18,** 285, 297
infrared, **12A,** 440, 451–460
photometric, **12A,** 112–120
pressure gauge, **18,** 504, 507, 509, 512, 514, 555, 592, 606, 610
of radioactive neutron sources, **5B,** 565–567
Raman scattering diagnostic system, **18,** 429
Calorie, **1,** 93, 260, 261
Calorimeter
continuous flow, **1,** 285, 296
detectors, simulation, **28,** 447
differential, **5B,** 282–283
electric, **1,** 270, 272, 273–274
ice, **1,** 266
Joly's steam, **1,** 266, 274
measurement of microwave power, **2,** 737–738
measurement of power, **1,** 468
metal block, **1,** 265, 273
Nernst, **1,** 266
vacuum, **1,** 266f
water, **1,** 263–265, 283, 284
X-ray, **5B,** 541–543
Calorimetric determination of work function, **6B,** 134
Calorimetric method, of energy-loss measurements, **17,** 48–49
Calorimetry, **1,** 94, 260–286; **19,** 192; *see also* Differential scanning calorimetry
applied to heat transfer measurement, **18,** 664, 670–679
capacitance calorimeter, **18,** 672–674
tangential conduction error, **18,** 670–671
Calutron source, *see* Arc plasma discharge source
CAMAC data acquisition system, **2B,** 173, 178
CAMAC standard, **2A,** 43
Camera
electron diffraction, **3,** 338–341
lenses, **25,** 74, 95
multiplate, for measuring differential cross-section, **5B,** 426–428
Camera tubes, **2,** 773–782; **2B,** 345–355
applications of, **2B,** 352–355
choice of, **2B,** 353
classification of, **2B,** 345
color cameras as, **2B,** 354–355
general principles of, **2B,** 345
image dissector type, **2B,** 351–352
image orthicon type, **2B,** 347
isocon-mode return beam, **2B,** 349
output circuit of, **2B,** 348–349
photoconductive targets in, **2B,** 349–350
photoelectric effects in, **2B,** 349–351
photoemitters as, **2B,** 350–351
return beam signal in, **2B,** 348–349
spectral response requirements of, **2B,** 353
storage mechanism in, **2B,** 345–348
storage targets of, **2B,** 350–351
target action in, **2B,** 346–348
vidicon type, **2B,** 347
Canada balsam, **1,** 379
Canal mechanism of spark formation, **18,** 696
Candela, **18,** 689
Candle, **1,** 389
Cantilevered reed, **1,** 111
Capacitance value, **8,** 7
Capacitance, **1,** 434–438, 466, 490
bridge, **1,** 67
cells, **13B,** 136
cylindrical geometry, **16C,** 384
definition, **16C,** 383
junction, **5A,** 270–271, 272
manometer, **7A,** 395–396; **14,** 49–51, *see also* Manometer
measured by bridge method, **16C,** 398
measurement in resonant circuit, **16C,** 406
methods, **22,** 17, 19
contact potential, **22,** 17, 18, 19
Fermi-level, **22,** 17

Kelvin-Zisman (vibrating capacitor), **22,** 17
reference surface, **22,** 19
variation circuit, **6B,** 16–17
Capacitative energy, **8,** 132
Capacitor(s), **1,** 436–438, 489, 496, 498; **2,** 42–44
analyzer, **22,** 220
coupling, vs. transformer coupling, **2A,** 282
electrolytic, **2A,** 37–38
in electronic circuits, **2A,** 36–38
equivalent circuit of, **2A,** 36
filter, **2,** 211
in integrated circuits, **2A,** 119–120
losses, **1,** 497
parallel plate, **1,** 495ff
temperature coefficient of, **2A,** 44
thermometry and, **11,** 538
Capillary correction to manometer, **18,** 507
Capillary discharge, **31,** 59–60
Capillary viscometer, **16A,** 19; **16C,** 45, 47
Capsule construction, **8,** 247
Capsule gauge for pressure measurement, **18,** 511
Capture cross section, neutron, **3,** 349–350
Capture efficiency, *see* Pumping speed, efficiency
Capture probability, *see* Pumping speed, efficiency
Capture pump, **14,** 249
Carbon, **22,** 154, 176, 184
arc, **1,** 373; **3,** 77
electron scattering, **5A,** 323ff
cross section, **3,** 337
physical properties, **14,** 383–384
tetrachloride
electron diffraction, **3,** 344–347
radial distribution function, **3,** 342
thermometer, **1,** 248
vapor pressure, **14,** 347, 350
boiling point, **14,** 350
Carbon arc, **8,** 27; **13B,** 5
Carbon arc sources, **25,** 248
Carbon bolometer, **10,** 169, 170
Carbon dioxide, **19,** 202
atmospheric concentration, **12B,** 170
electrical discharge in, **15B,** 70–71
from glass, ultraviolet irradiation, **14,** 329
ionization gauge production, **14,** 70
laser(s), **9A,** 82; **10,**61–64; **15B,** 83, 91; **29C,** 4, 148
amplifiers, **15B,** 91
physical properties, **14,** 421, 422
Raman scattering, **10,** 422, 423
in vibrational energy transfer, **15B,** 72
vibrational modes, **10,** 323, 324
Carbon disulphide, **23A,** 465, 481
Carbon film
for electron microscope specimen support, **16B,** 213
replica technique, **6A,** 180–181
Carbon foil. in beam-foil experiments, **17,** 532, 536–537
Carbon monoxide, **22,** 154
absorption coefficient of, **12B,** 169–171
atmospheric concentration, **12B,** 170
chemical laser, **15B,** 105–111
laser, **15B,** 83
microwave lines of, **12B,** 171
on Pd, **22,** 95
on Pd(100), **22,** 84
on Pd(111), **22,** 88
on Pd(210), **22,** 91
on Ru(001), **22,** 66, 113, 117
on W(100), **22,** 324
Carbon monoxide-carbon monoxide collisions, **15B,** 106
Carbon rod furnace, **13B,** 161
Carbon, reflectance spectra, **31,** 223–224
Carbon–13 NMR
application to polymer solid state studies, **16A,** 343–368
branching determination in polyethylene, **16A,** 285–288
probe for molecular dynamics, **16A,** 302–319
summary of information obtainable on polymers, **16A,** 368
Carbon–13 nucleus
spin-lattice relaxation rate, **16A,** 251
abundance in polymers, **16A,** 253, 326
Carbon-hydrogen-metal interactions, **23A,** 353
Carbonizing, steel, **14,** 357
Carburization of cathodes, **4A,** 26
Carcinotron, *see* Backward wave oscillator
Carcinotrons, **2,** 727; **2B,** 272–273
Cardinal points, **1,** 335
of a lens, **25,** 25
Carey-Foster bridge, **1,** 464
CARN technique, in thin films, **11,** 656
Carnot cycle, **1,** 239

Carr-Purcell sequence, *see* Measuring spin-spin relaxation time T_2
Carrier free isotope, **21,** 20, 21, 23, 25, 42
Carrier(s)
frequency, **2A,** 248
injection, **2,** 119
ion transporting, **20,** 530–535
majority, **2,** 114
modulation, in telemetering, **2B,** 248–249
systems, in feedback control, **2B,** 442–444
CARS (coherent anti-Stokes Raman scattering), **18,** 431–433, 489; **20,** 159–161; *see also* Coherent anti-Stokes Raman spectroscopy
Cascade emission, **3A,** 336
Cascade multiplier circuit, **2A,** 210
Cascade tripler, **2A,** 208–210
Cascading, **7A,** 46, 63, 133, 134; **13B,** 236
Cascading, effect of, on decay curves, 572–575, 581
Cascode circuit, **4A,** 335, 337
Cascode stage, **2,** 411–412, 816
Case hardening, *see* Carbonizing
Casimir's function, **3,** 18
Casimir-Polder force, **29B,** 341
Cassegrain antenna, **12B,** 13–14, 48
Cassels shower detector, **5A,** 664–665
Cassiopeia A, **12C,** 173
Casting of polymer films, from solution, **16C,** 57, 457
Catalyst surface, **22,** 326
Catalysts, **23C,** 469
Catalytic conversion gas purifiers, **4B,** 324
Catalytic hydrolysis of polyesters, **16B,** 274
Catalytic surfaces, **23B,** 63
Cataphoretic flow, in cadmium ion lasers, **15A,** 134–137
Catephoretic segregation (noble gases), **4B,** 323
Cathode
Ba-Sr oxide, **14,** 368
follower, **2,** 215
sheath, **14,** 448
sputtering
of films, **6A,** 129–130
of thin-surface layers, **7B,** 244–245
thoriated tungsten, **14,** 368
tungsten, **14,** 36.8
types, **2,** 48–51
dispenser, **2,** 50
monolayer emitter, **2,** 49
oxide emitter, **2,** 49–50
pure metal, **2,** 48–49
Cathode erosion
electron beam excitation, **31,** 76
hollow cathode, **31,** 69–71
Penning discharge, **31,** 76
Cathode ray oscilloscope, **14,** 82
Cathode-ray devices, **2,** 764–784; **2B,** 335–338
camera tubes as, **2B,** 345–355
electron-ray indicator tubes, **2B,** 335–336
image dissector, **2B,** 351–352
storage tubes as, **2B,** 355–358
Cathode-ray indicators, **2,** 570
Cathode-ray oscilloscope, *see* Oscilloscope
Cathode-ray tube, *see also* Oscilloscope
applications of, **2B,** 341–345
cathodochromic, **2B,** 358
for character displays, **2B,** 344
for computer readout and displays, **2B,** 344–345
defined, **2B,** 336
displays, in pulse amplitude measurements, **2B,** 119
electrostatic deflection in, **2B,** 338
flying-spot, **2B,** 342
operation of, **2B,** 337
phosphor screen in, **2B,** 339
principles of, **2B,** 336–345
in pulse amplitude measurements, **2B,** 130–132
radar applications of, **2B,** 343
screen materials used in, **2B,** 340
as TV picture tube, **2B,** 342–343
Cathodes
dispenser, **2A,** 150–151
oxide-coated, **2A,** 149–150
thermionic, **11,** 451
thermoionic, *see also* Matrix cathodes, Oxide cathodes
hexaboride, **4A,** 20
"hot" electron, **4A,** 34, 38
low temperature, **4A,** 33–35
metal, **4A,** 29–33, 35
rare earth oxide, **4A,** 23
thin film, **4A,** 28–29, 38
thoria, **4A,** 19
thoriated, **4A,** 25–27, 35
types and shapes of, **2A,** 151–152
Cathodochromic tubes, **2B,** 358
Cathodoluminescence, **6B,** 294, 301–331
efficiency, **6B,** 307

Catoptric optics, **25,** 169
Cauchois photographic x-ray spectrometer, **5A,** 599–601, 605
Cauchy formula, **9B,** 12
Cauchy function, to describe X-ray line shape, **16B,** 143, 145
Cauchy relation, **1,** 362
Causality principle, **24A,** 47
Caustic curve, **31,** 166–167
Cavalier's theorem, **1,** 60
Cavendish gravitational experiment, **23A,** 333
Cavitation, **1,** 215–216
  applications of, **19,** 403–411
    activity measures, **19,** 404–405
    cell disruption, **19,** 406
    cleaning, **19,** 405–406
    depolymerization, **19,** 406
    electroplating, **19,** 405
    emulsification and dispersion, **19,** 406
    gas diffusion and heat conduction, **19,** 405
  and atomization, **19,** 337
  definitions, **19,** 357
    cavitation streamer, **19,** 381
    stable cavitation, **19,** 358
    transient cavitation, **19,** 358
  dimensional analysis, **18,** 843, 848
  and electric-discharge machining, **19,** 352
  and emulsification, **19,** 328–329
  inception of, *see* Thresholds for cavitation
  as periodic process, **19,** 327
  prediction, **19,** 398–403
Cavitation dynamics, *see also* Bubble motion
  bubble motion, equations, **19,** 373–374
  nondimensional variables, **19,** 373
Cavitation inhibition, **19,** 406–408
  by degassing, **19,** 406
  by filtering, **19,** 407
  by increased pressure, **19,** 407
  paradoxes, **19,** 408
  by prepressurization, **19,** 407
  by wetting, using surfactants, **19,** 407
Cavitation nuclei, **19,** 365–367, 371–372
  motes, **19,** 360, 362
  skin model, **19,** 366
Cavities, **3,** 26–28; **10,** 124–126, 156
  materials for, **11,** 563–564
Cavity, **8,** 129–134, 138, 139, 141, *see also* Transmission cavity.
  resonator, *see* Microwave cavity resonator.
  tuning, **8,** 129
  walls, **8,** 137
Cavity, microwave resonant coupling, **3,** 476
  for double resonance, **3,** 479–480
  for high frequency modulation, **3,** 479–480
  impedance, **3,** 467
  low temperature, **3,** 478–479
  modes, **3,** 474–476
  quality, **3,** 473
  reflection coefficient, **3,** 490–491
Cavity, microwave, **6B,** 238
Cavity, resonant, **1,** 479, 485
Cavity methods for afterglow measurements, **7B,** 87–96
Cavity modes, tunnel junctions and, **11,** 229–233
Cavity radiation detector, **4A,** 449
Cavity resonators, **13B,** 161
CBEA, *see* Constrained binary encounter approximation
CBW, *see* Constant bandwidth
CCD, *see* Charge-coupled device
CCL, *see* Color-center laser
$(CD_3)_4NMnCl_3$, *see* TMMC
Cd sheets, **8,** 236
$CeAl_2$, **23C,** 198
CeB_6, **23C,** 148
Celacol film, **11,** 692
Cell(s)
  equivalent points, **3,** 272
  electrolytic, **1,** 452
  Fourier representations, **3,** 286–289
  Golay, **1,** 393, 401
  Kerr, **1,** 386, 397
  long path, **3,** 171–172
  multiple-pass absorption, **3,** 171–172
  photoconductive, **1,** 402
  photoemissive, **1,** 391
  photovoltaic, **2,** 758–759
  Raman spectra, **3,** 131–136
Cellulation during spherulitic growth, **16B,** 345
Cellulose
  chain conformation in crystalline regions, **16B,** 270, 272–273
  selective degradation by hydrolysis, **16B,** 269–273
  triacetate, single crystals, selective degradation, **16B,** 273
  water absorption, **16C,** 319
Celsius scale, **1,** 32, 238
CEM, *see* Channel electron multiplier

Cement, **14,** 414–417
Cements, **23C,** 446
  for optical components, **12A,** 370
Center of mass, **8,** 265, 266
Center of perspective, **25,** 80
Center-of-mass angles, **8,** 269
Center-of-mass coordinate, **8,** 277
Center-of-mass solid angle, **8,** 253, 258, 265
Center-of-mass system, **8,** 257, 258, 262, 278, 279
Center-of-mass system, in inelastic kinematics, **17,** 159, 161–163, 174
  scattering cross section, **17,** 164, 166
Center-of-mass velocity, **8,** 262
Centigrade scale, **1,** 32, 238
Central field region, **8,** 78
Central limit theorem, **28,** 26–27, 252
Central peaks, **23C,** 59
Central processing unit, **2B,** 167
Central-mass-image broadenings, **8,** 183
Centrifugal acceleration, **1,** 63
Centrifugal barrier factor, **8,** 279
Centrifugal distortion, **3A,** 72
Centrifugal force, **18,** 803
Centroid of counting time interval, **21,** 68
Centroid shift method, **5B,** 321–322
Ceramic capacitor spark light source, **18,** 700
Ceramic-to-meal seals
  Kovar, **14,** 379
  using platinum, **14,** 373
  using zirconium hydride, **14,** 365
Ceramics, **4B,** 272; **14,** 396–406; **19,** 30, 32, 52–55, *see also* specific ceramic materials
  boron-nitride, **14,** 405–406
  dielectric constant, common ceramics, **14,** 400
  Lava, **14,** 404–405
  Macor, **14,** 404, 481
  silicate, **14,** 398–401
  pure-oxide, **14,** 401–404
  in ultrahigh vacuum systems, **14,** 481
Čerenkov
  counters, **2B,** 80; **5A,** 10, 162–194
    velocity measurement, **5A,** 454–460
  light, **12A,** 40, 317, 343–350
  monitors, **5B,** 502
  radiation; **7A,** 177; **10,** 28, 300
    diffraction, **5A,** 176
    energy loss due to, **5A,** 8–10
    pulse time, **5A,** 192
  shower detector response to positrons, **5A,** 663
  spectrometer, **5B,** 531
  waves, **9A,** 238
Cerous magnesium nitrate (CMN), **11,** 503–511
  in thermometry, **11,** 503–511
CeSb, **23C,** 146
Cesium, **12C,** 133
  activation of negative-electron-affinity semiconductor photocathode, **29A,** 15–17
  ampoules, **11,** 113
  beam decay, **5B,** 67–68
  channels, **11,** 113
  iodide (T1), **12A,** 379
    scintillators, **5A,** 413ff, 618–619
  in photoemission measurements, **11,** 110–112
  resonant frequency, **12C,** 210
  restoration of photocathode efficiency, **29A,** 21–22
  sources, **29A,** 15
  thermal emission of ions, **29A,** 158–159
Cesium frequency standards, **8,** 163
Cesium-beam clock, **1,** 72; **12C,** 178
Cesium-beam resonator, **12C,** 208, 219–221
CFFT2, **12C,** 292, 294, 318
CFS, *see* Constant final state
Chain drive, **14,** 450
Chain folding, **16A,** 8, 397
Chain folds, infrared studies, **16A,** 129
Chain reaction, fission, **5B,** 591
Chain reactions, in chemical kinetics, **15B,** 283
Chain-folding model, **16B,** 342, 343–344
Chalcogenide glasses, **29C,** 375
Chamber
  bubble, **5A,** 203–207
  cloud, **5A,** 194–201
  diffusion, **5A,** 201–203
  discharge, **5A,** 281–288
  expansion, **5A,** 197–201
  luminescent solid, **5A,** 159–162
  spark, **5A,** 281–288
  tracks
    noise, **5A,** 381–382
    reconstruction in space, **5A,** 380–381
Channel
  conductance, **6B,** 141–142
  counts, analog storage, **2,** 611–612
  effect, **6B,** 104
  electron multiplier(s), **2B,** 310; **17,** 540–541, 544–547; **22,** 286

flow metering, **18,** 332–336
layer, **1,** 213
plate, **12A,** 239
in pulse amplitude measurements, **2B,** 85–86
waveguides
fabrication, **29C,** 384, 386
on glass substrates, **29C,** 383–384
width
errors, in pulse amplitude measurements, **2B,** 99–101
stability, in pulse amplitude measurements, **2B,** 81
Channel mass resolution, **8,** 261
Channel width, **8,** 196
Channel-plate electron multiplier, **22,** 140, 184
Chevron, **22,** 287
Channeled fraction of beam, **21,** 295, 308, 327
Channeled spectrum, **26,** 28
Channeling in penetration studies, **7B,** 245–264
Channeling of atomic beams, **5B,** 76
Channeling, apparatus
accelerator system, **21,** 297
basic, **21,** 278
beam lines, **21,** 298
current integration, **21,** 297, 299
detectors, *see* Detectors, for channeling
experimental techniques, **21,** 297–303
goniometer, **21,** 298
target
chamber, **21,** 298–299
preparation, **21,** 302, 303
vacuum system, **21,** 298, 345
Channeling, applications
absorbed atoms, **21,** 345, 350, 354–358
amorphous layers, **21,** 303, 304, 308
dislocations, bubbles and voids, **21,** 315–319
disorder, **21,** 303, 311–313
displacement of host crystal atoms, **21,** 308–322
displacement of solute atoms, **21,** 323, 333–335
effect of films, **21,** 304, 305
interstitial atoms, **21,** 277, 278, 333–339
lattice defects, **21,** 275, 283, 297, 303, 322, 345
location of solute atoms, **21,** 322
ordered alloys, **21,** 329
phase changes and precipitations, **21,** 281, 321, 331, 333
polycrystalline layers, **21,** 305–308
radiation damage, **21,** 311–316, 339
random sites, **21,** 345
sensitivity, **21,** 291, 290
substitutional impurities, **21,** 277, 328–339
surface studies, *see* Surface studies, channeling
vacancy trapping, **21,** 343, 344
Channeling, characteristic angles, **21,** 283
critical angle, **21,** 287, 289, 303, 304
Lindhardt angle, **21,** 283, 290
Channeling, definition, **21,** 283
Channeling, surface peak, **21,** 301, 347
Channeling, theory
Bohr potential, **21,** 286
computer simulation, **21,** 289, 291, 292, 349, 352, 353
continuum
model, **21,** 283
potentials, **21,** 276, 283–285
flux distribution of channeled ions, **21,** 291
Lindhardt potential, **21,** 284–286
Molière potential, **21,** 284–286
particle trajectories, **21,** 286–289
screening radius, **21,** 276, 283, 284, 290
string potential, **21,** 291
thermal vibrations, **21,** 286
Thomas-Fermi potential, **21,** 276
Channeling, yield
aligned, **21,** 289
angular dependence, **21,** 289, 293, 301, 306, 307, 348
from solutes, **21,** 323, 324
calculation of, **21,** 289, 290
from displaced atoms, **21,** 293
in double alignment, **21,** 290
minimum, **21,** 289, 290
amorphous layer, **21,** 304
normalized, **21,** 276, 289–295, 327
Channels, **20,** 447, 472, 473, 484, 497, 536–539
noise in, **20,** 447, 497
single-channel recording, **20,** 484
single channels in bilayers, **20,** 536–539
single channels in excitable membranes, **20,** 447, 472, 473
Channeltron multiplier, **22,** 320
Chaos, **23B,** 575
Chapman-Enskog equation, **1,** 140
Chapman-Jougute condition, **18,** 838
Character recognition, **26,** 197

Characteristic acoustic impedance, *see* Impedance
Characteristic function parameters, **20,** 368, 369
Characteristic loss, **22,** 196, 229
Characteristic matrix for thin films, **25,** 274
Characteristic time, **22,** 430
Charcoal, activated, **1,** 180; **14,** 237, 243, 244
Charge, *see* Electrostatic charge
  bound, **1,** 489
  determination, **5A,** 289–307
  distribution of nuclei, **5A,** 325, 327
  free, **1,** 489
  measurement, **1,** 451–452
  to mass ratio, **1,** 64
Charge (transfer) exchange, **7A,** 410–428
  Aston band method, **7A,** 423–426, 472
  beam attenuation in transverse fields, **7A,** 413–416
  condenser method of collecting slow ions, **7A,** 417–421
  dependence upon excited state population, **7A,** 426–428
  equilibrium method of measurement in gases, **7A,** 411–412
  growth curve method, **7A,** 421–423
  with mass analysis, **7A,** 419–421
Charge coupled devices (CCDs), **29B,** 365–366, 419–420
Charge density wave (CDW), **27,** 350, 385
  gap, **27,** 432, 434, 447
Charge density waves in
  1T-TaS_2, **27,** 358
  1T-$TaSe_2$, **27,** 358, 374
  1T-$TiSe_2$, **27,** 381
  1T-VSe_2, **27,** 359
  2H-$NbSe_2$, **27,** 386
  2H-TaS_2, **27,** 391
  2H-$TaSe_2$, **27,** 388
  4Hb-TaS_2, **27,** 393, 400
  4Hb-$TaSe_2$, **27,** 393
  $NbSe_3$, **27,** 402, 405
  $TaSe_3$, **27,** 402, 405, 416
Charge detection microscopy, **27,** 90
Charge exchange, **4B,** 252–254; **17,** 384–385; *see also* ; Charge transfer Electron capture; Electron loss
  Auger spectroscopy, **17,** 482, 488–489, 500–503
  fundamental processes, **17,** 78–83
  lithiumlike atoms, **17,** 509–510
  probabilities, **17,** 108
Charge measurement, **2,** 574–576; **2B,** 61–63
Charge on molecule, **4B,** 9
Charge rearrangement, by merging beams, **7A,** 478
Charge screening, **30,** 49
Charge state, **17,** 189–190
  heavy-ion-atom collision, **17,** 151, 253, 256–257, 272–278
  high-velocity ions in matter, **17,** 73–148
    atomic transitions and charge equilibrium in solid, **17,** 139–144
    density effects, **17,** 131–135
    electron capture, **17,** 89–110
    electron loss, **17,** 110–118
    equilibrium charge-state distribution, **17,** 118–124
    experimental techniques and data analysis, **17,** 83–89
    historical background and review, **17,** 73–78
    radiative electron capture, **17,** 144–148
  identification, **17,** 552–554
  postfoil charge distributions in beam-foil interactions, **17,** 551–552
Charge transfer
  role of energy defect, **29B,** 105
  as source of
    metastable atoms/ions, **29B,** 105
    fast neutrals, **29B,** 55–57, 305
    negative ions, **29B,** 305
    Rydberg states, **29B,** 119–121
Charge-analyzed reaction products, coincidence experiments involving, **17,** 345–453
Charge-changing cross section, **17,** 79–82; *see also* Electron capture cross section; Electron loss cross section
  measurement of, **17,** 83–84
Charge-changing process, in ion-atom collision, **17,** 303–354
Charge-changing reaction, **7A,** 429, 430
Charge-coupled device (CCD), **2A,** 130–133
  array, as short-pulse detector, **29C,** 199, 212
  buried channel, **2A,** 136
  camera imaging of microchannel plate output, **29A,** 260–261
  gate system and, **2A,** 135
  gate system, **2A,** 135
  structure, two-line, **2A,** 136
  three-phase, **2A,** 132
  two-level gate structure in, **2A,** 136

Charge-state dependence
conditions for measuring, **17,** 280–284
in Coulomb ionization, **17,** 495–496
in electron promotion, **17,** 490–491
inner-shell cross sections, **17,** 279–302
projectile cross sections, **17,** 297–302
target x-ray production, **17,** 284–289
Charge-state fractions, **17,** 118–119, 121
density effect, **17,** 132–133
measurement of, **17,** 83–89
nonequilibrium and equilibrium conditions, **17,** 80–83
Charged particle optics program, three-dimensional
electrostatic lends system design, **29A,** 206–207
features, **29A,** 206
Charged particle oscillators, **31,** 38
Charged particle reactions, **23A,** 532
table, **23A,** 548, *see* Nuclear reactions
Charging, insulators, **14,** 353
Charles' law, **1,** 165; **14,** 40
Chattock magnetic potentiometer, **1,** 553, 555
CHCH
transition frequencies of, **12C,** 26, 27
CHCN
transition frequencies of, **12C,** 26
Chebyshev response, **2A,** 334
Chebyshev's inequality, **28,** 25
Chemi-ionization, in excitation of bounds states, **15B,** 227
Chemical cell, reversible, **1,** 286
Chemical composition measurement, *see* Composition measurement
Chemical crystallography, **23A,** 339
Chemical deposition of films, **6A,** 130–132
Chemical energy surfaces, **15B,** 271–276
Chemical etching in thin surface removal, **7B,** 243
Chemical kinetics, **3B,** 638
bimolecular relaxation in, **15B,** 280
chain reactions in, **15B,** 283
diffusion-controlled reactions in, **15B,** 281
excitation processes in, **15B,** 278–279
in laser-selective chemistry, **15B,** 278–283
study in shock tube, **18,** 656–659, 792–795
surface-catalyzed reactions in, **15B,** 280–281
termolecular reactions in, **15B,** 280–281
unimolecular decay in, **15B,** 279–280
Chemical laser systems, operation of as supersonic chemical lasers, **15B,** 103–104
Chemical lasers, **15B,** 95–96, *see also* Chemically pumped lasers; Purely chemical lasers
in double-resonance experiments, **15B,** 118
efficient visible, **15B,** 139–140
new, **15B,** 120–140
operation of at short wavelengths, **15B,** 121–125
photodissociation in, **15B,** 111
purely, *see* Purely chemical lasers
rotational, *see* Rotational chemical lasers
rotational disequilibrium in, **15B,** 117
supersonic, *see* Supersonic chemical lasers
Chemical potential, **1,** 319, 325–327; **16C,** 286; **22,** 3
microscopy, **27,** 86, 87
Chemical processing in isotope production, **21,** 22–25
of irradiated material, **21,** 23
of target material, **21,** 25
of unirradiated material, **21,** 24
Chemical reactions, nature of with respect to laser action, **15B,** 99–100
Chemical release of atoms, **4A,** 178
Chemical shielding in NMR, **3B,** 481
Chemical shift, **16A,** 256–259; **20,** 23, 24, 39; **22,** 223, 224
anisotropy in, **16A,** 328–331
anisotropy relaxation, **21,** 450–451, 480–482
NMR, **21,** 398, 400–402
shielding sensor, **21,** 402
Chemical techniques for radioisotope sources, *see* Atomic beams of radioisotopes
Chemical vapor deposited (CVD) films, **2A,** 105
Chemical vapor deposition, of thin films, **11,** 645–648; *see also* Thin films
Chemically pumped lasers, **15B,** 94–142
Chemiluminescence, **6B,** 295
quenching of, **15B,** 133
use composition measurement, **18,** 641–643
Chemiluminescent method of flame study, **7B,** 214–215
Chemisorbed layer, **22,** 364
Chemisorbed species, **22,** 366
Chemisorbed systems, **23B,** 63

Chemisorption, **4B,** 324; **14,** 25, 26, 28, 231; **22,** 131, 154, 155, 157, 169, 181, 182, 365, 454
Chemistry studies, high temperature, **3,** 558–565
Chemorheology, **16C,** 7
Chevron feedthrough, **14,** 305
Chevron gasket, **14,** 305
Chi-square
  estimate of, **5B,** 804–806
  function, **1,** 5, 6, 22; **3,** 713; **5B,** 811; **6A,** 5
  test(s), **2,** 5; **28,** 188–192
    composite hypothesis, **28,** 189–190
    Pearson's, **5B,** 786–789
    simple hypothesis, **28,** 189
    two-sample tests, **28,** 191
Chief ray, **25,** 37
Child-Langmuir law, **2,** 52
Chirality, **23C,** 136
Chirp slowing, **29B,** 132–133
Chlorination method, **8,** 179
Chlorine, **19,** 220; **22,** 154, 155, 157, 158, 169, 181, 182
  atmospheric concentration, **12B,** 170
CHNH, transition frequencies of, **12C,** 27
CHO, transition frequencies of, **12C,** 26
CHOH, transition frequencies of, **12C,** 26
Choke, filter, **2,** 193–194, 210
Choke-input filters, **2A,** 192
Choked flow, **18,** 330, 772
Cholesteric liquid crystals, **29C,** 356, 357
Chopper(s), **1,** 451; **2A,** 300; **23A,** 211
  amplifier, **2A,** 300–305
  circuits, **2A,** 304
  fast, **5A,** 529–533
  Jacrat, **5A,** 503
  mechanical
    applications, **5A,** 504–509
    for pulsed neutrons, **5A,** 499–513
  neutron, **16A,** 220
    speed of rotation, **5A,** 500–501
  slow, fast neutron background, **5A,** 507–508
Chopper-stabilized amplifier, **4A,** 345–346
Chopping secondary mirror, **12A,** 446–447, 450
CHORDIS, *see* Cold and hot reflex discharge ion source
Christian filter, **23A,** 265
Christiansen filters, **1,** 401; **13B,** 25
Chromatic aberration(s), **1,** 331; **25,** 30
  coefficient, **16B,** 202
  in electron microscopy, **16B,** 190, 201–203
Chromatic aberration, **8,** 184
Chromatic dispersion in gratings, **26,** 72
Chromatography
  for composition in sampled fluid, **18,** 617–621
  gel permeation, *see* Gel permeation chromatography
  liquid, **16A,** 42
Chromium oxide film, stainless steel, **14,** 355
Chromium, reflectance spectra, **31,** 218
Chrono-interferometer, **18,** 551, 607
Chronograph, Le Boulangé, **1,** 75
Chronointerferometer, **7B,** 172–173
Chronophotography
  calibration of camera, **18,** 84–86
  camera requirements, **18,** 79–83
  compared with other velocimeters, **18,** 64–66
  dark and bright field illumination, **18,** 76–79
  data analysis, **18,** 86–87
  definition, **18,** 64, 66
  directional information, method, **18,** 67
  error analysis, **18,** 87–89
  illustration of system design, **18,** 89–93
  interrupted illumination, **18,** 67–76
  measuring volume, **18,** 83–84
  rotating flow apparatus, **18,** 818–819
  system elements, **18,** 67
Chronotron, **5A,** 548; **5B,** 316–317
  principle, **1,** 73
  vernier, **4B,** 221
CHS
  transition frequencies of, **12C,** 26
Chudley-Elliott model, **23B,** 132, 151, 201
Cilia, **16A,** 129, 394; **16B,** 342, 344, 346
Cin»-interferometry, plasma applications of, **9B,** 46–47
Cinematography. high speed, **18,** 726, 732–739
Cinnamic acid, **30,** 418
Circle of confusion, at a point image, **16B,** 200
Circle of least confusion, **1,** 332, 333
Circuit boards, mounting and wiring of, **2A,** 42
Circuit elements, **2A,** 45–146, *see also* Semiconductor circuit elements
  ideal, **2A,** 1
  in passive linear circuits and networks, **2A,** 1–7

Circuit trouble, causes and prevention of, **2B,** 218–219
circuit, **8,** 124
Circuits, *see also* Networks
  coincidence, **2,** 335–341
    biased-diode pickoff, **2,** 338–340
  components, **1,** 461–467, 481–486
  components of, **2A,** 33–39
  count rate, **2,** 351–356
  dc, **2,** 22
  digilog, **2,** 622
  diode permission, **2,** 340–341
    using multielement tubes, **2,** 341
  distributed, **1,** 475–488
  electronic, *see* Electronic circuits
  elements
    distributed, electromechanical devices, **2,** 40–41
    passive, **2,** 17–46
    semiconductor, **2,** 113–176
    transmission line, **2,** 39
  equivalent, **1,** 461ff
  frequency selective, **1,** 473
  gas-tube, **2,** 298–307
  high-frequency, wiring of, **2,** 46
  integrated, **2A,** 33
  linear transistor, **2,** 307–326
  low-noise, **2,** 815–817
  lumped, **1,** 459–475
  microwave, **2,** 705–742
  nonlinear, **2,** 327–451
  nonregenerative, **2,** 328–357
  rectifier, **2,** 177–211
  regenerative, **2,** 357–393
  scaling, **5B,** 795–799
    constant number of counts, **5B,** 796–797
    constant time interval, **5B,** 795–796
    varying time and counts, **5B,** 797–799
  semiconductor integrated, **2,** 171–174
  simple resonant, **2,** 22–24
  sweep, **2,** 394–412
  transistor switching, **2,** 412–451
  trigger, **2,** 371–382
  voltage reset, **2,** 787–789
Circular aperture, diffracting, **26,** 58, 66
Circular array antenna, **12B,** 24
Circular dichroism, **12A,** 397
Circular loop
  field, **8,** 61–62
  pair, magnetic field calculation, **8,** 77
Circular polarization, **1,** 377; **12A,** 395; **26,** 110; **31,** 7–10
  piezo-optical effect, **31,** 252
  by reflection, **31,** 246–251
  synchrotron radiation, **29C,** 30–33
  by transmission, **31,** 252–253
Circular polarizer(s), **10,** 127, 128; **26,** 124
Circular stop, **8,** 36
Circular waveguide, **6B,** 22
Circular, in ESR spectrometer, **2B,** 149
Circularly polarized light, **11,** 700
Circulator, **8,** 141
CIRCUS, digital computer program, **2A,** 286
CIS, *see* Constant initial state
CISS, *see* Constant-initial-state spectroscopy
Cl, see Chlorine
Clamp circuits, diode, **2,** 421
Clamping circuits, **2A,** 385–389
Clapeyron equation, **3,** 560
Clark Lake Radio Observatory, **12B,** 21
Clark's cell, **1,** 286
Clark's spherical magnet, **1,** 527
Clarke slug, **11,** 209, 260–262
  applications of, **11,** 296–299
Classical absorption, **19,** 186
Classical approximation, **23A,** 40
Classical probability, **28,** 6
Classical theory of e-m fields, **13A,** 32ff
Clausing factor, **1,** 185, 186; **14,** 17
Clausing formulation, **14,** 17
Clausius-Clapeyron equation, **1,** 281, 320, 325; **22,** 439
Clausius-Mosotti equation, **3,** 641
Clausius-Mosotti relation, **13B,** 171; **18,** 348
Clean Ge surfaces and alloys
  scanning tunneling microscopy of, **27,** 151
Clean semiconductor surfaces
  scanning tunneling microscopy of, **27,** 151
CLEAN, **12C,** 151
Cleaning, **14,** 441–448; **19,** 405–406
  abrasive, **14,** 500
  ceramics, **14,** 397
  chemical etch, **14,** 501
  etches, **6A,** 156–160
  laser-assisted, **30,** 357–359
  pumps, **14,** 502–503
  solvents, **14,** 499–500
  storage, **14,** 500
  thick layers, **14,** 501
  of vacuum systems, **14,** 498–503
Cleanup, electrical, **14,** 216

Cleavage planes, in photoemission measurements, **11,** 103
Clebish-Gordon coefficient, **3,** 652; **11,** 521–522
Climb, **6A,** 329–330
Clipper, defined, **2A,** 369
Clipping circuits, in pulse preamplifier, **2B,** 107
Clock transition, in logic circuits, **2A,** 415
Clocks
  atomic, **1,** 71–72
  cesium-beam, **1,** 72
  electronic, **5B,** 308–320
  microwave absorption, **1,** 71–72
  quartz-crystal, **1,** 70
Close encounter processes, **21,** 300
Closed-loop control, **2B,** 375
Closed-loop gain, **2B,** 377–378
Closed-loop transfer function, **2B,** 421
Cloud chamber measurement of mass, **5B,** 35–37
Cloud chambers, **5A,** 194–201
  drop count, **5A,** 14
  drop formation, **5A,** 446
  energy loss measurement in gases, **5A,** 22, 29–30
  ionization of atoms in, **5A,** 444–445
  momentum measurement, **5A,** 375–388
  range measurement, **5A,** 436–438
  solid state, **5A,** 280
  track ionization, photometric measurement, **5A,** 449–451
Cloverleaf pattern, in light scattered from thin polymer films, **16B,** 367
Cluster, scanning tunneling microscopy of, **27,** 266, 271, 273
Clustering, **23B,** 99
  kinetics, **23B,** 103
Clusters, **30,** 2, 212, 428
  desorption from, **30,** 74–75
  generation of beams, **30,** 4, 401–402, 628–629
  model calculations, **30,** 116
CMA, *see* Cylindrical-mirror analyzer
CMC, *see* Critical micellar concentration
CMN, *see* Cerous magnesium nitrate
CMOS, *see* Complementary MOS
CMR (common-mode rejection ratio), **2A,** 294–295
CMRR, *see* Common-mode rejection ratio
CN, transition frequencies of, **12C,** 27
CO, transition frequencies of, **12C,** 26
CO_2, *see* Carbon dioxide
CO_2 laser, **30,** 334
Co_2-Ar plasma, **7A,** 122
Coadding, **12A,** 524
Coagulation, transformation and bleaching of centers, **6B,** 332–333
Coal mine dust explosions, **18,** 796
Coatings, **29C,** 347–348
Coaxial cable, **8,** 113
Coaxial chokes, in ac measurement, **2B,** 75
Coaxial flashlamps, for short-pulse dye lasers, **15A,** 319–321
Coaxial line, **8,** 113, 118, 119, 121, 134
Coaxial solenoids, **8,** 59
Coaxial type-N female connector, **8,** 107
Cobalt, de Haas-Van Alphen effect, and, **11,** 61–63
Cobalt, magnetization, **14,** 358
Cockcroft-Walton accelerator, **5B,** 581–584; **17,** 4–5, 533; *see also* Electrostatic accelerator
Coddington equation, spherical reflector, **31,** 159
Coddington's equations, **25,** 17
Coded reference beams, for color holograms, **26,** 182
Coefficients
  absorption, **1,** 233, 334
  accommodation, **1,** 160, 298, 299
  aerodynamic force, **1,** 152
  attenuation, **1,** 221, 222
    scattering, **1,** 394
  of capacitance, **1,** 435
  demagnetization, **1,** 533
  of determination, **28,** 262
  of diffusion, **1,** 139, 141
  of drag, **1,** 148, 151
  extinction, **1,** 334
  of finesse, **25,** 218
  of friction, **1,** 98, 99
  of heat transfer, **1,** 137, 138, 300–301
  of induction, **1,** 435
  internal conversion, **5A,** 583
  lift, **1,** 150
  loss, of gases, **1,** 156
  Peltier, **1,** 507
  of potential, **1,** 435
  reflection, **1,** 479
  of resistivity, temperature, **1,** 245

of specular reflection, **1,** 161
temperature, **1,** 457–458
of thermal accommodation, **1,** 161
Thomson, **1,** 508, 510
transducer, **1,** 216
virial, **1,** 314
X-ray mass absorption, **1,** 416
Coercive force, **1,** 550; **6B,** 220–221
CoF_2, **23C,** 20, 28
antiferromagnetic resonance, **10,** 411
Coherence, **1,** 356; **13A,** 50, 75
in beam-foil experiments, **17,** 530, 591–591
function, **12C,** 175; **18,** 118–131; *see also* Heterodyne efficiency
lateral, **18,** 325
spatial, **18,** 406, 707, 710
temporal, **18,** 125, 129, 707, 710
weak periodic signals, **28,** 267–268
Coherence length, **25,** 258; **26,** 6, 41
definition, **18,** 711
light source, **18,** 130–132, 715
measurement, **18,** 711
Coherence time, **28,** 368
definition, **18,** 130, 711
measurement, **18,** 711
in VLBI systems, **12C,** 241
Coherent anti-Stokes Raman scattering (CARS), **30,** 116–118
Coherent anti-Stokes Raman spectroscopy, **15B,** 258
Coherent detection, **10,** 227, 228
Coherent emission, **10,** 10, 18–81
Coherent illumination, **8,** 41–44
Coherent population trapping, **29B,** 157
Coherent scattering, **21,** 148, 150, 152, 196, 203, 206–211; **23A,** 20
vibrational modes, **21,** 172
Coherent state, 236, 237
Coherent transient effects, **15B,** 282–284
Cohesive energy, **30,** 30
density, **16C,** 286
Coil ampere turns, **8,** 85
Coil volume, **8,** 91, 92
Coils, **1,** 525–528
flip, **1,** 522–523
search, **1,** 527
Coincidence arrangement, **8,** 260
Coincidence circuit, fast, **5A,** 439
Coincidence counting, *see* Angular correlation (of annihilation radiation); Lifetime (of positron), **21,** 52, 58, 59
Coincidence experiments, in ion-atom collision study, **17,** 188, 303–354
broadening effects, **17,** 170–171
charged-analyzed reaction products, **17,** 345–354
detectors, **17,** 313–316
inelastic energy-loss determination, **17,** 150–153, 155–159, 162, 184–188
inner-shell vacancy production, **17,** 304–344
intermediate regions, **17,** 337–344
molecular regime, **17,** 330–337
point projectiles, **17,** 321- 330
single-encounter experiment, **17,** 306–310
target-detector arrangements, **17,** 316–317
timing considerations, **17,** 311–313
Coincidence measurements, in ion-atom collisions, **7A,** 446–447
Coincidence techniques, **7A,** 134
Coincidence(s)
accidental, rate of, **5B,** 292
delayed-, measurement of, **5B,** 316
measurement analysis, **5B,** 320–327
method to measure specific activities, **5B,** 291
rate, radiation, **5B,** 132, 136–137
Coincident circuit, **8,** 261
Cokriging, **28,** 399
Cold and hot reflex discharge ion source, properties, **29A,** 115
Cold cathode discharge gauge, **14,** 64–66, *see also* Ionization gauge
pumping action, **14,** 65
Cold cathode glow discharge, **31,** 39
Cold finger, **8,** 219
Cold mirror, **25,** 296
Cold neutrons, **21,** 161, 164, 216
Cold-cathode discharge, **13A,** 215
Cold-drawing, **16C,** 138–140
Cole-Cole diagrams, **13B,** 94, 175
Cole-Cole plot, **16C,** 390–392
Collapse of transient cavity, **19,** 393–400, *see also* Bubble motion
acoustical emission, **19,** 394–395
jet production, **19,** 396–400
shocks versus jets for damage production, **19,** 396
sonoluminescence and sonochemistry, **19,** 395
violence of collapse, **19,** 393, 402

Collection of radioactive atom beams, **5B,** 76–78
Collection power, **25,** 195
Collective effects, **30,** 592
Collector current, **8,** 181
Collector diffusion isolation, **2A,** 111
Collector electrodes, potential charge of, **5A,** 90, 95
Collector plate, **30,** 599
Collector probe, **6B,** 85–86
Colligative properties of polymer solution, **16A,** 22
Collimated neutrons, **8,** 264
Collimation, **22,** 143
Collimation of X rays, **16B,** 41, 156–159
  slit-smearing effects at small angles, **16B,** 164–167
    correction for, **16B,** 177–183
Collimation of x rays, **20,** 347–349
Collimator, **8,** 207
  geometry, **8,** 207
  slit width, **8,** 207
Collimator, off-axis paraboloid, **3,** 97–99
Collimator-baffle system for x-ray spectrometer, **5A,** 601–602
Collimators, **23A,** 202
Collision
  alignment, **3,** 600
  broadening, **3A,** 111; **13A,** 23
    EPR, **20,** 104–109
  cascades, **22,** 329, 330, 333
  chambers, **7A,** 28
  cross section, **14,** 9
  deactivation measurement, **7B,** 220
  diameter, **5A,** 289
  frequency, **9B,** 110
  light, **4B,** 139, 141, 146–147
  molecular, **14,** 6
  processes, **9A,** 118
  radius, **5A,** 290
  strength, definition of, **9A,** 138
  time, **5A,** 7
Collision chamber, **8,** 34, 35, 37
Collision cross section
  for attachment, **7B,** 56
    measurement of, **7B,** 57
  of atoms, **4A,** 158
Collision cross sections from swarm experiments, **7B,** 45–48
  fractional energy loss per collision *f*, **7B,** 50–54
  quantitative descriptions, **7B,** 56
Collisional absorption, **29C,** 9, 10
Collisional broadening, **13B,** 38; **29B,** 257
  of lines, **12B,** 146–147
Collisional detachment, **7A,** 187–188
Collisional dissociation, **7A,** 466–472
Collisional drift instabilities, **9A,** 345–394
  alkali-metal plasmas for study of, **9A,** 362
  amplitudes of, **9A,** 379–383
    hysteresis, **9A,** 379
    soft onset, **9A,** 379
    turbulence, **9A,** 381
  anomalous losses, **9A,** 393
  critical fluctuations, **9A,** 346, 375–379
    damping rate, **9A,** 379
    thermal type, **9A,** 375–376
  Doppler shifts, **9A,** 367
  drift-wave study, **9A,** 345
  experimental aspects, **9A,** 362–365
  fluctuation-dissipation theorem, **9A,** 356
  fluid equations, **9A,** 349
  free-energy reservoirs, **9A,** 369
  growth rates, **9A,** 347
  identification of, **9A,** 365–370
  ion sheaths, **9A,** 354
  ionization potential, **9A,** 362
  Landau damping, **9A,** 346
  Langmuir probes in study of, **9A,** 365
  localization, **9A,** 366
  mode coupling, **9A,** 358
  mode stabilizations, **9A,** 366, 370–375
    diamagnetic velocity, **9A,** 374
  onset, **9A,** 360, 379
  overstable wave, **9A,** 352
  phase velocity, **9A,** 367
  pyrometer in diagnostics of, **9A,** 364
  Q machine, **9A,** 345, 362
  radial plasma transport from, **9A,** 383–392
    diffusion, **9A,** 383
      coefficient, **9A,** 387
    Fick's law, **9A,** 387
  resonance-fluorescence method, **9A,** 365
  by surface ionization, **9A,** 362
  theory of, **9A,** 349–362
    critical fluctuations, **9A,** 356–358
    linear theory, **9A,** 349–356
    amplitude master equation and turbulence, **9A,** 358–362
  turbulence, **9A,** 346
  vorticity, **9A,** 350
  work function, **9A,** 362

Collisional quenching, **29B,** 230–231
in purely chemical lasers, **15B,** 113
Collisionless multiphoton dissociation, **15B,** 268
Collisionless shock waves in laboratory plasmas, **9A,** 213–274
anomalous friction between electrons and ions, **9A,** 273
anomalous resistivity, **9A,** 238
bow shock, **9A,** 243
breakdown of a gas, **9A,** 254
Čerenkov type, **9A,** 238
characteristics of, **9A,** 222
charge-exchange collisions, **9A,** 242
collisionless dissipation, **9A,** 238
collisionless plasma, **9A,** 218
compression ratio of, **9A,** 215
compression waves, **9A,** 226
critical angle, **9A,** 249
critical density of, **9A,** 268
dispersion length, **9A,** 227
effective collision frequency, **9A,** 238, 271
electrostatic turbulence, **9A,** 238
expansion waves, **9A,** 225
experimental methods for, **9A,** 241–264
plasma formation, **9A,** 252–255
experimental results, **9A,** 264–274
macroscopic properties of θ-pinch waves, **9A,** 265–266
on waves with oscillatory structure, **9A,** 266–274
formation of, **9A,** 247
gas dynamic type, **9A,** 213
theoretical models of, **9A,** 220–226
gaseous discharges, **9A,** 255
generation of, **9A,** 243–246
hydromagnetic type, **9A,** 217
ion acoustic instability, **9A,** 238, 239
ionization
in electric fields, **9A,** 253
rate, **9A,** 253
large amplitude type, **9A,** 226–241
oblique type, **9A,** 240–241
"whistler" mode, **9A,** 240, 241, 270
phase-plane, **9A,** 229
phase mixing, **9A,** 237
photoionization, **9A,** 252
cross section, **9A,** 253
precursor type, **9A,** 270
probe techniques, **9A,** 258–264
dielectric waveguides, **9A,** 263
magnetic type, **9A,** 258–261
microwave type, **9A,** 261–264
resistive type, **9A,** 241
shock structure, **9A,** 217, 238
shock thickness, **9A,** 218
shock wave theory, **9A,** 220
singular points, **9A,** 234
solitary type, **9A,** 232, 267
in space, **9A,** 218
strong shocks, **9A,** 215
theoretical models for formation and structure of, **9A,** 220–241
θ-pinch and z-pinch experiments, **9A,** 244, 246, 255
cylindrical shocks from, **9A,** 255
fused-silica tube for, **9A,** 255
two-stream instability, **9A,** 238–239
wave-wave interactions, **9A,** 238
weak shocks, **9A,** 213
Collisions between counter-circulating beams, **5B,** 686–689
Collisions, in ablation plume, **30,** 494–499
Collisions, ion-atom, **13A,** 200
Collisions, ionizing, total number, **5A,** 33
Collisions, molecular, **1,** 183
Colloid technique, in magnetic domain studies, **11,** 687–692
Colloidal centers, **6B,** 330–332
Colloidal solutions, **23B,** 489
interparticle structures, **23B,** 524
intraparticle structure, **23B,** 514
Color cameras, **2B,** 354
Color center formation processes, **15B,** 13–22
additive coloration in, **15B,** 14–15
F center aggregation in, **15B,** 16–17
F_2^+ formation in, **15B,** 20–22
two-photon process dynamics and, **15B,** 18–20
U center in, **15B,** 17–18
Color center lasers, **15B,** 1–54
additive coloration in, **15B,**14–16
color center formation processes, **15B,** 13–22
construction and performance of cw laser using F_A(II) centers, **15B,** 28–37
definition in, **15B,** 1
vs. dye lasers, **15B,** 1
F centers in, **15B,** 3–4
F_A(II) centers in, **15B,** 4–7
F_B(II) centers in, **15B,** 7–9
F_2^+ centers in, **15B,** 9
gain formulas for, **15B,** 24–25

laser cavities with highly concentrated modal beam in, **15B,** 25–28
optical gain in, **15B,** 22–25
piecewise interferometric generation of gratings in, **15B,** 51–54
recent developments in F_2^+ center lasers and, **15B,** 41–51
relaxed excited state in, **15B,** 4
systems with pairs of spins in, **15B,** 12–13
timing schemes for, **15B,** 30–32
two-photon coloration in, **15B,** 18–20
Color center physics, **15B,** 2–13
color center, **6B,** 333–334
decay law, **6B,** 308–309
fast decay, **6B,** 311–313
fundamental concepts, **6B,** 293–297
growth and decay, **6B,** 308–313
modulated exciton sources, **6B,** 302–304
slow decay, **6B,** 309–311
spectral measurements, **6B,** 304–306
Color centers, **6B,** 324–334
coagulation and bleaching, **6B,** 332–333
determination of absorption spectrum, **6B,** 325–326
electrolytic coloration, **6B,** 328
luminescence, **6B,** 333–334
magnetic resonance measurements, **6B,** 333
methods of study, **6B,** 333–334
photoconductivity, **6B,** 333
photoelectric emission, **6B,** 333
production of, **6B,** 326–332
special types, **6B,** 330–332
Color correcting filters, **25,** 294
Color holograms, **26,** 180
Color indices, infrared, **12A,** 451, 456–460
Color interferometry, **18,** 742–743
Color production in silicate glasses, **6A,** 141–142
Color schlieren, **18,** 367, 370
Color-center laser (CCL), **29C,** 304
Coloration by ionizing radiation, **6B,** 328–330
Coma, **1,** 332; **25,** 42
definition, **31,** 155, 168
non-symmetrical systems, **31,** 168–170
in special-purpose resonators, **15A,** 352–353
Coma correction
circular mirror, **31,** 171–172
toroidal optics, **31,** 173–175
Combination bands in near infrared, **3,** 108–109
Combinatorial methods, **1,** 18
Combining laws, **3,** 236
Combustion
bomb, **1,** 284
driver, shock tube, **18,** 788
heat of, **1,** 262, 284ff
Commensurate overlayer, **22,** 278
Commensurate-incommensurate transitions, **23B,** 36
Common-emitter configuration, **2A,** 73–74
Common-emitter transistor amplifier stage, **2A,** 280
admittance and transconductance values for, **2A,** 271
cascading of, **2A,** 277–278
input capacitance of, **2A,** 279
Common-emitter transistor switch, **2A,** 77
Common-mode feedback, **2A,** 295
Common-mode rejection ratio (CMRR), **2A,** 294, 384; **2B,** 74
Common-path interferometers, **26,** 19
Communication network information capacity, **2,** 695–696
Commutation equation, **10,** 3, 401
Commutation factor, **2,** 111–112
Commutator bracket, **10,** 4
Compact source arc, **25,** 249
Mercury, **25,** 250
Mercury-Xenon, **25,** 250
Xenon, **25,** 250
Comparator(s), **2A,** 382–385; **4B,** 93; **13A,** 339
Abbé, **1,** 54
admittance, **1,** 487
amplifier, **2,** 445–446
common-mode pulse response of, **2A,** 384
voltage, transistor, **2,** 443–446
Schmidt-trigger, **2,** 444–445
Comparison method, **8,** 249, 251, 252
Compass needle, **1,** 524
Compatibility between polymers, **16C,** 275
Compatibility in polymer blends, **16B,** 335
Compensated impedance, **8,** 111, 113
Compensated interferometer, **26,** 11
Compensating circuit, **8,** 113
Compensating coils, **8,** 88
Compensating eyepiece, **25,** 132
Compensating plate, **26,** 11
Compensating retardation plate, **26,** 143
Compensation, **2,** 476–488
networks, **2,** 483–488
Compensator(s), **2B,** 421; **12A,** 401, 404; **26,** 143

closed vs. open-loop performance in, **2B,** 410–414
defined, **2B,** 410
in feedback control systems, **2B,** 410–425
objectives of, **2B,** 410
optical, **1,** 382–383
series, **2B,** 415–421
series vs. feedback, **2B,** 411
Competing anisotropy, **23C,** 53
COMPLEMENT function, **2A,** 410
Complementary MOS (CMOS), **2A,** 124–126, 139
Si-gate, **2A,** 138, 144
Complex coherency, **28,** 341
Complex conductivity of electrons, in afterglow studies, **7B,** 85, 86
Complex conductivity of plasma, **10,** 268–271
Complex conjugate, **8,** 5
Complex dielectric constant, **10,** 351–358
Complex effective dielectric constant of plasma, **10,** 271
Complex index of refraction, **31,** 227
Complex microwave conductivity, **4A,** 385
Complex notation, in network analysis, **2A,** 3–4
Complex permeability, **6B,** 217
Complex plane representation of polarization, **26,** 118
Complex refractive index, **6B,** 280–281
Complex refractive indices of plasma, **10,** 272–279
Complex transfer functions, **2B,** 156
Compliance
complex, **19,** 141, 175–177
creep, **19,** 141
equilibrium, **19,** 141
retardational, **19,** 141
Component failure, **14,** 455
Component mounting, in electronic circuits, **2A,** 42
Components of neutron beam instruments, **23A,** 199
Composite holography, **26,** 184
Composite hypothesis, **28,** 187–188
chi-squared tests, **28,** 189–190
Composite resonator, *see* Resonators
Composition analysis, **22,** 338
Composition areas of glass formation, **6A,** 140–143
Composition measurement
absorbed radiation by *in situ* fluid, **18,** 630–637
absorption spectrophotometry of sampled fluid, **18,** 621–630
analysis of emitted radiation by *in situ* fluid, **18,** 637–643
analysis of sampled fluids, **18,** 616–630
classification of methods, **18,** 613–616
electron beam fluorescence, **18,** 434, 445
of emulsion, **5A,** 215
mass spectrometer, **18,** 645–661
method of description, **18,** 611
methods, **18,** 611–661
molecular, **14,** 276
sampling methods, **18,** 617–617
species concentration by molecular scattering, **18,** 408–433, 643–645
Compound events, probability, **28,** 12–13
Compound microscope, **1,** 341; **25,** 159
Compound nucleus, **4A,** 203
Compound-grating antenna, **12B,** 23
Compressibility, **1,** 101; **19,** 4–7
adiabatic, of a liquid, **3,** 701
dynamic, **1,** 208
effects of, **1,** 128–130
fluid, **1,** 120–122
frozen, **19,** 182
of iron, **6A,** 419
isentropic, **19,** 183
relaxing, **19,** 183
Compressible flow field, density by light refraction, **18,** 346
Compressible flow, in wind tunnel, **18,** 759–761
Compression of materials, **6A,** 298
Compression ratio
lobe pump, **14,** 130, 132
rotary oil sealed pump, **14,** 114
Compression seal, **14,** 284
Compressional wave velocities, **24A,** 2
Compressive force, **1,** 67
Compressive strength, **16C,** 125
Compton
effect, **5A,** 622, 671, 672; **18,** 407, *see also* Gamma rays scattering in PIXE backgrounds
electrometer, **1,** 441
electron kinetic energy, **5A,** 80
polarimeter, **5B,** 139–142
scattering, **1,** 415; **4B,** 12; **5A,** 79–83, 631–632; **24B,** 549–550
correction in X-ray diffraction, **16B,** 65
cross section, **5A,** 80–82

from polarized ferromagnets, **5B,** 264
spectrometer, two-crystal, **5A,** 629
Compton continuum, **8,** 204, 218, 219
intrinsic efficiency, **8,** 216
Compton cross section, **8,** 228
Compton electron, **8,** 205
Compton peak, **8,** 221
Compton scattered qaunta, **8,** 223
Compton scattering events, **8,** 220, 225
Compton spectrometer, **8,** 211, 213, 219–225, *see also* Anti-Compton spectrometer; High resolution semiconductor Compton spectrometer; Semiconductor Compton spectrometer.
performance, discussion, **8,** 225
resolution, **8,** 220–222
response, nonlinearity, **8,** 222–223
spurious peaks, **8,** 223–225
Compton steps, *see* Double Compton steps.
Compton total linear attenuation coefficient, **8,** 215
Compton-scattered secondary gamma rays, **8,** 220
Computational design, **1,** 20
Computer control
in neutron scattering, **21,** 165
radiometer use of, **12C,** 20–21
Computer hologram, **26,** 189
Computer methods in NMR, **3B,** 570
Computer programs, use in radio astronomy, **12C,** 277–295
Computer reconstruction of holograms, **19,** 555–559
Computer simulation of plasmas in study of instabilities, **9A,** 302–304
Computer software, *see* Software
Computer systems, **2B,** 166–168
Computer tape drives, **2B,** 123
Computer-aided design, in integrated circuits, **2A,** 138
Computerized transmission tomography, **19,** 563–580
curved rays, **19,** 576–580
data acquisition, **19,** 580–589
history of, **19,** 563–565
multiple frequency, **19,** 571–573
notation, **19,** 565–568
straight ray, **19,** 568–576
Computers, **2,** 637–647; *see also* Analog computers, Digital computers
analog, **2,** 638–646
in counting systems, **21,** 55, 56, 139–143
digital, **2,** 646–647
Computing, *see* Statistical computing
Con Flat, **14,** 485
Concave grating, **1,** 368; **25,** 215
mounts, **13A,** 291
Concentration, **14,** 276
profiles, Gaussian, **21,** 40
Concentration power, **8,** 157
Concentrational sensitivity in impurity determination, **6A,** 29–30
limit, **6A,** 30
Concentric hemispherical analyzer (CHA), **30,** 319
Condensate, **14,** 233
Condensation, **14,** 25, 26; **19,** 5–7
coefficient, **14,** 231, 232, 244, 523
polymerization, **16A,** 4
polymers, end-group analysis, **16A,** 69
pump, *se* Diffusion Pump
pumping, **1,** 180
rate, **14,** 232
(vapors), **4B,** 319
Condensed multilayer, **22,** 508
Condenser discharge,
method of resistance measurement, **6B,** 45–46
Condenser lens, **1,** 364
Condenser lenses, in an electron microscope, **16B,** 207
Condenser microphone detectors, **13B,** 114
Condensers, **25,** 99
cylindrical and spherical, **5B,** 228–231
Conditional distribution, **28,** 15
Conditional probability, **28,** 11–12
Conditioning plot, **28,** 510–512
Conditioning, **22,** 369
Conductance, **14,** 13, 18, 276
aperture, **14,** 15
calculations, **14,** 14
formulas, **14,** 16
gas flow, **1,** 163, 164
orifice, **1,** 184
tube, **1,** 184ff
Conducting pastes, in magnetoresistance studies, **11,** 13
Conduction, **1,** 134
band, **12A,** 8
current density, **1,** 501
of heat, **1,** 287–299
Conductivity, **1,** 432

atmospheric, **1,** 504–505
cell, **6B,** 67–68
definition, **6B,** 33
electrical, **14,** 363, *see also specific materials*
laser modification of, **30,** 360–361
and mobility, **6A,** 17
of plasma, **2,** 81
in rotating coordinates, **10,** 271
thermal, **1,** 135ff, 287ff
Conductivity measurements
on dialectrics, **6B,** 1–31
general methods, **6B,** 33–40
on metals and semiconductors, **6B,** 32–71
ohmmeter and voltmeter-ammeter measurements, **6B,** 33, 53–54
sample preparation, **6B,** 49–53
Conductor material, **8,** 84
Conductors, **1,** 433
forces on, **1,** 518–521
torque on, **1,** 519
Cone formation, in laser ablation, **30,** 508–511
Cone-and-plate geometry, in shear viscosity measurements, **16C,** 49
Confidence bands, simultaneous, quantile-quantile plots with, **28,** 197–200
Confidence interval, **1,** 7; **2,** 6; **3,** 714; **28,** 178m 213
application, **28,** 169–170
approximate, based on normal-theory method, **28,** 229
interpreting, **28,** 167–168
likelihood-based, **28,** 225, 226
limited failure population model, **28,** 236–238
relation to statistical tests, **28,** 181
single parameter, **28,** 165–167
simulation-based, **28,** 242
simultaneous, **28,** 261
Weibull, **28,** 233–235
Confidence limits for mean, **28,** 264
Confidence rectangles, **28,** 174–178
Bonferroni method, **28,** 175–176
grain boundary triple point, frequencies, **28,** 174–175
Confidence regions, **28,** 173–174, 213
approximate likelihood-based, **28,** 225–226
elliptical, **28,** 178
normal-theory, **28,** 228–229
Weibull, **28,** 233–235
Configuration
of crystallizable polymers, **16B,** 340
coordinate, **30,** 40, 49–52, 59
definition, **16B,** 18
interaction, **3B,** 930; **4B,** 109
core polarization, **4B,** 109
Configuration of a chain molecule, **16A,** 4
NMR studies, **16A,** 288–294
Configurations, molecular, **3,** 181–182
Confining materials, **24A,** 85
Confluent beams, **7A,** 477
Confocal resonant cavities, **29C,** 366
Confocal scanning microscope, **25,** 174
Conformation
definition, **16B,** 19
helical nomenclature, **16B,** 20
Conformation of a polymer chain
in concentrated solution, **16A,** 501
infrared and Raman studies, **16A,** 134–148
from magnetic shielding constant, **16A,** 259
in melt or amorphous solid, **16A,** 5, 498–501
NMR studies, **16A,** 299–301
randomly coiled, **16A,** 4
in partially crystalline solid, **16A,** 503–506
in solution, **16A,** 18
Conformational change
in uniform polypeptides, **19,** 233
volume changes for, **19,** 203
Conformational order, **16A,** 146–147
Conformations, **20,** 137–150
lipid, **20,** 137–142
nucleic acid, **20,** 143–145
polysaccharide, **20,** 145
protein, **20,** 146–150
Confusion noise
from weak sources, **12C,** 16
Congruent transfer, **30,** 479
Conic constant, **25,** 53
Conical scanning, **12A,** 215–219
Conical spiral fees, **12B,** 76, 79
Conifgurational effects, in beam-foil experiments, **17,** 596–598
Conjugate image in holography, **26,** 170
Connector(s), **29C,** 172–173
components, **14,** 294–296
design, **14,** 294, 295
flexible, **14,** 300
gas, vacuum, water, **14,** 294, 295
Conoscope, **6B,** 266–268
Conoscopic mode, **25,** 173

Conrad probe, **18,** 254
Conservation of energy, **1,** 307; **23A,** 8
Conservation of momentum, **23A,** 8
Conservative freezing methods, **6A,** 53–55
Conservative motion of dislocations, **6A,** 329
Consistency, **1,** 9–11
Constant bandwidth (CBW), in telemetering, **2B,** 240
Constant current anemometer, hot-wire or hot-film
  basic circuity, **18,** 277
  calibration, **18,** 285
  compensation, **18,** 283
  square wave test, **18,** 285
Constant final state, **22,** 172, 173, 174, 175, 176, 177, 179, 183, 185
Constant initial state, **22,** 172, 177, 179, 180, 181, 182, 183, 185
Constant rate of strain, **16C,** 12
  radical generation during, **16C,** 207–210
Constant strain-rate apparatus, **6A,** 318
Constant stress apparatus, **6A,** 318–319
Constant temperature anemometer, hot-wire or hot-film
  basic circuitry, **18,** 290–292
  calibration, **18,** 297
  characteristic frequency, **18,** 295
  cutoff frequency, **18,** 292–293, 301
  damping coefficient, **18,** 295
  higher-order system response, **18,** 300
  linearization of signal, **18,** 302
  offset voltage, **18,** 291
  square wave test, **18,** 297
  unbalance parameter, **18,** 292, 294, 296,
Constant voltage generator, **8,** 13
Constant-deviation spectrometers, **25,** 208
Constant-final-state energy, **22,** 173
Constant-final-state spectroscopy, **22,** 173
Constant-frequency cyclotron, **17,** 13
Constant-initial-state mode, **22,** 177
Constant-initial-state spectroscopy, **22,** 177, 179
Constant-Q method, **23A,** 406
Constant-temperature bath technique for crystallization in thin films, **16B,** 359
Constants, fundamental, **1,** 35–43, 50–51
Constitutive relations, **19,** 34–37
Constrained binary encounter approximation, **17,** 201–202, 242, 244
Constraint, **1,** 307
Constricted glow discharge, **31,** 41
Contact area, real, **16C,** 431
Contact electrification, **16C,** 428–435
Contact formation technique, **6B,** 87–91
Contact front, **30,** 245
Contact potential microscopy, **27,** 92
Contact potential, **6B,** 140–141; **7A,** 40, 53; **22,** 12, 17, 18
  difference, **6B,** 131–133
  relation to charge density, **16C,** 430
Contactor control systems, **2,** 515–517
Contacts, for resistance measurement, **6B,** 50–52
Contamination, **14,** 257, 278, 299
  neutron spectrum, **21,** 163, 164, 179
  oil, **14,**124–126
  from organic molecules, in an electron microscope, **16B,** 208, 221
  particulate, **14,** 501, 502
  from pumps, **14,** 499
  of source, in beam-foil experiments, **17,** 84, 536
Continous wave (CW) radar, **1,** 77
Continuity equation, **1,** 144
Continuous channel multiplier, in deep space plasma measurements, **9B,** 72
Continuous time averaging, in nuclear magnetic resonance, **2B,** 142–143
Continuous wave (CW), **19,** 3, *see also* Standing wave
Continuous wave (CW) composite oscillator systems, **11,** 397–398
Continuous wave (CW) dye laser(s), **15A,** 325–355
  alternative geometries for, **15A,** 351–353
  astigmatic resonator for, **15A,** 343–347
  coma in, **15A,** 352
  defined, **15A,** 325–326
  dye cell astigmatism in, **15A,** 333–347
  dye system for, **15A,** 353–355
  efficiency of, **15A,** 331–332
  geometries of, **15A,** 326
  input photons and, **15A,** 333
  mode area for, **15A,** 341–351
  mode radius or waist for, **15A,** 339–340
  resonator for, **15A,** 326, 336–338
Continuous wave (CW) ion laser, **15A,** 71–74
  operating pressures for, **15A,** 89
  output power variation in, **15A,** 77
Continuous wave (CW) lasers
  dye lasers, **29C,** 45–46, 138
    alignment, **29C,** 71–73

astigmatism in, **29C,** 58–59
dyes, **29C,** 66–71
flowing dye cell, **29C,** 59–60
laser resonator in, **29C,** 56–58
mirror transmission, **29C,** 60–66
theory, **29C,** 45–56
triplet quenching, **29C,** 60
error sources, **29C,** 198
single spatial-mode semiconductor lasers, **29C,** 78, 79
Continuous wave (CW) techniques, theory of, **19,** 112–118, 134
propagating wave model, **19,** 112–114
sensitivity enhancement factors, **19,** 114–116
Continuous-channel electron multiplier, **17,** 436–439
Continuous-strip magnetic electron multiplier, **17,** 545
Continuum, **13A,** 21, 206
Continuum emission coefficients, electric arc use for, **7B,** 348–349
Continuum generation, by picosecond pulses, **15B,** 196–197
Continuum regime flow, **1,** 143
Continuum signals, **12B,** 202
Continuum sources
measurements of, **12C,** 7–18
Contour mapping
mechanical plotter programs for, **12C,** 296
nonmechanical plotter programs for, **12C,** 299
Contour slice images, **22,** 421
Contours, 3*d*-electron eigenfunction, **3,** 224, 225
CONTR, **12C,** 296
continuous contours, **12C,** 322–324
contours in segments, **12C,** 325–326
Contraction, thermal, **14,** 441
Contrast
amorphous-crystalline, in an electron microscope, **16B,** 189
defocusing, **16B,** 219
factor, **23C,** 430
matching, **23B,** 502
photographic film, **1,** 411
in small-angle scattering, **20,** 372–375
topographical, imaging, **16B,** 222–224
variation, **20,** 378–381; **23C,** 374
Control
chart, **1,** 5, 9; **2,** 4–5; **3,** 712–713
circuits
high-temperature, **2,** 304
thyratron, **2,** 300–303
feedback, see Feedback control systems
rods and drives, reactor, **5B,** 603
system
closed-loop, **2,** 454
contactor, **2,** 515–517
defined, **2B,** 375
Control elements, **8,** 249, 250
Controlled leaks for radioactive gases, **4A,** 233
Controlled rectifier circuits, **2A,** 224–241
applications of, **2A,** 230–241
elements of, **2A,** 228–230
inductive load and, **2A,** 233
motor speed control circuit and, **2A,** 235–238
relay service in, **2A,** 240–242
reverse inductive load in, **2A,** 233
single-phase full-wave rectifier in, **2A,** 230–233
single-wye, **2A,** 234
Controlled source electromagnetics
data acquisition, **24B,** 352–357
exploration method, **24B,** 346–349
FFM and TEM systems, **24B,** 352–354
frequency bands, **24B,** 357
source configurations, **24B,** 354–357
sources of noise in exploration, **24B,** 361–363
Convection, **1,** 134, 135, 137, 299–301
boreholes, **24B,** 202–206
currents, **1,** 294, 296
of heat at surface, **18,** 667
role in hot-wire and hot-film anemometer, **18,** 269
seafloor, **24B,** 257–259
Conventional photon sources, **22,** 141, 147
Conversion, **8,** 177
analog-to-digital, **2,** 621–622
coefficients
correlation function, **5B,** 152–162
from β and γ ray intensities, **5B,** 154–157
for mixture of two multipolarities, **5B,** 160
special methods, **5B,** 157–160
Y-shell, **5B,** 152
data, use of, **5B,** 160–162
digital-to-analog, **2,** 622–623
efficiency of diodes, **10,** 21–23

external, **5A,** 586–594
factors, **1,** 30, 31, 34, 49
pressure units, **14,** 2
frequency, **2,** 729–730
internal, **5A,** 296–297, 594–599
ratios, **5A,** 595
of units, **18,** 823–825
Converter(s)
analog-to-digital, **2A,** 416–419
balanced-, measurement of photon total beam, **5B,** 544
dc to dc, **2,** 177, 200–203
digital-to-analog, **2A,** 419–421
γ-ray, **5B,** 156
photoelectric, **5A,** 586ff
in rectifier circuits, **2A,** 211–214
rms to dc, **2B,** 65–67
successive approximation, **2B,** 53
time-to-height, **5B,** 312, 313
time-to-pulse height, **5A,** 440
torque, **1,** 95
tubes, **2,** 65
voltage-to-frequency, **2B,** 53–54
Convolution approximation, **21,** 210; **23A,** 45; **23B,** 151, 202
Convolution integral, **2B,** 381; **12A,** 499
solution of, **12C,** 2–7
Cook's distance, **28,** 272–273
Cooke-Yarborough meter, **2B,** 10
Coolant (water or LN) failure, **14,** 456
Coolant passages, **8,** 84
Coolant, reactor, **5B,** 602
Cooling
magnetic, **5B,** 186–190
in vacuum, **14,** 352
very low temperature, **5B,** 175–176
Cooling transitions, **30,** 50–51
cooling, **1,** 317
induction, **1,** 518, 521–523
leakage factor, **1,** 559
materials, **1,** 550, 551–559
media, **1,** 431
pole, **1,** 33
potentiometer, **1,** 553, 555
susceptibility, **1,** 537–546
systems, **1,** 317, 318
temperature, **1,** 259
Cooperative diffusion, **22,** 392
Cooperative process, **19,** 198, 231, 232
Coordinate method, **5A,** 378–379
Coordinates, rotating, **3,** 362–363
Copolymer sequence distribution
NMR studies, **16A,** 294–299
using fluorescent probe, **16A,** 437
Copolymer(s), **16A,** 3; *see also* Block copolymers
analysis by GPC, **16A,** 63
Copper, **22,** 170
activation measurement of relative photon beam, **5B,** 546–547
alloys, **14,** 379–382, 481
boiling point, **14,** 350
electrical conductivity, **14,** 363
hydrogen solubility, **14,** 354
ion laser, **15A,** 158–162
Kα X radiation, **16B,** 42, 47, 136–138, 161
laser lines, singly ionized, **15A,** 159
magnetoresistance of, **11,** 7
melting point, **14,** 347, 350
oxygen free high conductivity, **14,** 360–362
physical properties, **14,** 360–363
sputtering yield, **14,** 551, 553
thermal conductivity, **14,** 356
in ultrahigh vacuum systems, **14,** 480
vapor laser, performance of, **15A,** 44–45
vapor pressure, **14,** 347, 350
Copper, magneto resistance effect, **8,** 84
Copper, reflectance spectra, **31,** 211–212
Copper-phthalocyanine, **22,** 370
Corbino effect, **6B,** 146, 147, 155
Core
drivers, transistor, **2,** 433–443
linear, **2,** 442–443
electron binding energy, **22,** 215
magnetic, pulse spectrometer, **2,** 618
Core hole
excitation, **22,** 199
lifetime, **22,** 191, 200
Core-excited state, beam-foil interactions, **17,** 551, 560
Corioli's force, **9A,** 439
Coriolis coupling, **10,** 326, 327
Coriolis effects, **3A,** 62
Coriolis force, **18,** 803; **23A,** 296
Corner frequency, **2B,** 391
Cornu prism, **1,** 365, 383; **13A,** 285
Cornu spiral, **26,** 56
Corona devices, **2A,** 167–169
Corona discharge, **14,** 309; **16C,** 482
bars, **2A,** 167
Corona equilibrium model, **9A,** 123–124

Copper-phthalocyanine, **22,** 370
Corbino effect, **6B,** 146, 147, 155
Core
  drivers, transistor, **2,** 433–443
    linear, **2,** 442–443
  electron binding energy, **22,** 215
  magnetic, pulse spectrometer, **2,** 618
Core hole
  excitation, **22,** 199
  lifetime, **22,** 191, 200
Core-excited state, beam-foil interactions, **17,** 551, 560
Corioli's force, **9A,** 439
Coriolis coupling, **10,** 326, 327
Coriolis effects, **3A,** 62
Coriolis force, **18,** 803; **23A,** 296
Corner frequency, **2B,** 391
Cornu prism, **1,** 365, 383; **13A,** 285
Cornu spiral, **26,** 56
Corona devices, **2A,** 167–169
Corona discharge, **14,** 309; **16C,** 482
  bars, **2A,** 167
Corona equilibrium model, **9A,** 123–124
Corona high-voltage regulator tubes, **2A,** 167–169
Corona regulators, **2A,** 168
Coronal diagnostics spectrometer (CDS), 72
  calibration, **31,** 135, 138
Correcting coils, **8,** 88
Correcting solenoids, **8,** 88
Corrections for beam spreading, **19,** 71–76, 239, 260–262, 276, 292
Correlation
  angular, **5B,** 129–151
    determination of spins, **5B,** 131–137
  coefficient, **1,** 13, 14; **3,** 717; **28,** 23
  of conversion electrons, **5B,** 137–138
  directional, between two successive γ-rays, **5B,** 133–135
  experimental tests, **5B,** 148–149
  factor in diffusion, **21,** 70, 447
  function, **5B,** 150–151; **16B,** 175–177; **21,** 175
    conversion coefficients, **5B,** 152–162
    neutron scattering, **21,** 195–206
    pair-correlation, σ_{coh}, **21,** 197, 210
    self-correlation, σ_{inc}, **21,** 197, 200
    space-time, **23A,** 30; **23B,** 245
    theoretical directional, **5B,** 132–136
    VLBI use of, **12C,** 182, 196
  γ-γ, procedure in, **5B,** 149–150
  lengths, **23C,** 153
  range, **23C,** 5
  signals, **12B,** 218–219
    many-channel, **12B,** 219
    two-channel, **12B,** 218
  spectroscopy, **13B,** 28
  techniques, **1,** 229
  time, **3,** 382; **20,** 13–16; *see also* Jump frequency determination
    distribution, **16A,** 312–319
    effective, for segmental motion, **16A,** 303
    for liquids and solutions, **16A,** 248
    rotational
      effect of polymer crystallinity, **16A,** 473
      for nitroxide spin probe, **16A,** 462
      in polymer solutions, **16A,** 476
Correlations between jumps, **23B,** 149
Correlator(s), **20,** 314; *see also* Autocorrelation spectrometers
  logic, **12B,** 288–293
Corresponding states, principle of, **1,** 314
Corrosion by flux, **14,** 426
Corrugation amplitude, **27,** 227, 278
  anomalous, **27,** 289
  of metal surface structure, **27,** 278
Cross-sectional imaging, **27,** 261
  of GaAs/AlGaAs superlattices, **27,** 263
Coseismic strain, **24B,** 414
Cospectrum, **28,** 341
Cost, **14,** 84
  factor, **14,** 245
Costas loop, in telemetering, **2B,** 232
Coster-Kronig transition, **22,** 192
COT, *see* Cyclo-octatetraene
Cotton-Mouton effect, **1,** 388; **26,** 153
Cottrell cloud, **6A,** 327
Couette flow, **1,** 125, 132
Couette geometry, in shear measurements, **16C,** 27, 49, 52
Couette viscometer, **18,** 797
Coulomb
  approach, to inner-shell vacancy production, **17,** 321–330
  approximation, **7A,** 119
  deflection, **17,** 487
    low projectile velocities, **17,** 327–328
    in target ionization, **17,** 205–207, 236–237
  electron capture, **17,** 92

energy difference between two nuclei, **5B,** 363–364
energy of homogenously charged sphere, **5A,** 332
excitation; **13A,** 245
  by passing charged particles, **5B,** 446–448
excited gamma rays, **5A,** 613–616
factor, **4A,** 203
ionization or excitation, **17,** 200–207, 211, 253, 384–386, 482–483, 486–488
  alignment mechanism, **17,** 481
  Auger electron measurements, **17,** 492–500
  lithiumlike atoms, **17,** 509
  relativistic electron wavefunctions, **17,** 211
potential(s), **11,** 149; **17,** 163–164; **22,** 311
scattering; **5B,** 233; **17,** 31–33
  deviations, **5A,** 309
  probability for elastic, **5A,** 389

Coulomb energy, **8,** 259
Coulomb radius, **8,** 259
Coulomb-Born forumula for electric dipole transitions, **9A,** 135
Count distributions, photographic representation, **2,** 612–615
Counter
  electronic, **1,** 470–471
  Geiger, **1,** 412
  multistable magnetic, **2,** 442
  photon, **1,** 408
  proportional, **1,** 413
  revolution, **1,** 86
  ring, **2,** 439–442
  scintillation, **1,** 413
  time interval gated electronic, **2,** 542–543
  tube instruments, **6A,** 217–219
Counters, **13A,** 150
  alpha particle, **5A,** 107
  boron trifluoride proportional, **5A,** 533–534
    jitter time, **5A,** 534
  bulk, flux and energy measurement with, **5A,** 471–475
  Čerenkov, **5A,** 10, 162–194
    differential isochronous self-collimating, **5A,** 457
    focusing, **5A,** 166–186, 456–460
    gas, **5A,** 181, 184, 185, 457–460
    liquid, **5A,** 456–457
    non-focusing, **5A,** 186–189, 455–456
  cloud chamber control, **5A,** 200
  divide-by-*N*, **2B,** 6
  in frequency measurement, **2B,** 19
  gas-filled, **5A,** 110–120
  Geiger, **5A,** 118–120
    efficiency, **5A,** 120
    low pressure, **5A,** 33–34
  helium–3 proportional, **5A,** 573ff
  hydrogen-filled proportional, **5A,** 471–473
  mechanical, **2B,** 6
  proportional, **5A,** 108–118, 471–473, 533–534, 573ff
    ionization loss measurement in, **5A,** 30–34
    linearity of response, **5A,** 117
    space charge limitation, **5A,** 111
    time of response, **5A,** 113–115
  scintillation, **5A,** 120–159
  self-quenching, **5A,** 119
  solid for neutrons, **5A,** 473–475
  surface barrier, **5A,** 266
  synchronous, **2B,** 4–6
  tandem, background counts in, **5A,** 467
  telescope, **5A,** 466
  types of, **2B,** 3–4
  velocity interval selection, **5A,** 183, 186
  wall effect, **5A,** 573–574
Counters, particle, **5B,** 78–80
  alpha proportional, **5B,** 286–288
  beta ray 4π, **5B,** 288–290
  BF_3, **5B,** 375, 393
  Čerenkov differential high pressure, **5B,** 758–760
  characteristics, **5B,** 503
  gas ionizing, **5B,** 308
  gas recoil, **5B,** 375–376
  Geiger-Muller, **5B,** 442
  nonpolarizable, **5B,** 793
  polarizable, **5B,** 792–793
  radioactivity, **5B,** 283–284
  scintillation, **5B,** 307, 376
    NaI (Tl), **5B,** 445
    telescope geometry, **5B,** 503–505
  solid semiconductor ionization, **5B,** 308
  telescopes, **5B,** 441
    relative monitor, **5B,** 02
Counters, *see* BF3 counters
Counting, **2,** 519–535; **2B,** 1–14
  absolute, **21,** 61, 66
  amplification stability, **21,** 51
  background, **21,** 66, 67
  counting-rate meters in, **2B,** 9–12

countrate and resolution for Doppler broadening, **21,** 97
differential, **21,** 64
digital, **1,** 72
counters in, **2B,** 3–6
displays and recordings in, **2B,** 6–9
instabilities, *see* Instabilities
integral, **21,** 52, 64
loss(es), **2B,** 12–14
determination of, **2,** 522–523; **2B,** 13–14
input-stage, **2B,** 13
minimum efficient counting time, **21,** 67
pileup, **21,** 100
pulse, decay time, line-width, pileup, **21,** 51
sum peaks, **21,** 62
rate
mean, **5B,** 799–800
recoil, **5A,** 462–463
relative, **21,** 64
resolution or dead time, **21,** 65, 68–70
single-channel, effects of resolving time, **5B,** 792–794
statistics, **21,** 64, 66–67
systems, **2B,** 1
Counting efficiency, **8,** 195
Counting experiments, *see* Nuclear counting experiments
Counting rate, **8,** 8, 188, 193, 196
density, **8,** 196
measurements, *see* Backgrounds presence
Counting situations, evaluation, **8,** 193–194
Counting statistics, **8,** 187
Counting-rate meters, **2,** 523–525; **5B,** 799–802
analog, **2B,** 11–12
average of continuous readings, **5B,** 801–802
average of *n* independent readings, **5B,** 800–801
digital, **2B,** 12
Coupled accelerator, *see* Multiaccelerator
Coupled cavity interferometry, **9B,** 31–35
application to pulsed plasmas, **9B,** 32–33
frequency response, **9B,** 33–35
sensitivity of interfermometer for, **9B,** 32
Coupled reactions, **19,** 209
Coupled waves, **1,** 113
Couplers, directional, **1,** 484
Coupling iris, **8,** 137
Coupling reactance, **8,** 135
Coupling, **14,** 278, *see also* Microwave cavity/waveguide
constants, quadrupole, **3,** 523
efficiency, single-mode fiber, **29C,** 379–380
flexible, **14,** 301
interionic, effect on nuclear orientation, **5B,** 199–200
network, for superconducting resonators, **11,** 581–584
phenomena, of whistlers, **9B,** 124–126
quadrupole, **5B,** 147–148
Covalent bonding, **16A,** 2
Covan, **5A,** 228
Covariance, **28,** 22–25
function, **28,** 379–380
matrix, **28,** 382–383
Coverage measurement
absolute, **22,** 441, 442, 443
during programmed desorption, **22,** 451
relative, **22,** 441
Cowling conductivity, **9B,** 177
CPO–3D, *see* Charged particle optics program, three-dimensional
Cr, **23C,** 184
Cr_2O_3 antiferromagnetic resonance, **10,** 411
Crab-nebula pulsar, **12A,** 332
Crack propagation
chemical environment, **24A,** 136–137
double cantilever beam, **24A,** 138–140
double torsion, **24A,** 134–138
In situ measurements, **24A,** 146–152
load relaxation, **24A,** 136
notched bending beam, **24A,** 142–146
scanning electron microscope observations, **24A,** 146–150
stress intensity factor, **24A,** 134–135, 138–142
subcritical crack growth, **24A,** 134
velocity, **24A,** 135–138
Crack, definition, **16C,** 234
Cracking process, **8,** 179
Cramer–von Mises test, **28,** 201–202
Cranz-Schardin camera, **18,** 737
Crater effects, **22,** 338
Craze
definition, **16C,** 233–235
front, in solvent penetration, **16C,** 323
growth, **16C,** 252–255
internal structure, **16C,** 243–245
optical methods of study, **16C,** 237–242
porosity, **16C,** 240
retracted thickness, **16C,** 243
thickness, **16C,** 235, 240–245
velocity measurement, 253–254

Craze-opening displacement, **16C,** 243, 248
Crazing
creep equation, **16C,** 265
criteria for, **16C,** 249
effect of pressure on, **16C,** 116
environmental, **16C,** 255–262
molecular weight dependence, **16C,** 236
Creep, **1,** 108, 109; **14,** 258; **23B,** 117, *see also Surface migration*
compliance, **16C,** 4
in crazing polymers, **16C,** 265–268
response, **19,** 139, 141–143
rupture, molecular model, **16C,** 219–220
strength, **14,** 289
temperature, **14,** 286
thermal, **1,** 160, 169
velocity, terminal, **16C,** 5
CRESU, *see* Reaction kinetics in uniform supersonic flow technique
Crinkly foil, **7A,** 236
Critical angle(s), **1,** 329; **21,** 287, 289, 303, 304; **23A,** 268, 150
Critical angle, **31,** 263
interference filters, **31,** 328
mirror (CAM), **31,** 328–340
Critical correlation estimation, local, **28,** 463–464
Critical electron exposure, **22,** 509
Critical flow
liquid in channel, **18,** 333
nozzle throat, **18,** 330, 772
Critical fluctuations, **23C,** 1, 14
Critical illumination, **25,** 152
Critical magnetic fields, **6B,** 72–73
Critical micellar concentration, **23B,** 490
Critical nucleus, work of formation, **16B,** 392
Critical points, **1,** 243
Critical pressure, *see* Breakdown pressure
Critical resolved shear stress, **6A,** 316
Critical scattering, **23A,** 87
dynamics, **23C,** 24
Critical strain, for environmental crazing, **16C,** 256–258
Critical temperature for superconductivity, **10,** 76, 77, 432
Critical value, **28,** 180
Critical wavelength, **22,** 143
Critical-angle refractometers, **25,** 187
CRO, **8,** 126
Cross amplitude spectrum, **28,** 341
Cross antenna, **12B,** 21
Cross covariance sequence, **28,** 340–341
Cross entropy, **28,** 412
Cross periodogram, **28,** 343
Cross section
absorption, **5A,** 504; **23A,** 525, 527, 530
total, for gamma rays, **5A,** 87–89
for antineutrino bombardment of protons, **5A,** 686
capture, **5A,** 41; **23C,** 253
neutron, **3,** 349–350
carbon-electron scattering, **3,** 337
coherent scattering, **23A,** 21, 526, 530
Compton scattering, differential, **5A,** 80–82
for D(*d,n*) He^3 reaction, **5A,** 522–523
definition, **7A,** 3
determinations in static gases, **7A,** 390ff
delta ray ejection, **5A,** 55
differential, **23A,** 4, 197, 523
double differential, **23A,** 12, 17, 197
electron
at 1808 , **7A,** 9
differential in angle, **7A,** 5, 87, *see also* other permutations (including Electron scattering, etc.)
and energy, **7A,** 8, 72
for excitation in heavy particle collisions, **7A,** 455–466
for H^3 (*d,n*) He^3 reaction, **5A,** 524–525
incoherent scattering, **23A,** 21, 526, 530
ion-molecule reaction, **3,** 583
loss, **5A,** 41
macroscopic, **23C,** 407
table, **23C,** 410
magnetic, **23C,** 160
elastic, **23C,** 79
ion, **23C,** 76
measurements
absolute determination, **7A,** 91
calibration, **7A,** 39, *see also* Normalization of cross section measurements
critical elements, **7A,** 11
systematic errors, **7A,** 72
Mott scattering, **5A,** 55
for neutral pion production, **5A,** 339
neutron
measurement in kev region, **5A,** 553–554
plus He^3 reaction, **5A,** 571–572
proton, **5A,** 488–493
scattering, **5A,** 466, 484
non-spin-flip, **23C,** 94
nuclear coherent elastic, **23C,** 77
nuclear scattering of electrons, **5A,** 67, 69

pair production, **5A,** 83–87
for photoelectric effect, **5A,** 78–79
reference, **7A,** 294
Rutherford differential scattering, **5A,** 291
for secondary electron production, **5A,** 63, 65
single crystal, instruments, **23A,** 228
spin-flip, **23C,** 94
Thomson scattering, **5A,** 77
total integrated scattering, **23A,** 13, 197
total atom-atom, **7A,** 294, 295
total, of lead for 10 Ú neutrons, **5A,** 504, 507
total scattering, **23A,** 21, 525
by electrons, **7A,** 3, 38, 87
instruments, **23A,** 218
Cross section, *see* Pair production cross section
Cross sections, table, **23A,** 533
Cross spectral density function, **28,** 340
Cross spectrum
multitaper estimator, **28,** 344–346
Cross-correlation function
clipped signals, **12C,** 233
interferometer use of, **12C,** 145
VLBI use of, **12C,** 182
Cross-correlation spectrometry, **12B,** 107
Cross-correlation, **29C,** 198–203, 220–223, 224
Cross-disperser, **12A,** 475, 483
Cross-grating antenna, **12B,** 22
Cross-linked network, **16A,** 3
ESR spin probe studies, **16A,** 473
small-angle neutron scattering studies, **16A,** 502
Cross-polarization of spins, **16A,** 335–338
Cross-power spectrum, **12C,** 234
Cross-section, **22,** 244, 303, 304, 311, 330
capture, neutron, **5B,** 407–411
determination by magnetic reaction-particle analyzers, **5B,** 434ff
desorption, **22,** 322
differential
elastic, **5B,** 472–482
interaction, **5B,** 411–472
evaluation of errors, **5B,** 424–426
magnetic analyzers, **5B,** 431–440
sources of charged particles, **5B,** 413–416
thick target, **5B,** 436–439
thin target, **5B,** 435–436
for neutron-induced reactions, **5B,** 448–460
scattering, **16A,** 481–485
elastic scattering, **22,** 244
for elastic scattering of photons by nuclei, **5B,** 517–518
for electron-electron scattering, **5B,** 241–243
electron-induced surface process, **22,** 498
electron-stimulated desorption, **22,** 466, 500, 510
electron-stimulated desorption, adsorbed species, **22,** 501
for γ-induced reactions, **5B,** 461–472
inelastic scattering, **22,** 244
ion scattering, **22,** 303, 304, 330
low-energy electron scattering, **22,** 244
for Mott scattering, **5B,** 250
photoemission, **22,** 484
photon-stimulated desorption, **22,** 466
for positron-electron annihilation, **16A,** 377
for production of thin-target bremsstrahlung, **5B,** 520–524
scattering, **16A,** 208, 209
total
for Compton effect, **5B,** 267
interaction, **5B,** 366–396
back-angle, **5B,** 370
monoenergetic fast neutrons, **5B,** 366–388
using polyenergetic neutrons, **5B,** 388–396
measurements
with fast neutrons, **5B,** 385–388
with slow neutrons, **5B,** 394–396
Cross-semivariogram, **28,** 399
Cross-spectrum function, **12C,** 144, 148, 231
Cross-validation, **28,** 500
Crossed beam technique for measurement of relaxation times, **7B,** 24
Crossed field undulator, **29C,** 32, 33
Crossed spectra, **6B,** 305
Crossed-beam(s)
choice of angle between, **7A,** 296
electron-neutral experiments, **7A,** 67–68
general scattering equations, **7A,** 68–72
for particulate control, **30,** 513
photo-detachment, method for, **7A,** 181ff
Crosslinking by radiation, **16B,** 278, 352
location of, in polyethylene single crystals, **16B,** 282–286
Crossover frequency, **2B,** 413
Crowbar short-circuit devices, **2A,** 182–183
Crown glass, **29C,** 343

Crucibles, **6A,** 89, 92–95
Crustal deformation, **24B,** 410–411
Cryogenic magnetometer, **24B,** 328
Cryogenic microwave techniques, **11,** 584
Cryogenic pump, *see* Cryopump, Cryopumping
Cryogenics, **24A,** 361–363
Cryogens, **12A,** 429–432
Cryopump, **14,** 244
  in ultrahigh vacuum systems, **14,** 482, 483, 489
Cryopumping, **14,** 159, 231–235
Cryosar, **2,** 124–125
Cryoscopy, for molecular weight measurement, **16A,** 26
Cryosorption pump, **14,** 232, 237, 241, 242
Cryosorption roughing pump, **14,** 239
Cryostat design, **6B,** 156–160
Cryostats, **12A,** 420, 429–432, 438, 460
Cryotrap, **14,** 159, 234
Crystal
  analyzer spectrometers, **23A,** 250
  and space lattices, **6A,** 189–191
  axes of rotation, **6A,** 200–201
  center of inversion, **6A,** 200
  clock, **12C,** 179
  curved
    focusing spectrometer, **5A,** 599–616
    methods of bending, **5A,** 606–607
  definition, **6A,** 189
  detectors; *see* Diode detectors
    in ESR spectrometer, **2B,** 149
    for microwaves, **2B,** 283
  diffracted intensity formula, **6A,** 199
  diffraction,
    measurement of neutron momentum, **5A,** 566–570
    spectrometry, **5A,** 612–613
  diode detectors, **13B,** 111
  energy loss in, **17,** 70
  field splitting, **10,** 388–394
  growth from aqueous solution, **5B,** 184–186
  habit, **6A,** 100
  image, **3,** 286–287
  imperfections, detection by X-rays, **6A,** 241–244
  lattice(s), **6A,** 201
    vibrations, **10,** 358–372
  monochromator, **29C,** 38–39
    neutron, **5A,** 569–570
  mounts for electroluminescence, **6B,** 318
  optics, **6B,** 264–271
  orientation by optical methods, **6A,** 160–163
  periodicity of scattering power, **6A,** 194
  plane spacing, **5A,** 567
  polarizers, **26,** 138
  properties, **13A,** 161
  pulling, **6A,** 93–96
  quality testing, **23C,** 456
  reflection planes, **6A,** 201–202
  retardation plates, **26,** 142
  spectrometer, **17,** 372–374, 376–377, 386–388
  symmetry, **6A,** 200–202
    effect on diffracted rays, **6A,** 202–203
  systems, **3,** 269ff; **6A,** 15, 201; **16B,** 2–5
  types, **26,** 135
  unit cell of, **6A,** 190
Crystal distortions (defects)
  of the first kind, **16B,** 100
    effect on X-ray diffraction pattern, **16B,** 102–105
  of the second kind, **16B,** 100–102, 128
    effect on Z-ray diffraction pattern, **16B,** 105–111
    crystallite size effects, **16B,** 143–149
    typical corrections for, **16B,** 149–153
Crystal growth
  by addition of solute, **6A,** 104–108
  control of crystallization, **6A,** 87
  in a crucible, **6A,** 89–93
  without a crucible, **6A,** 96–98
  dislocations, **6A,** 332–335
  by electrolysis, **6A,** 108
  floating zone technique, **6A,** 96
  by interdiffusion, **6A,** 108
  from melt, **6A,** 86–98
  miscellaneous techniques, **6A,** 108–109
  from pure vapor, **6A,** 111–113
  rates
    impurity content, **6A,** 95
    magnitudes, **6A,** 87
    in solutions, **6A,** 99, 100
  by secondary recrystallization, **6A,** 119
  single crystals by primary grain growth, **6A,** 118
  in the solid phase, **6A,** 116–121
  from solution, **6A,** 98–109
  from solution by changing temperature, **6A,** 101–103
  from solution by evaporation of solvent, **6A,** 103–104
  in a temperature gradient, **6A,** 119–121

thermal gradients, **6A,** 87–89
from vapor phase, **6A,** 109–116
from a vapor phase reaction, **6A,** 113–116
Verneuil method, **6A,** 96
Crystal neutron filters, thickness, **8,** 194
Crystal optics, history, **31,** 401–402
Crystal structure
determination, **6A,** 187–282
moving-film methods, **6A,** 208–214
standard photographic methods for single crystals, **6A,** 206–208
of X-ray diffraction, **16B,** 68–98
of polypropylene, effect on melting behavior, **16B,** 304
Crystal-controlled oscillators, **12C,** 212–214
Crystal-field excitations, **23A,** 83; **23C,** 195
Crystalline polymers, **16A,** 7–8
Crystalline solids, **23A,** 370
Crystallinity, *see* Polycrystalline medium, description of; Grains
effect on electrical conduction, **16C,** 495
on gas transport properties, **16C,** 328–329
index, **16B,** 117–118
determination by X-ray diffraction, **16B,** 118–127
of polymers
determination by infrared, **16A,** 132–134
estimated from positron lifetime, **16A,** 394, 398
Crystallite shape, effect on X-ray line broadening, **16B,** 133–135
Crystallite size determination, **23C,** 445
Crystallite size, by wide-angle X-ray diffraction, **16B,** 129–153
Crystallization
Avrami theory, **16B,** 347–348
in bulk-type specimens, **16B,** 369–375
density balance technique, **16B,** 372–374
density gradient technique, **16B,** 374
dilatometric techniques, **16B,** 370–372
sample preparation, **16B,** 369–370
configurational requirements for, **16B,** 340
DSC studies, **16B,** 336
growth rate
effect of molecular weight, **16B,** 364
equation for, **16B,** 361
homogeneous nucleation theory, **16B,** 349
kinetic growth theory, **16B,** 348–349
primary, **16B,** 371
rate, experimental methods, **16B,** 352–385
secondary, **16B,** 303, 371
self-seeding technique, **16B,** 381–383
from solution, **16B,** 381–384
kinetics, **16B,** 383
in thin film specimens, **16B,** 354–369
analysis of data, **16B,** 361–365
depolarization technique, **16B,** 365–366
dielectric measurements, **16B,** 369
infrared absorption technique, **16B,** 368
light scattering technique, **16B,** 366–368
optical microscopic studies, **16B,** 356–365
under pressure, **16B,** 385
various polymers, supercooling dependence, **16B,** 363
Crystallography, **3,** 268–275; **16B,** 2–25; *see also* Chemical crystallography
X-ray, **1,** 421–427
Crystals
biaxial, **1,** 379
birefringent, **11,** 699
imperfections, **1,** 425–428
light propagation in, **11,** 700–702
quartz, **2,** 284–285
small-amplitude waves in, **11,** 376–381
for solid state lasers, **15A,** 188–191
uniaxial, **1,** 379
Cs adsorbed
GaAs(110)
spectroscopy of, **27,** 123
Cs^+ ionization, **7A,** 114
CS, transition frequencies of, **12C,** 27
Cs_3Sb, **12A,** 19–21
$CsCoCl_3$, **23C,** 139
CSEM, *see* Controlled source electromagnetics
CsI crystal scintillator, **7A,** 406, 468
CsI transmission, **10,** 95
$CsMnBr_3{\cdot}2D_2O$, **23C,** 38
$CsNa_2KSb$, **12A,** 22
$CsNiF_3$, **23C,** 228
Cu plume, **30,** 259–261, 270, 285
Cu Sil, **14,** 481
Cu(100)
scanning tunneling microscopy of, **27,** 303
Cu(110)
scanning tunneling microscopy of, **27,** 300
Cu(111)
scanning tunneling microscopy of, **27,** 297
Cu-Ni alloys, **23B,** 102
Cubic crystal system
definition, **16B,** 4
interplanar spacing formulae, **16B,** 26
reciprocal lattice relationships, **16B,** 24

Cubic crystals, **26,** 135
Cubic lattice, definition, **16B,** 5
$CuCl_2$ * $2H_20$ antiferromagnetic resonance, **10,** 411
CuH emission as a measure of relative [H], **7B,** 212
Culgoora Radio Heliograph, **12B,** 24
Cumulative distribution, definition, **28,** 14
Cumulative distribution function, **28,** 212, 217–218
  completely specified, aseessing fit, **28,** 201–202
  statistical computing, **28,** 491
  Weibull, **28,** 231
Cunsolo method, in ion drift velocity measurements, **11,** 456–458
Curie, *see* Radiation units
  formula, **2B,** 134
    for nuclear magnetization, **5B,** 107
  method, **1,** 538ff, 544, 548
  point
    definition, **24A,** 315, 330
    measurement, **24A,** 418–423
  temperature, **1,** 549; **6B,** 194, 197–200
Curie's law, **6B,** 172–173
Curie-Weiss law, **1,** 258; **6B,** 174; **10,** 375; **11,** 502–505, 517
Curing reaction, DSC studies, **16B,** 334
Current, **1,** 33
  alternating, **1,** 459–488
  channeling, **24B,** 318–320, 323, 341, 364–365
  convection, **1,** 294, 296
  dark, **1,** 392
    multiplier phototube, **2,** 749–752
  diffusion, **2,** 118
  direct, **1,** 447–459
  drift, **2,** 115–118
  focusing, **24B,** 308–310
  gain, common emitter, **2,** 149
  imaging tunneling spectroscopy, **27,** 129, 159
  integrator; *see* Ammeter
    circuit, **5B,** 418
  ionization chamber, **5A,** 109–110
  leakage, switching diode, **2,** 161
  maximum
    average, **2,** 110–111
    peak anode, **2,** 111
    surge anode, **2,** 111
  measurement, **1,** 449–451; **2,** 558–583; **7A,** 191–194; *see also* Ammeter
    ac, **2,** 576–583
    dc, **2,** 558–576, 650–651
    procedures, in equipment testing, **2B,** 190–191
  noise, **10,** 221, 222
  photomultiplier, **5A,** 129
  regulators, **2,** 489–503**2A,** 242–243; **2B,** 458–460
    ac, **2,** 491–493
    dc, **2,** 501–503
  sources, **2,** 18–19
Current density, **1,** 431, 432
  in afterglow studies, **7B,** 79
  atmosphere, **1,** 504
  conduction, **1,** 501
Current loop, **8,** 82
  field components calculation, **8,** 74–75
  pair, *see* Filamentary current loop pair
  radius a, **8,** 70–75
  scalar potential, **8,** 71–73
Current loop, magnetic fields, **8,** 67–75
  Biot-Savart Law, usage, **8,** 67–69
  intensity calculation, **8,** 71
Current loops, pair, **8,** 77–80
Current-dependent destruction mechanism, **15A,** 107
Curvature array
  intrinsic, **28,** 260
  small, **28,** 263
Curvature compensation, **5A,** 377–378
Curvature of field correction, **31,** 172
Curvature of field, **31,** 156–158
Curved crystal spectrometer, **17,** 387
  single cylindrical curved-crystal, **17,** 365–369
  single-plane, **17,** 361–365
Curved crystal X-ray monochromator, **16B,** 49
Curved electrode, **6B,** 27
Curved-plate analyzer(s)
  balanced type, **9B,** 63
  comparison of instruments, **9B,** 68–70
  coupled energy-angular response, **9B,** 65–67
  crossed-field velocity selector, **9B,** 68
  for deep space plasma studies, **9B,** 62–70
  energy-angle asymmetry ("skewing"), **9B,** 65
  extraneous effects, **9B,** 68
  geometries of, **9B,** 63–64
  types of, **9B,** 64
  unbalanced type, **9B,** 63

Curves
comparison, **5A,** 376
isoenergetic, **5A,** 656–657
Cut-weld joint, **14,** 434–436
Cutoff
plasma, **10,** 275, 280
waveguide, **10,** 116–119, 121
Cutoff wavelength, **31,** 305
temperature effect, 317–320
Cutting methods, **6A,** 148, 152
cw (continuous wave) chemical lasers, **15B,** 95–96
operation of at 5000, **15B,** 123
pumping of electronic transition in, **15B,** 111–112
cw (continuous wave) CO chemical lasers, **15B,** 106–110
cw (continuous wave) dye lasers, *see also* cw chemical lasers
modal beam and, **15B,** 25
passively mode-locked, **15B,** 190, 200
cw (continuous wave) HF(DF) chemical lasers, **15B,** 106
cw (continuous wave) lasers, **15B,** 32, *see also* cw chemical lasers; cw dye lasers
crystal preparation for, **15B,** 32–33
construction and performance of, using H_A(II) centers, **15B,** 28–37
excitation-deexcitation rate for, **15B,** 67
F_A(II) center stability in, **15B,** 33–34
performance of, **15B,** 34–37
tuning elements in, **15B,** 30–32
cw (continuous wave) Nd:YAG (neodymium-yttrium-aluminum garnet) laser, **15B,** 45
cw color center lasers, **15B,** 1, *see also* cw chemical lasers
focused beam cavity designs in, **15B,** 26
schematic of, **15B,** 29
cw dye laser systems, **15B,** 191–192
cw laser, *see* Continuous wave laser
CW lasers, *see* Continuous wave lasers
cw long-pulse operation, lower-state bottlenecking and, **15B,** 73
cw photoionization experiments, **15B,** 41–45
CW, *see* Continuous wave
Cyanide ion, **19,** 202
Cyanogen, electron-induced dissociation, **3,** 551–552
Cyclic laser, **15A,** 34–55
ideal, **15A,** 34
metal vapor, *see* Cyclic metal vapor lasers
noble gas, **15A,** 48–53
Cyclic metal vapor lasers
first oscillations of, **15A,** 38–40
properties of, **15A,** 36
Cyclo-octatetraene, in flashlamp-excited laser systems, **15A,** 359
Cyclotron, **5B,** 630–634; **8,** 189; **17,** 12–14, 533, 535
accelerator, *see* Electrodynamic accelerator
azimuthally varying field, **5B,** 683–686
electron, **5B,** 648–651
frequency, **5A,** 367
of electron, **1,** 45
"frequency-modulated," **5B,** 645–651
instability, **9A,** 300–302
irradiation, **6A,** 359
mass measurements, in indium, **11,** 60
"sector-focused," **5B,** 678–686
Cyclotron radiation, **10,** 259–264, 295–299
harmonic, **10,** 297–299
cadmium, **10,** 250, 251
copper, **10,** 250, 251
detection, **10,** 188, 201
free electrons, **10,** 77–81, 259–263
InSb, **10,** 128, 128, 187, 188
maser, see Electron cyclotron maser
semiconductors, **10,** 249–251
spectrometer-detector, **10,** 188–201
Cyclotron resonance, **6B,** 241–242; **13B,** 95
(ions), **7A,** 383–390
of electron, **4B,** 18–22
resonance frequency, **1,** 44
Cylindrical analyzer, **17,** 462–463
Cylindrical cavity resonator, **6B,** 19–20
Cylindrical curved-plate analyzers, **9B,** 64
Cylindrical film X-ray camera, **16B,** 56
Cylindrical mirror analyzer (CMA), **30,** 320
Cylindrical resonator, **19,** 187
Cylindrical source coordinates, **8,** 98
Cylindrical-deflector analyzer, 1278 , **22,** 34, 35, 139
Cylindrical-mirror analyzer, **17,** 454–458, 467, 544, 546; **22,** 136, 159, 169, 221, 474, 476, 477, 494
Cylindrical-sector field analyzer, **17,** 178, 180
Cystallinity
determination, comparison of methods, **16B,** 326
determination by calorimetry (DSC), **16B,** 316–326
determination by X-ray diffraction, **16B,** 114–127
nature of, **16B,** 99

Cytidine monophosphate, **19,** 202
Czerny-Turner monochromator, **7A,** 167–168; **10,** 84, 137; **17,** 539
Czerny-Turner mounting, **3,** 159
Czerny-Turner spectrometers, **13B,** 23
Czochralski technique, **11,** 57

D

D'Arsonval instruments, **1,** 449; **2,** 558–559; **2B,** 49–50
*D**, **10,** 224, 225
d-electron configuration, matrix elements, **3,** 227
D-field detachment, **7A,** 209
D.C. bridge methods, **6B,** 40–43, 58–59
D/A (digital-to-analog) converters, **2A,** 419–421
D2O laser, *see* Deuterium dioxide laser
Daly-type converters, **22,** 320
Damage
  desorption, **22,** 321
  electron-beam-induced, **22,** 505
  electron bombardment, **22,** 500
  ion beam, **22,** 321
  radiation, **22,** 173, 333
  sputtering, **22,** 321
  threshold, **29C,** 346; **30,** 573, 574
Damping
  eddy-current, **1,** 541
  effects, in ultrasonic studies, **11,** 425–426
  ratio, in feedback response system, **2B,** 387
  sound, **1,** 219
Damping of seismometer, *see also* Pendulum, damped
  critical, **24B,** 26, 52
  damping ratio, **24B,** 26
  resistive, **24B,** 29
Damping time, **8,** 167
Dark count noise, of photoelectric detectors, **17,** 545
Dark current, **12A,** 37–43, 243–248
  in image tubes, **12A,** 243, 248
  in photomultipliers, **12A,** 37–45
Dark discharge, **7B,** 63
Dark field illumination, **25,** 172
Dark field microscope, **1,** 342
Dark spontaneous force optical trap, **29B,** 140
Dark-field image, in electron microscopy
  description, **16B,** 194–195
  mode of operation, **16B,** 211–212
Dark-trace tube, **2B,** 357
Darling-Dennison resonance, **3,** 67
Darlington amplifier transistors, **2A,** 79
Data
  censored
    assessing distributional assumptions, **28,** 204–206
    interval-censored, **28,** 214–216
  evaluation of, **1,** 48
  preprocessing, **28,** 467
  presentation, techniques for, **12C,** 296–307
  prewhitening, **28,** 305, 320, 325–328
  processing techniques, **7A,** 64, 77, 277, 292–299
  recording of, **1,** 1; **2,** 526–535
  rejection of, **1,** 9; **2,** 7–8; **3,** 715–716; **5B,** 813–814; **6A,** 7
  statistical characterization, **5B,** 771–780
  tapering, **28,** 300, 320–325
  time series with missing, **28,** 305–309
  windowing, **28,** 297–302
  windows, frequency transforms, **28,** 299
Data analysis
  chronophotography, **18,** 86
  interferometry, **18,** 205–398
  positrons
    angular correlation, *see* Angular correlation (of annihilation radiation)
    Doppler broadening, *see* Doppler broadening (of annihilation radiation)
    lifetime, *see* Lifetime (of positron)
    vacancy formation enthalpy, **21,** 89
    vacancy-impurity binding, **21,** 92
    voids, **21,** 93, 94
Data handling, **21,** 387–391
  Fourier transforms, **21,** 388–391
  integration, **21,** 387–391
  sample and holders, **21,** 387
Data systems, **12A,** 307, 519
  for Fourier spectrometers, **12A,** 519
  for television, **12A,** 307
Dating
  of large ages, potassium-argon method, **5B,** 300
  radiocarbon, **5B,** 301–302
Davidson-Cole function, **16C,** 392
Day
  sidereal, **1,** 69
  solar, **1,** 69
Dayem bridge, **11,** 255
Dayhem bridge, **24A,** 353–354
DBR, *see* Distributed Bragg reflector

Dc circuit analysis, **2A,** 7
dc excess, versus periodic strength, **28,** 361–364
dc fields, superconductor shielding and, **11,** 612–617
Dc measurements, **2B,** 48–61
analog indicators in, **2B,** 49–52
conversion methods in, **2B,** 53
digital indicators in, **2B,** 52–58
digital voltmeter performance in, **2B,** 56–58
dual slope integration in, **2B,** 55
electronic amplifiers in, **2B,** 62–63
in equipment testing, **2B,** 215–216
null detectors in, **2B,** 59–67
potentiometers in, **2B,** 58–59
Dc motors, in servomechanisms, **2B,** 437–440
Dc reinserter, **2A,** 385
Dc restorer, **2A,** 385
techniques, **2B,** 79
Dc signals, spurious, **2B,** 60
dc supercurrent, **11,** 212–217
external noise and, **11,** 247–248
Dc tests, circuit requirements for, **2B,** 187–189, *see also* Equipment testing
Dc-to-ac converters, **2A,** 211–214
DCE, *see* Direct Coulomb excitation
DCN
transition frequencies of, **12C,** 27
de Broglie wavelength, **6A,** 246–247; **22,** 358, 379; **29B,** 342–343
De Broglie waves, **23A,** 259
De Haas-Van Alphen effect, **11,** 9, 24, 33–34
amplitude factor in, **11,** 37–39
amplitude measurements in, **11,** 54
Bessel functions in, **11,** 44, 55
block diagram for, **11,** 47–48
in cobalt, **11,** 61–63
computer analysis in, **11,** 55–56
Dingle temperature in, **11,** 55
experimental methods for, **11,** 33–64
Fermi surface and, **11,** 58–60
in ferromagnetic metals, **11,** 61–64
field dependence in, **11,** 54–55
field modulation technique in, **11,** 44–56
frequencies in, **11,** 35–37
frequency measurements in, **11,** 51–52
frequency selectivity in, **11,** 46–47
impedance matching and filtering in, **11,** 49–50
for lead, **11,** 56–60
magnetocrystalline anisotropy in, **11,** 40–41
modulation field in, **11,** 50–51
for nickel, **11,** 63–64
notation used in, **11,** 36
oscillation frequency in, **11,** 37
pickup coil for, **11,** 48–49
preamplifier in, **11,** 50
relative frequency measurements in, **11,** 53
rotation oscillations in, **11,** 53
spin-splitting factor in, **11,** 39
temperature dependence in, **11,** 38, 54
vector relations in, **11,** 40–44
de Haas-van Alphen effect, **6B,** 190–191
De Jong-Bouman method, **6A,** 212–213
De Morgan's theorems, **2A,** 411
Dead angle, **8,** 219
Dead solid angle, **8,** 219
Dead time
see also Counting in equipment testing, **2B,** 208
correction, **8,** 189
losses, **8,** 189
Dead volume, **8,** 194
Dead weight pressure gage, **18,** 512
Dead-band circuit, **2A,** 395
Deamplification of energy distribution in merging beams, **7A,** 479
Deborah number, **16A,** 7
diffusional, **16C,** 336
deBroglie wavelength, **1,** 47
Debye absorption, **13B,** 92
Debye dielectric model, **1,** 495
Debye equation(s), **3,** 640; **13B,** 171, 174; **16C,** 388
Debye length, **9A,** 3; **19,** 205
of a plasma, **9B,** 49
Debye ring, **16B,** 51, 53
Debye-Hückel law, **19,** 205
Debye-Scherrer cone, **23C,** 85
Debye-Scherrer lines, **1,** 423
Debye-Scherrer method, **6A,** 214–216
Debye-Scherrer rings, **16C,** 301
Debye-Scherrer, powder camera, for X rays, **16B,** 52, 59
Debye-Smoluchowski theory, **19,** 225
Debye-Waller effect, **22,** 213, 235
Debye-Waller factor, **5A,** 567; **16A,** 214, 233–234; **16B,** 112, 140; **20,** 240; **21,** 149, 150, 152, 206, 207, 108, 218; **22,** 274, 275, 296, 297; **23A,** 24, 395; **31,** 276
fast ions, **23B,** 197
hydrogen, **23B,** 138
Decade scalar, **2B,** 3

Decanting, in rotary oil sealed pumps, **14,** 122–124
Decay
amplitude, free, for single 90? pulse, **3,** 434
direct observation of, **5B,** 280–285
double beta, **5B,** 293–297
of free neutron, **5B,** 293–295
lifetime, radioactive, **5B,** 275–279
modes for mesons and hyperons, **5B,** 846
number of
per angular interval, **5B,** 832
per unit energy interval, **5B,** 832
radioactive, **5B,** 359–360
law of, **5B,** 766
sources of photons, **5B,** 514–515
rates, **13A,** 100, 104, 115, 118
sequence, intermediate state of, reorientation effects, **5B,** 199–200
signal, free, **3,** 426
two-body kinetics of, **5B,** 831–833
Decay curve, **8,** 188
Decay mode, **8,** 208
Decay of luminescence, **6B,** 308–313
of photoconductivity, **6B,** 347–348
Decay process, **17,** 296, 299, 311
beam-foil studies, **17,** 529–530, 532, 570–590
doubly-excited states, **17,** 561
autoionizing, **17,** 568–570
Decay time
charge density, **1,** 433
emission, **1,** 396–397
fast emission, **3,** 256, 258
fluorescence
of anthracene, **5A,** 421, 422
of cesium iodide (Tl), **5A,** 414
scintillation, in orgain scintillators, **5A,** 425
see also Phosphor decay time, **8**
Deceleration lens system, of parallel-plate analyzer, **17,** 440–449
Dechanneling, **21,** 281, 289, 293–297
dechanneled fraction, **21,** 281, 295
by defects, **21,** 295, 315–319
diffusion calculation, **21,** 293, 294
lattice vibrations and electronic collisions, **21,** 294
rate, **21,** 281
Decibel, **18,** 508
notation, **1,** 200–202
Decision procedure, **28,** 179
Decoding circuits, **2A,** 414
Decompression and cavitation, **19,** 409
Decontamination, **21,** 37
Deconvolution, **13B,** 37; **19,** 586–573
Deconvolution of positron spectra
angular correlation, 2-D, *see* Angular correlation (of annihilation radiation)
Doppler broadening, *see* Doppler broadening (of annihilation radiation)
lifetime spectrum, *see* Lifetime (positron)
Deconvolution procedures, in X-ray diffraction line analysis, **16B,** 141
Decoupling, **16A,** 294
dipolar, **16A,** 326
experimental, **16A,** 339
of moderations, **23A,** 184
Deep Sea Drilling Project, **24B,** 243–246
Deep space plasma measurements, **9B,** 49–88
experiments involving, 1959–1968, **9B,** 54–55
of flux, **9B,** 71–73
AC methods, **9B,** 71–72
DC methods, **9B,** 71
counting methods, **9B,** 72–73
synchronous detector, **9B,** 71–72
hydrodynamic parameters, **9B,** 77
instrumentation for, **9B,** 52–76
calibration of, **9B,** 736
curved-plate analyzer, **9B,** 62–70
modulated-potential Faraday cup, **9B,** 53–62
methods of analysis, **9B,** 77–88
differential and integral behavior, **9B,** 80–82
distribution function by interpolation, **9B,** 85–88
equations for detector current, **9B,** 79
estimation of plasma parameters, **9B,** 83
measurement relation to particle distribution function, **9B,** 78–80
method of moments, **9B,** 83
model distributions, **9B,** 83–85
microscopic character of instruments, **9B,** 49–50
organization of energy and angle measurements, **9B,** 73–76
on spin-stabilized spacecraft, **9B,** 74–75
on triaxially stabilized spacecraft, **9B,** 75–76
plasma properties, **9B,** 49
plasma regions, **9B,** 50
spacecraft-plasma interaction problems and, **9B,** 51–52
Deep Tow System, **24B,** 96, 100

Deexcitation process, **17,** 149
Defect clusters, **23B,** 226
Defect diffusion, **13B,** 185
Defect dynamics, **21,** 182
  heavy impurities, band mode, **21,** 186
  light impurities, localized modes, **21,** 190
  mass defect parameter, **21,** 182, 184
  theory, **21,** 182–183
Defect structure, **30,** 529–531
Defect-induced desorption, **30,** 16, 56, 75–78
Defects, **22,** 247; **23B,** 85
  dynamic properties, **23B,** 120
  ellipsoidal, **23C,** 435
  instrumentation, **23B,** 93
  neutron scattering cross sections, **21,** 152
  nonspherical, **23C,** 434
  point, **23B,** 91
    isolated, **23B,** 96
  in polymers, **16A,** 397
    infrared and Raman studies, **16A,** 130–132
  spherical, **23C,** 431
Definition circle, **8,** 110, 115
Deflection analyzer, *see* Electrostatic deflection analyzer
Deflection pattern of molecular beam, **3,** 612–613
Deflection, elastic, **14,** 287
Defocusing, *see* Feed defocusing
  contrast, in electron microscope image, **16B,** 219
Deformation, **1,** 100–114
  elastic, **1,** 66–67, 100–107
  non-elastic, **1,** 107–109
  plastic, **14,** 287
Degassing, **1,** 188; *see also* Outgassing
Degenerate parametric amplifiers, **12B,** 227
Degradation of polymers, thermal analysis studies, **16B,** 335
Degradation, in semiconductor diode lasers, **15A,** 242–244
Degradation, oxidative, ESR spin probe study, **16A,** 477
Degree of polymerization, **16A,** 5; **16B,** 340
Degrees of freedom, **1,** 6
deHaas-Van Alphen effect, **1,** 543
Dehydrohalogenation, of poly(vinylidene chloride), **16B,** 252–253
Delay ambiguity function, **12C,** 147
Delay line pulse stretcher, **2A,** 391
Delay line(s), **1,** 482; **2,** 39–40; **2A,** 31–32
  acoustic, **1,** 112
Delay, *see* Interferometer delay
Delay-line clipper, in pulse spectrometer, **2B,** 109
Delay-line, in pulse spectrometer, **2B,** 126–127
Delayed coincidence techniques, **7A,** 134, 135
Delocalized positron, *see* Block state of positron
Delta method, **28,** 228
Delta rays, **5A,** 43
  angle of ejection, **5A,** 55, 56
  density, **5A,** 301ff
  ejection cross section, **5A,** 55
  frequency, **5A,** 301
  production, **5A,** 300
Deltamax, **5B,** 247
Demagnetization
  adiabatic, **1,** 542
  alternating field, **24A,** 398–402
  analysis of curves, **24A,** 403–410
  chemical, **24A,** 402
  factor(s), **1,** 533, 547; **6B,** 191–193
  low temperature, **24A,** 402
  stability index, **24A,** 408
  thermal, **24A,** 394–398
  Zijderveld diagrams, **24A,** 407–408
Dember effect, **6B,** 365
Demodulation, **2,** 263–265; **3,** 481–484
Denaturation of proteins, **19,** 203, 235
Dendritic crystals, melting behavior, **16B,** 313
Dense plasma focus, **9B,** 187–249
  apparatus for, **9B,** 190–194
    coaxial plasma focus, **9B,** 190
    low-inductance cables, **9B,** 192
    vacuum spark gap, **9B,** 193
      conditioning
  computer study, **9B,** 244–248
  development of, **9B,** 194–215
    acceleration phase, **9B,** 198–203
    breakdown phase, **9B,** 194–198
    collapse phase, **9B,** 203–204
    filamentary structures, **9B,** 196
    sheath resistance, **9B,** 204
    "snowplow" model, **9B,** 200
  dynamic behavior of current sheath, **9B,** 205–208
  energy partition in discharge, **9B,** 209–210
  impedance considerations, **9B,** 210–215
  plasma diagnostic measurements, **9B,** 215–249
    accelerated deuteron beam model, **9B,** 219
    deuterium-tritium experiment, **9B,** 222–224

electron temperature, **9B,** 229–233
neutron production, **9B,** 215–222
neutron scaling, **9B,** 224–225
spatial stability of focus, **9B,** 238–242
X-ray emission, **9B,** 225–229
theoretical study, **9B,** 242–244
Densities of single crystals, **6A,** 291
Densitometer, micro-, **1,** 391
Density, **1,** 68; **19,** 4–7
approximation, **28,** 216
potential problems, **28,** 216–218
balance, for crystallization studies, **16B,** 372–374
blob, **5A,** 245–263
bubble, **5A,** 451–453
current, **1,** 431, 432
delta ray, **5A,** 301ff
effect, **5A,** 7–8, 10, 13, 23–29, 231, 241; **17,** 131–134
on electrical strength, **16C,** 473
electron energy loss measurement, **5A,** 38–40
in gaseous targets, **17,** 131–134
in heavy ion-atom collisions, **17,** 75–76
in solid targets, **17,** 134–139
emulsion, **5A,** 215, 224
of final states, **22,** 175, 177, 179
fluid, **1,** 201–211
substitution method, **1,** 69
function, **3,** 296–298
spectral, **3,** 382
gap, **5A,** 245–263
in gas flow, **1,** 154
gas, probe, **1,** 171
grain, **5A,** 212, 243, 247, 300
hydrostatic, **1,** 68
of initial states, **22,** 144
of liquids, **1,** 118
Density gradient,
column, **16B,** 374; **16C,** 473
sedimentation of polymer solutions in, **16A,** 68
by Raman scattering, **18,** 424
Density matrix, **4B,** 227–229, 242–245; **15A,** 20–26; **29B,** 156
defined, **15A,** 20–22
describing polarization state, **5B,** 221–222
diagonalization, **4B,** 228, 243
elements, relaxation times for, **15A,** 24
equation, **15A,** 22–26
steady-state solution of, **15A,** 26
time dependency of, **15A,** 22–26
Density measurement(s)
beam absorption technique, **18,** 405–408, 705
electron beam excited radiation, **18,** 434–455
interferometer technique, **18,** 345–403
under pressure, **19,** 171
Raman scattering technique, **18,** 418–433
Rayleigh scattering technique, **18,** 414–418
schlieren method, **18,** 363
Density measurements of shock waves, **7B,** 171–177
Density of solids, **6A,** 283–291
flotation method, **6A,** 287–288
gradient method, **6A,** 288–289
high precision determination with cubic crystals, **6A,** 284
hydrostatic weighing, **6A,** 289–291
liquids for flotation method, **6A,** 287
mixing method, **6A,** 288
pressure method, **6A,** 288
temperature method, **6A,** 288
X-ray method of determination, **6A,** 284–286
molecular, **14,** 184
of occupied states, **22,** 132, 149, 150, 175, 183
operator, **15A,** 21
of photographic films, **1,** 411
photographic track, **5A,** 449–451
photometric, relationship to cascade energy, **5A,** 681
various polymers, **16C,** 72
space charge, **1,** 504
standard deviation, **3,** 300
of states, **22,** 145, 147, 150, 151, 154, 171, 173, 200, 202
one-phonon, **23A,** 28, 376
at stp, of gas, **1,** 119–120
of unoccupied states, **22,** 149, 181, 183
water, **6A,** 286
weighing, **6A,** 281–291
Depletion
layer, *see* Barrier layer
region, **5A,** 266, 267
width, **5A,** 270–271
Depolarization, **5B,** 248–249, 257–259; **8,** 255
of atoms, of ions, **4B,** 249–250, 252–253
corrections, **12A,** 382
factors, **3A,** 426
of radiation, **4B,** 146, 176, 179–183
spectra, **13B,** 167

ratio, **3,** 123; **20,** 127, 130
in Raman spectroscopy, **16A,** 87, 97
Depolarizers, **12A,** 379–384; **26,** 153
Depopulation, of lower laser levels, **15A,** 109
DEPP, *see* Double escape pair peak
Depth
analysis, **22,** 327, 347
of field, **25,** 167
of focus, **30,** 191
of modulation, **18,** 121
profiling, **22,** 313
scale, **22,** 410
Depth profile, **21,** 231–233, 244–250
channeling, damage and displaced atoms, **21,** 309, 320
PIXE, **21,** 256
RRA, **21,** 250
SCANIR, **21,** 268
SIMS, **21,** 266
sputtering, **21,** 263
Derivatives, *see* Even order derivatives
Derjaguin-Landau-Verwey-Overbeek theory, *see* DLVO theory
Describing function, **2B,** 428–436
analysis, **2B,** 429–431
of function generator, **2B,** 434
oscillation and, **2B,** 432
for relay servomechanisms, **2B,** 444
linearity and, **2B,** 428–431
technique, **2,** 516–517
Descriptive statistics, **28,** 156–158
Design
extreme ultraviolet spectroscopy, **31,** 146–147
hollow cathode, **31,** 65–70
Penning discharge, **31,** 74
wall stabilized arc, **31,** 46–48
Design concepts, **25,** 92
Design tips, **25,** 51
Design, geometric
dielectric barrier discharge, **31,** 44
Desorption, **14,** 30, 276; **22,** 321, 322, 356, 376, 385, 408
activation energy, **22,** 438, 456
Auger-stimulated, **22,** 470
characteristic time, **22,** 430
cross section, **22,** 322
by electrons, **14,** 32, 68, 69
experimental data treatment, **22,** 447
field, **22,** 408
of gases on glass, **14,** 396
images, **22,** 412
by ions, **14,** 32
isosteric heat of, **22,** 439
kinetics
chemical methods, **22,** 453
using gas measurement, **22,** 447, 449
kinetics parameter, **22,** 448
laser-induced, **30,** 15, 58–78
mechanisms, **22,** 467
model, **22,** 469
molecules, **30,** 62–65
neutral atoms, **30,** 140, 157–158
photon-stimulated (PSD), **30,** 72
from polycrystalline substrates, **22,** 427
positive ions, **30,** 142–149
rate, **14,** 232; **22,** 430, 457
states, **22,** 434
statistical thermodynamics of, **22,** 457
thermal, **14,** 30, 447
Detachment by electric and magnetic fields, **7A,** 208ff
Detailed balance condition, **23A,** 35
Detection
channel, **6B,** 238–239
crystal video, **5B,** 24–25
efficiency of BF_3 counter, **5A,** 533
of free radicals in photochemical reactions, **3,** 576, 577–579
modulation signal, **5B,** 90–92
of molecular beams, **3,** 594–596
of neutral particles, **21,** 237
of objects, **25,** 124
particle, of atomic beams, **5B,** 71–85
of photodisintegration processes, **5B,** 462ff
sensitivity in microwave spectrometer, **5B,** 89
superheterodyne, **3,** 36
synchronous, **26,** 279
of ultrasound, **19,** 37–40, *see also* Transducers
Detection efficiency, **8,** 223, *see also* Neutron detection efficiency
Detection probability, **8,** 187, 189
Detective quantum efficiency (DCE), **12A,** 223–225
Detectivity, **13B,** 12
Detectivity "*D* star," *see D**
Detectivity (D*), **12A,** 424
Detectivity *D*, **10,** 224, 225
Detector(s), **2A,** 336–340; **3,** 481–484; **13A,** 26; **14,** 81; **17,** 175, 311–317, 436–438; **22,** 283, 286; **23A,** 208, 217; **26,** 291; *see also* Photodetector

acoustical, **1,** 217–219, 225
antineutrino, **5A,** 687, 688
bolometer, **10,** 157, 168–172; **22,** 497
for bridge use, **6B,** 23–24
calibration, **5A,** 154–159
for channeling, **21,** 300–301
circuits, **2,** 265
comparison of, **3,** 493–495
Compton γ-ray, sensitivity of, **5B,** 204–206
Condenser microphone, **13B,** 114
crystal, **2,** 728–729; **3,** 28–29, 493
  diode, **13B,** 110, 111; *see also* Diode detectors
  noise spectrum, **3,** 488, 489, 492
cyclotron resonance, *see* Cyclotron resonance detection
used in differential cross-section measurements, **5B,** 426
diffuse, **16B,** 172–175
doped germanium, **10,** 171
for elastic-scattering experiments, **5B,** 477–479
for electronic timing, **5B,** 306–308
energy-sensitive, **5B,** 413–430
for epithermal neutrons, **5A,** 533–538
for EPR, *see* EPR, spectrometer design
far infrared, **13B,** 51
Faraday cup, **22,** 283, 287
for fast time-of-flight, **5A,** 548
figures of merit, **26,** 292
fluorescent screen, **22,** 283, 284, 287
gallium arsenide, **10,** 178
gamma-ray, **13A,** 121
germanium, lithium-oriented and hyperpure, **16A,** 386
Golay cell, **10,** 167–168, 178; **26,** 298
gradient, **1,** 225
ideal, **2B,** 490
InSb, **10,** 179, 188
instrumentation, **16B,** 153–161
intensity changes, during polymer halogenation, **16B,** 249
Josephson junction, **10,** 201–210
line-broadening, due to crystal disorders, **16B,** 171–172
liquid crystal, **26,** 302
luminescent, **26,** 302
in melt-crystallized polymers, **16B,** 163–164
microwave, **6B,** 238; **13B,** 109
for microwaves, **3B,** 566
modeling lamellar crystal morphology, **16B,** 167–171
multichannel spectrometer, **12B,** 272
multiplier, **22,** 474
neutron(s), **5B,** 375–376; **21,** 160
for neutrons, **3B,** 976
noise, **26,** 328
noise in, **2B,** 488
noise, in solid-state detectors, **9A,** 105
optical, **20,** 194, 195
for optical region, **13A,** 314
particle, solid state, **5B,** 482–483
phase sensitive, **3,** 438–439, 482–484, 666–668
phase-sensitive, **2,** 552–554
photo, **13B,** 6
photoconductive, **10,** 174–188; **13A,** 323; **26,** 311
photoelectric, **22,** 208
  cells, **13A,** 321
photon, **26,** 303
photon-drag, **26,** 327
Pirani, for molecular beams, **3,** 595
pneumatic, **13B,** 8, 51
polarization, **26,** 158
in polymer single crystal mats, **16B,** 161–163
positron-sensitive pulse, **22,** 287
proton recoil, **5A,** 107–108
pyroelectric, **10,** 172–174; **26,** 299
quantum, **13B,** 6; **26,** 303
quasi-ideal, **2B,** 490
of radiation, **3A,** 239
  infrared, **3A,** 192–197
recoil-proton, **5B,** 441, 453
regenerative, **11,** 302
sample geometry, **16B,** 160
scintillation, **5A,** 418, 419, 420, 537–538
  detectors, **21,** 50–60
semiconductor, **5A,** 265–280
  detectors, **21,** 50, 53–54, 60–61, 62
shower, **5A,** 652–668
slit-smearing, **16B,** 164–167
  correction for, **16B,** 177–183
slow neutron, **5A,** 139, 140
for slow neutron measurement, **5B,** 393
solid state, **13A,** 153; **22,** 210
  for neutrons, **5A,** 536–537
standing wave, **1,** 485
superheterodyne, **3,** 493–494
surface barrier, **22,** 210
surface ionization, **3,** 595; **22,** 497
synchronous, **22,** 285
thermal, **13A,** 325; **13B,** 6; **26,** 293, 303
time-of-flight energy analysis, **17,** 180–181

two-dimensional (area), **16A,** 385, 495–497
ultraviolet, **13A,** 241, 244
universal for molecular beams, **3,** 595
vacuum ultraviolet, **1,** 408
X-ray (or x-ray), **1,** 411–413; **13A,** 149
detectors, **16B,** 154–156
for X rays, **16B,** 42–46, 154–156
Detectors, **8,** 162, *see also* Ge(-Li) detectors; Ge-Li-drifted primary detectors; Semiconductor; Si(-Li) detectors; Spectroscopic system
angle, **8,** 197
aperture, *see* Finite width detector aperture
material, **8,** 222
pulse, **8,** 218
signal, **8,** 34
system, **8,** 256
Detectors, annihilation radiation
angular correlation, *see* Angular correlation (of annihilation radiation)
Doppler broadening, *see* Doppler broadening (of annihilation radiation)
lifetime, *see* Lifetime (of positron)
Detectors, infrared, **1,** 401–402; **3,** 80–85; **12A,** 417–432; **13B,** 10, 11, 13; **16A,** 80
bolometer, **12A,** 415–416, 418–425, 432
coherent, **12A,** 417
Cu:Ge, **12A,** 424–425
germanium bolometer, **12A,** 416–417, 420–426, 432
Golay cell, **12A,** 415, 419
Hg:Ge, **12A,** 424–425, 432
incoherent, **12A,** 418
InSb, **12A,** 424
paramagnetic, **12A,** 426
PbS, **12A,** 415, 424, 453
photo conductive, **12A,** 417–419, 424–425
photoconductive, **3,** 81–83
photovoltaic, **12A,** 417, 419, 424
pyroelectric, **12A,** 419
superconducting, **12A,** 420
thermal, **3,** 83–85; **12A,** 418–426
thermistor, **12A,** 419
thermocouple, **12A,** 415
Detergent cleaning, **14,** 446
Determination of crystal structures, **6A,** 228–237
Detonation, **18,** 838–839
Detonation wave, temperature measurement, **18,** 470
Detuned open, **8,** 138
Detuned short, **8,** 138
Deuterated polyethylene
infrared studies, **16A,** 127–129
small-angle neutron scattering studies, **16A,** 504–506
Deuterium cyanide, **23A,** 483
Deuterium dioxide laser, **10,** 48, 49
Deuterium fluoride lasers, **15B,** 95
Deuterium fluoride molecular constants, **3,** 52–53
Deuterium fluoride-carbon dioxide transfer chemical lasers, **15B,** 113
Deuteron response of NaI (Tl), **5A,** 412
Development of thick emulsions, **5A,** 217
Deviation
mean scattering, **5A,** 400
of multiply scattered particles, **5A,** 391ff
standard, **1,** 4; **5B,** 774–776, 803, 810
fractional, **5B,** 805
Device noise models, **2B,** 469–481
Devices, active, **2,** 31–33
Devil's staircase, **23A,** 499
Dewar, **8,** 19–21
Dewars, **12A,** 429–432, 460
for Mössbauer spectroscopy, **11,** 352–356
Dexter-Klick-Russell processes, **30,** 52
DFB, *see* Distributed feedback
DFG, *see* Difference-frequency generation
DFT, **12C,** 285, 289
dHvA, *see* De Haas-Van Alphen effect
Di-methoxy-azoxy-benzene, *see* PAA
Diagnostics, **28,** 266–277
ACE transformation method, **28,** 273–274
ill-conditioning, **28,** 274–276
Larsen and McCleary plots, **28,** 269–270
model inadequacy, **28,** 266
multicollinearity, **28,** 274–277
nonconstant variance, **28,** 267–268
normal probability plot, **28,** 268–269
outliers, **28,** 267
plot residual versus
each regressor, **28,** 269
fitted value, **28,** 266–268
plots to detect influential observations, **28,** 270–273
plotting variables in pairs, **28,** 269
Diamagnetic anisotropy, **6B,** 246–247
Diamagnetic atomic susceptibility, **6B,** 192
Diamagnetic resonance, **6B,** 241–242
Diamagnetism, **6A,** 18; **6B,** 171, *see also* Magnetic susceptibility
Diameter of droplet, **5A,** 447
Diameter, molecular, **14,** 9, 10

Diamond anvil cell, **16C,** 101
Diamond anvils
  Brillouin scattering, **24A,** 22
  sintered, **24A,** 247
Diamond, **10,** 96
Diaphragm gauge, **14,** 49–51, *see also* Manometer
Diaphragm pressure gauge, *see* Pressure gauge, diaphragm
Diathermal contact, **1,** 308–309
Diathermal wall, **1,** 236
Diatomic chain, **10,** 359–361
Diatomic molecule, chemical energy surfaces and, **15B,** 271–272
Diatomic molecules, **3A,** 62, 73, 130, 133, 287–293
Dichroic filter, 313
Dichroic filters, **12A,** 364
Dichroic mirrors, **26,** 33
Dichroic polarizer, **31,** 232
Dichroic polarizers, **26,** 156
Dichroic ratio, **16C,** 173–174
Dichroic ration, **16A,** 84, 96
Dichroism, **1,** 383; **20,** 204, 205, 208; **26,** 33
Dichromated gelatin for holography, **26,** 202
Dicke limit, **29B,** 267–268
Dicke receivers, **12B,** 214–217
Dicke switch, **12A,** 446
Dickenson shift, **3,** 371
Dielectric, **1,** 488–501
  dispersion, **1,** 490
  energy dissipation, **1,** 493
  materials, **1,** 500
  polarization, **1,** 488, 491–493
  strength, **1,** 499
  susceptibility, **1,** 490
  theory, **1,** 491
Dielectric barrier discharge, **31,** 41–45
Dielectric beam splitters, **29C,** 349
Dielectric breakdown, *see* Electric breakdown
Dielectric cell, **6B,** 43–44
Dielectric constant(s), **1,** 488–489, 493ff, 499ff; **6B,** 1; **13B,** 92, 134; *see also* Complex dielectric constant
  complex, **16C,** 385
    relaxation theory, **16C,** 386–393
  definition, **16C,** 383
  dense media, **10,** 352
  derivation from transmission line equations, **16C,** 414
  dilute media, **10,** 352
  measurement, **3,** 641–645
    bridge method, **3,** 642–643
    resonance method, **3,** 643–645
  optical, **30,** 22
  plasma, **10,** 271
  polymer, **10,** 97
  static, **30,** 22
Dielectric constant, **8,** 113, 159
Dielectric constant, calculation, **31,** 258
Dielectric description, of energy-loss problem, **17,** 33–34
Dielectric filters, **31,** 305–314
Dielectric function, **30,** 457–458
Dielectric isolation, **2A,** 108–110
  in integrated circuits, **2A,** 142
Dielectric loss factor
  conductance contribution, **16C,** 404
  definition, **16C,** 385
  derivation from transmission line equation, **16C,** 414
  measurement in resonant circuit, **16C,** 407
  temperature dependence, **16C,** 394
Dielectric loss, **1,** 491–493; **6B,** 48
Dielectric material, **8,** 159
Dielectric measurement(s)
  auxiliary test equipment, **6B,** 30–31
  of crystallization in thin polymer films, **16B,** 369
  distributed circuits, **16C,** 408–421
  free waves, **6B,** 22–23
  generators and detectors, **6B,** 23–24
  at high pressures, **6A,** 424–425
  intrinsic electric strength, **6B,** 24–31
  lumped circuits, **16C,** 395–408
  method of mixtures, **6B,** 7–8
  miscellaneous arrangements, **6B,** 7–8
  nonuniform electric field, **6B,** 5–7
  null devices, **6B,** 8–15
  procedures, **6B,** 1–31
  specimen configuration, **6B,** 3–8
  "spectrometer," **6B,** 23
  uniform electric field, **6B,** 3–5
Dielectric mirrors, **29C,** 347–348
Dielectric relaxation, **19,** 194
  phenomena, **1,** 493–495
  spectra, **13B,** 168, 175, 194
Dielectric susceptibility, **29C,** 231
Dielectric tensor and energy distribution, **26,** 133
Dielectric test cell, **16C,** 400
Dielectric vane, **8,** 145
Dielectrics,
  effective conductivity of, **6B,** 2

plastic film for, **2A,** 36
resistivity measurements, **6B,** 70–71
Diesselhorst potentiometer, **1,** 246
Difference amplifier, **2B,** 449
Difference Fourier synthesis, **20,** 259, 260
Difference frequency generation
in large-signal generation, **15B,** 160
parametric oscillation and, **15B,** 168
Difference synthesis, **3,** 297
Difference-frequency generation (DFG), **29C,** 248–252, 254
Differential amplifier, **2A,** 292–298
common-mode feedback in, **2A,** 295–296
common-mode rejection ratio for, **2A,** 294–295
Differential coefficient, *see* Secondary differential coefficient
Differential energy loss, **8,** 271
Differential equation models, **28,** 501–509
kinetics example, **28,** 502–509
model, data, and method, **28,** 502–504
S function, **28,** 504–509
problem description, **28,** 501–502
Differential field coefficients, **8,** 93
Differential pressure flowmeter, **18,** 324–331
Differential pressure gauge, **14,** 48, 49, 51
Differential pumping, **7A,** 5, 75; **17,** 52–53; **22,** 321
in scattering chamber, **17,** 175
in x-ray spectroscopy, **17,** 373
Differential refractometer, **16A,** 48; **25,** 185
Differential scanning calorimetry (DSC)
baseline correction, **16B,** 318–324
crystallinity determinations, **16B,** 316–326
glass transition determinations, **16B,** 327–331
isothermal crystallization studies, **16B,** 336–337
melting studies, **16B,** 294–316
polymerization kinetic studies, **16B,** 331–334
principles, **16B,** 289–291
Differential strain curve analysis, **24B,** 404
Differential synthesis, **3,** 298–299
Differential thermal analysis (DTA)
apparatus, **16B,** 288
equilibrium melting point determination, **16B,** 375–381
instrumental factors affecting curve, **16B,** 293
principles, **16B,** 287–289
sample factors affecting curve, **16B,** 293–294
sample holder (block), **16B,** 293
theory, **16B,** 291–292
Differential tunnel conductance, **11,** 126
Differential voltmeter, **2B,** 52
Differential-scattering cross section, **17,** 165, 167, 174
Differentiation of spectrum, **22,** 203
Differentiator, defined, **2A,** 314
Diffracting circular aperture, **26,** 58, 66
Diffracting rectangular aperture, **26,** 65
Diffracting slit, **26,** 57, 65
Diffracting straight edge, **26,** 57
Diffracting two-dimensional array of holes, **26,** 80
Diffraction, **1,** 346–349; **22,** 212, 372; **23A,** 8; **26,** 49; *see also* Acoustooptic measurement of sound wave, beam diffraction, beam propagation; Beam spreading; Electron diffraction; X-ray diffraction
of atoms, **29B,** 345–347
of Čerenkov radiation, **5A,** 176
contrast, in electron microscopy, **16B,** 190
effect on schlieren method, **18,** 364
effects, in infrared systems, **12A,** 436, 438–443
efficiency, **25,** 214
electron, **3,** 334–348
by extended scattering medium, **12C,** 124–125
of fast electrons, **6A,** 249–261
Fraunhofer, **23A,** 271
Fresnel, **23A,** 272
geometrical theory of, references for, **12B,** 63
grating, **1,** 367; **5B,** 53
interferometer, **18,** 380
images with aberrations, **26,** 83
limited lens, **25,** 166
measurement of neutron momentum, **5A,** 566–570
methods of molecular structure determination, **3,** 265–358
theory, **3,** 268–301
of molecular beams, **3,** 611–612
neutron, **3,** 348–358
optics, **29C,** 360–361
pattern, **22,** 267
phenomena at high energies, **5A,** 322
of radiation, general experimental arrangements, **6A,** 187–188
of short-wavelength radiation by matter, **6A,** 187–203

of slow electrons, **6A,** 261–264
from surfaces, **22,** 242
theory, **22,** 240
by thin layer, **12C,** 119–121
tomography, **19,** 578
X-ray, **1,** 418, 421ff; **3,** 301–334
cards, **1,** 424
Diffraction grating, **31,** 357–358, 370–375, 380
astigmatism, **31,** 367–368
beamline design, **31,** 28–30
dispersion, **31,** 364–365
efficiency, **31,** 370–375
equation, **31,** 357–358
focal curves, **31,** 365–367
history, **31,** 347–348
measure of efficiency, **31,** 194
ray tracing, **31,** 358–361
resolving power, **31,** 368–370
Diffraction gratings, **3,** 102–103; **3A,** 185, 203; **12A,** 469–474; **13A,** 156, 160, 227, 287; **29C,** 34, 164–167, 361; *see also* Grating
for atoms, **29B,** 345
optical
blazed, **29B,** 375–376
curved, **29B,** 377
echelle, **29B,** 376–377
general considerations, **29B,** 371–375
transmission, **29B,** 375
reflection gratings, **12A,** 473
transmission gratings, **12A,** 474
Diffraction-limited point light source, **18,** 710
Diffractive focusing with a zone plate, **31,** 293–295
Diffractometer(s), **3,** 319–321; **23A,** 219
components, **22,** 280
four-circle, **20,** 251, 252
high-Q, **23A,** 226
high-resolution, **23A,** 223
single-crystal, **23A,** 230, 241
sources of instrument broadening, **22,** 260
X-ray counter, **1,** 423
Diffuse illumination, **25,** 152
Diffuse scattering, **23A,** 26, 55, 344; **23B,** 197; **23C,** 464
elastic, **23B,** 97, 107
instruments, **23A,** 228
magnetic, **23A,** 72
Diffuse scattering of neutrons, **21,** 147, 154, 158, 161–167
elastic, **21,** 148
lattice distortion studies, **21,** 169
limit of sensitivity, vacancies, **21,** 161
spectrometer, **21,** 161–163
thermal, **21,** 148
time-of-flight, **21,** 161
Diffused contacts, **6B,** 51
Diffused isolation, **2A,** 113–114
Diffused metal-oxide-semiconductor (DMOS) transistor, **2A,** 89–90
Diffuseness parameter, **5A,** 316, 321
Diffuser
orifice flowmeter, **18,** 326
wind tunnel, **18,** 758, 760, 772, 784
Diffusion, **1,** 139–142; **6A,** 364–373; **19,** 192, 217; **22,** 364, 378, 384, 389; *see also* NMR, measurement of diffusion coefficient; *see also* Permeation; *see also* Self-diffusion
activation energy, **22,** 365
alkali in orthoclase, **24A,** 100
of Au on Au(111), **27,** 292
autoradiographic and surface counting methods, **6A,** 371
autoradiography measurement, **21,** 48–49
BaF_2 or CaF_2, **21,** 208
back, **1,** 178
^{45}Ca in melts, **24A,** 103–104
calculated from permeability time lag method, **16C,** 357–361
cerium in obsidian, **24A,** 109
chambers, **5A,** 201–203
of chemisorbed layer, **22,** 364
coefficient of, **1,** 139, 141
concentration dependent, Fickian, **16C,** 332, 339
constant, **22,** 364
ambipolar, **2,** 83
controlled freezing-in, **6A,** 45–46, 66–70
cooperative, **22,** 392
^{51}Cr in MgO, **24A,** 108–109
current, **2,** 118
dual mode, **16C,** 333
of electrons and ions, **5A,** 94, 100–101
equation, **6A,** 72–73
exchange
isotropic, **24A,** 100
medium, **24A,** 99–101
technique, **24A,** 93
Fick's second law, **24A,** 92–94, 107
flame technique, **7B,** 207–208
flames, **7B,** 206
in fluids, **23A,** 44

forced, **1,** 142
of gases, theory, **16C,** 325
in glass, **14,** 323, *see also* Permeability
grain boundary, **21,** 48
H_2O in obsidian, **24A,** 109
of helium, in glass, **14,** 321
hydrogen, **21,** 194, 203–206, 211
in metals, **23B,** 132
of impurities, **6A,** 72–85
impurity control methods, variations, **6A,** 85
jump model, **21,** 198–206
junction, **6B,** 88
^{42}K in K_2O-SrO-SiO_2
Kirkendall markers, **21,** 235
laser assisted, **30,** 365
length, **2,** 118
measurement, **6B,** 105–107
from liquid-solid interface, **6A,** 75–77, 80–81
application to semiconductor, **6A,** 80
^{25}Mg in silicate garnets, **24A,** 104
molecular, **3,** 433–434
Ni in olivine, **24A,** 112
nondilute, **23B,** 150
non-Fickian, **16C,** 334–337, 348
of O on Cu(110), **27,** 302
p-n junction method, **6A,** 371–372
radiation enhanced, **21,** 231
in solids, **23A,** 46
pressure, **1,** 141
profile
autoradiography, **24A,** 110–111
electron microprobe, **24A,** 111–112
ion backscattering, **24A,** 118–120
nuclear reactions, **24A,** 112–118
secondary ion mass spectroscopy, **24A,** 120–124
serial sectioning, **24A,** 106–109
radiation assay techniques, **6A,** 369–371
rate, **14,** 29
residence time, **21,** 199, 205
sectioning method, **6A,** 366–368
into a slab, **6A,** 74–75
from a solid-solid interface, **6A,** 78–79, 83–84
specimens, preparation, **6A,** 364–366
^{30}Si in forsterite, **24A,** 104
^{85}Sr in diopside, **24A,** 119–120
Sr in orthoclase and microcline
surface diffusion activation energy, **22,** 395
thermal, **1,** 141
thick *vs* thin sources, **24A,** 101–106
thin films, **21,** 233–236
tracer, **24A,** 92
track widening by, **5A,** 381
from a vapor-solid interface, **6A,** 77–78, 81–83
volume, **24A,** 92
of water molecules, **23C,** 467
welding, **14,** 287
without a phase boundary, **6A,** 79, 84
Zr in obsidian, **24A,** 102
Diffusion coefficient(s), **7A,** 377–378; **21,** 40, 70; **22,** 364, 394; **23B,** 142, 483
in afterglow studies, **7B,** 79–80
chemical, **23B,** 152
definition, **16C,** 318
of gases in polymers, empirical relations, **16C,** 327
lateral, **20,** 217, 316
Diffusion measurement
by chromatographic method, **16C,** 355
in electron swarms, **7B,** 32–34
surface preparation, **24A,** 94–99
tracer emplacement, **24A,** 99–106
Diffusion pumps, **1,** 176–180; **14,** 141–182, 241, 247
backstreaming, **14,** 154, 155, 158, 159, 166–175
boiler, **14,** 142, 143, 154–157
design, **14,** 175–176
booster-diffusion pumps, **14,** 180–181
cleaning, **14,** 502
cold cap, **14,** 143, 167, 176, 177
comparison, ion pump, **14,** 144
construction, **14,** 142–143
cooling, **14,** 143, 179
discharge pressure, **14,** 154–155
ejector stage, **14,** 143, 144, 164, 165
foreline booster pumps, **14,** 163–165
forepressure, **14,** 154–158
forepump backstreaming, **14,** 164, 165
forevacuum pump
effect on performance, **14,** 145, 150
requirements, **14,** 155–156
glass, **14,** 320
high pressure operation, **14,** 149
maximum throughout, **14,** 145
valve operation, **14,** 467
mercury, **14,** 181
operating procedures, **14,** 155–159, 169, 170
overload, **14,** 458

performance versus power input, **14,** 156, 173–175
power requirement, **14,** 151, 152, 174, 176
pumping efficiency, **14,** 147, 150
pumping fluids, **14,** 160, 161 ,165, 171, 172, 417, 418
purification, **14,** 165
pumping mechanism, **14,** 141–145
pumping speed, **14,** 145–152, 167–168, 173, 174
pump size, determination, **14,** 152
pressure range, **14,** 144, 149
pressure ratio, **14,** 143, 144, 155, 156, 158, 159, 174
pressure stability, **14,** 157, 175, 177, 178
residual gases, **14,** 161–163
safety factors, **14,** 155–156
selection, **14,** 458–459
in series, **14,** 144, 160
size effects, **14,** 149–150
stages, **14,** 144
system design, **14,** 155, 156, 169, 170
throughput, **14,** 147–154, 175
speed relations, **14,** 147–154
tolerable forepressure, **14,** 154–158, 175
ultimate pressure, **14,** 158–165, 175
in ultrahigh vacuum systems, **14,** 482–483, 489
valve operation, **14,** 466–467
vapor and gas distributions, **14,** 144–146, 150, 154
Diffusion-controlled crystal growth, **16B,** 345
Diffusion-controlled reactions, in chemical kinetics, **15B,** 281
Diffusivity, **6A,** 386; **22,** 395
thermal, **1,** 136, 137, 288, 294
Digit, binary, **2,** 687
Digital circuits, testing of, **2B,** 209
Digital computer, **2B,** 165–169
application of, **2B,** 169
development of, **2B,** 165–166
first, **2B,** 166
Digital counter, **2B,** 3–6
Digital filter, **28,** 283–284
design, **28,** 292–296
Digital integrated circuits, **2A,** 138–140, *see also* Integrated circuits
Digital methods in Fourier analysis, **6A,** 235–236
Digital multiplier, in equipment testing, **2B,** 185
Digital phase meter, **2B,** 43, 46
Digital processing system, **2B,** 166
Digital raw-data recording, **2B,** 123
Digital systems, logic design of, **2A,** 409
Digital tape recording systems, **2B,** 123
Digital-analog computers, in pulse spectrometry, **2B,** 129
Digital-to-analog (D/A) converters, **2A,** 419–421
Digitometer, **1,** 57
Dilatational pressure gauge, *see* Pressure gauge, dilatational
Dilation, **1,** 127
Dilatometer, **24B,** 430–432, *see also* Strainmeter
for crystallization studies, **16B,** 370–372
in high-pressure measurements, **16C,** 99–100
measurements, **6A,** 418–420
Dilute flame method, **7B,** 206–207
Dilute magnetic solids, **6B,** 171–174
Dilution
experiment, extrapolated, **3,** 410–411
heat of, **1,** 262
paramagnetic, **3,** 496
Dimensional analysis
examples, **18,** 832–842
mathematical foundations, **18,** 821–828
nature, **18,** 821
Dimensional and dimensionless, **18,** quantities, **18,** 822–825
Dimensionless homogeneity, **18,** 826
Dimensionless numbers in fluid dynamics, **18,** 829–831
Dimensionless parameters, **8,** 95
Dimensions, **18,** 822–823
Dimer structure
scanning tunneling microscopy of, **27,** 176
Dimers
in effusive beams, **29B,** 14
in superheated heat pipes, **29B,** 72
water vapor, **12B,** 156, 173
Dimethyladenine, **19,** 202
Dimethylamine laser emission, **10,** 48–50, *see also* HCN laser
Dinactin, **19,** 202
Dingle temperature, **11,** 38, 54–59
Diode clipping circuits, **2A,** 372, 385–387
Diode detectors, **10,** 158–166
Diode function generator, in analog computer, **2B,** 157–158
Diode harmonic generator, **10,** 18–23
Diode injection laser, **13B,** 290
Diode laser array, **29C,** 90

Diode lasers, **10,** 64–69; **29C,** 77–78, 149–150, 260
  amplitude noise, **29C,** 83
  beam quality, **29C,** 82–83
  current sources, **29C,** 90–93
  electronics, **29C,** 89–95
  extended-cavity diode lasers, **29C,** 83, 84–89, 93, 95, 99
  extended wavelength coverage, **29C,** 99–100
  frequency conversion, **29C,** 232
  frequency noise and stabilization, **29C,** 97–99
  laser degradation, **29C,** 84
  optical feedback, **29C,** 83–84, 85
  output power, **29C,** 80–82
  temperature controllers, **29C,** 93–95
  tuning, **29C,** 78–80
Diode limiter, **2A,** 370
Diode phase detectors, **2B,** 44
Diode plus stretcher, **2A,** 390
Diode-transistor logic (DTL), **2A,** 138
Diodes, **2,** 52–57
  avalanche injection, **2,** 125–126
  backward, **2,** 138–140
  barrier layer, **2,** 135
  behavior, Llewellyn and Peterson determination, **2,** 55–56
  characteristics of, **2A,** 152
  cut-off frequency, **2,** 137–138
  gas, **2,** 357–363
  light-production, **2,** 141
  limiters, **2,** 328–335
  low current characteristics, **2,** 56–57
  nonlinear capacitance, **2,** 136–138; **2A,** 366–367
  as nonlinear device, **2A,** 362–367
  parametric, **2,** 136–138
  p-n junction, **2,** 134–135
  point contact, **2,** 133–134
  Schottky effect in, **2A,** 152–153
  semiconductor, **2,** 330
  single-junction, **2,** 127–141
  space charge limited, **2,** 52
  step recovery, **2,** 140–141
  switching, **2,** 160–162
  temperature limited, **2,** 52
  transient time effects in, **2A,** 154–155
  transit-time effects, **2,** 54–56
  tunnel, **2,** 138–140; *see also* Tunnel diode
  vacuum, **2,** 205
  vacuum-tube, **2A,** 152–154
  voltage regulation, **2,** 135–136
  whisker, **2,** 133–134
  Zener, **2,** 135–136
Diopter, **1,** 336
Diphenyl-picryl-hydrazyl, **2B,** 147
Diplexer circuits or networks, **2B,** 286
Dipmeter, **24B,** 402–403
Dipolar (multipolar) scattering, **22,** 46, 56, 60
Dipolar order, **23C,** 247
Dipolar relaxation in lab frame (T_{1D}), **21,** 456, 460–465
  strong collision theory for heteronuclear systems, **21,** 461–465
Dipolar relaxation in rotating frame ($T_{1D\rho}$), **21,** 465–469
  spin locking, ADRF sequence for, **21,** 467
  strong collision theory for, **21,** 466–467
Dipolar spin temperature, **23C,** 271
Dipolar systems, **23C,** 33
Dipolar theory, **22,** 56
Dipole, **13A,** 115, *see also* Multipoles
  approximation, **23C,** 99
  arrays, **12B,** 98–101, 116, 117
  attraction, **22,** 383
  energy, field-induced, **22,** 380, 384
  feeds, **12B,** 71, 78
  two parallel, **12B,** 79
  magnetic
    Larmor precession, **3,** 617–618
    moment, **1,** 535
  moment(s), **3B,** 778; **16A,** 396; **22,** 360
    electric, **3,** 607, 648–676; **10,** 310, 313, 314
      from dielectric properties, **3,** 648–650
      molecular, **3,** 670–677
    induced, **3,** 118, 119, 638
    matrix element, **10,** 304, 341
    measurement
      by molecular beam method, **3,** 668–670
      by Stark effect, **3,** 668–669
    permanent, **3,** 649
      molecular, **3,** 637
    precessing magnetic, components, **3,** 378
    rotating system, **3,** 60
  permanent, **1,** 488
  radiation pattern, **26,** 98
  relaxation, magnetic, **3,** 375–384
  selection rule, **22,** 230
  transition moment, **3,** 245
Dipole-dipole array, **24B,** 296–299, 303–305
Dipole-dipole interaction(s), **22,** 62, 64

nuclear, **23C,** 269
transition probability, **3,** 678
Dipole fields, **8,** 63–64
Dipole moment, **8,** 64
Dipole pattern, **31,** 2–3
Dipole-normal selection rule, **22,** 47
Dipole-quadrupole interactions, transition probability, **3,** 679
Dipping refractometer, **25,** 188
Dirac comb, **8,** 47; **12A,** 503
Dirac δ function, **4A,** 322–323
Dirac delta function, **2B,** 380
Dirac electron, **4B,** 23, 31
Dirac function, **8,** 5
Dirac moment, **1,** 43
Dirac notation, **10,** 304
Dirac relativistic theory, **17,** 587
Direct Coulomb excitation, **17,** 482–483, 486–488
Direct Coulomb ionization, **17,** 200–207, 482–483, 486–487
Auger electron measurements, **17,** 492–500
Direct current, **1,** 447–459
measurement, **2,** 558–576; *see also* Ammeter
Direct process model calculations, **8,** 277
Direct spectral estimate, data taper, **28,** 324
Direct spectral estimator, **28,** 321
Direct spectrum analysis, in LDV
illustrations of use, **18,** 220–227
image converter use, **18,** 208–209
method, **18,** 194–227
streak camera use, **18,** 217–218
synchronous detection use, **18,** 205–208
Direct stripping process, **8,** 277, 280
Direct viewing prism, **25,** 210
Direct-coupled amplifier, **2A,** 289–305
shunt feedback bias scheme for, **2A,** 291
two-stage, **2A,** 288–292
Direct-current voltage and current regulators, **2A,** 243–244, *see also* Dc (*adj.*)
Direction indices, crystallographic, **16B,** 6
Directional ambiguity in LDV, removal of, **18,** 186–190
Directional coupler switch/modulator, **29C,** 389
Directional couplers, **8,** 124, 127, 135, 137; **29C,** 382, 383
in microwave circuits, **2B,** 260
Directional radiancy expression, **8,** 29
Directional reflectance, measurement of, **26,** 274
Directivity function, **12B,** 2
Directivity index, **1,** 207–208
Dirichlet conditions, **2B,** 382 n
Disappearance-potential spectroscopy, **22,** 213
Disappearing-phase transformations, **22,** 279
Discharge
atomic beam source, **7A,** 88
chambers, **5A,** 281–288
coefficient, flowmeter, **18,** 328–330, 335
electrodeless, **3,** 167–168
electron-energy-exchange, **3,** 168
light sources, classification of, **4A,** 293
plasma, **14,** 310
sources, **25,** 247, 251
tubes, **3,** 167–169
for atomic beam, **4A,** 221
helium-ferric-bromide, **3,** 176
hollow cathode, **1,** 374; **3,** 176; **5B,** 51–52
hydrogen, **1,** 376
Discharges, *see* Electric breakdown
cold-cathode, **2,** 83–87
corona, **2,** 85
glow, **2,** 83–87
Langmuir mode, **2,** 92
Discrete emission, light sources, **4A,** 316–318
Discrete prolate spheroidal sequence tapers, **28,** 323
zeroth-order, **28,** 336–337
Discrete semiconductor devices, **2A,** 47–102, *see also* Transistor(s)
backward diode as, **2A,** 67
barrier rectifier diodes as, **2A,** 59–60
bulk semiconductor properties in, **2A,** 48–51
examples of, **2A,** 100–102
Hall effect devices as, **2A,** 49–50
multiregion devices as, **2A,** 92–99
nonlinear resistors as, **2A,** 50
piezoresistors and, **2A,** 51
P-N junction diodes and, **2A,** 51–57
symbols for, **2A,** 99–100
timing diodes as, **2A,** 63–67
transistors as, **2A,** 67–92
varactor diodes as, **2A,** 64–65
Discrimination level, **8,** 188
Discriminator circuits, **8,** 137
Discriminator(s), **2,** 371–382; **24B,** 43
chain, pulse spectrometer, **2,** 608–610
channel, pulse spectrometer, **2,** 607–608
Kandiah, **2,** 380–382
pulse shape, **5A,** 537
voltage, transistor, **2,** 443–446
Disequilibrium

in new chemical lasers, **15B,** 120
thermodynamic analysis of, **15B,** 97–100
with intense infrared radiation, **15B,** 295–296
Disintegration rate, radioactive α-source, **5B,** 568–570
Dislocation effects, in ultrasonic studies, **11,** 421–428
Dislocation line, **6A,** 323
Dislocation loop, square, **6A,** 327–329
Dislocation theory, **1,** 109
Dislocations, **6A,** 321–357
based on growth phenomena, **6A,** 341–343
basic properties, **6A,** 327–329
decoration methods, **6A,** 348–352
edge, **6A,** 323–325
etchants suitable for revealing, **6A,** 344–346
experimental methods of detecting, **6A,** 341–356
intuitive definitions, **6A,** 323–324
methods on etch phenomena, **6A,** 343–347
mobility, **6A,** 324–325
multiplication of, **6A,** 330–332
motion of, **6A,** 329–330
optical and electron-optical methods, **6A,** 347–348
origin of, **6A,** 356–357
positrons, **21,** 83, 84
prismatic, **6A,** 328–329
screw, **6A,** 323
two-dimensional, **23B,** 46
X-ray methods, **6A,** 352–355
Disorder in crystalline polymers, **16B,** 98–114
analysis in terms of helical transforms, **16B,** 113–114
calculation of scattered X-ray intensities, **16B,** 111–112
one-dimensional models used in small-angle X-ray scattering, **16B,** 167–171
Disorder in solids, **23A,** 56
Disordered systems, **21,** 94
Dispenser cathodes, **2A,** 150–151; *see also* Matrix cathodes
Dispersing instruments, **3,** 156–161
Dispersing systems, vacuum ultraviolet, **1,** 404, 405
Dispersion, **1,** 331, 345; **3,** 711–712; **5A,** 345, 352, 357; **19,** 41, 183, 187, *see also* Velocity measurements; **31,** 257–259, 364–365
of air, **3,** 172–173
angular, **1,** 363; **12A,** 463, 470
angular and linear, **4B,** 66–67
anomalous, **1,** 494
of beam, in beam-foil experiments, **17,** 577
classical theory, **10,** 351–357
curve(s), **13B,** 15, 192; **16A,** 100–104
determination of, **16A,** 106–116
of data, **6A,** 3
diatomic chain, **10,** 359–363
energies, additivity of, **3,** 235
grating, **1,** 367
index
binomial, **5B,** 790
Poisson, **5B,** 788–791
law of mass spectrometer, **5B,** 5–6
linear, **12A,** 464
prism, **1,** 337, 338
in pulsar signals, **12C,** 80–81
relation, **9B,** 6, 9–11
relations for dielectrics, **10,** 94, 350–355
relations for temperate plasma, **10,** 273–278
relations for warm plasma, **10,** 278–279
spectrometers, **25,** 200
of surface waves, **19,** 500–501, 506, 529–531
surfaces, **23A,** 315
width, **5A,** 177
Dispersion-compensated electrostatic analyzer, **22,** 217
Dispersive analyzer, **22,** 218
Dispersive counter, **17,** 358–359
Dispersive detector, **17,** 376
Dispersive devices, *see* Gratings, Prisms, and Interferometers
Dispersive spectrographs, **3A,** 203
Dispersive spectrometers, **3A,** 177
Displacement, **1,** 74–75
angular, **1,** 86
energy, **6A,** 361–362
free air, **1,** 173
particle, **19,** 4–6, 498
potential, **19,** 5
variables, **1,** 312
vector, electric, **1,** 489ff
Display analyzers, **22,** 140
Display storage tube, **2B,** 355–357
Displays
digital, **2B,** 6–9
driving circuits for, **2B,** 8
recording of, **2B,** 8–9
seven-segment, **2B,** 7
visual, **2B,** 6–8
Dissector, image, **2,** 781–782

Dissipation, **1,** 199; **8,** 10
factor, **1,** 493
definition, **16C,** 386
measurement by bridge method, **16C,** 398
Dissipative regulators, **2B,** 452–456
Dissociation, **7A,** 466–472; **14,** 26
by electric and magnetic fields, **7A,** 208ff., 212–214
energies, bond, **3,** 546–552
heat of, **1,** 262
multiphoton, **15B,** 265–268
Dissociation of molecules, **4A,** 178–183, 220–221
by discharge tube, **4A,** 181, 220–221
by radio-frequency discharge, **4A,** 182–183
thermal, **4A,** 179–221
Dissociative capture, **7A,** 467
Dissociative conjugation, **7A,** 81
Dissociative photoionization, **7A,** 163–166
Distillation
(gases), **4B,** 319
use in purification, **6A,** 22
Distorted wave approximation, **7A,** 84
Distorted-wave method, for electron capture, **17,** 90, 92
Distortion, **25,** 42; **31,** 153, **31,** 170, 172
emulsion, **5A,** 228
straggling, microscopic, **5A,** 228
of solid, measurement by laser speckle, **18,** 713
track, **5A,** 220
vacuum tube oscillator, **2,** 273–274
Distortional polarization, **10,** 347
Distortions, geometric, of electron microscope image, **16B,** 203–205
Distributed Bragg reflector (DBR), **29C,** 78
Distributed circuit measurement, **1,** 499–501
Distributed constant networks
electromechanical devices in, **2A,** 32–33
Distributed feedback (DFB), **29C,** 78
laser, **15A,** 247–249
generation and performance of, using F_A(II) centers, **15B,** 37–41
grating for, **15B,** 38
performance of, **15B,** 39–41
principle of, **15B,** 37
Distributed volume dipoles, **8,** 63
Distribution coefficient, **6A,** 42–50, 70
Distribution function, confidence bands, **28,** 176–177
Distribution function, normal, **1,** 4
Distribution, Gaussian, **2,** 4
Distributional assumptions, assessment, **28,** 187–209
chi-squared tests, **28,** 188–192
extensions to censored data, **28,** 204–206
formal test procedures, **28,** 201–204
completely specified cdf, assessing fit, **28,** 201–202
normality tests, **28,** 203–204
parametric families, testing for, **28,** 202–203
goodness-of-fit problem, **28,** 187–188
quantile-quantile plots, **28,** 192–200
two-sample comparisons, **28,** 207–209
Distributions
binomial, **5B,** 761–762
composite, **5B,** 780–786
frequency, **5B,** 761–771
Gaussian, **5B,** 810
interval, **5B,** 767–771
multinomial, **5B,** 762–763
normal, **5B,** 763–764, 810
parent, **5B,** 809
Poisson, **5B,** 764–767
Divalent ions, **19,** 193
Division
of amplitude, **26,** 4
of wave front, **26,** 1
Division, of amplitude or wave front, LDV configuration, **18,** 124
DKDP, **23C,** 33
DLVO theory, **23B,** 495
DMOS (diffused MOS transistor), **2A,** 89
DNA, **22,** 423
Do-it-yourself optical engineers, **25,** 3
DODCI, as absorber in flashlamp dye laser, **15B,** 188–190
Dolezalek electrometer, **1,** 440–441
Dolfus polarimeter, **12A,** 398
Domain
dimensions in clock copolymers, **16C,** 280
morphology, **16C,** 280–281, 295–296
Domain walls, **23A,** 273; **23C,** 140
Domains
ferromagnetic, **1,** 546
magnetic, **23C,** 82
nuclear, **23C,** 276
Dominant likelihood situation, **28,** 407
Dominant mode in waveguide, *see* Principle mode in waveguide
Donors per unit volume, **5A,** 270
Doorstopper gage, **24B,** 386–389

DOP, **12C,** 283, 308–314
Dopant tunneling microscopy, **27,** 90
Doping in PLD, **30,** 519, 521–523
Doping of tunnel junction, **16A,** 155
Doppler
ambiguity, *see* Ambiguity noise
bursts, **18,** 154, 175, 180, 190
interferometry, **26,** 42
line width, **12C,** 29
linewidth, **10,** 345
shift measurement of recoil nuclei stopping, **5B,** 336–337
velocimeter, acoustic, **18,** 317–318
Doppler broadening (of annihilation radiation), **21,** 79, 80, 95, 96, 98, 99–106
data analysis, **21,** 101–104
deconvolution, **21,** 101
detectors, **21,** 97–98
electronics, **21,** 99–101
energy resolution, **21,** 99, 100
lineshape, **21,** 86, 89, 90, 101–103
defect specific, **21,** 103
parameter, generalized, **21,** 103
pileup, **21,** 100
sample-detector configuration, **21,** 105, 107
source contribution, **21,** 104
stabilization, **21,** 139–143
Doppler broadening, **4B,** 61, 121, 123, 183; **5B,** 50; **13A,** 23; **13B,** 219; **17,** 67, 516, 518, 524
beam-foil interactions, **17,** 545, 554–557, 569
of cross-section resonance, **5B,** 395
in electron-positron annihilation energy spectrum, **16A,** 385–387
half-breadth, **3,** 10
Lamb dip and, **15B,** 233–234
of microwave lines, **12B,** 146
problem of, **15B,** 232
residual, **15B,** 240, 247
two-photon absorption experiment and, **15B,** 238–239
Doppler effect, **1,** 76–77, 215; **9A,** 163; **26,** 42
history of, **12C,** 277
for radiation, **7A,** 48
reference time for, **12C,** 280–282
relativistic, **12C,** 278
spectrometers, **11,** 316, 342
velocity components in, **12C,** 282
Doppler estimates, **8,** 161
Doppler shift, **13A,** 14
beam-foil interaction, **17,** 545, 554–557, 569
from fast atoms, **7A,** 463
formulas, **18,** 99–104, 342
of photons emitted by fast-moving ions, **17,** 369, 373
Doppler width; *see also* Thermal Doppler width
discussion, **29B,** 132, 158–159, 255–257
reduction in fast beams, **29B,** 303
second-order, **29B,** 264
Doppler-broadened line, hole burning in, **15B,** 234
Doppler-free laser spectroscopy, **15B,** 232–248
laser-induced line narrowing in, **15B,** 243–248
on-resonance two-photon spectrum in, **15B,** 247–248
saturation spectroscopy in, **15B,** 233–238
ac Stark shifts and, **15B,** 242–243
experimental setup for, **15B,** 240–241
Doppler-free spectroscopy
atomic recoil, **29B,** 264
by two-photon absorption, **29B,** 265–267
in fast beams, **29B,** 307
line narrowing techniques, **29B,** 257–260
of trapped particles in Dicke limit, **29B,** 267–670
using Ramsey fringes, **29B,** 260–265
Doppler-shift attenuation method of lifetime measurements, **17,** 65
Doppler-tune spectrometer
Doppler-tuned spectrometer, **17,** 369–371, 388
Double alignment, channeling, **21,** 281, 290, 291, 331
Double Compton steps, **8,** 216
Double crystal spectrometer, **6A,** 354
Double escape pair peak (DEPP) intrinsic efficiencies, **8,** 215
Double exposure holographic interferometry, **26,** 193
Double Gauss lens, **25,** 84
Double Helmholtz arrangement, **8,** 79
Double ion chamber method, **7A,** 146–148, 460
Double pulse method for measurement of spectrum and autocorrelation function of radiowaves, **9B,** 150
Double refraction, see Birefringence
Double resonance, **3A,** 113, 116; **7A,** 140; **16A,** 264, 326, 451; **20,** 32–38
technique, **4B,** 176, 181, 188, 206–208; **6B,** 243; **7B,** 1
Double scattering, **21,** 164

Double-beam spectrophometer, **6B,** 304–305
Double-doping, *see* Simultaneous segregation of two impurities
Double-electron transition, **17,** 344
Double-escape pair-peaks, **8,** 219, 220
Double-focusing mass spectroscopy, **22,** 328, 340
Double-humped distribution instability, **9A,** 285–291
  resonance heating, **9A,** 286
  runaway distribution, **9A,** 286
Double-pass interferometer, **26,** 37
Double-passed AOM, **29C,** 114
Double-pulse experiments, **30,** 100–101
Double-sideband receivers, **12B,** 212
Doublet separation, **3,** 13
Doubling, 1-type, **3,** 13
Doubly mode-locked systems, **15B,** 192
Doubly resonant oscillator, **15B,** 161
  conversion efficiency of, **15B,** 170–171
  risetime for, **15B,** 169
  threshold pump power density for, **15B,** 168
Doubly-excited state, **17,** 107
  autoionizing decay, **17,** 568–570
  beam-foil interactions, **17,** 549, 551, 560–564
Dow cell, for permeability, **16C,** 363–364
Downscattering, in phonon-neutron collisions, **16A,** 210
DPD, *see* Waveform processing, direct phase determination
DPH radical, **3,** 496–497
Drag, **1,** 147ff
  coefficient, sphere, **18,** 8, 10–15, 759, 800
  coefficient of, **1,** 148, 151
  force, **1,** 148
    measurement, **18,** 768–769
  form, **1,** 150
  skin friction, **1,** 150
Drain-gate feedback biasing, **2A,** 265–266
Draw ratio, **16C,** 140
Dried-droplet method, **30,** 422
Drift
  amplifier, **1,** 450
  instabilities, **9A,** 285, 311–313, 416
  mobility measurement, **6B,** 91–95
  region, **22,** 405
  in time measurements, **2B,** 29
  tolerant circuit design, transistor, **2,** 449–451
  tube, **7A,** 200; **14,** 89
    structure, proton linear accelerator, **5B,** 639–640
  velocity, **5A,** 90, 91, 93; **7A,** 374, 376–378; **14,** 92
    of electron swarms, **7B,** 34
    negative ions, **5A,** 92
    zero, of dc amplifier, **2,** 236–237
Drift-scan measurements, **12C,** 7–10
Drilling mud, *see also* Spine-and-ribs cross plot
  effect on well logs, **24B,** 445, 447–448
  invasion into borehole wall, **24B,** 448
Drive failure, **14,** 256
Drive for turbo pump, **14,** 265–266
Driver, shift-register, **2,** 437–439
Drives
  constant-velocity, **11,** 318–319
  for Mössbauer spectrometers, **11,** 316–346
Driving circuits, for displays, **2B,** 8
Driving-point admittance, **2A,** 9
Driving-point impedance, **2A,** 9
DRO, **29C,** 264; *see also* Doubly resonant oscillator
Drop formation, **19,** 338–339
Drop method calorimeter, **6A,** 381, 383
Drop size distribution, **12B,** 178–179
  Laws and Parsons, **12B,** 181
  Marshal and Palmer, **12B,** 181
Droplet formation, **16B,** 388–389
Droplet technique, for homogeneous nucleation, **16B,** 385–396
Droplets, **30,** 187
Dropout, *see* Signal dropout
Dropping resistors, in equipment testing, **2B,** 191
Drude model of dispersion, **31,** 257
Drude polarimetric method, **6B,** 280–281
Drude theories, **11,** 68
Drum camera, **18,** 733
Drum, magnetic, pulse spectrometer, **2,** 617–618
Druyvesteyn distribution, **7B,** 41, 50
Dry stripping negative replica method, *see* Plastic replica technique
Drying, **19,** 342–345
DSA method, *see* Doppler-shift attenuation method
DSC, *see* Differential scanning calorimetry
DSDP, *see* Deep Sea Drilling Project
DT-cut quartz, **19,** 52
DTA, *see* Differential thermal analysis
DTL (diode-transistor logic), **2A,** 138
DTS, *see* Doppler-tuned spectrometer
Du Bridge's method, **6B,** 383

Dual channel oscilloscope, **8,** 149
Dual laterolog, *see* Laterlog
Dual polarity tunneling images of Ge(001), **27,** 238
Dual slope integration, in dc measurements, **2B,** 55
Dual-beam technique
measurements with, **12C,** 14
Dual-mode sorption and diffusion in glassy polymers, **16C,** 333
Duality, **2,** 22–24
Duc de Chalnes method, **25,** 190
Duct
conical, **14,** 20
layer, **1,** 213
rectangular, **14,** 20
Ducted flows, **1,** 144–147, 163–164
Ducting, of whistlers, **9B,** 95–98
Duopigatron, *see also* Arc plasma discharge source
filament lifetime, **29A,** 105
geometry, **29A,** 104–105
ion-beam source densities, **29A,** 105
single-aperture source, **29A,** 105
Duoplasmatron, *see also* Arc plasma discharge source
filaments, **29A,** 102–103
geometry, **29A,** 102–104
for hydrogen plasma production, **9A,** 23
ion extraction region, **29A,** 101–102
ion source, **22,** 347
potential distribution within the plasma discharge, **29A,** 102–103
Duplexers, **2,** 730–732
Duranickel, **14,** 378
Duration-bandwidth product, **28,** 323
Durbin-Watson test, **28,** 268, 276
Dust, acceleration by shock wave, **18,** 31–33, 795
Dy, **23C,** 213
Dye cell
astigmatism, in cw dye lasers, **15A,** 344–347
cross-sectional area of beam within, **15A,** 349–351
for flashlamp excitation, **15A,** 312–314
free jet, **15A,** 314–318
Dye flow system, for short-pulse lasers, **15A,** 324–325
Dye laser(s), **13B,** 305; **15A,** 251–275; **25,** 263; **29C,** 146, 151; *see also* Steady-state laser(s)
advantages of, **29C,** 146
amplified spontaneous emission, **29C,** 154–162
applications, **18,** 715–716, 746
continuous wave, *see* Continuous wave dye laser(s)
dye lasers, **29C,** 45–46, 50–73
cw. *see* cw (continuous wave) dye lasers
disadvantages of, **29C,** 146
dye cell length for, **15A,** 265
electronic energy levels and transitions in, **15A,** 253–254
first successful, **15A,** 337
flashlamp-excited, **15A,** 297–298
flashlamp plasma temperature as efficiency factor in, **15A,** 300
as four-level system, **15A,** 255
free-jet dye cell in, **15A,** 314–318
frequency and phase locking, **29C,** 128–134
frequency stabilization, **29C,** 299
gain, **29C,** 52–53
grazing-incidence dye laser, **29C,** 164–167
heat stored in, **15A,** 313
intense excitation from, **15A,** 314–318
intracavity spectral filters, **29C,** 164–169
laser devices and, **15A,** 269
laser dyes in, **15A,** 355–359
laser threshold relation in, **15A,** 258–263
levels, optical transitions between, **15A,** 253–254
linear flashlamps and, **15A,** 302–305
most outstanding characteristic of, **15A,** 264
multiple passes through the gain medium, **29C,** 162–164
nitrogen-laser-pumped, **15B,** 217
operation, high discharge plasma temperatures in, **15A,** 298–300
oscillation conditions for, **15A,** 257–259
oscillator-amplifier system, **15A,** 286
output power, **29C,** 54–56
output wavelength selectivity of, **15A,** 264–265
performance, self-absorption factors in, **15A,** 255–256
principles, **29C,** 46–50
pump lamp, **18,** 693–394
pumping of by mode-locked laser, **15B,** 175
rate equation for, **15A,** 257–258, 269–271
resonators for, **15A,** 321–324
resonator stability criteria for, **15A,** 338–343
rhodamine 6G type, **15A,** 256–261, 331, 336, 357

safety, **29C,** 157–158
self-absorption in, **15A,** 255
short-pulse, *see* Short-pulse dye laser(s)
short-pulse oscillator and, **15A,** 276–281
simulated emission cross section in, **15A,** 263–264
single longitudinal mode tunable dye laser, **29C,** 167–168
solvent characteristics and, **15A,** 265–267
specialized techniques with, **15A,** 323–324
Spectra-Physics model, **15A,** 356, 375
steady-state laser as, **15A,** 293–325
thermal limitations in, **15A,** 264–269
threshold
equation for, **15A,** 259
pump power, **29C,** 53–54
traveling-wave dye laser (TWDL), **29C,** 251
two classes of, **15A,** 252
uses of, **15A,** 251
vs color center laser, **15B,** 1
water as solvent for, **15A,** 268–269
wavelength range and intensity maximum for, **15A,** 258–259
Dye molecules
energy states of, **15A,** 253–257
steady-state behavior of, **15A,** 294–297
Dye solution jets, **15A,** 314–318
nozzles for, **15A,** 354
Dye, marker for flow visualization, **18,** 819
Dynamic form factor, *see* Scattering functions
Dynamic graphics, **28,** 512–516
Dynamic mechanical properties
definition, **16C,** 10–11
frequency limitation measurement, **16C,** 42
instrumentation, **16C,** 35–44
isochronal measurement, **16C,** 38
of NBS nonlinear test fluid No. 1, **16C,** 21
Dynamic method in anelasticity, **6A,** 304–307
Dynamic nuclear polarization, **23C,** 248
Dynamic pressure, **18,** 247, 503
Dynamic range, **22,** 284
of electrographic recording, **12A,** 272
of Fourier-transform spectrometers, **12A,** 523
of photographic recording, **12A,** 524
of television systems, **12A,** 295, 309
Dynamic response, *see* Frequency response
Dynamic screening, **22,** 231
model, of heavy-ion incident on solid target, **17,** 253–255, 266
Dynamic susceptibilities, **23B,** 308
Dynamic test for elastic constants, **6A,** 298–303
Dynamic tests, circuit requirements for, **2B,** 203–204
Dynamic viscosity, **19,** 141–142
Dynamical light scattering, **19,** 192
Dynamical matrix, **23A,** 23, 379
molecular, **23A,** 445
Dynamical similarity, **18,** 828–829
Dynamical theory, **23C,** 89
Dynamical theory of diffraction, **6A,** 258–261; **23A,** 313
dispersion relation, **23A,** 316
Dynamics
fluid, **1,** 126–172
gas, **1,** 126–172
of molecular beams, **3,** 611–612
observation of, **27,** 291
of adsorbates on metal surfaces, **27,** 298
Dynamitron, *see* Electrostatic accelerator
Dynamometers, **1,** 66, 94–97, 460
Dyne, **1,** 30
Dynodes, **12A,** 27–37
box-and-grid, **12A,** 28
linear focused, **12A,** 29
materials, **12A,** 31
Venetian blind, **12A,** 29
Dysprosium ethyl sulphate, **10,** 391–393

E

e beams, in rare gas halide lasers, **15B,** 91–93
e ray
interaction of with o ray, **15B,** 146
noncollinear matching of with o ray, **15B,** 157
E ray, **1,** 379
E-H gradient balance magnet, **7A,** 95
e/*h* constant, Josephson effects and, **11,** 273–274
Eagle grating mount, **1,** 369, 370
Eagle mount, **3,** 159
Eagle mounting, **13A,** 293
EAPFS, *see* Extended appearance-potential fine structure
Earth rotation
nonuniform rate of, **12C,** 182
VLBI measurement of, **12C,** 261–276
Earth tides, **12C,** 264; **24B,** 143–146
VLBI measurement of, **12C,** 275
Earthquake prediction
dilatancy diffusion model, **24A,** 5
VLBI use in, **12C,** 276

Easy axes of magnetization, **10,** 400
Ebert arrangement, **3,** 101–102
Ebert-Fastie arrangement, **25,** 216
Ebert-Fastie spectrometer(s), **12A,** 479; **13A,** 235; **13B,** 23
EBIS, *see* Electron-beam ion source
EBIT, *see* Electron-beam ion trap
EBS vidicon, **12A,** 281
Ebulliometry, molecular weight measurement by, **16A,** 26
ECAP (Electronic Circuit Analysis Program), **2A,** 286
ECDL, *see* Extended-cavity diode laser
Echelette grating, **10,** 134, 135
  filters, **10,** 138–140
Echelle grating, **8,** 162
Echelle(s), **1,** 369; **3,** 156; **12A,** 470, 474; **13A,** 299; **26,** 79
  spectrometers, **12A,** 482–489
Echellette grating, **1,** 368, 369, 400
Echellettes, **13B,** 16, 58; **26,** 79
Echelon(s), **1,** 372–373; **13A,** 301; **20,** 174–176; **26,** 31
Echo amplitude, **3,** 426
  damping term, **3,** 399
Echo sounders
  beamwidth, **24B,** 87–90
  conversion of traveltime to depth, **24B,** 94
  design considerations, **24B,** 87–97
  frequency of operation, **24B,** 87, 95
  multibeam, **24B,** 90–94
  pulse length, **24B,** 87
  recording, **24B,** 88
  resolution, **24B,** 88–89
  single beam, **24B,** 94–96
Echo trains, of whistlers, **9B,** 92–93
  duct of, **9B,** 9
  electron cyclotron frequency, **9B,** 93
  earth-ionosphere wave-guide, **9B,** 93
Echo-overlap method
  cycle-for-cycle matching of pulses, **19,** 95–97
  in ultrasonic velocity measurement, **19,** 97–99
Echoes
  rotary, **3,** 400
  plasma wave type, **9A,** 27–36
  spin, **3,** 395–399
Eckersley law, in ducting of whistlers, **9B,** 95
ECL (emitter-coupled logic), **2A,** 138
ECRIS, *see* Electron-cyclotron resonance ion source
EDC, *see* Energy distribution curve
Eddy current(s), **1,** 74; **6B,** 216–218
  damping, **1,** 541
  dynamometer, **1,** 96–97
  induction, **24B,** 351
  loss, **1,** 556
EDGDL, *see* Electric discharge gas-dynamic laser
Edge dislocation, **6A,** 323–325
Edge filters, **29C,** 350
Edge waves, **26,** 52
EDJDOS, *see* Energy distribution of the joint density of states
Edlén dispersion formula, **3,** 173
EDSAC, **2B,** 166
Edser-Butler fringes, **26,** 28
EDVAC, **2B,** 166
EELS, *see* Electron energy loss spectroscopy
Effective area, definition of, **12B,** 3
Effective counts, **8,** 192
Effective dielectric constant of plasma, **10,** 271
Effective magnetic moment of atom, **4B,** 236
  variation of, **4B,** 236
Effective mass, **6A,** 17; **10,** 241–243
Effective pathlength of a cavity, **10,** 156
Effective wavelength, **12A,** 154–160, 399
Effective wavelengths, infrared, **12A,** 452–456, 459
  higher-order corrections, **12A,** 454, 459
Efficiency
  aperture, **12B,** 5
    in paraboloidal reflectors, **12B,** 30–33
  counting of boron-loaded liquid scintillation counters, **5A,** 535–536
  detection
    of boron trifluoride counter, **5A,** 533
    of helium–3 counter, **5A,** 576
  discharge chambers, **5A,** 284–285
  gamma ray telescopes, **5A,** 669–670
  Geiger counter, **5A,** 120
  ionization, **5A,** 103–105
  of phosphors, **6B,** 306–308
  photoluminescence, **1,** 397–398
  pumping, **14,** 17
  quantum, **5A,** 125ff, 134
  radiation, **12B,** 3
  relative scintillation
    of gas scintillators, **5A,** 433
    of sodium iodide (Tl), **5A,** 413
  scintillation
    absolute, **5A,** 421
    vs particle mass, **5A,** 416–417

scintillator, **5A,** 126–128, 618
simulation, **28,** 451–452
uniform field pair spectrometer, **5A,** 641–652
Effusion, **30,** 242–243
rate of molecular beam, **3,** 592
source, **7A,** 71, 74–75, 232
Egerton-Powling burner, **7B,** 191–192
Eigenfunctions
electronic, **3,** 192
harmonic oscillator, **3,** 252
Eigenvector fitting, **23A,** 472
Eikonal equation, **9B,** 20; **19,** 569
Einstein coefficient of absorption, **10,** 341
Einstein mobility relation, **6B,** 94
Einstein probability coefficients, **4B,** 215–216
Einstein relation, **7A,** 377–378
Einstein transition probabilities, **10,** 46, 341
Einstein-Bohr controversy, **23A,** 332
Einstein-Debye theory (light scattering), **20,** 363
Ejector pump, **14,** 112, 113, 136–138
diffusion pump comparison, **14,** 145
with liquid piston pump, **14,** 136
oil, **14,** 180–181
Ekman boundary layer, **18,** 805–806, 810, 817
Elastic
deformation, **1,** 66–67, 100–107
hysteresis, **1,** 108
limit, **1,** 107
modulus, **1,** 100, 101, 110
Elastic coefficients, **6A,** 294
wave equation and, **11,** 439–441
Elastic constants
calculation of, **11,** 411–414
central potentials and, **11,** 414–418
determination from measurement of diffuse X-ray reflections, **6A,** 302–303
dynamic test methods, **6A,** 298–303
by high-pressure measurements, **16C,** 108–112
noncentral potentials in, **11,** 418–419
under pressure, **6A,** 420–421
pulse ultrasonic method of determination, **6A,** 301–302
significance of, **11,** 410–411
static test methods, **6A,** 296–298
static-lattice contributions to, **11,** 414–419
temperature dependence of, **11,** 419–421
ultrasonic measurements of, **1,** 107; **11,** 410–421
by ultrasonic measurements, **16C,** 79–84
of various polymers, **16C,** 81
Elastic differential cross section, neutral-neutral, **7A,** 304–310
Elastic differential scattering
large angles, **7A,** 316–320
small angles, **7A,** 315–316
Elastic energy loss, *see* Nuclear energy loss
Elastic incoherent structure factor, **23B,** 557
Elastic instabilities, **23C,** 35
Elastic moduli, **6A,** 294
anisotropy, **24A,** 4
bulk, **19,** 4–5
complex, **24A,** 47
complex shear, **19,** 140
dispersion, **24A,** 37
instantaneous elastic shear, **19,** 138, 146, 172–173, 175
measurement
Brillouin spectroscopy, **24A,** 13–23
thermal diffuse scattering of x rays, **24A,** 23–28
ultrasonics, **24A,** 2–13
pore fluids, effect on, **24A,** 5
pressure derivatives, **24A,** 10
of pyrolytic graphite, **19,** 293
in scattering formulas, **19,** 242–244, 270
shear, **19,** 139–141
of steel bar, equivalent, **19,** 294
elevated temperature and pressure, **24A,** 4–5
Elastic neutral-neutral collisions, theoretical survey, **7A,** 310–315
Elastic properties of solids, **6A,** 291–321
Elastic scattering, **21,** 149–151; **22,** 244; **23A,** 10, 24
cross sections, electric-arc use in studies of, **7B,** 351–353
Elastic total cross sections, neutral-neutral, **7A,** 279–292
Elastic waves, *see also* Free oscillations
amplitude, **24B,** 11–12
body, **24B,** 12
frequency range, **24B,** 11
surface, **24B,** 12
Elastic yield, **22,** 213
Elasticity, **1,** 100
Elasto-optical effect, **6B,** 272–273
Elastomer, **14,** 243, 406–411, *see also specific substances*
gasket, **14,** 279–281, 407, 408
joint, **14,** 279
Neoprene, **14,** 408–410

outgassing, **14,** 479
periodic inspection, **14,** 503
properties, **14,** 408
silicone-based, **14,** 408
silicone-based, **14,** 282, 410, 411
soft rubber, **14,** 408, 409
Viton, **14,** 411
Elastomer (rubbers)
fluorosilicone, permeability, **16C,** 368
fracture studies by ESR, **16C,** 213–217
low temperature ductility, **16C,** 216
permeation of gases, historic, **16C,** 316–320
Elbow, cylindrical, **14,** 19
Elbows, microwave, **1,** 482
ELDOR, *see* NMR, double resonance techniques
Electret, **16C,** 439, 468
Electric and magnetic fields, **24B,** 306–307, 317–318, *see also* Maxwell's equations
Electric arcs, **7B,** 307–353
Abel inversion process, **7B,** 315–319
applications of, **7B,** 347–353
in atomic transition probabilities, **7B,** 347–348
in continuum emission coefficients, **7B,** 348–349
in elastic scattering cross sections, **7B,** 351–353
in line-broadening parameters, **7B,** 349–351
arc sources, **7B,** 307–315
fluid-stabilized types, **7B,** 309–312
high-current, controlled atmospheric types, **7B,** 309
low-current atmospheric types, **7B,** 308–309
plasmas, local thermal equilibrium (LTE) in, **7B,** 341–347
temperature and density determinations in, **7B,** 319–341
absolute line intensities, **7B,** 330–333
Fowler-Milne method, **7B,** 333–334
Electric breakdown field, in air at STP, **16C,** 429
Electric dipole decay process, **17,** 579, 583–584
Electric dipole interaction, of linearly polarized monochromatic wave, **15A,** 25
Electric dipole moment, limit
of Cs atom, **4B,** 44–46
of electron
measurement, by atomic beam, **4B,** 44–46
by (*g*–2) experiment, **4B,** 43–44
by high energy scattering, **4B,** 44
of neutron, **4B,** 43
of positron, **4B,** 47
Electric dipole moments, **3B,** 778
elementary particles, **3B,** 843
Electric dipole selection rules, in multielectron transitions, **17,** 396–397
Electric dipole transition, in doubly-excited states, **17,** 561–564
Electric discharge gas-dynamic laser, **15B,** 79, 82–83
multiatmosphere operation of, **15B,** 80
Electric discharge, in pure CO_2, **15B,** 71
Electric displacement field, **16C,** 381
Electric field(s)
at an interface, **16C,** 433
definition, **16C,** 381
gradient, in Mössbauer spectroscopy, **11,** 314
intensity, **1,** 518
polarization as function of, **15B,** 144
polarization dependence on, **15B,** 143
time dependent, **16C,** 384
Electric precession of electron, **4B,** 35
Electric quadrupole decay process, **17,** 583
Electric quadrupole moment, nuclear, **4B,** 100, 103–104
Electric resonance, **3B,** 750
spectroscopy, **3A,** 355
Electric strength, *see* Electric breakdown
Electric units, conversion, **1,** 432
Electric-resonance optothermal spectrometer
description, **29B,** 282–284
applications, **29B,** 284–287
Electrical breakdown
by discharges
artificial void specimens, **16C,** 460
experimental, **16C,** 481
theory, **16C,** 450–451
cryogenic methods, **16C,** 476–478
electrode application, **16C,** 462–464
industrial tests, **16C,** 488
in nonuniform fields, **16C,** 448
in single crystals, **16C,** 459
intrinsic, **16C,** 444–445
prestressing effect, **16C,** 466–469
space charge effects, **16C,** 467–469
spark development, phototechniques, **16C,** 478–481
specimen preparation, **16C,** 451–462
strength, hydrogen, **14,** 421

thermal
experimental, **16C,** 475
impulse, **16C,** 447
steady state, **16C,** 446
theory, **16C,** 446–447
time lag measurements, **16C,** 470–471
voltage modes, **16C,** 464–470
Electrical circuit, dielectric barrier discharge, **31,** 42
Electrical conduction in polymers, high-field measurements, **16C,** 489–497
Electrical conductivity, **14,** 363, *see also specific materials*
Electrical matching, *see* Impedance matching
Electrical methods of impurity detection, **6A,** 37
Electrical properties of matter, **13B,** 134
Electrical properties of solids, **6B,** 1–143
Electrical propulsion
criteria for, **9B,** 251–252
as a plasma physics problem, **9B,** 269–289
azimuthal electric field, **9B,** 288
electric field measurement, **9B,** 277
ion flux measurements, **9B,** 282
MPD arc parameters, **9B,** 270
potential contour map, **9B,** 280
preheated cathode, **9B,** 271
Rogowski coils, **9B,** 272–276
plasma problems in, **9B,** 257–289
as a problem in magnetohydrodynamics, **9B,** 255–269
systems for,
categories of, **9B,** 254–255
plasma parameters in, **9B,** 252–254
Electrical resistance measurements at high pressure, quasi hydrostatic techniques, **6A,** 426–428
Electrical resistivity
apparent resistivity, **24B,** 295, 504
arrays, **24B,** 296, 301, 303–305
departure curves, **24B,** 507–511
dependence on pore fluids, **24B,** 460–463
effect of clays, **24B,** 284–285
effect of temperature, **24B,** 283
formation factor, **24B,** 460
measurement
boreholes, **24B,** 502–530
surface arrays, **24B,** 291–302
pressure dependence, **16C,** 113
properties of earth materials, **24B,** 283
surveys, principles, **24B,** 292–299
Electrical variables, **1,** 306
Electricity, **1,** 431–515
alternating current, **1,** 459–488
atmospheric, **1,** 501–506
direct current, **1,** 447–459
electrostatics, **1,** 431–447
thermoelectricity, **1,** 506–515
Electro optical effect, **6B,** 273–275
Electro-optic
effect, **26,** 147
modulators, **26,** 147
Electro-optical shutter, **18,** 727–732
Electro-optics, **1,** 385–388
Electroacoustic instruments, **1,** 210
Electroacoustic transducers, **1,** 216
Electroacoustic wave, *see* Bleustein-Gulyaev wave
Electrochemical cells, **23B,** 193
Electrochemical equivalent, **1,** 41
of silver, **4B,** 4
Electrochemical potential, **22,** 2
Electrode attachment, **1,** 496–497
Electrode configurations, in dielectric measurements, **6B,** 3–8
Electrode excitations, in normal-metal tunneling, **11,** 169–170
Electrode polarization, **13B,** 140; **24B,** 285–287, *see also* Induced polarization, mechanisms
Electrode-semiconductor-electrode tunnel junctions, **11,** 187
Electrodeless discharge, **13A,** 261
Electrodeposition, of thin films, **11,** 648–650
Electrodes
application, **16C,** 453, 462–464
guard, **5A,** 99
Electrodynamic accelerator, *see also* Storage ring
cyclotron accelerator
development, **29A,** 312–313
isochronous cyclotron, **29A,** 313–314
maximum energy, **29A,** 313
principle, **29A,** 312
superconducting cyclotron, **29A,** 314
principle, **29A,** 306–307
radiofrequency linear accelerator
applications, **29A,** 314–315
principle, **29A,** 314
synchrotron accelerator, **29A,** 315–316
Electrodynamic instruments, **2B,** 64–65
Electrodynamometer, **1,** 458, 467
Electrographic image tubes, **12A,** 252–276
Electrojet, equatorial, scattering from, **9B,** 176–186

Electroluminescence, **6B,** 294–295, 315–321
cells, **6B,** 316–318
crystal mounts, **6B,** 318
methods of observation, **6B,** 319–320
properties of, **6B,** 315–316
surface effects, **6B,** 320–321
Electroluminescent phosphors, **6B,** 299–300
Electrolysis, **6A,** 23
Electrolyte conduction, **24B,** 283
Electrolytes, resistance measurements, **6B,** 67–70
Electrolytic capacitors, **2A,** 37–38
Electrolytic cell, **1,** 452
Electrolytic coloration, **6B,** 328
Electrolytic electrodes, **16C,** 464, 496
Electrolytic peeling of thin surfaces, **7B,** 243–244
Electrolytic saw, **11,** 10
Electrolytic solutions, **6B,** 67–69
Electrolytic tank, **1,** 445–447; **4A,** 103–105, 112–113, 138
Electromagnetic absorption methods, in flame studies, **7B,** 229
Electromagnetic anemometer, **18,** 318–321
Electromagnetic description of light, **26,** 108
Electromagnetic engines, **9B,** 255
Electromagnetic fields, **1,** 517–524; **13A,** 32, 38, 61
Electromagnetic flowmeter, **18,** 337, 340
Electromagnetic lens, **16B,** 196–197
Electromagnetic measurement of mass, **1,** 64–66
Electromagnetic phenomena, Maxwell's equations for, **15B,** 250
Electromagnetic problem in microwave scattering, **9A,** 41–48
Electromagnetic radiation, **8,** 165
Electromagnetic shock tubes, **7B,** 148; **9A,** 173–207
auxiliary components, **9A,** 188–204
blast wave drive, **9A,** 175
cables for, **9A,** 197–199
coaxial type, **9A,** 197
Rogowski probes, **9A,** 199
transmission line effect, **9A,** 198
capacitors, **9A,** 181
castor oil use in, **9A,** 184
dielectric in, **9A,** 185
header for, **9A,** 184
capacitor bank, **9A,** 178
mandril use in, **9A,** 184
pulse-forming type, **9A,** 184–185
dielectric materials for, **9A,** 199–200
dielectric strength of, **9A,** 199
dielectric switches, **9A,** 189
solid type, **9A,** 189
electrodes for, **9A,** 191
energy sources of, **9A,** 179–186
experimental components of, **9A,** 179–207
gaseous dielectrics for, **9A,** 200–201
electronegative gases, **9A,** 200
ignition, **9A,** 189
instrumentation for, **9A,** 205–207
electric field probes, **9A,** 205
high-speed photography, **9A,** 206
image converter camera, **9A,** 206
Kerr cell, **9A,** 206
magnetic field probes, **9A,** 205
phasography, **9A,** 207
photodiodes, **9A,** 206
pressure probes, **9A,** 205–206
shadowgraph and schlieren techniques, **9A,** 207
jitter time, **9A,** 192
liquid dielectrics for, **9A,** 201
low-inductance switch, **9A,** 190–191
power supplies, **9A,** 194–195
monocyclic circuit, **9A,** 194
saturable reactor, **9A,** 194
pressurerized gas switch, **9A,** 190
pulse transformer, **9A,** 193
shielding and grounding for, **9A,** 195–197
screen room, **9A,** 195
shock jump equations, **9A,** 177
snowplow model, **9A,** 176
solid dielectrics for, **9A,** 201–204
surface resistivity, **9A,** 204
strip line transformer, **9A,** 194
switch triggering, **9A,** 194
switches for, **9A,** 189
system performance and experimental results, **9A,** 208–212
of coaxial shock tubes, **9A,** 208–211
of theta pinch shock waves, **9A,** 208
of Z-pinch shock waves, **9A,** 211
thyratron, **9A,** 193
trigger time, **9A,** 192
types of, **9A,** 174–175
coaxial type, **9A,** 186
inverse pinch, **9A,** 188
theta-pinch, **9A,** 188, 208
z-pinch type, **9A,** 188
vacuum spark gap switch, **9A,** 190
Electromagnetic spectrum, **13A,** 2

Electromagnetic theory of light, **1,** 376
Electromagnetic units, **1,** 34
Electromagnetic wave equations
boundary conditions for earth, **24B,** 275–277
derivation, **24B,** 270–272
impedances, **24B,** 277–283, 315–316, 320–327, 333–335
solutions, **24B,** 272–275
Electromagnetic wave propagation in well logging, **24B,** 491
Electromagnetic wave, polarization wave and, **15B,** 143
Electromagnets, **1,** 528–530
Electromagnets, characteristics of, **11,** 595
Electromechanical breakdown
experimental, **16C,** 473–475
suppression of, **16C,** 475
theory, **16C,** 447
Electromechanical coupling constant, **19,** 37, 54
for surface waves, **19,** 502, 507
Electromechanical relays, **2A,** 378
Electrometer, **2,** 570; **7A,** 23, 104; **14,** 84
amplifier, **2A,** 298–300
tubes, **2,** 570–573; **4A,** 338–339
vibrating-reed, **2,** 573; **7A,** 23, 104
Electrometer circuits
DuBridge-Brown, **4A,** 339–340
Miller, **4A,** 341
multistage amplifier, **4A,** 341–342
noise in, **4A,** 346–347
Wynn-Williams, **4A,** 340
Electrometer probe, for electrostatic charge measurement, **16C,** 426
Electrometers, **1,** 438ff, 460; **6B,** 38–40
and electroscopes, ultimate sensitivity, **4A,** 347–348
Electromigration
definition, **28,** 156
failure time
confidence band, **28,** 176–177
linear regression, **28,** 163
natural logs, **28,** 155–156
statistical model, **28,** 180
Electron accelerator
Cock-Walton accelerator, **29A,** 307
Dynamitron, **29A,** 312
principle, **29A,** 306
Van de Graaff accelerator
principle, **29A,** 307
resonant coherent excitation experiments, **29A,** 309, 311–312
sectional view, **29A,** 309–310
tandem accelerator, **29A,** 307, 309
Electron affinities/affinity, **3B,** 619; **12A,** 11
affinity
difference method, **3,** 557–558
mass spectrometric method, **3,** 552–558
Electron attachment, **4A,** 399; **5A,** 92, 100–101
in oxygen, **5A,** 101
Electron avalanche, **14,** 220; **30,** 87, 95, 588
definition of, **7B,** 66–67
Electron beam, **8,** 36, 129, 131, 132
in cathode-ray tube, **2B,** 338
densitometers, in shockwave studies, **7B,** 177–178
energy analyzers, **7A,** 15–17
energy measurements, **7A,** 40, 52, 53
gas density probe, **1,** 171
ionizing, **14,** 82
pulsing technique, **7A,** 109–113
Schlieren system, **1,** 172
space charge effects, **7A,** 52, 53, 102, 103
welding, **14,** 437, 439
contamination, **14,** 438
Electron beam excitation, **31,** 76–80
Electron beam fluorescence
beam generation, **18,** 452
beam spreading, **18,** 452
calibration of system, **18,** 450–451
chemical composition measurement, **18,** 434, 445
compared with laser light scattering technique, **18,** 435
density measurements, **18,** 450–451
flow visualization, **18,** 399, 437, 453
general description, **18,** 434–438
intensity relation to gas density, **18,** 441–450
role of gas motion, **18,** 438, 446–447
role of secondary electrons, **18,** 443–445
selection rules, **18,** 438–441
temperature measurement, **18,** 489–497
Electron beam method, for work function measurement, **22,** 14
adsorption, **22,** 15
application, **22,** 14
contact potential, **22,** 15
Electron beam-plasma interaction, **7A,** 96
Electron beams, fundamental limitations
Liouville's theorem, **4A,** 86
space charge, **4A,** 85–86, 89–90, 99–101, 105
optimum launching angle, **4A,** 86, 130
Electron beams, in beam-plasma interactions, **9A,** 130

Electron bolometer effect, **4A,** 462
Electron bombardment, **14,** 447; **16C,** 437
detector, **7A,** 272, 352
outgassing, of glass, **14,** 327–328
in photoemission measurements, **11,** 104
Electron bunching, **10,** 27, 28, 30–33, 35, 39, 42–44; **31,** 1, 25
description, **31,** 124
Electron capture cross section, **17,** 75, 79, 96, 143–144
density effects and excited states, **17,** 132–133
experimental results, **17,** 97–104
K shell, **17,** 140
multiple capture, **17,** 104–105
probabilities, **17,** 108
simple collision system, **17,** 91
Electron capture, **17,** 89–110, 139–142, 384, 482, 488–489; **21,** 2–4; *see also* Charge exchange; Radiative electron capture
beam-foil interactions, **17,** 549–550
charge exchange probabilities, **17,** 108
in complex systems, **17,** 94–97
into continuum states, **17,** 109–110
in heavy-collision systems, **17,** 75, 78
detector, **14,** 513
into high quantum states, **17,** 108–109
hydrogenic states, **17,** 566
from K-shell of heavy targets, **17,** 350–354
and loss at very low energies, **5A,** 41–42
multiple, **17,** 104–107
orbital by Ar^{37}, **5A,** 693–697
in PIXE backgrounds, **21,** 254–255
process, **3,** 535
in simple collision system, **17,** 89–94
transitions, **5A,** 296
Electron charge
by electrolysis, **4B,** 4
by oil drop, **4B,** 3
by shot effect, **4B,** 4
measurement by charge on cloud, **4B,** 3
limit on range of values, **4B,** 11
Electron collimating magnetic field, **8,** 185
Electron collision
frequency, derivation of, **7B,** 86
helium metastable destruction by, **15A,** 108–109
ions excited by, **15A,** 59–97
neutral atom lasers excited by, **15A,** 33–59
in noble gas ion lasers, **15A,** 82–83
Electron cyclotron, **5B,** 648–651
frequency, **1,** 45; **9B,** 9
maser, **10,** 77–81
resonance, *see* Cyclotron resonance
spectrometer-detector, *see* Cyclotron resonance spectrometer-detector
Electron cyclotron resonance (ECR), **31,** 52–54
Electron density
distribution, Fourier series representation, **16B,** 38–39
mapping, **23A,** 363
maps, **20,** 256–258
measurement, **18,** 698, 704, 705, 742, 750–753, 794
of a plasma, **29C,** 8–10, 11
in plasmas, measurement of, **4A,** 386–388
Electron detector(s), **7B,** 282–284; **17,** 314
continuous-channel electron multiplier, **17,** 436–439
Electron diffraction; **3,** 334–348
accuracy of results, **3,** 347–348
analysis of results, **3,** 341–347
apparatus and procedures, **3,** 338–341
by crystals, **6A,** 246–247
methods, **6A,** 246–278
mode of operation, in electron microscope, **16B,** 211–212
patterns, **6A,** 273–278; **16B,** 192–194
polycrystalline specimen, **6A,** 276–278
several crystals, **6A,** 276
single crystal, **6A,** 273
thick single crystal, **6A,** 273–275
very thin single crystal, **6A,** 275–276
small-angle, **16B,** 222
Electron emission
methods, for work function measurement, **22,** 5
field-electron emission, **22,** 8
photoelectric emission, **22,** 7
thermionic method, **22,** 5
platinum, **14,** 373
secondary, in tetrodes, **2,** 63–64
secondary, zirconium, **14,** 365
tantalum, **14,** 370
tungsten, **14,** 368
Electron energy analyzer(s), **7B,** 279–282
construction, **29A,** 227–228
efficiency and electron energy, **29A,** 303–304
electrostatic deflection devices
applications, **29A,** 215
configurations, **29A,** 213–214
energy resolution, **29A,** 217–218, 228
figure of merit and luminosity, **29A,** 218–220

focusing conditions, **29A,** 216–217
fringing field minimization, **29A,** 220–221, 227
input lens
conifugurations, **29A,** 223
design, **29A,** 224–225
multielement lenses, **29A,** 225
optimization of voltages, **29A,** 226
operating modes, **29A,** 221–223
operating voltages, **29A,** 222
preretardation, **29A,** 215
principle, **29A,** 214–215
electrostatic shielding, **29A,** 220–221, 227–228
imaging, **29A,** 226–227
magnetic field, maximum tolerable in construction, **29A,** 227–228
multichannel energy detection, **29A,** 226–227
reflection/transmissionn device, **29A,** 213
retarding grid devices, **29A,** 212
time-of-flight devices
components, **29A,** 210
design, **29A,** 211
detection system response time, **29A,** 211
sources, **29A,** 209–210
space-charge effects, **29A,** 210
Electron energy, **8,** 166
Electron energy loss spectroscopy (EELS), **30,** 76, 291
Electron flow in electrostatic fields, **4A,** 97–100, 109–115
basic patterns
curvilinear, **4A,** 109–111
hollow, **4A,** 110–111
rectilinear, **4A,** 98–99
equations of, **4A,** 97
trajectories
by analog methods, **4A,** 111–113, 138–139
by digital computer, **4A,** 113–116
Electron gun(s), **7B,** 278–279; **22,** 206, 281, 366, 476, 486
in cathode-ray tube, **2B,** 337
conventional design, **7A,** 65
EBF system, **18,** 451–452
field emission source, **22,** 282
high current type, **4A,** 96–123
analysis by digital computer, **4A,** 113–120
anode configurations, **4A,** 104–107, 111–116
cathode boundary conditions, **4A,** 97
convergent beam
beam compression, **4A,** 101
cylindrical, **4A,** 101–107
hollow, **4A,** 108, 121
strip, **4A,** 107
crossed field, **4A,** 120–121
cylindrical beam, parallel, **4A,** 101
design by analog methods, **4A,** 103–105, 112–113
focusing
electrostatic, **4A,** 100–101
magnetic, **4A,** 105
magnetron injection, **4A,** 121–123
strip beam, parallel, **4A,** 101
thermal velocity effects, **4A,** 108
low current type
brightness limitation, **4A,** 90–92
cathode ray tube gun, **4A,** 93
multistage, very low energy, **4A,** 94–95
triode gun, **4A,** 90–92
low-energy, **22,** 282
magnetically focused, **22,** 282
for RHEED, **22,** 283
Electron impact excitation, to Rydberg states, **29B,** 117–119
Electron impact ionization, **31,** 40
Electron impact technique, **7B,** 1
Electron interception, **8,** 134
Electron lens, **7A,** 8
Electron loss, **17,** 75, 78, 110–118
multiple, **17,** 116–118
projectile ionization, **17,** 113–114
target ionization, **17,** 110–113
Electron loss cross section, **17,** 75, 79, 98–100
density effects and excited states, **17,** 132–133
experimental results, **17,** 114–115
multiple loss, **17,** 105–106
probabilities, **17,** 108
Electron mass
measurement of by Compton scattering, **4B,** 12
by e^+-e^- annihilation, **4B,** 12, 13
by spectra, **4B,** 12
Electron microscopes, **6A,** 183–186
Electron microscopy, **20,** 391–444; **22,** 511
aberrations, **20,** 392, 395, 398, 405, 444
absorption coefficient, **20,** 396
back projection, **20,** 413, 414
beam damage, **16B,** 190
bright field image, **16B,** 190–192, 211–212
dark field image, **16B,** 194, 211–212

dark-field methods, **20,** 433–438
diffraction pattern mode, **16B,** 211–212
electron optics, **16B,** 195–205
focusing, **16B,** 217–220
high voltage, **20,** 441–444
hollow cone problem, **20,** 416–418
hydration, **20,** 430–433
image degrading factors (aberrations), **16B,** 199–205
image restoration, **20,** 404, 405
instrumentation, **16B,** 206–212
isolation technique, **6A,** 181
low-loss, **16B,** 227–228
in magnetic domain studies, **11,** 726–734
magnification calibration, **16B,** 221
observation of single polymer molecules, **16B,** 228
of polymeric alloys, **16C,** 291–299
radiation damage, **20,** 418–430
relocation of the same region, **6A,** 181–182
replica method for, **11,** 692–693
resolution, **16B,** 220–221; **20,** 392–395
scanning, **16B,** 224–227; **11,** 734–735
scanning-transmission, **16B,** 227
shot noise, **20,** 427–430
specimen preparation, **16B,** 212–217
strong phase object approximation, **20,** 399
to study crazes, **16C,** 242–245
three-dimensional reconstruction, **20,** 409–418
topographical contrast imaging, **16B,** 222–224
transmission techniques, **6A,** 176–177
weak phase angle approximation, **20,** 397–399
Electron monochromators, **7A,** 4, 54, 65
Electron multiplier phototube, **4B,** 87–90
cooling, **4B,** 89
dark current, **4B,** 89
photon efficiency, **4B,** 89
windowless, **4B,** 88
Electron multiplier, **17,** 436–439
detector, **7B,** 282–283
Electron multipliers, **7A,** 23–28, 104, 193–194, 364, 401–405; **11,** 99
calibration, **7A,** 26
continuous channel, **7A,** 24, 405
magnetic, **7A,** 24, 85, 403–405
operating modes, **7A,** 25
activation, **7A,** 28
Electron multipliers for ion detection, **4A,** 355–360
circuits, **4A,** 359–360
continuous dynode, **4A,** 357–359
crossed-field, **4A,** 357–358
hollow tube, **4A,** 359
multiple dynode, **4A,** 355–357
Electron optics, **4A,** 135–148
comparison of electron and light optics, **4A,** 136–137
geometrical, **4A,** 135–145
electrostatic lenses, **4A,** 141–142
fields
aberrations in, **4A,** 143–144
classes of, **4A,** 140–144
methods of estimating, **4A,** 138
magnetic lenses, **4A,** 142, 144
optical elements, properties of, **4A,** 139–140
trajectories, methods of estimating, **4A,** 138–139
physical electron optics, **4A,** 145–148
electron diffraction, **4A,** 147–148
electron polarization, **4A,** 148
electron scattering, **4A,** 148
free electron interference, **4A,** 146–147
refractive index, **4A,** 135–136, 145
Electron paramagnetic resonance (EPR), **2B,** 133; **3B,** 497; **21,** 405–428; *see also* EPR
defect models, **21,** 423
g-tensor theory, **3B,** 504
as materials science tool, examples of, **21,** 425–428
minimum number of detectable spins, **21,** 407, 408
representative studies, **21,** 423–428
spectra, **21,** 405, 406
spectrometer, **20,** 57
design, **21,** 407–423
block diagrams, **21,** 410–413
bolometer detector, **21,** 415–416
cavities, **21,** 418–423
cryogenics and variable temperatures, **21,** 418–423
crystal diode detector, **21,** 415–416
detectors, **21,** 415–416
frequency stabilization, **21,** 414, 415
homodyne spectrometer, **21,** 412–413
matching devices, **21,** 422–423
microwave power sources, **21,** 414
modulation, **21,** 416–418
noise figure, **21,** 408, 409
operating frequency 413

simple spectrometer, **21,** 410–412
superheterodyne spectrometer, **21,** 413
Electron plasma echoes
mechanism for, **9A,** 29–32
observation of, **9A,** 35–36
theory of, **9A,** 31–32
types of, **9A,** 27–29
Electron plasma frequency, **9B,** 9
Electron plasma waves, **9A,** 4–6
observation of, **9A,** 18–27
Electron polarimeter, *see also* Mott polarimeter
accuracy, **29A,** 248–249
calibration
calculation of Mott asymmetry, **29A,** 232, 235, 242–244
double-scattering measurements, **29A,** 244
measurement of analyzing power, **29A,** 232, 235, 244
use of electron with known polarization, **29A,** 244–245
classes, **29A,** 231–232, 248
degree of polarization, **29A,** 231–232
Mott scattering error sources
background, **29A,** 247
extrapolation error, **29A,** 245
instrumental asymmetry, **29A,** 245–247
optical electron polarimetry
error sources, **29A,** 247
instrumentation, **29A,** 242
theory, **29A,** 234, 242
polarized low-energy electron diffraction polarimeter, **29A,** 240
secondary electron polarimeter, **29A,** 242
selection for experiments, **29A,** 247–250
Electron promotion, **17,** 482–486
Auger electron production, **17,** 490–492
Electron range effects, **8,** 215
Electron scattering, *see* Inelastic electron scattering
angular resolving power, **7A,** 11–14, 93
by an atom, **16B,** 187
by atoms, **3,** 335–338
Doppler broadening, **7A,** 43
elastic, **7A,** 2
forward, **7A,** 9, 10, 39
inelastic, **7A,** 2
multiple, **16B,** 190
phase change, **3,** 347–348
pressure dependence, **7A,** 39, *see also* Multiple scattering
resonances, **7A,** 40, 53, 88
Electron source (gun), **16B,** 206–207
Electron sources, monoenergetic, *see also* Energy monochromators
cool cathodes, **4A,** 125
low temperature, field emission, **4A,** 126
photoelectric, **4A,** 127
Electron spectroscopy, **22,** 218
Electron spectroscopy for chemical analysis (ESCA), **30,** 318
Electron spin resonance (ESR), **2B,** 147–151; **3,** 441–500; **3B,** 497
energy levels, **3,** 443, 445–453
magnetic fields for, **3,** 484–490
vs nuclear magnetic resonance, **2B,** 147
quantum mechanical description, **3,** 442–456
rf and microwave components, **3,** 469–480
sensitivity of apparatus, **3,** 490–498
for species concentrations, **18,** 634–637
spectrographs for, **3,** 464–469
spectrometer(s), **2B,** 148; **3B,** 554–572
circulator in, **2B,** 149
steady-state methods in, **2B,** 147–150
techniques, **6B,** 234–241
transient methods in, **2B,** 150–151
transition frequency, **3,** 465
Electron spin resonance (ESR) spectroscopy
apparatus, **16C,** 188–191
instrumentation, **16A,** 452–456
paramagnetic (spin) probes in polymer, **16A,** 457–465
limitations, applied to fracture studies, **16C,** 225
principles, **16C,** 186–187
sensitivity threshold, **16C,** 189
signal line shape and width, **16A,** 456
spectra of nitroxide spin probe, **16A,** 459–462
spin probe applications, **16A,** 470–478
theory, **16A,** 442–452
Electron stimulated desorption, **29B,** 61–63
Electron storage ring, **8,** 165–168
Electron swarms, **7B,** 29–77
attachment and ionization occurring simultaneously, **7B,** 74–75
characteristic variables, **7B,** 29–31
collision cross sections by, **7B,** 45
cross sections for rotational excitation by, **7B,** 48–49
diffusion coefficient in, **7B,** 38
diffusion measurement in, **7B,** 32–34
coefficient of diffusion, **7B,** 33–34

mean free path for momentum transfer, **7B,** 38
direct-current breakdown, **7B,** 75–77
D/μ measurement, **7B,** 37
experimental values, **7B,** 41–44
vs. E/*p*, **7B,** 41
drift velocity in, **7B,** 31–33
experimental values of, **7B,** 45
electron attachment and electron affinity studies, **7B,** 55–56
measurement at low E/p_8, **7B,** 57–60
electron detachment from negative ions, **7B,** 60–61
electron drift speeds, measurement, **7B,** 34–36
electron shutter method, **7B,** 35
light source pulsing, **7B,** 36
voltage pulsing process, **7B,** 36
electron energies above room temperature, **7B,** 48–49
E/N, definition of, **7B,** 31, 64
E/*p* and *pd*, definitions, **7B,** 30–31
ionization by electron collisions, **7B,** 61–69
kinetic theory of diffusion in, **7B,** 37–40
kinetic theory of drift velocity and mobility, **7B,** 39–40
loss of kinetic energy of electrons in gases, **7B,** 49–51
mobility in, **7B,** 32, 46
observables of experiments on, **7B,** 31
Electron temperature, **18,** 464, 472, 475, 705
and beam-foil interactions, **17,** 550
optimizing of, **15A,** 112
Electron transfer, **17,** 321, 341
Electron transit angle, **8,** 129
Electron trap(s), **14,** 310; **30,** 141, 148, 149, 150–152
Electron tube technology, **8,** 132
Electron tubes, **10,** 27–29, 30–45
Electron tunneling, **22,** 350, 378
in solids, **11,** 123–197, *see also* Normal-metal tunneling; Semiconductor tunneling; Tunneling; Tunnel junction
Electron(s)
accelerator, high energy as neutron velocity spectrometer, **5A,** 527ff
atomic, screening of nuclear charge, **5A,** 294–295
atomic collisions of, **7B,** 78–124
in flames, **7B,** 198
measurement, **7B,** 227–229
Auger, *see* Auger electron
anomalous magnetic moment, **3,** 629–630
backscattering on organic scintillators, **5A,** 425–426
beam deflection, **2,** 768
behavior in gases, **5A,** 91–94
bremsstrahlung, **5A,** 56–61
configurations, **4B,** 90, 91, 96, 108, 109
conversion, directional correlation of, **5B,** 137–138
de Broglie wavelength, **3,** 334
desorption, **14,** 447
diamagnetic effect, **1,** 44
diffusion, **5A,** 94, 100–101
distribution function, **3,** 529
elastic scattering, **5A,** 322–329
from hydrogen beam, **3,** 600
energy, internal conversion, **5A,** 594
loss measurement study of density effect, **5A,** 38–40
energy loss, **7A,** 8–9
energy scale calibration, **3,** 536–537
escape of at the surface, **11,** 80–81
field emission of, **11,** 452–453
gun, **2,** 767
"hot," **2,** 114
impact ionization of ions
confinement time, **29A,** 172
theory, **29A,** 170–172
impact method for positive ions, **3,** 527–532
initiated avalanche, **5A,** 118
initiated showers in lead, **5A,** 661–662
injection of from thermionic cathodes, **11,** 451–452
ionization loss of, **5A,** 18–19
lenses, **2,** 767–768
linear accelerator, **5B,** 642–645
longitudinal polarization, **5A,** 3
magnetic moment, **1,** 43, 45–46; **3,** 442, 491
maximum kinetic energy of, **11,** 68
mean free path in gases, **5A,** 113
mean range, **5A,** 62
mobility, **2,** 81
monoenergetic, impact method, **3,** 532–535
motion in electromagnetic field, **5B,** 222–233
motion of in magnetic field, **11,** 2
multiplier tubes, **2,** 66
nuclear double resonance, **5B,** 127–129
open and closed orbits of, **11,** 2–3, 6

pairs
energy measurement, **5A,** 676–678
ionization measurement, **5A,** 678–681
opening angle, **5A,** 677–678
particle-wave concept, **16B,** 186
passage through matter, **5A,** 56–72
penetration depth, **22,** 505
photoelectric injection of in liquid helium, **11,** 448–449
photoinjection of, **11,** 122
polarization, **5B,** 214–274
conversion, **5B,** 253–255
detection, **5B,** 239–264
longitudinal, **5B,** 256
non-relativistic, **5B,** 219
relativistic, **5B,** 219–220
transverse, **5B,** 252–256
polarized target, **5B,** 243
process, **30,** 18, 176, 225, 232–234, 272, 282, 413
production of secondaries, **5A,** 63–71
properties, table of, **4B,** 2
pulse height energy relationship in anthracene, **5A,** 423–425
radiation by, **5A,** 56–61
reflection of, **7A,** 22, 23, 52
relaxation, dipolar, **3,** 428
response of semiconductor detector, **5A,** 276
at rest, polarization of, **5B,** 217–219
secondaries of x-rays, energies, **5A,** 582
secondary, **7A,** 22, 52
selector, Marmet-Kerwin, **3,** 533
sensitive emulsions, **5A,** 212
spectrum, mu-meson decay, **5A,** 155
spin
dependent cross section, **5A,** 64ff
magnetic quantum number, **3,** 447, 449
system, net power absorbed, **3,** 454
spin-polarized sources, **29A,** 1, 30–32
synchrotron, **5B,** 653–656, 676–678
total absorption counter, **5A,** 189–191
trajectories, solenoidal spectrometers, **5A,** 366
velvet, **7A,** 23
volt, **5A,** 497
Electron-beam deposition, **29C,** 348
Electron-beam evaporation system, **11,** 635–636
Electron-beam ion source
design, **29A,** 172–173
excitation, **29A,** 172
fractional abundance of ion charge states, **29A,** 175
operation, **29A,** 174
performance, **29A,** 174–177
Electron-beam ion trap
apparatus, **29A,** 392–393
beam
collection, **29A,** 395
energy, **29A,** 177, 393–395
radius, **29A,** 394–395
charge state
range, **29A,** 391
selection, **29A,** 399–400
compactness, **29A,** 177
development, **29A,** 393
electron-ion collision cross-section measurement, **29A,** 410, 412
dielectric recombination, **29A,** 414–417
excitation of an X-ray line, **29A,** 412
ionization measurement techniques, **29A,** 413–414
excitation of highly charged ions, **29A,** 417–418
ion heating and cooling, **29A,** 398–399
ion injection methods, **29A,** 395–398
ion trapping, **29A,** 397–398
metal vapor vacuum arc source, **29A,** 397–398
operating parameters, **29A,** 393–394
performance, **29A,** 176, 391–392
super 200-keV instrument
apparatus, **29A,** 400–401
X-ray polarization, **29A,** 406, 408
X-ray spectroscopy, **29A,** 402
lifetime measurements, **29A,** 408, 410
transition energy measurements
above 2 keV, **29A,** 402, 404–405
below 2 keV, **29A,** 405–406
Electron-beam trap (EBIT), **31,** 54–56
Electron-beam-induced damage, **22,** 505
Electron-crystal interactions, EPR, **21,** 406–407
electrostatic interaction, **21,** 407
hyperfine interaction, **21,** 406
spin-orbit interactions, **21,** 407
superhyperfine interaction, **21,** 406
Electron-cyclotron resonance ion source
comparison with other sources, **29A,** 126–127, 184
design, **29A,** 178
development, **29A,** 1223, 177–178
electron velocity distribution, **29A,** 179–180

gas pressure, **29A,** 181
ion beam intensity, **29A,** 182–183
ionization mechanism, **29A,** 124–125
lifetime, **29A,** 127
low-charge-state sources, **29A,** 125–127
low-emittance, high-intensity source, **29A,** 129
magnetic field and microwave frequency relationship, **29A,** 179
maximum plasma electron density, **29A,** 183
microwave
frequency, **29A,** 183
power, **29A,** 182
multicusp-field source, **29A,** 127
operation, **29A,** 178–182
performance, **29A,** 182–185
properties, **29A,** 123–124
Electron-electron scattering, **5A,** 63–71
detection, **5B,** 244ff
in photoemission, **11,** 77–79
polarization, dependence, **5B,** 239–249
Electron-energy analyzers, **22,** 134, 182
Electron-energy distribution, **22,** 11
Electron-energy loss spectrometer, **22,** 33
Electron-energy loss spectroscopy, **22,** 26, 28, 66, 75, 344
resolution, **22,** 38
sensitivity, **22,** 30
tabulation of results, **22,** 121
Electron-hole pairs, **30,** 176
Electron-impact ion source, **22,** 313
Electron-impact ionization process, **3B,** 585–590
Electron-induced desorption, **22,** 209
Electron-ion crossed beam experiments, **7A,** 97
Electron-ion radiative recombination, by merging beams, **7A,** 478, 485
Electron-ion recombination, afterglow studies of, **7B,** 78
Electron-ion source, **7A,** 363–370
Electron-lattice coupling, **30,** 30–32
Electron-nuclear double resonance (ENDOR), **3,** 498–500; **3B,** 506
Electron-nuclear interactions, **21,** 398–403
chemical shift, **21,** 398, 400–402
Knight shift, **21,** 398–400
Electron-phonon coupling, **30,** 81
constant, **11,** 149
Electron-phonon interaction, tunneling as probe of, **11,** 149–150
Electron-phonon scattering, **11,** 79
Electron-plasmon scattering, **11,** 79
Electron-positron charge difference, **4B,** 11
Electron-positron mass difference
measurement of by e^{+}-e^{-} annihilation, **4B,** 13, 14
by mass spectrometer, **4B,** 14–16
Electron-positron pair ionization loss, **5A,** 36–37
Electron-proton charge difference measurment of by gas efflux, **4B,** 6–9
by molecular beam, **4B,** 9–11
by suspended sphere, **4B,** 11
by volatile liquid, **4B,** 9
Electron-solid interactions, **7B,** 275–305
analyzers for, **7B,** 279–282
detectors for, **7B,** 282–284
electron scattering in, **7B,** 277–278
experimental strategy, **7B,** 287–289
guns for, **7B,** 278–279
specimen preparation, **7B,** 284–287
Electron-spin resonance (ESR), in flame kinetic studies, **7B,** 209–210
Electron-stimulated adsorption, **22,** 511, 513
Electron-stimulated desorption, **22,** 367, 446, 467, 471, 477, 487, 493
Electron-stimulated-desorption ion angular distribution, **22,** 467, 471, 477, 485, 493
Electronic amplifiers, in dc measurements, **2B,** 62–63, *see also* Amplifier(s)
Electronic angular momentum of molecules, **10,** 328–338
Electronic band structure by X-ray spectroscopy, **6B,** 281–292
see also X-ray spectroscopy
Electronic charge, **8,** 271
Electronic Circuit Analysis Program (ECAP), **2A,** 286
Electronic circuits
capacitors in, **2A,** 36–38
component mounting in, **2A,** 42–43
construction techniques in, **2A,** 39–44
electrolytic capacitors for, **2A,** 37–38
grounding considerations in, **2A,** 41–42
inductors and transformers in, **2A,** 38–39
passive components in, **2A,** 33–39
plastic film dielectrics in, **2A,** 37
shielding and skin effect in, **2A,** 39–41
skin depth in, **2A,** 40
temperature considerations in, **2A,** 43–44
Electronic collisions, **8,** 35
Electronic components, **2A,** 33–39

resistors in, **2A,** 34–36
standards, tolerances, and preferred values in, **2A,** 34
Electronic counter, time-interval gated, **2B,** 24
Electronic deactivation, **15B,** 285
Electronic energy loss, *see* Inelastic energy loss
Electronic energy states, **3A,** 248–259
Electronic excitation, **22,** 331
relaxation and, **15B,** 284–286
selective chemistry by, **15B,** 288–293
Electronic feedback systems, **2B,** 445–459, *see also* Feedback control systems
operational amplifiers in, **2B,** 446–452
Electronic multiplier, **2B,** 66
Electronic noise, **2B,** 461–500, *see also* Noise
Electronic nuclear double resonance (ENDOR), **2B,** 151; **3,** 498–500; **3B,** 506, 508; **5B,** 127–129
line shape, **3,** 500
Electronic polarization, **10,** 347, 348
Electronic potential surfaces, nonadiabatic transients between, **15B,** 101
Electronic quadrupole moments, **3B,** 786, 788
Electronic spectroscopy, **3A,** 203–301
Electronic sputtering, **30,** 176
Electronic structure, synchrotron radiation studies, **29C,** 40
Electronic units, **8,** 231
Electronic-rotation interactions, **15B,** 271
Electronically excited excimer molecule, **15B,** 139
Electronics for scanning tunneling microscopy, **27,** 60
Electronics of infrared photometers, **12A,** 432
Electrooptic effect, lithium niobate, **29C,** 387–389
Electrooptic modulation, **19,** 488–490
Electrooptic modulator (EOM), **29C,** 109, 122–124, 129
Electrooptic shutter, **15A,** 27
Electrooptical shutter, **16C,** 478
Electrophoresis, **6A,** 26
Electrophotography, **6B,** 356–357
Electrophotoluminescence, **6B,** 295
Electroplating, **14,** 357, 443
Electropolishing, **14,** 442, 443
Electrostatic analyzer, **17,** 175, 178–180, 183–185, 439–463; **22,** 219
Electrostatic charge
decay, **16C,** 440–442
density, **16C,** 424
generated with ionizing radiation, **16C,** 435
methods of measurement, **16C,** 424–428
migration, **16C,** 439
surface, maximum in air, **16C,** 429
Electrostatic field, *see* Homogeneous electrostatic field
Electrostatic focusing lens, **17,** 442
Electrostatic force microscopy, **27,** 90
Electrostatic interactions, **4B,** 108, 109
Electrostatic lens
aberrations
significance, **29A,** 190
spherical aberration effects, **29A,** 199–200
types, **29A,** 199
alignment, **29A,** 204
asymptotic trajectory of particle, determination, **29A,** 194
circular aperture lens, **29A,** 195–196
combining lenses, **29A,** 201–204
computer simulation in design
advantages, **29A,** 204–205
CPO–3D, **29A,** 206–207
SIMION, **29A,** 205–206
cylindrical lens, **29A,** 195–196
data sources for various electrode configurations, **29A,** 195
design rules, **29A,** 201
disc of least confusion, **29A,** 200–201
electron energy analyzer input lens
configurations, **29A,** 223
design, **29A,** 224–225
multielement lenses, **29A,** 225
optimization of voltages, **29A,** 226
four-cylinder lens, **29A,** 199
materials, **29A,** 204
optical lens analogy, **29A,** 189–191
thick lens representation, **29A,** 193–195
three-cylinder lens
focal properties, **29A,** 197–198
voltage, **29A,** 197–199
voltage
measurement, **29A,** 192
ratio, **29A,** 201
Electrostatic potential, **22,** 155
Electrostatic stress, **22,** 417
Electrostatic time-of-flight spectrometer, **8,** 175
Electrostatics, **1,** 431–447
field, **1,** 433, 443–447
force, **1,** 439
voltmeter, **1,** 460

Electrothermal engines, **9B,** 254
Elemental analysis, **30,** 224
Elementary particles, electric dipole moments, **3B,** 843
Elementary step, in chemical reaction, **19,** 207
Elements, identification of new, **5A,** 295–296
Eliashburg equations, **11,** 149
Ellipsoid
of inertia, **10,** 303
magnetic measurement with, **1,** 533–534
magnetic moment, **1,** 535
moments deduced from external fields, **1,** 534–536
in uniform magnetic field, **1,** 532–533
Ellipsoidal mirror analyzer, **22,** 492
Ellipsometer, **25,** 179
Ellipsometry, **31,** 234
Elliptic flow equation, **18,** 760
Elliptic integrals, **8,** 67
Elliptic polarization,
azimuth and phase angle measurements, **6B,** 258–261
measurement of, **6B,** 258–262
principle angle of incidence and principle azimuth measurements, **6B,** 261–262
Elliptical polarization, **1,** 377; **26,** 113
Elongation flow, **16C,** 143
Elongation viscosity
definition, **16C,** 12
measurement of, **16C,** 25
EMAT, 69, 518–519, *see also* Surface acoustic wave transducers
Embossing of holograms, **26,** 206
Embrittlement
nickel, **14,** 358
steel, **14,** 357
EMF, *see* Electron beam fluorescence
Emission
absolute gamma ray, **5A,** 640–641
angular distribution
α-particle, **5B,** 206–207
β-particle, **5B,** 209–210
γ-ray, **5B,** 207–209
charged particle from charged particle induced reactions, **5B,** 413–440
cross section, **7A,** 455
current density, maximum, **2,** 47–48
decay time, **1,** 396–397
induced, **1,** 488
intensity, in spectroscopy, **7B,** 102
intensity per molecule, **3,** 242
microscope, **6A,** 185
neutral particle from charged particle induced reaction, **5B,** 440–448
secondary, multiplier, **14,** 82, 84
secondary electron
from Faraday cups, **5B,** 491
monitor, **5B,** 501
thermionic, **2,** 47–51
Emission angle, **8,** 253
Emission methods, for work function measurement, **22,** 4
electronic band structure effect, **22,** 5
patch field, **22,** 5
temperature dependence, **22,** 5
Emission spectra, **13A,** 20, 87
in infrared, **13B,** 3
laser plasma source, **31,** 86–87
Emission spectroscopy, **6A,** 30
Emission spectrum, **22,** 197
iron, **3,** 175
of scintillator, **5A,** 126, 131–132
Emission techniques, for work function measurement, **22,** 20
band structure, **22,** 20
field emission, **22,** 20
free-electron limit, **22,** 20
photoemission, **22,** 20
thermionic emission, **22,** 20
Emission-yield studies, **17,** 191–192
Emissivity, **1,** 255, 302, 303; **8,** 28, 33; **12A,** 435–439, 442–443, 449
Emittance, **1,** 335
incandescent surface, **1,** 376
ion beam, **29A,** 71–73
of light sources, **18,** 465, 689
Emitted characteristic radiation, for velocity measurement, **18,** 341–345
Emitted current density, **22,** 6
Emitted electrons, **22,** 15
energy spread, **22,** 15
Fermi-energy, **22,** 15
Emitter follower, **2,** 808–809
Emitter-collector delay time, **2A,** 75
Emitter-coupled logic, **2A,** 138
Emitter-follower, in light-scattering optical systems, **9A,** 104
Emitters, thermionic, in gas tubes, **2,** 94–95
Empirical curves, **8,** 272
Empty magnification, **25,** 168
Emulsification, **19,** 328–332
Emulsion, photographic, **5A,** 208–264

charge determination of particles, **5A,** 298–307
-cloud chamber method, **5A,** 680
diluted, **5A,** 215
electron sensitive, **5A,** 212
gamma ray measurement, **5A,** 676–682
grain count, **5A,** 14
grain density, **5A,** 35
ionization
loss in, **5A,** 34–37
measurement in, **5A,** 240–245
nuclear
momentum measurement, **5A,** 388–408
neutron flux and energy measurement, **5A,** 475–478
processing, **5A,** 216–224
range-energy relation, **5A,** 45
range
of particles, **5A,** 226–240
straggling, **5A,** 54
standard, scattering constant, **5A,** 405
tracks, opacity measurement, **5A,** 303–305
water content, **5A,** 222–224
Emulsion fog, **1,** 411
Enclosure stiffness, **8,** 14
End form, **22,** 385
End-group analysis, **16A,** 69–72; **16C,** 226
by infrared, **16B,** 251
End-to-end distance, root mean square, **16A,** 18
ENDOR, *see* Double resonance techniques; Electronic nuclear double resonance; NMR
Endothermic reaction, **1,** 262
Energetic-beam-assisted growth, **30,** 494–495
Energies/Energy, **1,** 31
agitation, **5A,** 92, 93
analysis, **22,** 314, 316, 340, 474
analyzer, **22,** 316
of annihilation photons, **5B,** 512
band gap, **4A,** 456
binding, **5A,** 297–298; **5B,** 1, 2
bond dissociation, **3,** 546–552
of characteristic x-ray quanta, **5A,** 294
charged particle, **5A,** 342
conservation of, **1,** 307
conversion
factor, **5A,** 124
to wavelength, **5A,** 607
correction, Stark effect, **3,** 657–658
critical, **5A,** 660
defect, **29B,** 105
deficit, **30,** 429
density, **19,** 7
determination, **5A,** 341–348
with ionization chambers, **5A,** 409–410
due to electric quadrupole moment, **5B,** 48
end point, **5B,** 359
equivalent of a mass of nuclear magnitude, **5B,** 339
to form a crack and a craze, **16C,** 271
to form an ion pair in gas, **5A,** 1, 43–44
function, internal, **1,** 307ff
gamma ray
determination with curved crystal spectrometer, **5A,** 599–616
determination with pair spectrometer, **5A,** 641–652
measured by absorption, **5A,** 671–676
interaction, nuclear quadrupole, **5B,** 97
kinetic, of cyclotron particles, **5B,** 631
kinetic, of fragment ions, **3,** 537–539
lost per revolution, betatron, **5B,** 630
due to magnetic moment, **5B,** 46–47
measurement(s), **1,** 459, 468–469
with cloud and bubble chambers, **5A,** 436–438
of electron pairs, **5A,** 676–678
by recoil methods, **5A,** 461–494
neutrino, **5A,** 683–692
neutron
determination with helium–3 spectrometer, **5A,** 570–581
instrumentation for measurement, **5A,** 466–481
interaction recoil, **5A,** 463–464
nuclear binding, change during fission, **5B,** 592–593
of a nuclear magnetic moment in a uniform field, **3,** 361
of π^0 meson, **5A,** 681
potential, function, **3,** 40, 43ff
reaction, of nuclear reaction, **5B,** 339–352
resolution, **22,** 303, 314, 317, 318
relation to momentum resolution, **5A,** 345
relationship with gamma ray energy, **5A,** 634ff
time-of-flight spectrometer, **5A,** 545
response, scintillator, **5A,** 128–129
of rigid symmetric top, **3,** 59
rotational, **3,** 11, 47
of rotational increase, **3,** 18
spread, **22,** 405

electron beam, **3,** 531–532
of filtered neutrons, **5A,** 509–510
Stark perturbation, **3,** 654, 655, 656, 659
storage rate per unit volume, **3,** 390
theorem, **26,** 70
threshold, **5B,** 340–341, 346–349
endoergic reaction, **5B,** 346–347
threshold for unstable particle production, **5B,** 833–838
total stored, in tuned circuit, **3,** 391
X-ray total beam, **5B,** 537–553
Energy dependence of dechanneling, **21,** 281, 315–319
of χ_{min}, **21,** 306
Energy distribution curve, **11,** 69; **22,** 130, 137, 140, 144, 147, 154, 159, 165, 167, 170, 173, 175, 179, 181, 183
ac method for obtaining, **11,** 92–97
block diagram of ac method in, **11,** 94
angular measurements for, **11,** 100–101
cesium channels and, **11,** 113–114
for copper, **11,** 71
future research in, **11,** 121–122
interpretation of, **22,** 147
linear sweep method in, **11,** 97
lowered electron affinity and, **11,** 110–112
measurement of, **11,** 84–102
photoemission chambers and, **11,** 116
sample preparation in, **11,** 102
temperature and, **11,** 101–102
time differentiation method in, **11,** 97–98
weak structure in, **11,** 96
Energy distribution of electrons from heavy particle ionizing collisions, **7A,** 449–454
Energy distribution of the joint density of states, **22,** 145, 146, 148, 172, 173, 177, 178, 180
Energy distribution of photoelectrons, **12A,** 12–14
Energy distribution of secondary electrons, **12A,** 15
Energy distribution, **22,** 137, 332, 473, 476
of collected electrons, **22,** 16
of electrons, **22,** 129
of field-emitted electron, **22,** 361
of field-ionized gas atoms, **22,** 380
in gratings, **26,** 75
of ions, **22,** 492
of neutrons scattered inelastically by aluminum, **5A,** 512
of neutrons scattered by one photon, **5A,** 513
in pair production, **5A,** 84, 95
of photoelectrons, **22,** 130, 164
of secondary particles, **22,** 332
Energy gain, **8,** 170, *see also* Multimodulator system; Single modulator
Energy gap spectroscopy
of charge density waves (CDW), **27,** 351, 354, 386, 421
Energy gap spectroscopy of
1T-TaS_2, **27,** 363
1T-$TaSe_2$, **27,** 376–378
1T-TiS_2, **27,** 383
1T-$TiSe_2$, **27,** 383
1T-VSe_2, **27,** 378
2H-TaS_2, **27,** 389, 393
2H-$TaSe_2$, **27,** 389
4Hb-TaS_2, **27,** 401
4Hb-$TaSe_2$, **27,** 400
$NbSe_3$, **27,** 418
$TaSe_3$, **27,** 418
Energy gap, in superconductors, **11,** 135–140
Energy gap, *see* Superconductor energy gap
Energy level(s); **3,** 443, 445–453
asymmetric top molecules, **3,** 14
atoms in a magnetic field, **3,** 603
broadening, **3,** 378
diagram, "one-electron jump," **6B,** 282
diagrams, **3,** 448–453
diatomic molecules, **3,** 12
diatomic rotator, **3,** 46, 47, 49
dye lasers, **29C,** 46–50, 60
of ENDOR, **3,** 499
HF harmonic oscillator, **3,** 44
molecules in electric field, **3,** 606
nuclear, from reaction energies, **5B,** 352–366
nuclear quadrupole, **3,** 502–507
in organic molecules, **5A,** 123
population, **3,** 454
rotational, of symmetric-top molecule, **5B,** 20–21
in solid state lasers, **15A,** 171–176
splitting in atoms, **4B,** 109–110
spacing of compound nucleus, **5A,** 538, 541–544
symmetric top molecules, **3,** 13
in thermal equilibrium, **15A,** 32
vibrating rotator, **3,** 52
Energy loss
anomalous, **1,** 556
due to Čerenkov radiation, **5A,** 8–10
electrons at very low energies, **5A,** 42–43

fluctuations, **5A,** 17–18
in gases, cloud chamber measurement, **5A,** 22, 29–30
per ion pair, **5A,** 103–105
in liquid scintillators, **5A,** 143–146
of Li^7 ions in emulsion, **5A,** 239
maximum, in single event, **5A,** 448
particle, **5A,** 449
smallest measurable in ionization, **5A,** 409
per unit length, **5A,** 230–231
Energy loss processes, **8,** 275
Energy loss, of high-velocity in matter, **17,** 25–72
experimental methods, **17,** 46–68
mechanisms of, **17,** 26–30
stopping cross sections, **17,** 68–72
theories, **17,** 30–46
Energy meter, as short-pulse detector, **29C,** 199
Energy monochromators for electron beams, **4A,** 127–134
barrier filter types, **4A,** 132–134
deflection filters, **4A,** 127–132, 143
effect of space charge, **4A,** 129–131
"energy aberrations," **4A,** 129
sector fields, **4A,** 143
image force filter, **4A,** 134
klystron bunching, **4A,** 127
scattering resonance filter, **4A,** 134
time of flight filter, **4A,** 134
use of axial fields, **4A,** 142–143
Energy resolutions, **8,** 213
Energy spectra
of atoms in plume, **30,** 523
of electron emission, **30,** 14–16
of positive ions, **30,** 155–164
Energy spectrometer, **8,** 255
Energy spectrum of sputtered atoms, **21,** 258–261
Energy straggling, **17,** 70–71
Energy to depth conversion
in nuclear reaction analyses, **21,** 247
in Rutherford backscattering, **21,** 227–230
Energy transfer
between antenna pigments, **20,** 182, 183
heavy particles to electrons, **5A,** 300
kinetic, **5A,** 290
maximum, **5A,** 4, 8
processes, **5A,** 133
in projectile-atom collision, **17,** 197–198
Energy transmittance of a monochromator, **10,** 136
Energy wave-number characteristic, **23A,** 384
Energy-calibrated magnetic analyzer, **17,** 536
Energy-loss measurements, **17,** 46–68
backscattering, **17,** 58–65
gaseous target transmission, **17,** 51–53
inelastic, in simple collisions, **17,** 149–192
inverted Doppler-shift attenuation method, **17,** 65–68
supported films, **17,** 53–54
thin foil transmissions, **17,** 47–51
thin layer on solid-state detector, **17,** 54–58
Energy-window analyzers, **22,** 136
Enforced dipole transitions, **13B,** 41
Engan's heterodyne acoustooptic technique, **19,** 490–492
Enhanced nuclear moments, **23C,** 281
Enhancement-mode device, MOS transistor as, **2A,** 87
ENIAC, **2B,** 166
Enthalpy function, **1,** 261, 262; **30,** 181
Entrainment, diffusion pumps, **14,** 141, 142
Entrance aperture, **14,** 15
Entrance slit, *see* Monochromator
Entrance window, **1,** 334
Entropies, mathematical, of continuous processes, **2,** 688, 689–690
Entropy, **1,** 321, 322, 324, 327
molar, **5B,** 212–214
Shannon-Jaynes, **28,** 411, 412–415
Envelope delay, **28,** 341
Envelope function, **20,** 403, 404
Enveloping curvature, **31,** 367–368
Environmental crazing
critical strain rate, **16C,** 260
effect of hydrostatic pressure, **16C,** 260
in gases, **16C,** 257–259
in liquids, **16C,** 256
Enzymes, **20,** 41–44
EOM, *see* Electrooptic modulator
Eötvös effect, **24B,** 130
Epitaxial growth, **14,** 563; in thin films, **11,** 621
Epitaxy of
Ag on Au(111), **27,** 296
Au on Ni(110), **27,** 295
Ge on Si(001), **27,** 247
Ni on Au(111), **27,** 296
Epithermal energy range, **8,** 242
Epithermal neutron detection, **24B,** 540–541
Epoch folding, **28,** 354–355
coherent signal, **28,** 368
Epoxies, in magnetoresistance, measurements, **11,** 13
Epoxy, **14,** 243, *see also* Sealant

Epoxy resin(s), *see* Polyepoxide
curing studies by DSC, **16B,** 334
electron tunneling spectrum, **16A,** 166
EPR, *see* Electron paramagnetic resonance
Epstein strips, **1,** 551, 552
Equal-thickness fringes, **26,** 8
Equation of radiation transfer, **10,** 290
Equation(s) of state, **18,** 612; **30,** 115
of a polymer, **16C,** 84–86
Equator of X-ray fiber pattern, **16B,** 55
Equilibrium, **1,** 303
associativity of, **1,** 309
pseudo-, state, **1,** 305, 306
radioactive, **5B,** 275–276
site, **23A,** 23
states, **1,** 305ff, 327–328
thermal, **1,** 308
Equilibrium charge state distributions, **17,** 80–83, 118–124
experimental results, **17,** 118–122
semiempirical descriptions of, **17,** 122–124
Equilibrium charge state, average, **17,** 125–131
experimental, **17,** 127–128
Lamb-Bohr criterion, **17,** 125–127
semiempirical relationships, **17,** 129–131
Equilibrium excitation, of fast ions in solids, **17,** 142–144
Equilibrium melting point, **16B,** 296, 350, 362, 375–381
Equilibrium position determinations, **22,** 291
Equipment specifications, in testing, **2B,** 212–215
Equipment testing, **2B,** 170–219
advanced measuring instruments in, **2B,** 184–186
amplifier noise testing in, **2B,** 212
analog and digital meters in, **2B,** 179–180
and automatic testing during experiments, **2B,** 172–173
bandwidth specifications in, **2B,** 191
basic test instrumentation in, **2B,** 176–178
calibration problems in, **2B,** 186
characteristic curve tracers in, **2B,** 177
circuit equations for significant dynamic test in, **2B,** 203–205
circuit trouble and prevention in, **2B,** 218–219
classification of tests in, **2B,** 170–176
clip-on probe measurements in, **2B,** 191
component testers in, **2B,** 176–178
counters, timers, and frequency meters in, **2B,** 184–185
current-measuring procedures in, **2B,** 190–191
dc measurements in, **2B,** 215–216
dead time in, **2B,** 208
dynamic, **2B,** 202–212
environmental, **2B,** 174–176
equipment specifications and, **2B,** 213–215
fast pulse discrimination in, **2B,** 210
fast pulse generators in, **2B,** 181–183
fault isolation by semiautomatic techniques in, **2B,** 174
fault tracking in, **2B,** 171–172
frequency domain measurements in, **2B,** 205–207
general-purpose instrumentation in, **2B,** 178
high-frequency input configurations in, **2B,** 196–197
input characteristics in, **2B,** 195–197
integrated circuit testers in, **2B,** 177–178
internal adjustments and calibrations in, **2B,** 171
internal circuit probing in, **2B,** 204
large signal response in, **2B,** 208
loading corrections in, **2B,** 204–207
marginal testing in, **2B,** 171
mixed circuits in, **2B,** 210
multimeters in, **2B,** 179–180
multiple signal input in, **2B,** 197
multivibrator in, **2B,** 188
negative feedback systems in, **2B,** 208–209
ohmmeter in, **2B,** 176
oscilloscopes in, **2B,** 184–187, 191–202
permanent dropping resistors in, **2B,** 191
power supplies in, **2B,** 178–179
quadrupole parameters in, **2B,** 202–203
recovery time in, **2B,** 208
RLC bridges in, **2B,** 176–177
routine checks and adjustments in, **2B,** 170
sampling technique in, **2B,** 199–201
signal-flow strategies in, **2B,** 216–218
signal generators in, **2B,** 180–183
small-signal response in, **2B,** 207
square wave train generator for, **2B,** 183
steady-state conditions in, **2B,** 186–191
switching circuits in, **2B,** 209–211
test pulser in, **2B,** 183
time domain measurements in, **2B,** 206–207
trigger and sweep facilities in, **2B,** 197–198
trouble-shooting procedures in, **2B,** 212–219
voltage measuring procedures in, **2B,** 189–190
Equipment, slewing rate in, **2B,** 208

Equivalent circuit models, of tunnel junctions, **11,** 268–273
Equivalent circuit of point-contact diode, **10,** 162
Equivalent circuits
  in amplifier stages, **2A,** 267–272
  of triodes, **2A,** 157
Equivalent circuits (transducer), **19,** 40–51
  KLM, **19,** 47–51
  Mason, **19,** 44–47
Equivalent degrees of freedom, **28,** 331
Equivalent electrical generator, **2B,** 489
Equivalent noise conductance, **2B,** 500
Equivalent noise resistance, **2B,** 498; **4A,** 332–333; **10,** 225, 226
Equivalent surface conductance, **18,** 667
Equivalent width, **7A,** 125; **29B,** 248–249
Equivocation, **2,** 693
Er, **23C,** 208
Er:fiber amplifier (EDFA), **29C,** 175
Erbium ethyl sulphate, **10,** 394
$ErFe_2$, **23C,** 214
Erfle eyepiece, **25,** 168
ErIG ferrimagnetic exchange resonance, **10,** 419
EROS, *see* Electric-resonance optothermal spectrometer
Erosion rate, **14,** 222, 224
Erosion, electrode, by ion bombardment, **2,** 95–97
Error analysis
  chronophotography, **18,** 87–89
  particle tracking methods, **18,** 50
Error coefficients, **8,** 98–100, **8,** 102
  envelopes, **8,** 85; in feedback system response, **2B,** 389
Error contour plots, **8,** 85
Error functions, definitions, **18,** 666, 668
Error signal, **29C,** 106, 115–120, 281–286
  in Mössbauer spectroscopy, **11,** 326–327
Error(s), **6A,** 2; **28,** 245–246
  accidental, **1,** 3, 9; **2,** 2; **5B,** 808; **6A,** 2
  compounding of, **1,** 12
  of computation, **1,** 19–24; **2,** 13–15; **3,** 721–723; **5B,** 819–821; **6A,** 12–14
  definition, **28,** 1
  determination in scattering measurements, **5A,** 407–408
  of direct measurement(s), **1,** 6–9; **2,** 6–7; **3,** 714–715; **5B,** 812–813; **6A,** 5–7
  drift, in gas pressure measurements, **1,** 170
  in drop counting, **5A,** 449
  estimation, **28,** 452
  generation of, **1,** 20
  grain, **5A,** 396
  in indirect measurements, **2,** 8–10
  of indirect measurement, **3,** 716–718
  instrumental, **28,** 30–31
  interpolation, **1,** 21
  of linear equation solving, **1,** 19
  magnitude of, **1,** 1
  matrix, **28,** 23
  maximum possible, **28,** 3
  periodic, **1,** 2
  probable, **5B,** 778
  propagation of, **1,** 12–14, 20; **3,** 716–718; **5B,** 783–784; **6A,** 8–9
    in indirect measurements, **5B,** 814–816
  random, **1,** 9
  in range measurements, **5A,** 437
  round off, **1,** 21; **5B,** 819; **6A,** 1
  standard, **1,** 4; **5B,** 776–778, 803, 810; **28,** 257
  statistical view, **28,** 252–253
  systematic, **1,** 2–3, 6, 11; **2,** 2; **5B,** 808; **6A,** 2; **28,** 28–29
  in track density measurements, **5A,** 451
  truncation, **1,** 20; **5B,** 819
Erythemal ultraviolet, **1,** 395
ESA, *see* Electron-stimulated adsorption
ESA, *see* Spherical energy analyzer
Escape depth, **22,** 127, 130, 131, 152, 179
ESD, *see* Electron-stimulated desorption
ESDIAD, *see* Electron-stimulated-desorption ion angular distribution
ESR, *see* Electron spin resonance
Estimator
  biased, **28,** 316
  robust, **28,** 390
Etalon, **7B,** 106; **9A,** 98; **12A,** 532; **13A,** 304, 306, 309, 314; **13B,** 304; **25,** 221; **26,** 41
Etalon, Fabry-Perot, Fizeau-Tolansky, **18,** 198, 202, 214, 708, 717, 740
Etalons, **3A,** 405
Etch, chemical, **14,** 442, *see also* Cleaning
Etches for imperfections, **6A,** 164–165
Etches for optical orientation determination, **6A,** 158–159
Etches for revealing imperfections, **6A,** 163–172
Etching
  chemical, **30,** 355–357

of electron microscope specimens, **16B,** 216, 240, 242–245
guide to literature, **6A,** 152
with plasma, **16B,** 241–245
with solvent, **16B,** 239–241
Etching methods, **6A,** 153–155
alloy, **6A,** 155
cathodic bombardment, **6A,** 154
chemical, **6A,** 153
electrolytic, **6A,** 153–154
solvation, **6A,** 154
thermal, **6A,** 154
Etendue, **25,** 195
(throughput), **12A,** 536
definition of, **9A,**91
Ethane, rotational Raman spectrum, **3,** 142
Ethyl acetate, **19,** 203
Ethylene, rotational Raman spectrum, **3,** 143
Ettingshausen effect, **6B,** 166, 169
Ettingshausen-Nernst effect, **6B,** 166
Euler, Leonhard, **28,** 3
Eulerian cradle, **3,** 321
EuO, **23C,** 30
EuS, **23C,** 30
EV (electronic-vibrational) event, **15B,** 64
Evaluation of measurement, **6A,** 1–14
Evanescent coupler, short-pulse dye laser and, **15A,** 289
Evaporated contacts, **6B,** 51
Evaporation of film in vacuum, **6A,** 123–129
Evaporation of neutrons, **23A,** 106
Evaporation rate, **14,** 522
Evaporative cooling, **29B,** 140–141
Even order derivatives, **8,** 86
Ewald construction, **16C,** 153; **22,** 241
grazing-angle x-ray diffraction, **22,** 241
LEED, **22,** 241
penetrating radiation, **22,** 241
RHEED, **22,** 241
Ewald sphere of reflection, **16C,** 154
for oriented fiber, **16C,** 156
Ewald sphere, **16B,** 34–35; **20,** 233, 234, 402; **23C,** 83
Ewald summation, **23A,** 385
Ewald transformation, **23A,** 455
EXAFS, *see* Extended-x-ray-absorption fine structure
Excess acoustic absorption, **19,** 187, *see also* Absorption
Excess carriers, **6B,** 84–87
collector measurements, **6B,** 85–86
contactless detection methods, **6B,** 86–87
detection by sample conductance, **6B,** 84–85
Excess conductance, **6B,** 84–85
Excess mass, **24B,** 156
Exchange degeneracy, **10,** 396
Exchange effects, spin relaxation, **3,** 459
Exchange energy, **10,** 398
Exchange field, **10,** 398–400
Exchange integral, **10,** 398
Exchange interaction, **10,** 395–400
Exchange reactions, dynamics, *see* Mass exchange reactions
Excimer formation, **16A,** 407
Excimer lasers, **29C,** 4, 147–148, 151, 259–260; *see also* Rare gas halide excimer lasers
Excitation, **17,** 78, 87, 149, 188–191, 384–485
after electron capture, **17,** 107
decay-time dependence, **5A,** 130
density effect and, **17,** 131–139
detection of excited states, **17,** 88
direct Coulomb, **17,** 482–482, 486–488
energy loss of high-velocity ions, **17,** 26
fast ions in solids, **17,** 142–144
inner shell, **17,** 116–117
mean excitation energy, **17,** 35–37
multiple, **17,** 135–139
potential, mean, **5A,** 10–14
Excitation cross section, electron beam, **18,** 438–447
Excitation energies, determination, **5B,** 354–363
Excitation functions, threshold measurement, **7A,** 63–64
Excitation in heavy particle collisions, **7A,** 455–466
optical techniques, **7A,** 456–466
Excitation methods for afterglows, **7B,** 82–85
Excitation of atoms
by discharge, **4A,** 184
by electron bombardment, **4A,** 184–187
by optical radiation, **4A,** 187
thermal, **4A,** 183–184
Excitation of ions by electrons, **7A,** 105–109
Excitation of x-rays by particles, **21,** 251–253
Excitation potential of atoms, **1,** 47
Excitation processes, in laser-selective chemistry, **15B,** 278–279
Excitation rate coefficient, **9A,** 136, 139
Excitation system, nuclear, **3,** 436
Excitation transfer, **7A,** 48, 62

Excited atom ionizing collisions, afterglow studies of, **7B,** 78
Excited state, of molecule
decay paths, **16A,** 406
lifetime, **16A,** 408
Excited states, **3B,** 810; **7A,** 214, 223, 426–428
Exciton, **6B,** 285
Excitons, **30,** 34–36, 584
magnetic, **23C,** 200
Exclusion, **6B,** 83–84
Exfoliation, exfoliational sputtering, **30,** 102, 176, 236
Exit channel, **8,** 274
Exit pupil, **25,** 168
Exit slit, *see* Monochromator
Exit window, **1,** 334
Exothermic reaction, **1,** 262
Expansion
adiabatic, temperature change during, **5A,** 197
chambers, **5A,** 197–201
isentropic, **1,** 194
strength, in shock waves, **1,** 194
thermal, **14,** 289, 440
thermometry, liquid, **1,** 243–244
time, cloud chamber, **5A,** 200
Expected values, **28,** 18–19
Experimental design, **28,** 184
Experimental difficulties, various, **19,** 134–135, 291–297, 520
Experimental light intensity, **8,** 36
Experimental procedures in holography, **26,** 199
Experimental science, probability, **28,** 1–4
Experimental targetry, **8,** 271–275
Experimental visibility, coefficient, **8,** 51
Explanatory variable, **28,** 220–221, 245
Exploding wires, **7A,** 156
Exploding-wire technique, **15B,** 137–138
Explosion diagnostics, **18,** X-ray flash, **18,** 408
Explosion hazard, **14,** 450
Explosive boiling, *see* Phase explosion
Explosive vaporization, **30,** 89
Exponents, **23C,** 6, 43
Exposure times, photographic, **18,** 692–702
Exposure-dose, **5B,** 537, 539–540
Extended appearance-potential fine structure, **22,** 232, 235, 236
Extended scattering, **20,** 380, 381
Extended x-ray absorption fine structure (EXAFS), **29C,** 40
Extended-cavity diode laser (ECDL), **29C,** 83, 84–89
control, **29C,** 99
optical coating, **29C,** 95
temperature stability, **29C,** 93
Extended-cavity laser, **29C,** 83, 84
Extended-chain crystals, **16B,** 347, 350, 385
melting behavior, **16B,** 308
Extended-range "grasshopper" monochromator, **22,** 144
Extended-x-ray-absorption structure, **22,** 297
Extensometer, **24B,** 381–382, *see also* Strainmeter
capacitance-type, **16C,** 121, 128
External enhancement cavity, **29C,** 239–243
Extinction, **23A,** 348; **23C,** 89
coefficient, **1,** 334; **6B,** 250–263; **13B,** 135
distance, **22,** 245
law criteria, **3,** 284–285
primary, **3,** 305–306; **23A,** 325
secondary, **3,** 306
Extinction, atmospheric, **12A,** 123–179, 453–456
error analysis, **12A,** 124–136
infrared, **12A,** 453–456
King's theory, **12A,** 155–160, 190, 456
variation with tie, **12A,** 140–145
Extra-atomic relaxation, **22,** 224
Extraction and adsorption, **6A,** 24
Extraordinary beam, **26,** 136
Extraordinary ray, **12A,** 372
Extraordinary wave in plasma, **10,** 275, 276
Extreme Ultraviolet (EUV)
design, **31,** 146–147
instrumentation, **31,** 133–135, 138
mirror optics, **31,** 145
Extrinsic Auger satellites, **22,** 229
Extrinsic electron traps, F_2^+ center production and, **15B,** 41–45
Extrinsic photoconductivity, *see* Impurity photoconductivity
Extrinsic photoconductors, **26,** 317
Extrinsic radiation, **6B,** 321
Eye, **25,** 120, *see also* Lens/eye; Unaided eye
circle, **25,** 131
relief, **25,** 61, 131
Eyepiece, Ramsden, **1,** 339
Eyepieces, **25,** 59,71, 127, 129, 131
Eyring theory of reaction rates, **22,** 457, 459, 461

F

F center, **15B,** 3–4
aggregation, in color center formation process, **15B,** 16–17
optical pumping cycle of, **15B,** 4, 23
F distribution ratio, **1,** 24
F ratio, **1,** 11, 18, 19
F tensor, **23A,** 447
F^- ion, **19,** 202
F-band, **6B,** 324–325
F-centers, **6B,** 324
F-like centers, optical pumping cycle of, **15B,** 23
F-like laser-active centers, **15B,** 2–12
F-number, **10,** 84, 137
F-statistic, **28,** 262
F-tensor approximation, **23A,** 487
F/number, **25,** 25
F1/F2, *see* In phase F1/F2
F_2^+ center lasers
future developments in, **15B,** 50–51
in hosts potassium fluoride and lithium fluoride, **15B,** 45–48
mode locking of, **15B,** 48–50
recent developments in, **15B,** 41–51
F_2^+ center, **15B,** 9–12
absorption and emission bands of, **15B,** 11
energy level diagram of, **15B,** 10
enhanced production of through two-step photoionization and extrinsic electron traps, **15B,** 41–45
light-to-light conversion efficiencies of, **15B,** 11
production of, **15B,** 20–22
F_2^+ formation process, **15B,** 20–22
F_A(II) centers, **15B,** 4–7
in distributed-feedback laser, **15B,** 37–39
luminescence decay time for, **15B,** 7
normal and relaxed ionic configuration of, **15B,** 5
self-absorption and, **15B,** 23
stability of in cw laser, **15B,** 33–34
tuning elements for, **15B,** 30–32
FA, *see* Flowing afterglow
Faber-Ziman formalism, **23B,** 251
Fabrication, **14,** 425–448
Fabry lens, **12A,** 103, 383, 386, 388, 392, 434
Fabry-Perot
Etalon(s), **1,** 370, 371; **15A,** 332; **25,** 191, 217; **26,** 31; **29B,** 379–383; **29C,** 64–66, 145, 350
use in laser, **18,** 708, 717, 740
filter, **10,** 105, 107; *see also* Direct spectrum analysis, in LDV
resonator, **10,** 50, 51, 125, 126
spectrometer(s), **10,** 156; **25,** 217
spherical, **25,** 220
Fabry-Perot cavity(ies), **15A,** 10–11, 27, 210; **29C,** 107
error signal deviation, **29C,** 115–120
mechanical stress, **29C,** 110–113
optical properties, **29C,** 107–110
phase locking, **29C,** 131
tunability vs. stability, **29C,** 113–115
Fabry-Perot instruments, **12A,** 531–569
adjustment of, **12A,** 6557, 561
angular scanning, **12A,** 538
basic properties, **12A,** 532–534
calibration of, **12A,** 561
comparisons with other instruments, **12A,** 564–569
coupling etalons, methods of, **12A,** 543
double-etalon spectrometer, **12A,** 555
matching to astronomical sources, **12A,** 551–553
matching to telescopes, **12A,** 549–551
mechanical scanning, **12A,** 538
multiple-etalon, **12A,** 540–548
PEPSIOS, **12A,** 542, 556
premonochromatization, **12A,** 535
pressure scanning, **12A,** 537
single-etalon nebular spectrometer, **12A,** 533
Fabry-Perot interferometer(s), **1,** 350, 351, 364; **3A,** 403; **7B,** 106; **9A,** 95, 97, 98; **13A,** 304; **13B,** 55, 84, 127; **15A,** 9; **16A,** 184–188; **18,** 105, 195–227, 708, 717, 740; **26,** 25; **29C,** 350, 362–366
applications to acoustooptics, **19,** 478–481
measurements on dissociative recombination, **7B,** 120
Face-centered cubic lattice, **16B,** 5
Facet damage, in semiconductor dye lasers, **15A,** 242
Factor-group modes, **16A,** 104–106, 119–120
Factorial, **5B,** 763
Fahrenheit scale, **1,** 32
Fahy permeammeter, **1,** 555
Fail-safe devices, **14,** 169–170
Failure rate, transistor, **2,** 160
Failure, definition, **28,** 156
FALP, *see* Glowing afterglow/Langmuir probe

FAMOS, *see* Floating gate avalanche-injection MOS
Far infrared, **12A,** 416–417, 423–424, 447, 460
  lasers, **15B,** 63, 76, *see also* Infrared radiation; Infrared transition
  sum frequency generation in, **15B,** 181–183
Far point of the eye, **25,** 120
Far sighted, **25,** 68
Far-Infrared Region, **13B,** 50
  astronomical applications, **13B,** 100
  detectors, **13B,** 51
  interferometry, *see* Fourier transform spectrscopy
  spectra of gases, **13B,** 88
  spectra of solids, **13B,** 90
Far-ultraviolet region, **13A,** 204
Faraday, **1,** 41
Faraday cage, **7B,** 282
Faraday constant, **1,** 39, 41–42; **4B,** 3
Faraday cup, **5B,** 416–417, 486–494; **7A,** 6, 18–22, 364, 397, 398; **21,** 298–299
Faraday cup detector, **22,** 283, 287
Faraday cup method, for electrostatic charge measurement, **16C,** 424
Faraday effect, **1,** 387–388; **6B,** 275–276; **7A,** 140; **11,** 704–705, 707–708; **26,** 153
Faraday law, **1,** 431, 449; **2B,** 361
Faraday method, **6B,** 178–179, 181, 184
Faraday mirrors, **29C,** 173–174
Faraday rotation, **11,** 703–704
  ionospheric, **12B,** 126–127
  plasma, **10,** 286, 287
  in pulsar signals, **12C,** 82
  semiconductors, **10,** 130, 257–259
Faraday rotation, in optical refractivity of plasmas, **9B,** 27–31
  magnitude of, **9B,** 27–29
  measurement of, **9B,** 29–31
Faraday rotator, **2,** 715
Faraday shutter, use in high speed photography, **18,** 731–732
Farady transition, **23B,** 191
Farvitron tube, **8,** 169
Fast atoms
  by charge exchange, **4A,** 187–190
  detection of, **4A,** 412–414
Fast beam sources
  charge exchange
    techniques, **29B,** 55–56
    beam putiry, **29B,** 57–58
  CW laser/arc jet, **29B,** 60–61
  electron stimulated surface desorption, **29B,** 61–63
  negative ion photodetachment
    H(^{2}S), **29B,** 50–51
    O(^{3}P), **29B,** 51–52
    other neutral species, **29B,** 53
  photodissociation, **29B,** 63
  plasma/surface neutralization, **29B,** 58–59
  pulsed laser-generated plasmas, **29B,** 59–60
  space charge considerations, **29B,** 51, 54
Fast collision, **17,** 31
Fast decay of luminescence, **6B,** 311–313
Fast flux generators, in equipment testing, **2B,** 181
Fast flyback operation, in Mössbauer spectroscopy, **11,** 326–331
Fast Fourier transform (FFT)
  accuracy of, **12C,** 293
  filtering method, **28,** 296–297
  fixed-indexing, **12C,** 290
  implementation of, **12C,** 295
  in-place, **12C,** 290
  memory requirements of, **12C,** 293
  multidimensional data, **12C,** 292
  natural, **12C,** 289
  power spectrum measurement, **29C,** 330
  principles of, **12C,** 284–287
  real data, **12C,** 291
  real symmetric data, **12C,** 293
  testing programs for, **12C,** 295
  types of, **12C,** 284
  zeros in data, **12C,** 291
Fast ion conductors, **23B,** 115, 187
  anharmonicity, **23B,** 211
  diffraction, **23B,** 205
  inelastic scattering, **23B,** 233
  neutron scattering methods, **23B,** 189
  quasi-elastic scattering, **23B,** 223
Fast luminous fronts, **18,** 696
Fast molecular beam source, **7A,** 252–254
Fast neutral beam detectors, **7A,** 350–352
Fast neutral beams, **7A,** 341ff
Fast neutral charge exchange beam source, **7A,** 344–349
Fast neutron reactions, *see* Nuclear reactions
Fast neutron spectroscopy, **8,** 263–264
  application, **8,** 253–255
Fast neutrons, **8,** 268
Fast response pressure gages, **18,** 576–610
Fast-atom bombardment (FAB), **30,** 415
Fast-beam spectroscopy

beam sources, **29B,** 304–305
kinematics, **29B,** 302–304
lifetime measurements, **29B,** 232–239, 312
metrology, **29B,** 318
molecules, **29B,** 314–317
negative ion, **29B,** 312–314, 317–318
optical, **29B,** 306–309
radio frequency and microwave, **29B,** 309–311
Fast-Fourier-transform infrared spectroscopy, **22,** 78
Fast-ion accelerator, **17,** 464–467
Fast-ion conductors, **23C,** 465
Fast-reaction studies, **3B,** 655
Fastie-Ebert mounting, **3,** 160
Fastpassage in NMR, **3B,** 516
Fatigue, **14,** 440; **23B,** 117
testing, **19,** 345–348
Fault isolation by semiautomatic techniques (FIST), **2B,** 174
F_B(II) center, **15B,** 7–9
normal and relaxed configuration of, **15B,** 8
FBA, *see* Accelerometer, force-balanced
FDT, *see* Flow drift tube
Fe on GaAs(110), spectroscopy of, **27,** 127, 130, 271
Fe, **23C,** 31, 179
Fe-Si, **23C,** 95
$FeCl_2$, **23C,** 37
Feed defocusing
axial, in paraboloidal reflectors, **12B,** 45
lateral, in paraboloidal reflectors, **12B,** 44
FEED, *see* Field-emission energy distribution
Feedback, **1,** 468; **2,** 453–518
advantages of, **2B,** 377
in broadband amplifiers, **2A,** 307–308
dc voltage regulators, **2,** 496–501
pulse stretcher and, **2A,** 393
in square-loop core drives, **2A,** 406
system, linear performance, **2,** 476–481
Feedback compensation
minor or inner loop, **2B,** 419
tachometer in, **2B,** 423
Feedback control systems, **2B,** 375–459, *see also* Electronic feedback systems, Feedback system response
advantages, **2,** 453–455
amplifiers and signal conditioners in, **2B,** 442
angular position transducers in, **2B,** 440–441
block diagrams in, **2B,** 376
Bode plot in, **2B,** 419–420, 424
carrier systems in, **2B,** 442–444
closed-loop gain in, **2B,** 377–378
closed- vs. open-loop performance in, **2B,** 410–414
compensation in, **2B,** 410–425
conditionally stable, **2B,** 435–436
description of, **2B,** 375–376
differential equations, **2,** 458
dynamic response, **2,** 458–472
gain or magnitude in frequency response, **2B,** 390
gain phase plot in, **2B,** 395–397
general techniques of, **2B,** 375–436
Laplace transforms and, **2B,** 381–385
lead and lag networks in, **2B,** 415–416
linear, time-invariant differential equations for, **2B,** 379–381
mathematical aspects of, **2B,** 379–385
nonlinear, **2B,** 425–436
Nyquist stability criterion in, **2B,** 401–405
open-loop gain in, **2B,** 377–378
partial fraction expansion in, **2B,** 383
phase margin of, **2B,** 414
phase-sensitive demodulation in, **2B,** 441
relay servomechanisms and, **2B,** 444–445
root-locus techniques in, **2B,** 405–410
Routh criterion in, **2B,** 398–401
series compensation in, **2B,** 411, 415–421
servomechanisms and, **2B,** 436–447
stability in, **2B,** 398–410
stable vs. unstable, **2B,** 379
superposition principle in, **2B,** 379
symbolology in, **2B,** 376–377
time-invariant, **2B,** 380
transfer function in, **2B,** 383
transient vs. frequency response in, **2B,** 395–397
unity feedback system and, **2B,** 381
Feedback control theory
scanning tunneling microscope design, **27,** 33
Feedback electronics, **27,** 61
Feedback function, **8,** 236
Feedback oscillation
experimental problems in scanning tunneling microscope operation, **27,** 73
Feedback regulators, **2B,** 452
Feedback relations, **8,** 233
Feedback stabilization, in transistor amplifiers, **2A,** 280–281

Feedback system response, **2B,** 385–397, *see also* Feedback control systems
  Bode plot in, **2B,** 391–394
  damping ratio in, **2B,** 387
  electronic, **2B,** 445–459
  error coefficients in, **2B,** 389
  frequency response in, **2B,** 445–446
  gain-phase plot in, **2B,** 392
  natural frequency in, **2B,** 387
  transient response in, **2B,** 385–388
Feeds, *see also specific types*
  characteristics, **12B,** 78–79
  use on paraboloidal reflectors, **12B,** 64–81
Feedthrough
  bakeable, **14,** 307
  electrical, **14,** 277, 278, 306–310
  high current, **14,** 308, 309
  high voltage, **14,** 309
  linear motion, **14,** 304–306
  mechanical motion, **14,** 278, 301–306
  rotary, **14,** 301–306
  sputter-proof, **14,** 310
  swing motion, **14,**305
FEEM, *see* Field-electron emission microscope
FeF_2, **23C,** 20, 27
  antiferromagnetic resonance, **10,** 411–414
  Raman scattering, **10,** 424, 425
Fejér's kernel, **28,** 317–319
Fellgett advantage, **10,** 141; **12A,** 404, 493; **13B,** 53; **25,** 198
FEM, *see* Field-emission microscope
FeP glasses, phonon density of states, **23B,** 294
Fermat's principle, **1,** 329; **18,** 352; **31,** 355–357
Fermi
  plateau, **5A,** 15, 18
  plot of allowed beta-decay electron spectrum, **5A,** 684
  type neutron velocity selector, **5A,** 502–503
  unit, **5A,** 311
Fermi energy, electron escape and, **11,** 80–81
Fermi formula, **4B,** 163
Fermi level, **2,** 114; **10,** 65, 159; **11,** 9; **12A,** 7; **22,** 19
  in copper, **11,** 72
  photoemission analyzer and, **11,** 91
  in semiconductor tunnel junctions, **11,** 180
Fermi pseudopotential, **23A,** 17, 263; **23C,** 243
Fermi resonance, **3A,** 66
Fermi sea, **22,** 8
  current density, **22,** 8
Fermi surface, **11,** 6, 29; **21,** 94
  changes in, **11,** 58–60
  of copper, **11,** 7
  electron motion and, **11,** 2
  Hall coefficient and, **11,** 4
  longitudinal magnetoresistance and, **11,** 30
  open and closed orbits on, **11,** 3
  umklapp, **21,** 116, 117
Fermi velocity, **11,** 550–551
Fermi-Dennison resonance, **3,** 67
Fermi-Dirac distribution, **10,** 64, 65, 243
Fermi-Thomas potential, screening angle, **5A,** 392
Fermions, 360-degree rotation, **23A,** 299
FERP, *see* Field emission retarding-potential method
Ferrimagnetic exchange, **10,** 416–419
Ferrimagnetic resonance, **10,** 416–419
Ferrimagnetic structures, **23C,** 129
Ferrimagnetism, **6B,** 176–177, 193–208
  body force method, **6B,** 204–208
  calibration and error, **6B,** 208
  experimental method, **6B,** 201–208
  induction method, **6B,** 201–202
  oscillating pendulum, **6B,** 204–205
  theory of, **6B,** 194–197
  torsion balance, **6B,** 207–208
  vibration system, **6B,** 203–204
Ferrite devices, in microwave circuits, **2B,** 261–263, 286
Ferrite harmonic generator, **10,** 27
Ferrites, study of in region of Curie temperature, **6B,** 198–200
Ferritin, **22,** 417, 421
Ferroelectric ceramics, *see* Ceramics
Ferroelectric crystals, **10,** 372–385
Ferromagnetic anisotropy, **6B,** 243–244
Ferromagnetic contamination, **1,** 543–545
Ferromagnetic domains, observation, **6B,** 277
Ferromagnetic equation of motion, **6B,** 215, 224–227
Ferromagnetic materials, **1,** 546–559
  Hall effect in, **6B,** 148–149
Ferromagnetic nickel, spin-dependent tunneling in, **11,** 154
Ferromagnetic resonance, **6B,** 240; **10,** 394–407
Ferromagnetic structures, **23C,** 119
  nuclear, **23C,** 273

Ferromagnetism, **6A,** 19; **6B,** 193–208
body force method, **6B,** 204–208
calibration and error, **6B,** 208
experimental methods, **6B,** 201–208
induction method, **6B,** 201–202
oscillating pendulum, **6B,** 204–205
theory of, **6B,** 193–194
torsion balance, **6B,** 207–208
vibration system, **6B,** 203–204
Ferromagnets, **23A,** 68; **23C,** 176
losses in, **6B,** 209–227
weak, **23C,** 191
Fery pyrometer, **1,** 253
FET, *see* Field effect transistor
FeTi, *see* Iron-titanium
Feynmann, R. P., **23B,** 330
FFT, *see* Fast Fourier transform
FFTS, *see* Fast-Fourier-transform infrared spectroscopy
Fiber couplers, **29C,** 376–377
Fiber diffraction, **20,** 275–288
Fiber lasers, **29C,** 171
cavity building, **29C,** 171–175
diagnostics, **29C,** 187–190l
modelocking, **29C,** 175–184
operating parameters, **29C,** 184–187
Fiber optics, **12A,** 205, 244
in light-scattering optical systems, **9A,** 94, 102
Fiber stress, maximum in flexural test, **16C,** 125
Fiber, X-ray patterns, **16B,** 56–60
indexing, **16B,** 71–76
Fiber-optic faceplate, **22,** 389
Fiberoptic delay line, **29C,** 376
Fiberoptics, **29C,** 392
components, **29C,** 376–381
optical fibers, **29C,** 371–376
Fibers, **16A,** 8–9
hard-elastic, **16C,** 145
natural, **16C,** 138
Fibrillar crystals, in spherulites, **16B,** 342
Fibrils, in a craze, **16C,** 233
Fick's law, **6A,** 364; **9A,** 387
of diffusion, **16C,** 317–318
Fickian diffusion, concentration-dependent, **16C,** 332, 346
Fiducial marks, in strain measurement, **16C,** 119
Field(s), *see* Electric field
crystal, **3,** 222–229
term splitting, **3,** 223
curvature, **1,** 333, 337
distribution, magnetic, **5A,** 354, 355
electromagnetic, **1,** 517–524
electrostatic, **1,** 433, 443–447
equations, **26,** 108
fringing, **5A,** 364
index, magnet, **5B,** 731
inhomogenous, axially symmetric magnetic, **5A,** 370–375
intensity, magnetic, **14,** 87
Lamb shielding, **3,** 366–367
local time average, **3,** 363–365
magnetic, **1,** 524–530
for cloud and bubble chambers, **5A,** 386
electron spin resonance, **3,** 484–490
measurement, **3,** 471
nuclear magnetic resonance measurement of, **5B,** 122–123
measurements, **26,** 266
oscillating electric and magnetic, **3,** 620–622
potentials, ligand, **3,** 223
randomly fluctuating local, **3,** 363
sound, **1,** 224–230
static, **1,** 490
stop, **1,** 334
strength, **14,** 87
sweep direction, **3,** 407
theory, ligand, **3,** 222
Van der Waals, **3,** 229–238
vector, electric, **1,** 489ff
Field absorption as visualization method, **18,** 389–392
Field components, **8,** 65, 67, 69, 77, 85
calculation, *see* Current loop
Field corrections, **8,** 79, *see also* Successive field corrections
Field desorption, **4A,** 45–46; **22,** 385, 398
mass spectrometry, **22,** 400
microscopy, **22,** 412
Field effect apparatus, **6B,** 139
Field effect transistor (FET), **2A,** 84–85; **2B,** 105–106; **4A,** 336–338
applications, **4A,** 336–337
bias circuits and, **2A,** 261–267
bias instability in, **2A,** 265
in dc amplifiers, **2B,** 51
equivalent circuits for, **2A,** 269
insulated gate type, **4A,** 337–338
noise in, **2B,** 105, 466, 479–481

noise figure for, **2A,** 329
noise sources, **4A,** 336–337
as nonlinear device, **2A,** 362
in nuclear magnetic resonance oscillators, **2B,** 140
in preamplifier circuit, **2B,** 105–106
Field effect transistors, resistance thermometers and, **11,** 499–501
Field emission, **14,** 64, 66
basic process, **4A,** 39–41
cathodes, **4A,** 44–45
devices, **4A,** 45–51
emitter fabrication, **4A,** 42
experimental verification, **4A,** 41–42
projection microscope, **4A,** 42–43
Field emission Field, **22,** 8, 20, 351, 356, 368
electron source, **22,** 206
energy distribution, **22,** 26
microscope, **22,** 21
Field emission ion sources, **11,** 452–453
Field emission retarding-potential method, **22,** 15
field emission cathode, **22,** 15
work function, **22,** 15
Field emission source, **22,** 282
Field evaporation, **22,** 385, 387, 397, 399, 404, 410; *see also* Liquid-metal ion source
brightness of sources, **29A,** 142–143
initiation, **22,** 404
ion current density, **29A,** 141
ion formation mechanism, **29A,** 140–141
microfocused beam applications, **29A,** 140, 143–144
Field homogeneity, **8,** 85, 86, 97, **8,** 101
calculation, *see* Finite rectangular cross-sectional area
Field ionization, **22,** 378, 401; *see also* Ion sources
basic method, **29B,** 121
ion current low-field ionization, **29A,** 136–138
lifetime transitions, **29A,** 136
mass spectrometry, **4A,** 47, 53
to measure populations of excited levels, **7A,** 223
microscopy, techniques, **4A,** 46–47, 51–53
positive ion formation, **29A,** 134
potential energy of atoms, **29A,** 135
in resonance ionization spectroscopy, **29B,** 175, 188
sources
applications, **29A,** 138
types, **29A,** 138–139
theory, **7A,** 210–211
tunneling mechanics, **29A,** 134–135
Field lens, **25,** 130; *see also* Fabry lens
Field mill, for electrostatic charge measurement, **16C,** 425
Field modulation, de Haas-Van Alphen effect in, **11,** 34, 44–56
Field of view, of infrared photometer, **12A,** 434, 439 441, 443–444, 450–451, 460
Field optics, **12A,** 424, 434–440, 444, 460
Field significance, **28,** 458
testing for
field statistics, **28,** 465–466
temperature correlation maps, **28,** 470–471
Field splitting
spin-dependent tunneling and, **11,** 154
strong magnetic, Mn^{2+}, **3,** 449–450
Field statistics, **28,** 457–478
assessment by Monte Carlo simulation, **28,** 462–466
critical area estimation, **28,** 465
field significance testing, **28,** 465–466
local critical correlation estimate, **28,** 463–464
random time series, **28,** 462–463
simulation of field statistics, **28,** 463
atmosphere temperature fields, **28,** 466–471
correlation statistic map calculation, **28,** 460–461
cross correlation between ozone and solar flux time series, **28,** 475–477
field significance, **28,** 458
finite data set lengths, temporal interdependence, **28,** 459
global ozone data fields, **28,** 471–475
higher dimensions, **28,** 478
known spatial degrees of freedom, **28,** 459–460
spatial interdependence, **28,** 459
time-lag correlation for TOMS, **28,** 457–458
Field stop(s), **25,** 37, 131, 171; **31,** 150–151
Field strength, **22,** 410
measurement, microwave, **2,** 741–742
measurements, in microwaves, **2B,** 298–300
Field-dependent lifetime, and Lamb shift, **17,** 589–590

Field-effect transistors, **12A,** 426, 432
Field-electron energy distribution, **22,** 362
Field-electron-emission microscope, **22,** 349, 353, 354, 359
  pressure in, **22,** 354
Field-electron-emission microscopy, **22,** 349, 363
  application, **22,** 363
Field-emission source, **16B,** 207
Field-emission-referenced electron spectroscopy, **22,** 226
Field-emitted electrons, **22,** 353
  energy distribution, **22,** 361
Field-induced dipole energy, **22,** 380, 384
Field-ion current, **22,** 402
Field-ion images, **22,** 379
Field-ion microscope, **22,** 376, 384, 385, 387, 392, 412
Field-ion source, **22,** 339, 401
  liquid metal, **22,** 403
Field-ion tomography, **22,** 421
Field-ionization ion source, **22,** 347
Field-ionization mass spectroscopy, **22,** 400
Field-quenching technique, Lamb shift measurements, **17,** 588–590
Field-widened interferometer, **25,** 227
  Fields, *see* Circular loops; Dipole fields; Finite cross-sectional area; Line currents; Solenoid pair
  calculation, *see* Axial symmetric systems
  deviation, **8,** 78
  distribution calculation, *see* Finite rectangular cross-sectional area
  intensity, **8,** 57, 58
  magnetized dipoles, influence, **8,** 65–66
  region, *see* Central field region
Figure of merit, **1,** 556
  of cavity, **3,** 27
  for detectors, **26,** 292
  pentode, **2,** 232
  tuned amplifier, **2,** 244
Figure of merit, electron energy analyzer
  configuration effects, **29A,** 218–219
  luminosity effects, **29A,** 218–220
Filament material, electron beam, **3,** 531–532
Filament transformers, **2A,** 222–223
Filament(s), **25,** 240
  coiled, **25,** 240, 244
  emissivity of, **25,** 240
  rhenium, **14,** 373
  ribbon, **25,** 240
  tungsten, **14,** 368
Filamentary crystals, *see* Whiskers
Filamentary current loop pair, **8,** 100
Filar micrometer eyepiece, **25,** 132
Filled aperture antennas, **12B,** 11–21
Filling gas, **25,** 241
Filling pressure, *see* BF3 counters
Film(s), *see* Thin films
  autoradiograph, **21,** 44–49
  characteristics, **6B,** 63–64
  conductivity, **6B,** 64–65
  doping in PLD, **30,** 475–572
  dosimeter, **21,** 35
  growth by PLD, **30,** 475–572
  magnetic, pulse spectrometer, **2,** 618
  nonabsorbing refractive index and thickness, **6B,** 279–280
  non-reflecting, **1,** 352
  optical definiton, **6B,** 279
  for phosphors, **6B,** 298–299
  reactivity, **14,** 197
  resistance measurements, **6B,** 63–65
  resistors, **2A,** 35
  resolution, **5A,** 448
  specimen, for electric breakdown, **16C,** 454–456
  speed, **1,** 411
  stoichiometry in PLD, **30,** 482–483, 520–521
  thickness, **6B,** 279
    measurement, **6B,** 64
  ultrathin
    preparation, **16C,** 456–459
    thickness measurement, **16C,** 458
Filter capacitors, **2A,** 223
Filter chokes, **2A,** 222–223
Filter radiometers, **26,** 257
Filter values, normalized, **2A,** 26
Filter-type spectrometers, **12C,** 22, *see also* Multichannel filter spectrometers
Filtering time series, **28,** 283–297
  bandpass filter, **28,** 295–296
  digital filter, **28,** 292–296
  direct Fourier analysis-resynthesis, **28,** 296–297
  fast Fourier transforms, **28,** 296–297
  frequency coefficients, **28,** 285
  frequency response, **28,** 284–291
    filter coefficients, **28,** 289
    finite difference operators, **28,** 290–291
    low-pass filter, **28,** 287

1-1-1 moving average filter, **28,** 286–289
periodogram, **28,** 285
spectral power densities, **28,** 285, 287–289
Gibbs phenomenon, **28,** 292
Lanczos window weights, **28,** 293–295
ringing, **28,** 296
total variance, **28,** 284–285
Filtering, **12A,** 513, 520
Fourier, **28,** 417
of rectifier output, **2,** 203–204
Filters, **1,** 220; **13B,** 25, 26, 29, 59, 84; **22,** 485; **25,** 282; *see also* Optical filters
absorption and reflection, **10,** 102, 110
active, **2A,** 28
band-pass, **10,** 105, 109
in beam-foil measurements, **17,** 538–539
boron, **5A,** 518
Christiansen, **1,** 401
design of, **2A,** 25–27
echelette grating, **10,** 138, 140
frequency response, **28,** 284–291
frequency transformations in, **2A,** 24–25
high pass, **1,** 477
hydrogen, using palladium, **14,** 373
image-parameter design in, **2A,** 23–24
infrared, **12A,** 415–416, 426–428
circular variable, **12A,** 427
dichroic beamsplitters, **12A,** 428
electronic, **12A,** 440
metal-mesh interference, **12A,** 428
scatter, **12A,** 427
thin-film interference, **12A,** 416, 426
interference, **1,** 355; **10,** 105, 107
ladder, **2A,** 27
low-pass, **10,** 102–104, 109, 110
metal grating, **10,** 105, 107
in microwave circuits, **2B,** 256–257
monochromators, **22,** 485
in network theory, **2A,** 23–28
neutron, **5A,** 507–509, 568–569
nickel foil, for X-ray monochromatization, **16B,** 47–48
optical, **25,** 273
performance, **25,** 289
prewhitening, **28,** 327–328
program, **25,** 276
Raman lamp, **3,** 129–131, 135
replica grating, **10,** 103, 104, 139
scattered radiation, **3,** 86
transient response of, **2A,** 27–28
transmission, **1,** 356, 485
wave, **2,** 33–37
waveguide, **10,** 121
Filtrage, **25,** 196
FIM, *see* Field-ion microscope
Final density of states, **22,** 172
Final-state correlation effects, **22,** 231
Final-state effects, **22,** 150, 169, 179
Final-state formation, **22,** 337
Fine interaction, in foil excitation process, **17,** 592, 595
Fine structure constant, **1,** 42–43
Fine structure, atomic, **4B,** 108–109, 131–150, *see also* Lamb shift
doublet splitting, **4B,** 113
Paschen-Back limit, **4B,** 112, 119
table of, **4B,** 132
Fine-control-rod position indicator, **8,** 236
Finesse of interference fringes, **26,** 27
Finesse, **10,** 86, 87; **12A,** 532, 539, 540; **18,** 197; **29B,** 380–381; **29C,** 362, 363
defect, **12A,** 539
instrumental, **12A,** 540
of interferometer, **9A,** 98
reflective, **12A,** 532
Finite amplitude acoustic, **24B,** 90
Finite amplitude effects, **19,** 229–303
Finite angular resolution, correction for in neutral-neutral collisions, **7A,** 287–289, 354–356
Finite conjugate lenses, **25,** 91
Finite cross-sectional area, solenoid field, **8,** 81–92, 99
exercise, **8,** 85–92
Finite difference operators, **28,** 290–291
Finite rectangular cross-sectional area, solenoid pair
field distribution/field homogeneity, calculation, **8,** 93
Finite rotational relaxation rate, collisions and, **15B,** 67
Finite temperature, **8,** 1
Finite width a slit source, **8,** 51
Finite width detector aperture, **8,** 50–53
Finite-amplitude waves, in solids, **11,** 387–389
Finite-difference methods, **1,** 18
Finite-size effects, **22,** 246
FIR, *see* Far infrared lasers

Firsov theory, of electronic energy loss, **17,** 39–40
First moment method, **5B,** 321
First order plate, **26,** 143
First sound in superfluid helium, **19,** 417–453
  Doppler shift of, **19,** 445, 446
  properties of, **19,** 421
  transducers for, **19,** 421–429
  velocity of, **19,** 419, 420
First-order autocorrelation, **29C,** 205
First-order desorption kinetics, **22,** 425, 448, 454, 459, 460, 462
First-order-Doppler effect, elimination of, **15B,** 232, *see also* Doppler broadening
First-principles calculations, of photon spectra, **23A,** 386, 457
Fisher information, **28,** 227
  matrix, **28,** 241
Fisher t function, **2,** 6–7
Fisher's *g* statistic, **28,** 339
Fisher's *t* function, **1,** 8; **3,** 715; **5B,** 812–813; *see also* Student's *t* function
Fissile materials, **8,** 239
Fission, *see* Nuclear reactions
  chambers, **5A,** 108–109
  counter, **5A,** 140
  cross-section, **5B,** 407–411
  fragments, **17,** 73–74, 118
    response of semiconductor detector, **5A,** 273–274
  neutron sources, **5B,** 562
  spectrum, **23A,** 106
  threshold, **5A,** 109
  tracks, **24A,** 110
Fission chambers, **8,** 249
Fission cross section, **8,** 240, 247
Fission density, **8,** 203
Fission isotopes, *see* Gamma-emitting fission isotopes
Fission product resonance absorption, **8,** 244
Fission products, **8,** 181, 203, 211
Fission yield, **8,** 210
FIST (fault isolation by semiautomatic techniques) project), **2B,** 174
Fit, goodness of, tests of, **5B,** 786–792
Five-kilometer Cambridge telescope, **12C,** 165–166
Fixed calibration pressures, **6A,** 415
Fixed points, thermometric, **1,** 255–257
Fixing of thick emulsion, **5A,** 217
Fizeau effect, **23A,** 303
Fizeau fringes, **9A,** 97, 102
Fizeau interferometer, **1,** 358, 359; **26,** 8
Fizeau velocity selector, **7A,** 255–257
Fizeau wavemeter
  calibration, **29C,** 335
  limitations and variations, **29C,** 335–337
  principles, **29C,** 331–332
  structure, **29C,** 332–333
  wavelength calculation with, **29C,** 333–335
Fizeau-Tolansky interferometer, use in LDV, **18,** 198–199, 214
Flame bands, hydrocarbon, **3,** 166
Flame composition
  by emission spectroscopy, **18,** 641–643
  by mass spectroscopy, **18,** 659–660
Flame front velocity, measure by schlieren method, **18,** 369
Flame intensity profiles, **7B,** 218
Flame kinetic studies, **7B,** 189–230
  experimental techniques, **7B,** 209–229
    optical studies, **7B,** 209–220
Flame photometry, [O] determination by, **7B,** 214
Flame temperature, **18,** 421–425, 466–470
Flames, **13A,** 260
Flames, collisional deactivation measurement, **7B,** 220
  composition of, **7B,** 190–209
  definition of, **7B,** 189
  diffusion type, **7B,** 206
  electromagnetic absorption methods for, **7B,** 229
  electrons in, measurement, **7B,** 227–229
  equilibria and nonequilibria in, **7B,** 194–195
  high-pressure types, **7B,** 205
  in high-temperature studies, **7B,** 189
  ions and electrons in, **7B,** 198–201
  isothermal type, **7B,** 218–220
  life-period method, **7B,** 208
  low-pressure type, **7B,** 202–205
  microprobe sampling, **7B,** 221–227
    scavenger methods, **7B,** 223–227
  premixed type, **7B,** 201–202
  radicals in, **7B,** 196
  rate constants from, **7B,** 230
  sources of, **7B,** 191–194
Flange, **14,** 277–282, 291, 293
  all-metal, **14,** 286
  dimensions, **14,** 279–281
  sexless, **14,** 282

Flanges, vacuum, **4B,** 283–286
Flappers, **29C,** 173
Flash coding, sheet beam, **2,** 624
Flash desorption, **22,** 428
Flash lamp
  characteristics, **18,** 692–695
  dye laser pump, **18,** 719
Flash photochemical reactions, **3,** 578–579
Flash photolysis, **3,** 161–166; **3A,** 233; **7A,** 149; **20,** 166, 167
Flash radiography, **18,** 408
Flash-discharge lamp, **25,** 254
Flashlamp arc, wall stabilization of, **15A,** 307
Flashlamp dye amplifiers, **15B,** 189
Flashlamp dye lasers, **15B,** 188–189
Flashlamp excitation, dye cell for, **15A,** 312–314
Flashlamp lasers, single pulse from, **15B,** 201
Flashlamp plasma temperature, **15A,** 298–302
  laser efficiency and, **15A,** 300
Flashlamp(s)
  ablating, **15A,** 305–307
  coaxial, **15A,** 319–321
  ionized gas in, **15A,** 304
  linear, **15A,** 302–305
  xenon-filled, **15A,** 302
Flashlamp-excited dye lasers, **15A,** 297–298
  dye solution jets in, **15A,** 314–318
  emission spectra in, **15A,** 301
Flashlamp-excited systems, oxygen and cyclo-octatetraene in, **15A,** 359
Flashlamp-pumped dye laser
  optimized, **15B,** 189
  pulse generation by, **15B,** 188
Flashlamp-pumped lasers, **15B,** 186
Flashover, **16C,** 445
Flat crystal spectrometers, **31,** 405–408
Flat jack, **24B,** 381–383
Flat radiated power response, **8,** 13
Flat spiral feed, **12B,** 77, 79
Flattening of spheroid, **24B,** 132
Flavanthrene, **22,** 372
Flavin mononucleotide, **19,** 202
Flavinyl-tryptophane peptides, **19,** 203
Fletcher-Munson contours, **1,** 218
Flexural oscillations, **6A,** 299
Flexure of a beam static test, **6A,** 297
Flicker noise, **2B,** 465–467; **10,** 224; **22,** 202, 259, 402
Flicker, in low-noise amplifier, **2A,** 329
Flight paths, **23A,** 199
Flight testing
  apparatus, **18,** 779–781
  heat transfer, **18,** 664
Flight time, **5A,** 545
  measurement, **5A,** 547–548
Flight-distance, direct, measurement of lifetime, **5B,** 335–336
Flip-flop (bistable multivibrator), **2A,** 398–401
Flip-flop circuits, **2,** 422–427; **2A,** 398–400
Flip-flop timing specifications, **2A,** 415–416
Flip-flops
  in frequency measurements, **2B,** 19
  in phase detection, **2B,** 46
Flipping ratio, **23A,** 68, 77; **23C,** 101, 264
Floating gate avalanche-injection MOS (FAMOS), **2A,** 127–128
Floating solid phase, **23B,** 38
Floating zone technique, **6A,** 96–98
Flop, **3,** 616
Flop-Out method, **3,** 616
Floquet's theorem, **4A,** 368
Florida, University of, **2B,** 466
Flory, P. J., **23C,** 299
Flotation, **1,** 69
Flow
  cycle, **14,** 440
  free molecular, **1,** 183–186; **14,** 11–13, 16
  gaseous, **14,** 11
  laminar, **14,** 12
  meters, **1,** 117
  phenomena, normal pressures, **1,** 143–157
  plastic, **14,** 288
  quantities, **14,**11
  rate, mass, **1,** 170
  regimes, **1,** 158ff
  turbulent, **14,** 12
  viscous, **14,** 12, 13
  volumetric, **1,** 172, 184; **14,** 11, 13
Flow disturbance
  by electron beam fluorescence diagnostics, **18,** 451, 453
  by hot-wire probe, **18,** 308
  tracer particles, **18,** 38, 41, 49–51
  by Pitot probe, **18,** 243, 250
  by Raman scattering diagnostics, **18,** 419
Flow drift tube
  apparatus, **29A,** 291–292
  ion-neutral reaction studies, **29A,** 291–292
Flow measurement, **14,** 51
Flow proportional counter, **17,** 357–358
Flow straightener, **18,** 326

Flow tracing particles
  advantages and disadvantages, **18,** 3–4, 97
  definition, **18,** 2, 6
  dynamic characteristics, **18,** 32–34, 221–222, 225
  effect on flow field (loading error), **18,** 38–41
  effect of sedimentation, **18,** 41–43
  equation of motion, **18,** 8
  generation and dispersal, **18,** 43–50
  hydrodynamic resistance, **18,** 8–15, 795
  interaction effects, **18,** 223–224
  light scattering, **18,** 52–60, 64
  limit of sensitivity in velocity measurement, **18,** 38–41
  location in measuring volume, effect on LDV, **18,** 126
  motion of, effect of size and density, **18,** 15–16
  optical characteristics, **18,** 51–60
  refractive index data, **18,** 53–54
  response time determination, **18,** 26–32, 795
  response time effect on turbulence measurements, **18,** 34–38
  selection of, illustration, **18,** 60–64
  size and density, measurement of, **18,** 16–26
  size effect on LDV performance, **18,** 124
  system for velocity measurement, **18,** 2–4, 6–7, 201–206, 235–240
  use in rotating flow apparatus, **18,** 818–819
Flow visualization
  electric glow discharge, **18,** 402
  electron beam fluorescence, **18,** 399, 437
  Hele-Shaw apparatus, **18,** 798–799
  high speed photography, **18,** 725–753
  infrared, **18,** 750–753
  interferometer, **18,** 377
  jet, **18,** 359
  light source, **18,** 694
  phase constant, **18,** 389
  radiation emission, **18,** 398
  rotating flow apparatus, **18,** 818–819
  schlieren, **18,** 365
  shadowgraph, **18,** 358
  shock waves, **18,** 365, 386
  smoke, **18,** 6–8, 770
  tuft, **18,** 241, 770
  wind tunnel, **18,** 769–771
Flowing afterglow
  apparatus, **29A,** 280–282
  applications, **29A,** 282–283
  ion detection, **29A,** 282
Flowing afterglow/Langmuir probe
  apparatus, **29A,** 283–285
  applications, **29A,** 283–284
  carrier gad, **29A,** 284
  electron attachment rate determination, **29A,** 284–285, 291
  ionic recombination coefficient determination, **29A,** 285
  operation, **29A,** 283
  spectroscopic experiments, **29A,** 294
Flowing afterglows, chemical aspects of, **7B,** 133–139
  mass spectrometry of, **7B,** 131–133
Flowing dye cell, **29C,** 59–60
Flowmeter
  acoustic, **18,** 337–340
  bundle of capillaries, **18,** 336–337
  calibration, **18,** 322
  definition, **18,** 241
  electromagnetic, **18,** 337, 340
  float meter, **18,** 331
  flume, **18,** 332, 334
  orifice, **18,** 324–330
  positive displacement, **18,** 323–324
  power loss, **18,** 328
  sonic nozzle, **18,** 330
  turbine, **18,** 324
  variable area, **18,** 331
  Venturi, **18,** 324, 331
  weir, **18,** 332–336
  wet-gas, **18,** 323
Fluctuating process, spectral resolution, **2B,** 496
Fluctuation dissipation theorem, **23A,** 35
Fluctuation spectrum, of plasmas, **9A,** 49, 61
Fluctuation theory of Blatt, **5A,** 263
Fluctuation(s)
  concentration, **23B,** 253
  dynamics of, **23C,** 28
  number density, **23B,** 253
  statistical, **14,** 83, 84
    in nuclear processes, **5B,** 761–806
Fluence, **30,** 174–175
Fluence dependence
  of ablation yields, **30,** 616
  of atom emission, **30,** 158
  of ion emission, **30,** 145
Fluid density, **1,** 210, 211
Fluid dynamic equations in rotating coordinate system, **18,** 802–806

Fluid dynamics, **1,** 126ff
Fluid flow, compressible, **1,** 128
Fluid properties, ultrasonic studies, **3,** 681–708
Fluid systems
  simple, **1,** 324
  two phase, **1,** 319
  two-variable, **1,** 316, 318
Fluid(s), **1,** 222–223; **23B,** 405
  compressible, mass flow rate, **1,** 155, 156
  computer simulation, **23B,** 455
  data analysis, **23B,** 419
  definition, **18,** 501
  dynamic phenomena, **1,** 126–197
  experimental techniques, **23B,** 414
  homogeneous additive, **1,** 318
  mechanics of, **1,** 115–198
  molecular, **23B,** 408
  nature of, **1,** 126–127
  normal modes, **23B,** 461
  pair approximation, **23B,** 405
  scattering function, **23B,** 411, 450
  static phenomena, **1,** 115–126
  structure factor, **23B,** 408, 445
Fluid-stabilized arcs, **7B,** 309–312
Flume, **18,** 332, 334
Fluorescence, **1,** 396–398; **3,** 116; **6B,** 294; **7A,** 166; **13A,** 145; **13B,** 334; **20,** 133, 134, 197–227
  analysis, X-ray, **1,** 417–419
  anisotropy, **20,** 200
  contours, **30,** 250–251
  correlation spectroscopy (FCS), **20,** 208, 209, 212–216
  decay, **22,** 495
  decay time
    of anthracene, **5A,** 421, 422
    of cesium iodide (Tl), **5A,** 414
  depolarization
    comparison with ESR spin probe techniques, **16A,** 479
    theory, **16A,** 413–426
  description, **16A,** 405–407
  dye laser, **18,** 715–716
  emission intensity, theory, **16A,** 418–421
  excitation, **3,** 165
  from impurities, problem in Raman spectroscopy, **16A,** 86–87
  infrared sensor, **18,** 672, 751
  laser-excited, **22,** 498
  lifetime measurements, **16A,** 441
  meaning, **18,** 411–412
  microscopy, **25,** 174
  photobleaching recovery (FPR), **20,** 208–216
  polarization, **20,** 200–208
    techniques, **16A,** 412–436
  quenching, **18,** 412
    techniques, **16A,** 407–412
  relation to resonance scattering, **18,** 413
  resonance, **1,** 397
  specific of sodium iodide (Tl), **5A,** 413
  spectrometers, **16A,** 439–441
  use for density, temperature, composition diagnostic, **18,** 410–414
Fluorescence yield, **17,** 191, 380, 382, 389, 392–395, 477–479
  argon L-shell, **17,** 347
  in gases, **17,** 422
  in high-charge collisions, **17,** 289–297, 300
  hypersatellites, **17,** 409
  K-shell ionization, **17,** 195
  lithiumlike atoms, **17,** 506, 586
  L-shell ionization, **17,** 197
  in multiple ionization, **17,** 207–209, 238–239, 503–506
  for thick target, **17,** 277–228
Fluorescent labels, **16A,** 435
  types, **16A,** 438
Fluorescent lacquer, visualize transition to turbulence, **18,** 771
Fluorescent lamps, **25,** 251
  instant start, **25,** 252
  preheated, **25,** 252
  rapid start, **25,** 252
Fluorescent probe
  geometry, **16A,** 414
  hydrophobic and hydrophilic, **16A,** 408
  types, **16A,** 437
Fluorescent radiation Doppler shift to measure velocity, **18,** 343
Fluorescent-screen detector, **22,** 283, 287
Fluoride fibers, **29C,** 375
Fluorine–19
  cross-polarization in carbon–13 NMR of PTFE, **16A,** 359
  NMR study of poly(chlorotrifluoroethylene-*alt*-isobutylene) conformation, **16A,** 299–300
Fluorite, **25,** 36
  structures, **23B,** 204, 226, 230
Fluorometer, **1,** 397

Fluorozirconate fibers, **29C,** 375
Flute instabilities, **9A,** 309, 399
Flux, **8,** 50; **12A,** 193, 465, 466; **21,** 17, 18, 25; **31,** 6, 60
absolute, in laser plasma, **31,** 85–86
alternating, **1,** 535
antineutrino, **5A,** 686
average primary neutron, at scatterer, **5B,** 454
ball, **1,** 523
brightness ratio, **31,** 21
charged particle, **5B,** 485–507
chemical, **14,** 426
decrease rate, **5B,** 367
deep space measurements of, **9B,** 71–73
depression, **8,** 245
at detector of doubly scattered neutrons, **5B,** 382
fast neutron, **21,** 19, 20
absolute measurement of, **5B,** 440–441
level, **8,** 247
linkage, **1,** 534–535
luminous, **1,** 388, 389
magnetic, **1,** 521ff
meter, **1,** 522, 523
neutron, instrumentaion for, **5A,** 466–481
reactor neutron, **5A,** 498
scale, for antenna calibration, **12B,** 84
thermal neutron, **21,** 19
time dependence, **8,** 251
transformer, **24A,** 359
transversely coherent, **31, 31,** 21
Flux calculation, **31,** 6
electron-beam excitation, **31,** 77
Flux densities/density
in deep space plasma measurements, **9B,** 80
of particles scattered from a target, **5B,** 502–507
units of, **12B,** 4
Flux density measurements
complex sources, **12C,** 7
Gaussian source-corrections in, **12C,** 5
small-diameters sources, **12C,** 3
uniform-disk corrections in, **12C,** 5
Flux distribution
in channeling, **21,** 291–293, 309
HFB reactor, **5B,** 601
in surface studies, **21,** 346
Flux-gate magnetometer, **2B,** 371
Fluxmeters, **5B,** 434–435
FM saturation spectroscopy, **29C,** 120
FM telemetering transmission, notch-noise test data for, **2B,** 236
FM/FM telemetering, **2B,** 220, 238
Focal curves, **31,** 365–367
Focal distance, **1,** 335, 336
Focal isolation, **13B,** 85
Focal length, solenoidal spectrometer, **5A,** 367
Focal points, **1,** 335
Focal shift, **25,** 42
Focal systems, **25,** 132
Focal-plane modulators, **12A,** 446
Focault knife edge test, **25,** 175
Focus, **12A,** 460
depth of, for cloud chamber measurements, **5A,** 447
first order directional, **14,** 88
Focused beam cavity designs, **15B,** 25–26
Focusing
action of a lens, **16B,** 196
action of magnetic lens, **5B,** 696–699
alternating-gradient, **5B,** 663–664, 693
axial, **5A,** 355
beam, **5B,** 692–747
Čerenkov counter, **5A,** 166–186
double, **5B,** 692ff
of two-magnet system, **5B,** 704–706
in electron microscopy, **16B,** 217–220
electrons for X-ray production, **16B,** 42
equations for wedge-shaped magnets, **5B,** 731–747
properties of magnetic fields, **5A,** 370–375
properties of two-magnet system, **5B,** 701–712
property, **14,**87
strong, **5B,** 693ff
with electrostatic lenses, **5B,** 708–709
system with momentum analysis, **5B,** 742–746
vertical, by fringing field of wedge-shaped magnet, **5B,** 732–734
an X-ray beam, **16B,** 157, 158
Focusing diffractometer, **6A,** 218–219
Focusing electrode design for electron guns, **4A,** 103–104, 116–120
Focusing, in image tubes, **12A,** 238–241
electrostatic, **12A,** 240, 246–249, 253–255
magnetic, **12A,** 240, 21, 243, 245–246, 255
proximity, **12A,** 239
Focusing, spectrometer, **23A,** 252
Foil seal, **14,** 280, 288

Foil target, **17,** 371–372, 387, 532, 536–537; *see also* Beam-foil experiments
- charge state fraction measurements, **17,** 84–86
- energy-loss measurements, **17,** 47–51
- spectral identification of emitters, **17,** 348–350
- thickening effects, **17,** 537, 577

Foil, magnetized ferromagnetic, polarized electrons from, **5B,** 243

Fold period
- deduced from GPC of degraded samples, **16B,** 263–266
- definition, **16B,** 162
- dependence on crystallization temperature, **16B,** 351
- from diffuse X-ray scattering, **16B,** 172–175
- distribution in cellulose triacetate, **16B,** 273
- effect of annealing on, **16B,** 300, 351
- estimated from longitudinal acoustic modes, **16A,** 124–125

Fold surface
- chlorination and bromination of, **16B,** 248
- nature of, **16B,** 128, 229, 238, 342, 344, 346
- selective oxidation with nitric acid and ozone, **16B,** 261
- thickness of, **16B,** 150

Fold surface free energy, **16B,** 349
- effect on melting point, **16B,** 352

Folded chains, **16C,** 219

Folded tandem accelerator, **17,** 9–10, 12

FONAR, *see* Imaging, NMR

Forbidden Auger transition, **17,** 561, 586–587

Forbidden decays, **13B,** 260

Forbidden electronic transitions, **3A,** 282
- excitation of, **3B,** 927

Forbidden radiative transition, **17,** 583–586

Forbidden transitions, **7A,** 118, 141

Forbidden zone, **22,** 383

Force, **1,** 25, 30, 31, 60–69
- on atom, **4B,** 236, 239
- coercive, **1,** 550
- on conductors, **1,** 518–521
- drag, **1,** 148
- electrostatic, **1,** 439
- equation, **1,** 431, 432
- gravitational, **1,** 439
- measurement of susceptibility, **1,** 537–542
- on molecule, **4B,** 10
- moment of, **1,** 84–86
- variables, **1,** 312

Force balance, aerodynamic model in wind tunnel, **18,** 768–770

Force constants, **23A,** 23, 373, 378
- from vibrational spectra, **16A,** 93–94

Force deflection, **14,** 288

Force gradient microscopy, **27,** 89

Forces, betatron, **5B,** 735

Forces, role of in STM
- theory of, **27,** 24

Forcing function, role in hot-wire and hot-film circuit response, **18,** 278

Fore optics, **12A,** 526

Forepressure, *see* Diffusion pump, forepressure

Forevacuum pump, **14,** 111–140, *see also* Pump backstreaming, **14,** 171, 467, 469
- with diffusion pump, **14,** 142–147, 154–158
- ejector, **14,** 136–138
- installation, **14,** 126–128
- liquid piston, **14,** 133–136
- lobe pump, **14,** 128–133
- manifold, **14,** 461
- oil-sealed rotary pumps, **14,** 417
- pressure range, **14,** 111–114
- rotary oil sealed pump, **14,** 114–128
- rough pumping, **14,** 141
- selection, **14,** 461
- trapping, **14,** 482
- types, **14,** 112
- volumetric efficiency, **14,** 113
- water vapor pumping, **14,** 121–124

Form factor
- cell, **23B,** 509
- magnetic, **23A,** 6
- atomic, **23A,** 61

Form factor, magnetic, **23C,** 75
- actinides, **23C,** 108
- measurement, **23C,** 100
- rare-earths, **23C,** 105
- theory, **23C,** 97

Formaldehyde
- absorption coefficient of, **12B,** 172
- atmospheric concentration, **12B,** 170
- molecule, decomposition of, **15B,**291
- in photodissociative selective chemistry, **15B,** 291–293
- rotational structure in singlet-triplet transition, **3,** 203, 204

Formate ion, **19,** 202

Formation, heat of, **1,** 262

Formic acid, **30,** 421
Forming gas, **14,** 422
FORS1, **12C,** 294, 320–321
Fortran, **12C,** 284
code for ODE solver, **28,** 506
Forward bias, in P-N junctions, **2A,** 53–56
Foucault
prism, **1,** 380, 402
method, in magnetic domain studies, **11,** 728–730
test, **1,** 336
Fountain analyzer, **22,** 221
Fountain spectrometer, **7A,** 7
Four mirror circular polarizer, **31,** 251
Four mirror polarizer, **26,** 124; **31,** 239–242
adjustable, **31,** 241–242
Four-point probe method, **6B,** 36–38
Four-point X-ray pattern, **16C,** 180
Four-pulse method for, measurement of spectrum and autocorrelation function of radio wave scattering, **9B,** 152
Four-terminal networks, noise in, **2B,** 499–500
Four-wave DFG, **29C,** 254
Four-wave frequency mixing, **29B,** 78–79
Four-wave mixing interactions, **29C,** 253–254
Four-wave SFG, **29C,** 254
FOUR1, **12C,** 290, 294, 316
FOURG, **12C,** 289, 294, 315
Fourier analysis, **1,** 473
Fourier components, **25,** 156, 167
Fourier computational methods, **6A,** 234–237
Fourier filtering, **28,** 417
Fourier frequency, independent, **28,** 366–368
Fourier heat conduction equation, **18,** 665
Fourier methods in X-ray diffraction, **6A,** 232–233
Fourier number, **18,** 668–669, 675
Fourier representation(s), **13A,** 55; **28,** 313
Fourier series, **1,** 199
for density of scattering power in crystal, **6A,** 197–199
Fourier spectrometer, scanning Michelson interferometer and, **29C,** 330
Fourier synthesis, **23B,** 208
Fourier transform, **8,** 41, 47, 49, 51; **12A,** 498–516, 529; **22,** 165; **25,** 148; **26,** 62, 69; *see also* Data handling, **21,** 389–391
holograms, **26,** 178
of quantum beat pattern, **17,** 594–595
Fourier transform infrared *see* Infrared spectrometer
Fourier transform ion cyclotron resonance mass spectrometer, *see* Mass spectrometer
Fourier transform microwave spectroscopy
instrumentation, **29B,** 287–292
applications, **29B,** 292–295
Fourier transform of atomic force microscopy (AFM) images, **27,** 369, 371, 376
Fourier transform spectra, for species concentrations, **18,** 629
Fourier transform spectrometer, **25,** 190, 222
Fourier transform spectroscopy, **13B,** 53, 60; *see also* Multiplex spectrometry
amplitude, **13B,** 73
noise, **13B,** 75
Fourier transformation of pulse NMR signal, **16A,** 269
Fourier transforms, **20,** 409–413
Fourier-transform mass spectroscopy (FTMS), **30,** 438
Fourier-transform spectrometer, **3A,** 177, 186
Fourth sound, **19,** 417–453
Doppler shift of, **19,** 419, 120
properties of, **19,** 421
transducers for, **19,** 434–435
velocity of, **19,** 419, 420
Fourth-generation sources, **31,** 24–25
Fowler function, **6B,** 380–381, 383
Fowler theory, **11,** 72–73
Fowler's coordinate method, **5A,** 397–399
Fowler-DuBridge equation, **4A,** 54–56
Fowler-Milne method, modified, **7B,** 333–334
Fowler-Nordheim equation, **22,** 350, 352, 353, 369
Fowler-Nordheim formula, **22,** 8
Fowler-Nordheim relation, current density calculation, **29A,** 131
Fowler-Nordheim slope, **22,** 359
Fowler-Nordheim theory, **22,** 351, 352
Fowler-Nordheim tunneling equation, **11,** 162
Fowler-Richardson equation, **4A,** 2
Fox-Smith adaptations, of Michelson interferometer, **15A,** 322
Fractal aggregates, **23C,** 437
Fractional height, **25,** 39
Fractionation of polymers, **16A,** 30–41
by column elution, **16A,** 37, 39
by extraction, **16A,** 35–36

by precipitation, **16A,** 32
by solubility, **16A,** 32, 35
Fractionation, **16B,** 353
Fracture, **30,** 92–98, 102–104, 592–594, 604
as related to crazing, **16C,** 270
modes, **24A,** 132–133
molecular mechanisms, **16C,** 217–225
precursory bulge, **24A,** 151
of solids, **6A,** 309–321
toughness, **24A,** 134, 143
Framing camera
application, **18,** 696, 733–734
light source, **18,** 705, 715
Franck-Condon
principle, quantum-mechanical, **3,** 247
transition, **3,** 526
Franck-Condon excitation, **22,** 468; **30,** 58
Franck-Condon factors, **3B,** 856; **15B,** 127, 134; **18,** 491–492
Franck-Condon ionization, **22,** 468
Frank and Tamm equation, **5A,** 8
Frank-Condon principle, **7A,** 468–469
Frank-Read source of dislocations, **6A,** 331–332
Franz-Wiedmann law, **4A,** 439
Fraunhofer
approximation, **12A,** 438
cemented doublet, **29C,** 351
diffraction, **1,** 346ff; **26,** 62; **8,** 156
region, **1,** 212, 224
hologram, **26,** 175
Free carrier
absorption, **10,** 181, 182
photoconductivity, **10,** 179–188
reflectivity minimum, **10,** 255–257
Free e-m fields, **13A,** 60
Free energy, **1,** 323, 326
Free energy of mixing, **16C,** 276
segmental, **16C,** 278
Free field, **1,** 231
Free flight apparatus, **18,** 779–781
Free induction decay (FID), **21,** 365
in nuclear magnetic resonance, **2B,** 146
Free liquid surface, ions and, **11,** 481–483
Free oscillations, **24B,** 11, 14
Free parameters, **8,** 191
Free piston technique, **6A,** 414–415
Free radical production; **3,** 543–544
by chemical reaction in discharge/flow system, **29B,** 90–92
by laser ablation, **29B,** 91
in low pressure flames, **3,** 569–571
by photolysis, **29B,** 89–90
in discharges
DC
conventional, **29B,** 87–88
hollow cathode, **29B,** 88–89
magnetic field enhancement, **29B,** 89
microwave, **29B,** 86–87
in supersonic beams
laser photolysis, **29B,** 39
corona discharge, **29B,** 40
flash pyrolysis, **29B,** 41
van der Waals clusters, **29B,** 92–93, 296
Free radicals, **3B,** 606; **23C,** 103
detection, **4A,** 414
in photochemical reactions, **3,** 576, 577–579
in thermal reactions, **3,** 566–577
ionization potentials, **3,** 543–546, 548
reactors, Eltenton's, **3,** 566ff
trapped, **3,** 230–234
Free rotations, **3B,** 984
Free spectral range (FSP), **12A,** 471, 475, 532; **25,** 198; **29C,** 107, 108
of interferometer, **9A,** 98
Free stream Mach number, **1,** 153
Free stream temperature, **1,** 153ff
Free volume, **16A,** 396
theory of gas diffusion, **16C,** 325
Free wave rf techniques, **13B,** 154
Free working distance, **25,** 165
Free-air correction, **24B,** 151
Free-carrier heating, **30,** 119, 588
Free-electron chambers, **5A,** 94
Free-electron lasers, **22,** 184
Free-electron wave, **22,** 164
Free-molecule flow, **18,** 763
Free-precession technique, in NMR measurements, **11,** 516
Free-space propagation methods, **7B,** 96–99
Free-space wavelength, **8,** 131, 150
Freedom, degrees, **8,** 3
Freeman source, *see* Arc plasma discharge source
Freeze-drying, or single crystal suspensions, **16B,** 259
Freezing-in equilibrium, **6A,** 44
Freezing-in of impurities, **6A,** 40
Frenkel defect, **6A,** 357

Frequency, **1,** 69–73, 221; **13A,** 2, *see also* Wavenumber
absorption, pure rotational, **5B,** 21
attenuation dependence on, **19,** 242–247, 267, 272–275, 280–284, 286–290
audible, **1,** 199
combination, **3,** 122
constant-, sources, **2,** 537–540
control, automatic, **2,** 507–508
crossover, **2B,** 146
cut-off, video amplifier, **2,** 320
cyclotron resonance, **1,** 44
of delta rays, **5A,** 301
difference between two nonequivalent nuclei, **3,** 404
dividers
electronic counter, **2,** 539–540
regenerative, **2,** 538–539
group, **3,** 58
gyromagnetic resonance, **1,** 44
for infinitesimal amplitudes, **3,** 43
infrasonic, **1,** 199
instability, klystron, **3,** 491–492
limitations, oscillator, **2,** 267–269
natural, **2,** 19–20
of operation, electron spin resonance, **3,** 473–474
of operation, proton linear accelerator, **5B,** 638
operational, for transducers, **19,** 257, 266–267
of oscillation, **1,** 199
overtone, **3,** 122
proton resonance, **1,** 523
quadrupole resonance, pressure effect, **3,** 522
resonance, **1,** 498
chemical shift of, **5B,** 121
resonant, **14,** 90
response, **1,** 225, 230, 231
revolution, synchrocyclotron, **5B,** 646
ringing, **3,** 165
in scattering formulas for attenuation, **19,** 242–247
separation of Stark component from main line, **3,** 668–669
symmetric top molecule, **3,** 13
transformations, **2,** 36
transition, diatomic molecule, **3,** 12
ultrasonic, **1,** 199
Frequency calibration, **8,** 147, 149, 151
Frequency chain, **29C,** 304
Frequency conversion, **2,** 729–730
in microwave circuits, **2B,** 284–285
nonlinear, *see* Nonlinear optical frequency conversion
Frequency counting as LDV signal processing technique, **18,** 174–175
Frequency deviation, **8,** 140
Frequency difference, **8,** 148
Frequency discrimination, **8,** 139
Frequency distribution, **8,** 3, 4
Frequency distributions, **5B,** 761–771
of elastic waves in solids, **5A,** 510
function, *see* Density of states
Frequency division, **2,** 538–540; **2B,** 18
Frequency domain response, **19,** 31–34, 56–65
Frequency domain signal processing in LDV, **18,** 161
Frequency doubling
intracavity, **29C,** 237–239
synchronously pumped, **29C,** 243
Frequency measurement(s), **1,** 469–474; **2,** 535–558; **2B,** 15–28; **3,** 471
in de Haas-Van Alphen effect, **11,** 51–52
harmonic generation in, **2B,** 20–21
heterodyne method in, **2B,** 27–28
measurement techniques in, **2B,** 22–29
microwave, **2,** 740–741; **3,** 30–31
multivibrators in, **2B,** 18
techniques, **2,** 540–543
time interval gated electronic counter in, **2B,** 24
ultra-high-frequency, **2B,** 26–28
wavemeters in, **2B,** 15–17
Frequency meters, in equipment testing, **2B,** 184–186
Frequency mixing, **13B,** 319, 321
with two inputs, **15B,** 174–181
Frequency modes, neutron scattering, *see* Band modes; Localized modes; Normal modes; Resonant modes
Frequency modulation, **2,** 265; **3,** 22; **2A,** 248–249
devices for, **2A,** 338
in seismic data transmission, **24B,** 35–41
well logging, **24B,** 198
Frequency multiplier(s), **2,** 263; **2B,** 370–372
klystron, **2,** 723–724
magnetic, **2,** 793–794
Frequency noise, diode lasers, **29C,** 97–99

Frequency resolved optical gating (FROG), **29C,** 215, 216–220
Frequency response, **8,** 153; *see also* Response time
calibrator for pressure gage, **18,** 555
defined, **2A,** 4–5
density measurement Rayleigh scattering, **18,** 418
diaphragm pressure gage, **18,** 562
in feedback control system, **2B,** 389
flow tracing particles, **18,** 32–34, 221, 225
function, **18,** 527
hot-wire and hot-film probe, **18,** 273, 278, 282, 293, 297
pressure bar gauge, **18,** 599
Raman scattering diagnostics, **18,** 420, 421
Rayleigh scattering diagnostics, **18,** 418
stub pressure gauge, **18,** 587
system, **2,** 467–469
time in, **2A,** 4–6
transistor current gain, **2,** 149–150
vane anemometer, **18,** 257–259
Frequency shift, **22,** 62
correlation with mass change, **5B,** 29–35
dipole-dipole interaction, **22,** 62, 64
isotopic, **5B,** 21–22
keying, **2B,** 232
local bonding variation, **22,** 62, 65
vibrational coupling, **22,** 62
Frequency shifting in LDV, **18,** 163–164, 185, 187–189, 193–194
Frequency spectrum, **8,** 199
Frequency stability
characterization of, **12C,** 198
frequency-domain measures of, **12C,** 200–202
oscillator, **2,** 270–272
time-domain measures of, **12C,** 200–202
Frequency stabilization, **1,** 488; **3,** 469–471; **29C,** 103–104, 299–301
of diode lasers, **29C,** 97–99
of dye lasers, **29C,** 299
of tunable lasers, **29C,** 106–107
design examples, **29C,** 126–134
loop filter, **29C,** 124–126
optical frequency references, **29C,** 107–121
transducers, **29C,** 121–124
Frequency standard, atomic
cesium beam, **4B,** 210–211
hydrogen beam maser, **4B,** 211–212
Rb optical pumping, **4B,** 186, 212–214
Frequency standards, **1,** 73, 469–470; **12C,** 198–227
bibliography of, **12C,** 227
Frequency switching, **12C,** 31–33, 49
Frequency synthesizers, **2B,** 21
Frequency uncertainty, **29C,** 301–303
Frequency variation circuit, **6B,** 17–18
Frequency-division mutliplexed telemetering, **2B,** 220, 235–242
FRESCA, *see* Field-emission-referenced electron spectroscopy
Fresnel
biprism, **26,** 3
diffraction, **1,** 346ff; **26,** 53
effect, **11,** 727
equations, **26,** 113
formula, **24A,** 293
fringes, in electron microscope image, **16B,** 218–219
hologram, **26,** 175
integrals, **26,** 56
rhomb, **1,** 381, 382; **12A,** 373, 398; **26,** 133
zone plate(s), **13B,** 84; **26,** 59
Fresnel equation, **31,** 265
Fresnel formulas, **31,** 273–274
Fresnel lens, **8,** 40; **31,** 289
Fresnel reflection coefficients, **31,** 280
Fresnel rhomb, **31,** 247–248
Fresnel zone plate lens, **8,** 153, **8,** 154
Friction, **1,** 97–100
factor, **1,** 145
internal, **1,** 110
Frictional sliding
cycling tests, **24A,** 171, 173
fault gouge, **24A,** 173
measurement, **24A,** 170–171
stick-slip, **24A,** 171
Friedel's law, **3,** 285, 286; **6A,** 202
Friedel-Craft acylation, **16B,** 255
Fringe amplitude
effects of noise on, **12C,** 238–247
Fringe anemometer, **18,** 109, *see also* Optical heterodyne detection
Fringe distortion methods, **18,** 369
Fringe interpretation, of LDV operation, **18,** 117
Fringe phase
VLBI measurement of, **12C,** 265–266
Fringe rate
VLBI measurement of, **12C,** 266

Fringe-frequency spectrum
VLBI measurement of, **12C,** 233
Fringe-visibility spectrum, **12C,** 234
Fringed-micelle model, **16A,** 7–8; **16B,** 342, 343
Fringes, **4B,** 76–77, 80; *see also* Interferometers, fringes in
Brewster, **1,** 350, 351, 359
of equal chromatic order, **26,** 28
of equal inclination, **26,** 5
of equal thickness, **26,** 4
Haidinger, **1,** 105, 350, 356
Fringes, *see* Young's interference fringes
observation, **8,** 43–44
pattern, **8,** 39
Fringing fields, **8,** 185
FROG, *see* Frequency resolved optical gating
Froome harmonic generator, *see* Arc harmonic generator
Froude number
definition, **18,** 830
role in ship fluid dynamics, **18,** 848
FRS, *see* Free spectral range
Fuel elements, reactor, **5B,** 601
Fuel irradiation capsule, **8,** 245–248
design consideration, **8,** 245–247
exercises, **8,** 248
results, **8,** 248
Full energy peak fractions, **8,** 216
Full-wave bridge circuit, **2A,** 199–201
Full-wave center-tapped circuit, **2A,** 197–201
Full-wave doubler, **2A,** 207
Full-wave rectifiers, **2A,** 197–207
with capacitive load, **2A,** 197–201
with inductive load, **2A,** 202–207
in nonlinear circuits, **2A,** 395
single-phase, **2A,** 230–233
Fume hoods, **21,** 32, 33
Function generator, in analog computer, **2B,** 162
Functional model, **28,** 278
Functions
chi square, **1,** 5, 22
describing, *see* Describing function
least squares approximation, **1,** 15
polynomial, **1,** 15
Student's t , **1,** 8, 23
Fundamental frequency, **28,** 339
Fundamental transverse mode operation, in heterojunction lasers, **15A,** 238–239
Fuoss-Kirkwood function, **16C,** 391
Furnace, **14,** 277, 278
Furnace, gas pressure apparatus, **24A,** 73–74
Furnaces for crystal growth, **6A,** 90–92
Fused fiber couplers, **29C,** 376–377
Fused junctions, **6A,** 65–66
Fused silica, **29C,** 344, 345–346
Fused silica (a-SiO_2), **30,** 93–96, 606–607
Fusion, **1,** 262; **6A,** 23
heat of, **1,** 283
Future in optics, **25,** 5
FXLR1, **12C,** 292, 294, 317

G

G (thermal conductance), **12A,**–424
g Factor, atomic, **4B,** 110, 114, 191–199, 201–208
relativistic corrections, **4B,** 197–198
tables of, **4B,** 193–194, 199, 204
g factor, electron, **20,** 83, 84
g factor, of electrons in various environments, **16A,** 446–448
G value, **22,** 507
g, **24B,** 128–129
g-factor (or g-value), **10,** 386, 388, 389, 391, 392
g-factor, **3,** 487; **5B,** 224; **6B,** 228
g-value, **1,** 45ff; *see also* g-factor
GaAlAs diode, **29C,** 150
GaAs surface reconstruction
scanning tunneling microscopy of, **27,** 254
GaAs surfaces
scanning tunneling microscopy of, **27,** 251, 255
GaAs(100)
scanning tunneling microscopy of, **27,** 110, 255
Cs adsorbed, **27,** 123
Fe adsorbed, **27,** 127, 130
General I-V characteristics, **27,** 118
O adsorbed, **27,** 110
Sb adsorbed, **27,** 136
surface buckling, **27,** 254, 258
GaAs(111)
scanning tunneling microscopy of, **27,** 261
GaAs, *see* Gallium arsenide
GaAs/AlGaAs superlattices
cross-sectional imaging of, **27,** 263
spectroscopy of, **27,** 264
Gabor plate, **26,** 61
Gage factor, resistance sensor, **18,** 538

Gages
  Alphatron, **1,** 168–169, 190
  Bayard-Alpert, **1,** 190
  block, **1,** 54
  heat conduction, **1,** 166ff
  ionization, **1,** 182, 189–191
  Knudsen, **1,** 191
  McLeod, **1,** 164–166, 191
  molecular, **1,** 191
  Philips, **1,** 190
  Pirani, **1,** 167–168, 191
  response time, **1,** 169
  strain, **1,** 56, 80, 102–103; **2,** 126–127
  thermocouple, **1,** 167
  unbounded strain, **1,** 74
  vacuum, response time, **1,** 191–192
  wire strain, **1,** 67
Gages, vacuum pressure
  effects on gas composition, **4B,** 309–311
  partial pressure, **4B,** 291, 305–309
    mass spectrometers, **4B,** 305–309
  pumping effects, **4B,** 300–301
  sensitivity, **4B,** 300
  total pressure, **4B,** 290–305
    Bayard-Alpert, **4B,** 292–294, 301, 304
    capacity, type, **4B,** 304
    inverted magnetron, **4B,** 299
    ionization, **4B,** 291–304
    Kreisman, **4B,** 299–300, 303
    Lafferty, **4B,** 296–298
    magnetron,
      cold cathode, **4B,** 299–301
      hot cathode, **4B,** 296
    modulated ionization, **4B,** 295–296
    orbitron ionization, **4B,** 296–297
    Penning, **4B,** 298–299
    Schuemann suppressor, **4B,** 294–295
    thermal conductivity, **4B,** 304
    triggered discharge, **4B,** 300
  X-ray limit, **4B,** 294, 298, 303, 306
Gain, **1,** 230, 231
  adjustment, **2,** 481–483
  antenna, **12B,** 3
  control, automatic, **2,** 506–507
  determination, **2,** 476–488
  formulas, in optical gain measurements, **15B,** 24–25
  intensity
    due to magnet system, **5B,** 710
    due to rectangular quadrupole, **5B,** 712–713
  modulation
    spectral-line measurements with, **12C,** 50
    use in receivers, **12B,** 216
  multiplier phototube, **2,** 749
  relative wideband, **2,** 698
Gain coefficients, saturated laser, **31,** 104
Gain-phase diagram, **2,** 468
Gain-phase plot, **2B,** 395
GaInPAs laser, **29C,** 80, 149
Gal, **24B,** 128
Galactic hydrogen spectra
  data presentation for, **12C,** 65
Galilean telescope, **25,** 69, 136
Galling, **1,** 98
Gallium, **22,** 403
  boiling point, **14,** 350
  melting point, **14,** 347, 350
  physical properties, **14,** 375, 377
  source, **22,** 403
  vapor pressure, **14,** 347, 350
Gallium arsenide (GaAs), **22,** 170, 171; **30,** 70–71, 293
  cesiated, **11,** 110–111
  chemical cleaning of, **11,** 182
  contaminants of, **11,** 106–107
  diodes, **2A,** 365; **10,** 20, 165, 166
  GaAs(001), **22,** 295
  GaAs(110), **22,** 291
  sputter etched, **22,** 292
Gallium arsenide (GaAs), negative-electron-affinity semiconductor photocathode
  applications, **29A,** 32
  comparison of polarized electron sources, **29A,** 30–32
  development, **29A,** 1–2
  diffusion length, **29A,** 4, 6
  electron escape, **29A,** 4–5
  electron gun
    brightness values, **29A,** 25–26
    configuration, **29A,** 27
    electron-optical considerations, **29A,** 24–25
    low-energy electron diffraction gun, **29A,** 27–28
    spin rotation, **29A,** 28, 30
    stability, **29A,** 26
  incident radiation, **29A,** 22–23
  optical spin orientation, **29A,** 3
  photocathode
    aniodization, **29A,** 12–13, 34
    heat cleaning, **29A,** 13–14

lifetime, **29A,** 21–22
limitations of response, **29A,** 20–21
material, **29A,** 11–12
photocurrent monitoring, **29A,** 15
surface activation, **29A,** 15–18
surface cleaning, **29A,** 12–15, 33–34
polarization
circular polarization of incident radiation, **29A,** 22
film growth, **29A,** 7, 9, 12
formula, **29A,** 18
yield curve, **29A,** 19
steps in photoemission, **29A,** 3
thickness and electron polarization, **29A,** 6–7, 9–10
Gallium indium oxide ($GaInO_3$), **30,** 552–553
Gallium phosphide (GaP), **30,** 69, 77–78, 159
Gallium selenide, **22,** 150, 165, 167, 180
Galvanomagnetic effects, **6B,** 145–170
Galvanometer(s), **1,** 449
amplitude and phase, **24B,** 34
attenuation factor, **24B,** 33
ballistic, **1,** 451
Clarke slug as, **11,** 298
coupled to pendulum, **24B,** 32–35
coupling factor, **24B,** 34
equations of motion, **24B,** 32–34
mirror scanners, **25,** 113
moving coil, **1,** 519
as null detectors, **2B,** 60
string, **1,** 459
use for signal magnification, **24B,** 32
vibration, **1,** 459
Games, nonalgebraic solutions, **28,** 443–444
Gamma activity, **8,** 231
Gamma distribution, generalized, **28,** 239
Gamma lines, **8,** 198, 227
Gamma quanta, **8,** 207, 222
Gamma radiation, on glass, **14,** 322
outgassing, **14,** 333–336
Gamma ray detectors, **8,** 269
Gamma rays, **5A,** 582–682; **8,** 207, 216, 272, 274, *see also* Compton-scattered secondary gamma rays; *see also* Angular correlation (of annihilation radiation); Doppler broadening (of annihilation radiation); Lifetime (of positron)
absorption, **21,** 8
cross section, total, **5A,** 87–89
longslit, **21,** 109
attenuation, **5A,** 76, 671ff
calibrated sources, **21,** 65
Compton effect, **21,** 7, 8, 61, 62, 66
Coulomb excited, **5A,** 613–616
counters, **5B,** 78–80
decay of nuclear states, **5B,** 357–359
detection
photographic film, **21,** 44, 60
pulse counters, **21,** 50, 61
emission, **5B,** 442–448; **21,** 2, 6
from neutron-induced reaction, **5B,** 459–460
energy determination, **5A,** 599–616
by absorption, **5A,** 671–676
energy spectrum, **21,** 6, 61
analysis, **21,** 62
flux measurement, **5B,** 527–528
heating, **21,** 20
from lead, **5A,** 330
mass absorption coefficient, **5A,** 88
mean free path in toluene, **5A,** 144
momentum determination, **5A,** 599–616
neutron capture, **5A,** 613–616
pair production, **21,** 6, 7, 8, 61, 66
penetration, **5A,** 76–89
photoelectric effect, **21,** 7, 61, 62, 66
photographic emulsion measurement, **5A,** 676–682
from π^0 meson decay, **5A,** 679–682
radioactive sources, **5B,** 572–580
recoil-energy loss in, **11,** 308
sample preparation, **21,** 61
scintillation spectrometry, **5A,** 616–641
sensitivity of inorganic scintillators, **5A,** 411
sources, **5A,** 605
composite, quantitative analysis, **5A,** 639
monoenergetic, **5A,** 675
size considerations, **5A,** 605–606
specific ionization, **21,** 8
telescopes, **5A,** 668–671
transmission through polarized absorber, **5B,** 267–269
width, total, **5A,** 544
Gamma spectroscopy, **8,** 203–211
exercises, **8,** 211
Gamma-emitting fission isotopes, **8,** 203
Gamma-gamma correlation, **13A,** 138
Gamma-ray photoelectric absorption, **24B,** 473–476, 559–563
Gamma-ray anisotropy, **11,** 529
Gamma-ray astronomy, **12A,** 315

Gamma-ray detectors, **24B,** 551–555, 569–581
Gamma-ray energy, **8,** 205
  determinations, **8,** 218
Gamma-ray formation density, **24B,** 545–559
Gamma-ray interactions, **24B,** 545–548
Gamma-ray radioactivity, **24B,** 450, 491
Gamma-ray regions, **13A,** 16, 115
  detectors, **13A,** 121
  spectrometers, **13A,** 131
Gamma-ray scattering, **24B,** 467–472
Gamma-ray spectra, **8,** 204, 208; **24B,** 491, 582; *see also* Multi-gamma-ray spectrum
Gamma-ray spectrometer, **8,** 226
Gamma-ray spectrometry
  natural source, **24B,** 489–491
  neutron excited, **24B,** 483–489, 563–581
Gamma-ray-induced reaction cross-sections, **5B,** 461–472
Gamma-rays, detected by atmospheric Čerenkov radiation, **12A,** 343–359
Gap
  density, **5A,** 245–263
  length, **5A,** 246ff
Gap isolation (voltage clamp), **20,** 486–504
Gardon heat transfer gage, **18,** 672–675
Gas
  Čerenkov counters, **5A,** 457–460
  constant, **1,** 38, 41
  density, **1,** 119–120
  dynamics, **1,** 127ff
  ideal, **1,** 128, 313
  kinetics, **3,** 589, 597–602
  law, **1,** 37, 116, 119, 128
  loss coefficient of, **1,** 156
  molecular weight of, **14,** 82
  multiplication, **5A,** 110–118
    statistics, **5A,** 117
    voltage dependence, **5A,** 112ff
  noble, **14,** 82
  phase, **14,** 276
  properties, **14,** 418–423
  rarefied, measurements in, **1,** 164–172
  residual, **14,** 82
  sampling system, molecular beam, **3,** 575, 581
  scintillator
    light wavelength, **5A,** 433
    noble, **5A,** 431–435
    purity, **5A,** 432–433
    spectrometer resolution, **5A,** 435
  Van der Waals, **1,** 122
  velocity of sound in, **1,** 277ff
Gas ballast, *see* Rotary oil-sealed pump
Gas chromatography, **4B,** 322
Gas diode laser, **25,** 267
Gas discharge(s), **2,** 357–363; **2A,** 165–166
  laser, **15B,** 82
Gas discharge column, **8,** 117
Gas doser, **22,** 487
Gas dynamics, **30,** 225, 230–231
Gas evolution measurement, **22,** 447, 449
Gas film bearing, **14,** 264
Gas flow
  conductance, **1,** 163, 164
  coolers, for Mössbauer spectroscopy, **11,** 356
  density in, **1,** 154
  in duct, **1,** 144–147
  dynamics, **1,** 126–172
  external, **1,** 144
  free molecule, **1,** 158, 159
  about immersed bodies, **1,** 147–152
  internal, **1,** 144
  measurement, *see* Flow measurement
  measurement of, **1,** 153–157
  in pipes, **1,** 144ff, 163
  Poiseuille's law of, **1,** 144
  pressure in, **1,** 164
  rarified, **1,** 157–172
  regime, slip, **1,** 158, 159–163
  regime transition, **1,** 158, 159
  schlieren technique in, **1,** 171
  similarity of, **1,** 142–143
  subsonic, **1,** 155
  temperature in, **1,** 170
  visualization, **1,** 171
Gas jet, **22,** 339
Gas kinetics, **3B,** 768
Gas laser interferometry, **9B,** 31–39
  couple cavity interferometry, **9B,** 31–39
  off-axis modes, **9B,** 37
  oscillating mirror, **9B,** 36–37
  rotating feedback mirror for, **9B,** 35–36
  without coupling of laser and plasma, **9B,** 38–39
Gas lasers, **10,** 45–64; **25,** 262; **29C,** 147–148, 151
  atomic and ionic, **15A,** 31–166l *see also* Ion lasers; Neutral-atom lasers
Gas laws, ideal, **14,** 6

Gas load, **14,** 304
Gas mantle, **3,** 76; **10,** 15, 16
Gas photoionization, in ultraviolet photoemission measurements, **11,** 83
Gas projectile, **17,** 415–416
Gas proportional counter, **17,** 86
Gas rectifiers, **2A,** 173–174
Gas regulators, **14,** 579
Gas sample analysis, **4B,** 322, 328
Gas target, **17,** 316–317, 373–374, 379
  angle resolution effects, **17,** 310
  average equilibrium chargers, **17,** 125–130
  charge state fraction measurements, **17,** 85, 87
  collisional quenching, **17,** 426–427
  density effect in, **17,** 131–134
  energy-loss measurements, **17,** 51–53
  energy straggling, **17,** 70–71
  isotropic composition of, and broadening effects, **17,** 170–171
  K-shell vacancy production, **17,** 335, 337, 342
  multielectron ions passing through, **17,** 136
  neon, **17,** 410–411
  outer-shell rearrangements in, **17,** 424–425
  scattering, **17,** 306
  spectra, **17,** 422–423
  x-ray production rate, **17,** 228
Gas thermometers, **6B,** 124
Gas transport, *see* Molecular transport
Gas tubes, **2A,** 165–188
  anode dissipation factor, **2A,** 188
  anode voltage drop, **2A,** 188
  arc devices, **2A,** 172–186
  arc drop, **2A,** 188
  average current in, **2A,** 187
  commutation factor, **2A,** 188
  corona devices and, **2A,** 167–169
  deionization time in, **2A,** 188
  glow devices, **2A,** 169–172
  hydrogen thyratrons, **2A,** 176–180
  ignitrons, **2A,** 180–182
  ionization time, **2A,** 188
  jitter in, **2A,** 188
  microwave, **2A,** 186
  peak anode current in, **2A,** 187
  peak forward voltage in, **2A,** 187
  phenomena of, **2A,** 165–167
  spark gaps and, **2A,** 183–184
  surge anode current in, **2A,** 187
  thyratrons, **2A,** 174–176
  tube and spark gap ratings in, **2A,** 186–188
  vacuum gaps and, **2A,** 184–185
  voltage regulator, **2A,** 169–171
Gas-discharge display, **2B,** 7
Gas-dynamic laser, **15B,** 63
Gas-filled counters, **5A,** 110–120
Gas-filled detector, **17,** 356–358
Gas-filled phototubes, **2B,** 307–309
  applications of, **2B,** 308
  characteristics of, **2B,** 307–308
  types and ratings of, **2B,** 308–309
Gas-filled rectifier, **2A,** 219–220
Gas-kinetic reference rate, **15B,** 65
Gas-stabilized arcs, **7B,** 309–312
GaSe, *see* Gallium selenide
Gaseous discharges, ion injection in, **11,** 453–454
Gaseous electronic devices, **2A,** 165
Gases
  electron mean free path in, **5A,** 113
  loss of kinetic energy of electrons in, **7B,** 49–51
  microwave spectroscopy in, **5B,** 93–100
  velocity of sound in, **3,** 681, 682–700
Gases as transmitting media, **31,** 326–329
Gasket(s)
  aluminum, **14,** 363
  creep, **14,** 287
  elastomer, **14,** 407–408, 479
  gold, **14,** 373
  indium, **14,** 375
    wire, **3,** 170
  metal, **14,** 287, 291
  rectangular, **14,** 285, 286, 288, 289
  square, **14,** 285–287, 289
Gaskets, vacuum
  aluminum, **4B,** 272, 283
  copper, **4B,** 283–285
  gold wire, **4B,** 283–284
  rubber, **4B,** 269, 285
Gasometer, variable volume, **1,** 155
Gate system, in integrated circuits, **2A,** 135–136
Gate valve, **14,** 296
Gated double-pulse superposition method
  phase shifts due to bonded transducers, **19,** 93
  in ultrasonic velocity measurements, **19,** 90–94
Gated integrator
  noise minimization, **29A,** 446

schematic, **29A,** 447
Gates, linear, **2A,** 377–382
Gating circuitry, in pulse amplitude measurements, **2B,** 112–114
Gating circuits, **2A,** 410–411
balanced linear, **2A,** 380
transistor, **2A,** 378–379
Gating currents, **20,** 447, 471, 472
Gating in linear circuits, **2,** 341–344
Gauche bonds, in polyethylene
in chain folds, **16A,** 129
chemical shift in NMR spectra, **16A,** 357
in melt, **16A,** 142
Gauche conformation, in polyethylene, infrared bands, **16B,** 251
Gauge transformations, **13A,** 32
Gaunt factor, **7A,** 84, 173; **9A,** 135, 162
Gauss, **1,** 518
equation, **25,** 17
laws, **1,** 431, 434
rule, **1,** 58
Gauss-Newton method, **28,** 255–256
Gaussian approximation, **23A,** 43
Gaussian beam(s), **25,** 107, 260; **29B,** 256–257; **30,** 173
velocity profile, **17,** 578
Gaussian coil, **16A,** 497
Gaussian distribution, **1,** 4; **3,** 712; **28,** 18
central limit theorem, **28,** 27
of sizes in semicrystalline polymers, **16B,** 169
Gaussian fitting
drift-scan use of, **12C,** 7
Gaussian function, to describe X-ray line shape, **16B,** 145
Gaussian laser beams, **30,** 179–180
Gaussian line profile, distortion by Brillouin scattering, **18,** 414–416
Gaussian line shape, **3,** 458
Gaussian noise time series, generation, **28,** 468
Gaussian optics, **25,** 22
Gaussian peak shapes, **8,** 216
Gaussian peaks, **8,** 196
Gaussian random number generator, **28,** 448
Gaussian shape, **8,** 215
Gaussian units, **1,** 431
Gaussian white noise process, **28,** 339
Gaydon-Wolfhard burner, **7B,** 192
Gd, **23C,** 105, 210
GDL, *see* Gas dynamic laser
GDMS, *see* Glow-discharge mass spectroscopy
GDSN, *see* Global Digital Seismic Network
Ge detector, **8,** 222, 223
Ge surfaces and alloys, scanning tunneling microscopy of, **27,** 235
Ge surfaces, scanning tunneling microscopy of, **27,** 225
Ge(-Li) detectors, **8,** 214, 215, 218, 220
Ge(001)
spectroscopy of, **27,** 237
dual polarity tunneling images of, **27,** 235
Ge(001)–2 × 1
scanning tunneling microscopy of, **27,** 235
Ge(111)
high temperature measurements of, **27,** 234
Ge(111) c2 × 8
scanning tunneling microscopy of, **27,** 227
spectroscopy of, **27,** 230
Ge-Li-drifted primary detectors, **8,** 213
Geiger (Geiger-Müller) counter, **16B,** 44, 45, 155
Geiger counter, **1,** 412; **7A,** 458, 465; **7B,** 282
Geiger-Müller, **8,** 188
Geiger-Müller counter(s), **5B,** 442; **21,** 34
Geissler tubes, **13A,** 265
Geks, **20,** 333–336
Gel permeation chromatography (GPC), **16A,** 41–65
band broadening, **16A,** 55
calibration, **16A,** 54–59
columns and packing, **16A,** 50
detectors, **16A,** 49, 63
high speed, **16A,** 52
high temperature, **16A,** 52
of hydrolyzed cellulose, **16B,** 272
instrumentation, **16A,** 48
molecular weight determination by, **16A,** 53, 60
of oxidatively degraded crystalline polymers, **16B,** 263
effect of crosslinking radiation, **16B,** 282–286
resolution, **16A,** 47, 51
theory of separation, **16A,** 46
Gen I image tubes, **12A,** 240, 247
Gen II image tubes, **12A,** 239
General relativity
tests of, **12C,** 270
Generalized oscillator strength, **7A,** 35–38, *see also* Oscillator strength

Generation of ultrasonic, **19,** 30–31, 37–40, *see also* Transducers
Generation-recombination noise, **2B,** 461, 464–465
Generator, electrostatic, **5B,** 584–590
  for elastic scattering measurements, **5B,** 476
Generators
  for bridge use, **6B,** 23–24
  broken-line, **2,** 408–409
  delay, **2,** 369–371
  function, **2,** 641–643
  harmonic, **2,** 540
  multipulse, **2,** 658
  sine-wave, **2,** 652–654
  square-wave, **2,** 654
  sweep, parabolic, **2,** 407–408
Gentisic acid, **30,** 418
Geodesy, VLBI applications in, **12C,** 261
Geodetic latitude, calculation of, **12C,** 257
Geodetic satellites
  cameras, **24B,** 166–168
  orbital characteristics, **24B,** 164–166
  orbit dynamics
    disturbing function, **24B,** 173
    equations of motion, **24B,** 170–173
    forced perturbations, **24B,** 175–177
    resonance, **24B,** 177–178
  radar altimetry, **24B,** 169, 181–185
  range rate, **24B,** 168–170
Geodetic techniques, **24B,** 409
Geodimeter, **1,** 37
Geoelectric section, **24B,** 346–348
Geoid, **24B,** 130, 146–148, 182
Geomagnetic surveys, **1,** 524
Geometric image, **8,** 42, 45
Geometric similarity, **18,** 828, 843, 844
Geometrical optics, **1,** 329–346
  definitions, **1,** 329–335
  instruments, **1,** 338–346
  optical systems, **1,** 335–338
  references for, **12B,** 61
Geometrical optics, **31,** 150–158
Geometry factor, **8,** 83
Geophysical apparatus, **18,** 801–819
Geophysical measurements, VLBU techniques for, **12C,** 274–276
Geostationary orbiting earth satellite, **24B,** 50
Geostatistical model, **28,** 378–381
Geostrophic flow, **18,** 803–806
Gerdien aspirator, **1,** 504–505
Germanate anomaly, **23B,** 278
Germanium, **22,** 155
  Auger recombination in, **15B,**208
  bolometer, **10,** 171
  diodes, **10,** 163–165
  Ge(111), **22,** 154, 155
  infrared detectors, **3,** 83
  n-type, resistivity, **2,** 115ff
  photoconductors, **10,** 178
  rectifiers, **2A,** 220
  resistors, in thermometry, **11,** 491–492
  spectrometer, **24B,** 580–581
  windows, **10,** 95, 96, 97
Germanium (lithium) detector, **17,** 67, 313, 358–361
Germicidal lamp, **25,** 253
GeSi(111) 5 × 5, scanning tunneling microscopy of, **27,** 239
Getter, **14,** 326–327
Getter pumping, **14,** 193
Getter-sputtering system, **11,** 641–642
Gettering, **1,** 180
  chemical, **14,** 216, 217
  tantalum, **14,** 372
  titanium, **14,** 366
  zirconium, **14,** 364
  zirconium hydride, **14,** 365
Getters
  bulk, **4B,** 326
  flash, **4B,** 324–326
Ghost lines in gratings, **26,** 75
Gibbs
  free energy, **1,** 323
  function, **1,** 286
  phenomenon, **28,** 292
  sampler, **28,** 427, 429–430
Girard grille spectrometer, **13B,** 85
Gladstone-Dale constant
  electron, **18,** 350
  ionized gas, **18,** 350
Gladstone-Dale relation, **18,** 348, 351
Glan laser polarizers, **29C,** 354
Glan-Foucault prism, **12A,** 369, 394; **26,** 141
Glan-Thompson polarizers, **29C,** 354
Glan-Thompson prism(s), **11,** 706, 720; **12A,** 369; **26,** 139; **26,** 141
Gland, **14,** 284
Glass, **4B,** 269–270; *see also specific composition*
  aluminosilicate, **4B,** 275
  alumino-silicate, **14,** 315

annealing point
common glasses, **14,** 315
definition, **14,** 314, 388
bleaching, **14,** 334
barium-alkali-silicate, composition, **14,** 315
borosilicate, composition, **14,** 317, 385
chemical resistance, **14,** 396
common, **14,** 315, 317
composition, **14,** 314–317
cracking, neutron irradiation, **14,** 333, 334, 336
gas permeability, **14,** 394
hard, **14,** 384
helium permeation, **14,** 337–343
hydrogen evolution, for common types, **14,** 331, 333
lead, composition, **14,** 385
lead-loaded, **5A,** 190
lime, composition, **14,** 385
MaCor, **14,** 481
mechanical properties, **14,** 386
melting point, **14,** 388
optical properties, common types, **14,** 394–396
outgassing, **4B,** 270; **14,** 396
nuclear radiation, **14,** 333–337
permeation constants, common glasses, **14,** 337–343
permeation of gases, **4B,** 270–271, 275
potash-soda-lead, composition, **14,** 315
properties, **14,** 385
silica, composition, **14,** 315
soda-lime
composition, **14,** 315
thermal outgassing, **14,** 323
softening point, **14,** 314, 388
common glasses, **14,**315
soft glasses, **14,** 384
solarization, **14,** 395
strain point, **14,** 314
common glasses, **14,** 315, 387
surface reactions, **4B,** 270
thermal conductivity, **14,** 390
thermal expansion, common glasses, **14,** 314–315, 386, 388, 389
thermal shock resistance, **14,** 389
in ultrahigh vacuum systems, **14,** 481
viscosity, common glasses, **14,** 314–315, 387, 388
wetting, metals, **14,** 316
working point, **14,** 314
common glasses, **14,** 315
Glass diffraction, **23A,** 226
Glass fiber, **14,** 307
glass-to-glass, **14,** 314, 316
glass-to-metal, **14,** 316–319
graded, **14,** 389
housekeeper seals, **14,** 317
tip-off, outgassing, **14,** 325–327
Glass transition, **16A,** 6–7
and anomalous transport of vapors, **16C,** 334–337
of block copolymers, **16C,** 311–314
by compressibility, **16C,** 106
effect on positron lifetime, **16A,** 394
paramagnetic probe studies, **16A,** 463, 471
of polymeric alloys, **16C,** 306–311
Rayleigh-Brillouin scattering studies, **16A,** 200–204
spin-lattice relaxation below, **16A,** 255
temperatures, **19,** 146
and ultrasonic measurements, **16C,** 88
various polymers, ESR study, **16A,** 466
viscosity at, **16C,** 6
Glass transmission, **10,** 94
Glass vacuum systems, **14,** 313–343
advantages, disadvantages, **14,** 313–316
components, **14,** 317, 320–321
diffusion pumps, **14,** 320
fabrication, **14,** 314–321
gauges, **14,** 320
helium permeation, **14,** 337–343
outgassing, **14,** 322–327
ultimate pressure, borosilicate glass, **14,** 321
water vapor, **14,** 323–327
Glass window, **14,** 311
Glass working, **14,** 316
Glass-ceramic, machinable, **14,** 307
Glass-transition temperature
determination by calorimetry (DSC), **16B,** 327–331
definition, **16B,** 328
effect of heating rate and sample size, **16B,** 329
effect on crystallization rate, **16B,** 361–363
effect on molecular weight, **16B,** 330–331
of various polymers, **16B,** 327
Glasses, **23B,** 243; **29C,** 343–346, 351, 375; *see also* Preparation of laboratory glass samples
for solid state lasers, **15A,** 191–193

Glassy compliance, **16C,** 4
Glide plane, **6A,** 323
  in crystal symmetry, **16B,** 9, 10
Global Digital Seismic Network, **24B,** 7
Global ozone data fields, field statistics, **28,** 471–475
Global Positioning System, **24B,** 93, 119, 168
Global Seismographic Newtork, **24B,** 71
Global significance, **28,** 458
Globar, **1,** 399; **3,** 76; **8,** 27; **10,** 15, 16; **13B,** 4
  source, **22,** 77
Globular protein solutions, **23B,** 528
Gloria, **24B,** 103
Glove boxes, **21,** 3
Glove-box-type operations, **8,** 177
Glove-box-type spectrometer, **8,** 178
Glow curves, **6B,** 313–315
Glow devices, **2A,** 169–172
  voltage reference tubes, **2A,** 171–172
  voltage regulator tubes as, **2A,** 169–171
Glow discharge polymerization, **16C,** 456–457
  of ethylene and benzene, **16A,** 167–168
Glow discharge scrubbing, **14,** 447, 448
Glow discharge(s), **1,** 374; **2,** 83–87; **14,** 310, 558–563
Glow discharges, **31,** 37–41
Glow indicator and counting tubes, **2A,** 172
Glow-discharge etching, **16B,** 242
Glow-discharge mass spectroscopy, **22,** 329
Glow-discharge sputtering, **11,** 638–642
Glue, **14,** 414–417
Glutamic acid, **19,** 222–225
Glycinate ion, **19,** 202
GOES, *see* Geostationary orbiting earth satellite
Golay cell, **1,** 393, 401; **3,** 85; **10,** 167–168; **13B,** 8, 51; **26,** 297
Golay thermal detector, **4A,** 445–448
Gold (Au), **30,** 83–85
Gold, **22,** 176, 403
  black, **7A,** 23
  chemical properties
    boiling point, **14,** 350
    melting point, **14,** 347, 350
    vapor pressure, **14,** 347, 350
  electrical conductivity, **14,** 363
  electron-electron scattering in, **11,** 78
  ion laser, **15A,** 164
  as mirror coating, **29C,** 347
  nuclei decoration, of electron microscope specimens, **16B,** 217, 234
  physical properties, **14,** 373–374, 376
  point, **1,** 253, 254
  wire seal, **14,** 290, 291
Gold, reflectance spectra, **31,** 209–210
Goldstack interferometer, **12C,** 267, 269
Goniometer, **22,** 288
  X-ray, **16B,** 61, 63
Goniometer. *See* reflectometer
Gordeyev integral, **9B,** 164
Gorski relaxation method, **23B,** 152
Gorter-Casimir model, **6B,** 76
Goulay method, **25,** 206
Gouy method, **1,** 538ff, 544; **6B,** 178–179
GPC, *see* Gel permeation chromatography
GPS, *see* Global Positioning System
Graded period multilayer filters, **31,** 325–338
Graded seals, glass, **14,** 389
Graded-index lenses, **29C,** 351–352
Gradiometers, **11,** 283–286
Graft copolymer, **16A,** 3
Grain
  density, **5A,** 300
    in high energy electron pairs, **5A,** 678–679
  energy, **5A,** 396
  scattering, *see* Scattering
Grain boundaries geometry of, **6A,** 336–339
Grain boundary
  definition, **6A,** 336
  pure tilt, **6A,** 336–338
  pure twist, **6A,** 336–339
Grain boundary triple points, **28,** 162, 166–167
  frequencies, confidence rectangle, **28,** 174–175
Grain size distribution
  attenuation dependence upon, **19,** 241, 247–248, 267, 271–274, 281–282
  calculation of, **19,** 248–250
  influence on effective grain volume, **19,** 247–248
  moments of, **19,** 247, 271
  in Nickel and Fe–30% Ni alloys, **19,** 271
  photomicrograph images used as data, **19,** 248–251, 269–271
  in polycrystalline materials, **19,** 247–251, 267–271
  transformation to find, **19,** 248–250

Grains
  anisotropy of, **19,** 241–244, 267, 270, 277–278, 281–290
  attenuation due to, **19,** 237–257, 267–291
  diameters in scattering regions, **19,** 242–247, 272–274, 281
  elastic anisotropy, *see* Grains, anisotropy of
  as inhomogeneities, **19,** 241–244
  in metal, **19,** 237–238, 240, 267, 267–290
  multiple phases within, **19,** 253–257, 277–289
  orientation, **19,** 238, 240–241, 275–277, 292–294
  per cluster, **5A,** 259ff
  scattering cross sections, **19,** 245
  shapes, **19,** 238, 269–271, 282–283
  single-crystal, **19,** 242, 268–277
  of single phase, *see* single crystal
  size distribution, **19,** 241, 247–251, 267, 271, 274, 282
  in steel, anisotropy of, **19,** 267, 284–290
  volume, effective, calculation of, **19,** 247–251, 271
  volume, effective, for scattering, **19,** 242–244, 247–248, 251, 268, 271
Graphecon, **2,** 783
  storage tube, **2B,** 356
Graphical methods, **28,** 509–510
Graphical ray tracing, **25,** 18
Graphics
  box plots, **28,** 510–511
  brush, **28,** 515–516
  conditioning plot, **28,** 510–512
  dynamic, **28,** 512–516
  identification, **28,** 513
  linking, **28,** 513, 516
  quantile-quantile plot, **28,** 510–511
  rotation, **28,** 515–516
  static, **28,** 510, 514
Graphite, **23B,** 4
  exfoliated, **23B,** 4
  intercalation compounds, **23A,** 431; **23B,** 70
  latent heat of sublimation, **3,** 560–562
  lattice dynamics, **23A,** 429
  STM of
    theory of, **27,** 14
  tube furnace, **7A,** 125
Grashof number, **1,** 301
  definition, **18,** 830
  hot-wire and hot-film convection, **18,** 269
Graticules, **25,** 131
Grating spectrometer(s), **12A,** 475–482; **13B,** 20, 57; **25,** 210
  for light-scattering optical systems, **9A,** 100–101, 102
Grating(s), **1,** 367–370; **3,** 137–138; **3A,** 185, 203; **4B,** 66, 68–70; **13A,** 227, 287; **13B,** 20
  angular distribution of light, **26,** 70
  aspherical, **13A,** 229
  blazed, **13B,** 16, 58; **26,** 158
  chromatic dispersion, **26,** 72
  concave, **13A,** 226
  crystal, **13A,** 156, 160
  diffraction, **3,** 102–103
  disperse radiometer, **26,** 261
  echellette, **13B,** 16, 58
  efficiency, **12A,** 472
  equation, **4B,** 66; **10,** 130; **12A,** 469; **25,** 221
  fabrication, **13A,** 297
  filter, **10,** 105–107, 139
  Gunn effect, **13B,** 108
  holographic, **13A,** 229; **13B,** 22
  instruments, **3,** 156–160
  lamellar, **13B,** 80
  lobes, use in array antennas, **12B,** 102–103
  monochromator(s), **3A,** 174, 180; **7A,** 461–462; **10,** 130–140; **29C,** 34–35
  mounts, **3,** 158–159; **13A,** 291, 293
  phase, **13A,** 231
  piecewise interferometric generation of, **15B,** 51–54
  polarizer(s), **10,** 105, 130; **26,** 157
  polyethylene, **10,** 103, 104, 139
  replica, **13B,** 16, 21
  resolving power, **3,** 157; **26,** 72
  spectrographs, **3A,** 207–213
Grating, *see* Echelle grating; Incoherent light illuminated grating
Gravimeter, **11,** 284
  absolute, **24B,** 133–135
  free-fall, **24B,** 134–135
  LaCoste-Romberg, **24B,** 137–139
  moving platform, **24B,** 138–142
  pendulum, **24B,** 133, 140
  relative, **24B,** 135–138
  superconducting, **24B,** 138
Gravimetric techniques
  for permeability, **16C,** 366
  for sorption, **16C,** 350–352

Gravitational and inertial mass, equivalence, **23A,** 260
Gravitational blue shift, **8,** 163, 164
Gravitational constant, **1,** 15
Gravitational mass, **1,** 60, 61–62
Gravitational red shift, **8,** 164
measurement, **8,** 161–162
Gravitational units, **1,** 30
Gravitational-radiation bursts, X-rays from, **12A,** 337
Gravity
acceleration of, **1,** 35–36, 66
anomaly, **1,** 36
international formula, **1,** 36
Newton's law of, **1,** 35
Gravity anomaly
Bouguer, **24B,** 152–153
downward continuation, **24B,** 154–155
interpretation, **24B,** 153–156
isostatic, **24B,** 152
monitoring, **24B,** 156–159
Gravity measurement, *see also* Gravimeter
at sea, **24B,** 138–141
from aircraft, **24B,** 141–142
in boreholes, **24B,** 155–156
satellite, **24B,** 163–164
Gray body equation, **24A,** 299
Gray-scale mapping
programs for, **12C,** 303–307
Gray-wedge technique, in pulse spectrometer, **2B,** 115–118
GRAYM, **12C,** 304, 331
GRAYMP, **12C,** 332–333
Grazing incidence spectrometer, **17,** 539–544
Grazing-incidence dye laser, **29C,** 164–167
Grease, **14,** 414–417
lubricated bearing, **14,** 263
vacuum, **14,** 320
Green Bank
three-element interferometer at, **12C,** 164–165
Green's function matrix, **21,** 175, 183
of defects, **21,** 183
Green's function, **30,** 349
dyadic, references for, **12B,** 61
Gregorian antenna, **12B,** 48–50
Greinacher cascade multiplier, **2,** 198–200
Greninger method for back-reflection, **6A,** 239
Grid
filter, *see* Grating filter
microchannel plate, **22,** 487, *see also* Microchannel plate detector, Channel-platepolarizer, *see* Grating polarizer
Grid-dip meters, **2B,** 16–17
Gridded lens, of parallel-plate analyzer, **17,** 443
Grids
for electron plasma wave study, **9A,** 18
for ion wave study, **9A,** 13
Grids, for specimen support, in electron microscopy, **16B,** 213
GRIN lenses, **29C,** 351–352
Grinding and lapping, **6A,** 149–151
Grinding in surface preparation, **6A,** 150
hand operations, **6A,** 150
machine operations, **6A,** 150
Groove, **14,** 281–285, 288, 291
Grooved moderators, **23A,** 184, 187
Groszkowski gauge, **14,** 72, *see also* Ionization gauge
Grotrian diagrams, **9A,** 165
Ground loop, **1,** 225
Ground loops, **8,** 135
Ground rod, **14,** 452
Ground state, **5B,** 352
potential surface, for Br-H-Cl system, **15B,** 274
Ground, in electronic system, **2A,** 41–42
Grounding, **14,** 451, 452
Group delay, **28,** 341
ionospheric, **12B,** 127–128
Group frequencies, **3A,** 147
Group theory, **23A,** 459; **23C,** 110
geometrical, **3,** 270ff
Group velocity, **9B,** 11
Groups, isochromatic, **3,** 395
Growth of luminescence, **6B,** 308–313
Grüneisen parameter(s), **6A,** 376; **16A,** 200; **16C,** 85, 112; **23A,** 475; **24A,** 222–224
GSN, *see* Global Seismographic Network
(g–2) Experiments
for electron, **4B,** 27–41
for positron, **4B,** 42–43
theory, **4B,** 30–36
Guard ring, **1,** 496; **16C,** 399, 400, 491
Guide tubes, *see* Neutron guides
Guide wavelength, **8,** 129, 130, 144, 149, *see also* Waveguide
Guided wave integrated optics, **29C,** 381–382, 392

channel waveguides in glass susbtrates, **29C,** 383–384, 392
lithium niobate, **29C,** 385–392
optical plumbing, **29C,** 382–383
polymer waveguides, **29C,** 383–384
quasi-phase-matching in, **29C,** 245
Guided wave propagation in afterglow studies, **7B,** 99–100
Guinier approximation, **23A,** 52; **23B,** 514; **23C,** 431
Guinier, A., **23C,** 381
Guinier-Preston zones, **23B,** 114, 125
Gunn devices, **2A,** 51
Gunn diode, **29B,** 277
Gunn-effect devices, **2B,** 148–149, 273, 277–278, 280
applications of, **2B,** 280–282
frequency stabilization and, **2B,** 281–282
modulation in, **2B,** 282
noise in, **2B,** 281
power generation capacities of, **2B,** 279
Gurevich effect, **6B,** 118
Gyration, radius of, **1,** 84
Gyrator, **2A,** 335–336
Gyromagnetic ratio, **16A,** 443
of intermediate state in a γ-γ cascade, **5B,** 144
nuclear, **5B,** 122–123
angular correlation determination, **5B,** 142–147
of proton, **4B,** 17–18
proton, **1,** 44; **5B,** 122–123
Gyroscope, **24B,** 2, 15
optical, **26,** 43

H

H^+, 2s capture, **7A,** 465
H^+, electron capture, **7A,** 464
H, in highly excited levels, **7A,** 214
H, resonant charge exchange by merging beams, **7A,** 483
H^- detachment, **7A,** 114
H^-, **7A,** 179–180, 183–184
H-adsorbed
Si(111) 7×7
scanning tunneling microscopy of, **27,** 194
H-D constant spacing model, **5A,** 252ff
H-test, **28,** 360
H-value, maximum, **28,** 371
H^2 (*d,n*) He^3 reaction, **5B,** 371
H_2^+ dissociation, **7A,** 466–469
by electrons, **7A,** 97
H_2^+ field dissociation, **7A,** 209, 212–214
H_2^+, by electron bombardment, **7A,** 469–470
H2, field ionization, **7A,** 209
H_20
laser, **10,** 48, 49
spectrum, **10,** 91–93, 320, 321
H_2O-D_2O mixtures, **23C,** 375
$H_2PO_4^-$ ion, **19,** 202
H^2 (*d,n*) He^3 reaction, **5B,** 371
H^3 (*d,n*) He^4 reaction, **5B,** 372
Hadamard spectrometer, **13B,** 87
Hagen-Poisseuille formula, **18,** 336, 798
Hagen-Rubens relation, **10,** 112
Haidinger fringes, **1,** 105, 350, 356, 371; **26,** 11
Half life, **21,** 9
isotope separation, **21,** 71
Half-cylindrical mirror analyzer, **17,** 457–458, 467
Half-guide wavelength, **8,** 138
Half-intensity bandwidth, **8,** 37
Half-life, **5B,** 275, 278; **8,** 208, 210, *see also* Nitrogen13
Half-power points, **8,** 147, 148
Half-shade device, **25,** 124
Half-tone hologram
Half-wave circuits, **2A,** 193–197
Half-wave doubler, **2A,** 208
Half-wave plate, **1,** 381; **12A,** 369, 373, 379, 389
Half-wave rectifiers, **2A,** 193–197
with capacitive load, **2A,** 193–195
in nonlinear circuits, **2A,** 395
with resistive or inductive load, **2A,** 195
6-phase, **2A,** 205
Half-waveplate, **29C,** 353–354
Half-width, **8,** 198
Halide ion, in color centers, **15B,** 3
Hall
effect devices, **2,** 123–124
mobility, **2,** 123
Hall coefficient, **11,** 4–5
magnetoresistance and, **11,** 26
Hall coefficient measurement, **6B,** 151–160
alternating current methods, **6B,** 153–154
circular symmetry, **6B,** 154–155
cryostat design, **6B,** 156–160
direct current methods, **6B,** 152–153
Hall effect, **6B,** 145–160; **11,** 1

in bismuth, **1,** 524
definitions, **6B,** 145–149
devices, **2A,** 49–50
effect of anisotropy, **6B,** 147–148
of inhomogeneities, **6B,** 150–151
ferromagnetic materials, **6B,** 148–149
field dependence of, **11,** 4
geometrical properties, **6B,** 149–150
in insulating photoconductors, **6B,** 156
magnetocaloric and thermoelectric disturbances, **6B,** 150
multiple band conduction, **6B,** 148
probes for, **11,** 743
simple model, **6B,** 146
use in impurity measurements, **6A,** 37
velocity distribution effect, **6B,** 146–147
Hall mobility, **6B,** 91; **9B,** 177
Hallow-cathode discharge, in cadmium-ion lasers, **15A,** 129–131
Halogen atoms, chemistry of, **15B,** 288
Halogen negative ions, **7A,** 179ff
Halogenation, of crystalline polymers, **16B,** 245–251
effect on small-angle X-ray scattering, **16B,** 249
Halperin-Nelson theory, **23B,** 46
Hamaker constant, **23B,** 496
Hamiltonian
of atom, **4B,** 108–109
for atom in a magnetic field, **3,** 602
electron spin resonance, **3,** 443
of ENDOR method, **3,** 499
for EPR systems, **21,** 406, 407, 433
for first-order quadrupole interaction of one nucleus, **3,** 651
formulation, **13A,** 60
for interaction of electric field with molecular dipole moment, **3,** 650
for magnetic interactions in molecules, **3,** 200
for molecule in presence of a magnetic field, **3,** 605
for prolate symmetric rotor, **3,** 202
for quadrupole interaction, **3,** 502, 504, 505
for rigid, rotating electric dipole in an electric field, **3,** 606
spin, **3,** 446–447
electron, **20,** 78, 84–90
spin-orbit, for system of electrons and nuclei, **3,** 207
for vibronic interactions, **3,** 192
Hamming, **12A,** 404
Handamard spectrometer, **29B,** 367, 416
Handedness, **13A,** 45
Hanle effect, **7B,** 4; **29B,** 332–335
lifetime measurements, **7A,** 137, 138
Hanning data taper, **28,** 321–322
Hanning window, **28,** 300–301
Haptens, spin labeled, **20,** 110–122
Hard axes of magnetization, **10,** 400
Hard sphere model, **1,** 132ff
Hard-soldering, **14,** 426
Hardness, indentation, **6A,** 307–309
Hardware platforms, **28,** 482
Harker sections, **3,** 288–289
Harmonic approximation, **23A,** 23, 372
Harmonic beam splitter, **29C,** 349
Harmonic conversion efficiency, **10,** 18
arc harmonic generator, **10,** 25, 26
diodes, **10,** 21–23
plasma, **10,** 26
Harmonic frequency, **28,** 339
Harmonic generation, **13B,** 128; *see also* Nonlinear effects
diodes, **10,** 18–23
electron beams, **10,** 27–29, 44
ferrites, **10,** 27
Josephson junction, **10,** 30
linear accelerators, **10,** 27, 28, 44
microtron, **10,** 28, 44
of microwave frequencies, **6B,** 236
nonlinear reactors, **10,** 18–23
nonlinear resistors, **10,** 18–23
optical mixing and, **15B,** 255–258
picosecond spectroscopy, **20,** 191, 192
plasma arcs, **10,** 23–26
plasmas, **10,** 26–30
Harmonic generator, crystal, **3,** 29
Harmonic mixing, **10,** 21–23, 165
Harmonic modulators, **2B,** 370–372
Harmonic oscillator
circuit configurations for, **2A,** 343–347
damped, **1,** 90
design of, **2A,** 342
oscillation frequency in, **2A,** 343–344
voltage control of, **2A,** 347–348
Harmonics spectrum, **8,** 199–200
Harmonics, in ac measurements, **2B,** 69–70
Harmonic motion, forced, **1,** 89–90
Harris instability, **9A,** 279–285
Hartmann dispersion formula, **3,** 174
Hartmann-Hahn condition, **16A,** 335

Hartree atomic units, **3B,** 948
Hartree-Fock approximation, **7A,** 119, 173
Hartree-Fock-Slater charge distribution, **17,** 41, 45
Hartshorn bridge, **6B,** 184–187, 188
Hartshorn-Ward circuit, **6B,** 16–17
Hat matrix, **28,** 225
Hay bridge, **1,** 463, 464
Haynes-Shockley drift length method, **6B,** 105
Haystack Radio Observatory, **12B,** 16
Hazard
- electrical, **14,** 451
- mechanical, **14,** 450
- thermal, **14,** 453

Hazards, nuclear reactor, **5B,** 605–606
HCl, **10,** 89, 310–313; **19,** 220
HCN
- laser, **10,** 48–61, 328
- maser, **10,** 50
- molecule, **10,** 310, 324, 325, 328
- transition frequencies of, **12C,** 27

HCNO, transition frequencies of, **12C,** 26
HCOOH, transition frequencies of, **12C,** 26
He^+, electron capture, **7A,** 463
He^+, ionization and excitation by electrons, **7A,** 97
He^- field detachment, **7A,** 209
Head (of fluid), meaning, **18,** 333
Heat, **1,** 235–328
- capacity, **1,** 261, 313, 320, 325
- of chemical reaction, **1,** 283–286
- combustion, **1,** 284ff
- flow in very low temperature cooling, **5B,** 175–176
- influx during magnetic cooling, **5B,** 182–183
- Joule, **1,** 246, 292, 507
- loss, rate of, **1,** 300
- mechanical equivalent of, **1,** 128, 261, 269–272
- Peltier, **1,** 506–507
- production rate due to radioactivity, **5B,** 286
- Thomson, **1,** 507–508
- of vaporization, **1,** 281ff

Heat capacity
- chemical contribution to, **19,** 195–207
  - frozen, **19,** 182
- relaxing, **19,** 199, 200

Heat capacity of solids, **6A,** 375–385
- at constant volume, **6A,** 375
- isothermal method, **6A,** 382–385
- method of mixtures, **6A,** 380–382
- methods, **6A,** 376–385

Heat cleaning, in photoemission measurements, **11,** 105–107
Heat conduction, **1,** 287–299; **8,** 247; **30,** 178–179, 239
- gages, **1,** 166ff
- relations, one-dimensional, **18,** 665–670

Heat flow, **1,** 287, 288ff, 293; **24A,** 272–273; *see also* Thermal conductivity
- calculation, **24B,** 216–217, 254–255
- corrections, **24B,** 191
- environmental disturbances, **24B,** 255–259
- intergrain, **1,** 214
- in shallow boreholes, **24B,** 218–219
- refraction, **24B,** 256–257

Heat of adsorption, **22,** 366, 456
Heat of fusion, of polyethylene single crystals
- effect of bromination, **16B,** 250
- effect of irradiation, **16B,** 315

Heat of polymerization, by DSC, **16B,** 332
Heat of reaction, **1,** 262
- from DTA curve, **16B,** 292

Heat of sublimination, **1,** 281
- latent, of graphite, **3,** 560–562

Heat pipes
- closed, **29B,** 68
- open-ended, **29B,** 69

Heat shield, **14,** 277, 278
Heat sink, **2A,** 217
Heat transfer, **1,** 134–139, 287–303; **23A,** 136; **30,** 177–178
- coefficient, definition, **18,** 459
- free molecule probe, **1,** 171
- hypersonic tunnel, **18,** 762
- radiation loss in arc-plasma tunnel, **18,** 674–677, 784

Heat transfer coefficient, **8,** 247
Heat transfer gage
- asymptotic type, **18,** 672–675
- balanced heat removal type, **18,** 664
- calorimeter type, **18,** 664
- capacitance calorimeter, **18,** 672–674
- construction and principles, **18,** 663–685
- Gardon type, **18,** 672–675
- high heat flux, **18,** 676–677, 681
- high temperature gas flows, **18,** 683–685
- infrared bolometer, **18,** 679
- membrane calorimeter, **18,** 664
- multilayer gage, **18,** 666
- radiation type, **18,** 683–686

sandwich type, **18,** 664
shock tube and shock tunnel, **18,** 666, 677–679, 682–685, 792
thick film type, **18,** 677–678
thin film type, **18,** 664, 679–683
thin membrane calorimeter, **18,** 675–675
use in arc-plasma tunnel, **18,** 784
use in free-flight model, **18,** 781

Heat transfer measurement
conceptual methods, **18,** 664
gages, **18,** 663–653
shock tube technique, **18,** 791
wind tunnel technique, **18,** 674, 778

Heat-pipe ovens
application in
collision cross section measurement, **29B,** 77
oscillator strength determination, **29B,** 72–74
vapor pressure measurement, **29B,** 71
concentric, **29B,** 74–77
correction for end effects, **29B,** 73–74
crossed arms, **29B,** 71–72
for mixtures of different vapors, **29B,** 79–82
metal vapor plasma production, **29B,** 74
operation, **29B,** 69–71
superheated, **29B,** 72
technical considerations, **29B,** 82

Heating
effect of radioactive substance, **5B,** 187–188
local, of gas target along incident beam path, **5B,** 423

Heavy impurities, *see* Band modes

Heavy ion
beam transport, **17,** 464
charge exchange phenomena, **17,** 74–75
charge-state fractions in solids, **17,** 272–278
Coulomb excitation, **17,** 384
ionization cross sections, **17,** 384
K-shell ionization, **17,** 250–272
Lamb shift, **17,** 587–590
multiple inner-shell vacancy production, **17,** 374–375, 377
projectile, **17,** 403–408
sources, in beam-foil experiments, **17,** 533, 535
target considerations in inner-shell ionization, **17,** 221–228

Heavy ion reactions, **8,** 265–266
application, **8,** 257–262

Heavy particle excitation, **9A,** 140

Heavy particle stripping process, **8,** 277, 280

Heavy target, electron capture from, **17,** 350–354

Heavy-ion accelerator, **17,** 16–19, 22, 74, 84

Heavy-ion-atom collision, **17,** 151–155, 500
charge dependence of atomic inner-shell cross sections, **17,** 270–302
electron loss, **17,** 117

Heisenberg models, three-dimensional, **23C,** 29

Heisenberg representation, **10,** 3

Hele-Shaw apparatus, **18,** 788–789

Helical conformations, analysis by X-ray diffraction, **16B,** 83–87

Helical magnetic structures, **23C,** 133

Helical point net system, **16B,** 20

Helical polymer chains, normal vibrations, **16A,** 108–109

Helical undulator, **29C,** 32

Helicity, **13A,** 45
electron, **5B,** 215, 219
determination using bremsstrahlung, **5B,** 259–262
of γ-rays
by Mössbauser effect, **5B,** 273–274
by transmission, **5B,** 267–269
persistence, **5B,** 215, 223
positron, from annihilation-in-flight, **5B,** 262
transfer efficiency, **5B,** 234
polarization measurement depending on, **5B,** 259–262

Helicopters, use in antenna calibration, **12B,** 113

HELIOS, **23A,** 130, 202

Helium 3–4 mixtures, **23B,** 385
excitations, **23B,** 388
scattering function, **23B,** 387

Helium 4, **11,** 469–470

Helium barrier thickness, **8,** 246

Helium gas, conductivity/purity, **8,** 247

Helium lines, **8,** 161

Helium metastables
creation of, **15A,** 104–105
destruction of, **15A,** 108

Helium, **14b,** 419–420; **23B,** 303
atmospheric concentration, **12B,** 170
diffusion
in glass, **14,** 321
through glass, **14,** 337–343
discharge, **7A,** 156
from glass, **14,** 322

neutron irradiation, **14,** 335–337
glass permeation, **14,** 394
ionizing potential of, **2A,** 178
isotopes, **11,** 468–469
laser emission, **10,** 48, 49
leak defector, **14,** 81
liquid, *see* Liquid helium
permeation, **14,** 515–516, 518
phase diagrams, **23B,** 303
in pump oil, **14,** 469
in rare gas halide lasers, **15B,** 90
resonances in, **7A,** 40
solubility, in H_2O, **14,** 419
turbulent superfluid, **11,** 479–481
whistlers, **9B,** 131–133
Helium–3, **23C,** 138
ion behavior in, **11,** 468–469
melting curve for, **11,** 535–536
neutron spectrometer, **5A,** 570–581
Helium–3, liquid, **23B,** 370
dynamic susceptibility, **23B,** 374
interactions, **23B,** 377
kinetic energy, **23B,** 383
paramagnon model, **23B,** 380
spin fluctuations, **23B,** 370
zero sound, **23B,** 370
Helium–3, solid, magnetic ordering, **23B,** 325
Helium–4, liquid, **23B,** 326
adsorbed on graphite, **23B,** 59
anomalous dispersion, **23B,** 337
condensate, **23B,** 331, 350
elementary excitations, **23B,** 328, 337
kinetic energy, **23B,** 355
linewidths, **23B,** 342
Monte Carlo calculation, **23B,** 357, 366
multiphonon scattering, **23B,** 348
pair correlations, **23B,** 359
scattering function, **23B,** 338
singular behavior, **23B,** 354
structure factor, **23B,** 359
superfluidity, **23B,** 304, 326
two-fluid model, **23B,** 342
Helium–4, solid, **23B,** 309
anharmonicity, **23B,** 324
interference terms, **23B,** 321
dynamics, **23B,** 310
kinetic energy, **23B,** 322
phonons, **23B,** 317
scattering function, **23B,** 315
short-range correlations, **23B,** 314
single-particle excitations, **23B,** 321
Helium-cadmium laser, **15A,** 128, *see also* Cadmium in laser; Ion lasers
noise in, **15A,** 137–138
other excitation methods for, **15A,** 138–142
Helium-fluorine laser, **15A,** 118–120
Helium-II
thermodynamic properties of, **11,** 463–464
"turbulent" state of, **11,** 479–481
Helium-iodine laser, **15A,** 146–149
Helium-like ions, **9A,** 126
Helium-mercury laser, **15A,** 125–126
Helium-neon lasers
with dye lasers, **29C,** 167
excitation processes in, **15A,** 104–110
frequency stabilization, **29C,** 126–128
iodine-stabilized, **29C,** 289–294, 307
noise in, **15A,** 137–138
number of, **15A,**31, 97–116
performance of, **15A,** 110–114
pulsed operation of, **15A,** 114–116
as reference laser, **29C,** 325
small-signal gain of, **15A,** 113
spectroscopy of, **15A,** 101–104
transitions of, **15A,** 103
uses, **29C,** 260
Helium-pulsed discharge, **15A,** 143
Helium-recoil counter-polarimeter, **5A,** 484–486
Helium-zinc laser, *see* Zinc ion laser
Helix feed, **12B,** 74, 79
Helix-coil transition, **19,** 203, 233
Helmer gauge, **14,** 72, 73, *see also* Ionization gauge
Helmholtz
coils, **1,** 526, 528
equation, **23A,** 260
formulation, **19,** 569
free energy, **1,** 323; **11,** 437
resonator and superfluid helium, **19,** 438–439
Helmholtz arrangement, **8,** 99, *see also* Double Helmholtz arrangement
Helmholtz combination, **8,** 79
Helmholtz field, **8,** 100
Helmholtz solution, **8,** 105
Helmholtz-Kirchhoff theorem, **26,** 50
Helmholtz-Lagrange law, **4A,** 86, 124, 144; **29B,** 54
lens configuration in electron energy analyzer, **29A,** 224–225
pencil angle determination, **29A,** 192, 202

Hemispherical curved-plate analyzer, **9B,** 64
Hemispherical deflector, 1808, **22,** 34, 35
  lens, **22,** 35, 40, 41
  resolution, **22,** 40, 41
Hemispherical power radiancy, **8,** 28
Hemoglobin, **20,** 257, 387
Hemoproteins, **20,** 150–154
Hencky strain, **16C,** 13
Henry's law of sorption, **16C,** 318
Heptode, **2,** 65
  tube, **2A,** 160
Hermanson Rule, **22,** 159, 161
Herpin matrix, **26,** 129
Herrick effect, **14,** 172–173
Hertz-Knudsen equation, **30,** 226
Hertzian polatization, **26,** 157
Herzberg-Teller effect, **3,** 192–193
Heterodyne acoustooptic techniques, **19,** 481–492, *see also* Superheterodyne acoustooptic techniques
  general principles, **19,** 481
  heterodyne systems, **19,** 482–487
  superheterodyne systems, **19,** 487–492
Heterodyne detection, *see* Optical heterodyne detection
Heterodyne efficiency, **18,** 119
Heterodyne interferometry, **9B,** 37–38
  principle of operation, **9B,** 37
  sensitivity, **9B,** 37–38
Heterodyne oscillator, in frequency measurements, **2B,** 22–23, *see also* Oscillator
Heterodyne spectroscopy, **13B,** 87
Heterogeneous nucleation, *see* Nucleation
Heterojunction laser(s), **15A,** 215–217; **25,** 268
  carrier confinement in, **15A,** 219–220
  emission range for, **15A,** 232–233
  fundamental transverse mode operation in, **15A,** 238–239
  material used in, **15A,** 230–234
  operation of, **15A,** 217–218
  radiation confinement in, **15A,** 220–222
Heterojunction, **2,** 120
  discontinuities, **22,** 155
Heterostructure interface
  Au/AlAs/GaAs
    ballististic electron emission microscopy of, **27,** 337, 339
Heusler alloys, **23C,** 176
Hewlett Packard frequency meter, **2B,** 18
Hexagonal crystal system
  definition, **16B,** 4
  interplanar spacing formulae, **16B,** 26
  reciprocal lattice relationships, **16B,** 24
Hexagonal lattice
  definition, **16B,** 5
  reciprocal, **16B,** 23
Hexamethylene tetramine
  electron diffraction curves, **3,** 342–345
  Laeu photograph, **3,** 324
  neutron diffraction study, **3,** 355–358
  precession photograph, **3,** 322
  rotation photograph, **3,** 314
  structure determination, **3,** 329–334
  zero-level Weissenberg photograph, **3,** 318
Hexamethylenetetramine, **23A,** 441
Hexatic phase, **23B,** 46
Hexode, **2,** 65
  tube, **2A,** 160
HF rotation spectrum, **10,** 312
HF(DF) chemical lasers, **15B,** 75, 105–111, *see also* Deuterium fluoride chemical laser; Hydrogen fluoride laser
  chemical reaction rate in, **15B,** 122
HFFT2, **12C,** 292, 294, 319
Hg^{197} decay scheme, **5B,** 159
Hg^{203} decay scheme, **5B,** 159
HI spectrum, **10,** 312, 314, 337
Hierarchical Bayes, **28,** 429
High energy diffraction,
  apparatus and techniques, **6A,** 264–273
  camera, **6A,** 268–269
  electron source, **6A,** 265
  high voltage systems, **6A,** 269–271
  optics, **6A,** 265–267
  precision and resolving power, **6A,** 271–272
  specimen manipulator, **6A,** 267–268
  specimen preparation, **6A,** 272–273
  vacuum systems, **6A,** 269
High frequency effects in semiconductors, **6B,** 110–114
  measurement technique, **6B,** 113–114
High frequency instabilities, **9A,** 275–304
  absolute and convective instabilities, **9A,** 278–279
  betatron oscillations, **9A,** 284
  computer simulation experiments, **9A,** 302–304
  "superparticles" in, **9A,** 302
  driving mechanisms for, **9A,** 276

electrostatic instabilities, **9A,** 275
free energy for, **9A,** 276
linear theory and dispersion relation, **9A,** 277–278
longitudinal instabilities, **9A,** 279–299
beam-beam type, **9A,** 296–299
double-humped type, **9A,** 285–291
drift type, **9A,** 285
Harris type, **9A,** 279–285
negative mass type, **9A,** 291–295
modified, **9A,** 295–296
microinstabilities, **9A,** 275
transverse instabilities, **9A,** 299–304
cyclotron type, **9A,** 300–302
Whistler type, **9A,** 299–300
Vlasov equation, **9A,** 277
High leverage point, **28,** 271
High magnetic fields, superconducting tunneling in, **11,** 153–156
High polymer, definition, **16A,** 3
High pressure bomb for optical measurements, **6A,** 431–432
High pressure investigations, **6A,** 407–437
High pressure methods, miscellaneous, **6A,** 436–437
High pressure techniques, for liquids, **19,** 170–172
High pressure tensile testing apparatus, **6A,** 433
High *Q*, *see Q*
High reflectivity mirrors, design, **31,** 279–281
High resolution semiconductor Compton spectrometer, **8,** 203
High speed recording methods, **18,** 725–753
High temperature gases, produced in shock-tube, **18,** 787–790
High temperature scanning tunneling microscopy, **27,** 208
High vacuum system
design, **14,** 169, 170, 457–464
valved pumping systems, **14,** 461–463
valveless systems, **14,** 463–464
High voltage systems for electron diffraction, **6A,** 269–271
High-beta plasma instabilities, **9A,** 395–449
anomalous diffusion, **9A,** 441
anomalous resistivity, **9A,** 434
Bohm diffusion, **9A,** 396, 441
classical diffusion in, **9A,** 441
collective scattering, **9A,** 435, 443
Corioli's force, **9A,** 439
cusp configuration, **9A,** 401, 441
definition of, **9A,** 396
diagnostic measurements, **9A,** 418–429
Abel inversion, **9A,** 426
bremsstrahlung, **9A,** 420
high-speed photography, **9A,** 420–421
laser-light scattering, **9A,** 428
roating mirror cameras, **9A,** 421
streak camera, **9A,** 421
Thomson scattering, **9A,** 419, 428
summary of, **9A,** 422–425
diffusion coefficient, **9A,** 435
drift type, **9A,** 399
dynamic stabilization, **9A,** 447–448
examples of, **9A,** 401–403
experimental investigation of, **9A,** 429–447
feedback stabilization, **9A,** 448
finite Larmor radius stabilization, **9A,** 402, 438
flutes, **9A,** 399
high-speed photography of, **9A,** 403
"interchange" instability, **9A,** 399
magnetohydrodynamic instability, **9A,** 395
MHD stability, **9A,** 436
microinstabilities, **9A,** 395–396, 434
plasma focus, **9A,** 401
plasma satellites, **9A,** 435
Rayleigh-Taylor instability, **9A,** 429, 436, 441, 443
resistive type, **9A,** 433
from reversed-field theta-pinches, **9A,** 440
rotational type, **9A,** 438
streaming type, **9A,** 434
tearing type, **9A,** 433
theory of, **9A,** 403–418
anomalous diffusion, **9A,** 417
bounce model, **9A,** 408, 413
drift instabilities, **9A,** 416
effective collision time, **9A,** 417
energy principle, **9A,** 405
enhanced resistivity, **9A,** 417
finite Larmor radius effects, **9A,** 412
gravitational mode, **9A,** 413
Kink modes, **9A,** 407
localized modes, **9A,** 406
magnetic shear, **9A,** 406
mirror instability, **9A,** 418
normal modes, **9A,** 405
Ohm's law, **9A,** 404
resistive instabilities, **9A,** 412
shear length, **9A,** 414

sheared systems, **9A,** 405
streaming instabilities, **9A,** 416
tearing mode, **9A,** 413
theta-pinch, **9A,** 401
toroidal systems, **9A,** 443–447
bumpy theta-pinch, **9A,** 443
classical diffusion, **9A,** 447
hardcore theta-pinch, **9A,** 444
high-beta stellarator, **9A,** 445
Kruskal-Shafranov limit, **9A,** 447
reversed field pinch, **9A,** 447
screw pinch configuration, **9A,** 446
shear, **9A,** 447
stabilization, **9A,** 444
Tokamak configuration, **9A,** 446
Zeta experiment, **9A,** 447
turbulent heating experiments, **9A,** 434
universal type, **9A,** 399
z-pinch, **9A,** 401
High-brightness source, nearly monochromatic electrons, **22,** 368
High-charge collision, fluorescence yields, **17,** 289–297
High-density gas jet target, for charge state fraction measurements, **17,** 85
High-energy beam-foil experiments, **17,** 533, 535–537
lifetime measurements, **17,** 577
High-energy particles, interactions, **28,** 444–445
High-field magnetoresistance, measurement of in metals, **11,** 1–31
High-frequency electrodeless discharge lamp, **7B,** 16
High-order harmonic generation, **31,** 109–115
disagreement with lowest order perturbation theory, **31,** 110–111
features from rare gases, **31,** 109–110
theory, **31,** 111–113
vs. x-ray lasers, **31,** 115
VUV beam properties, **31,** 113–115
High-pass waveguide filter, **10,** 121
High-pressure ion sources, **7A,** 363–370
High-pressure measurements
belt apparatus, **16C,** 100
diamond anvil cell, **16C,** 101–104
calibration, **16C,** 103
dilatometer, **16C,** 99
piston-cylinder device, **16C,** 94–99
calibration, **16C,** 98
types of equipment, **16C,** 92–104
High-pressure tunable lasers, **15B,** 79–82
High-*Q* superconductivity resonators
design and fabrication of, **11,** 567–570
frequency measurement and control in, **11,** 580–581
frequency stability in, **11,** 585–586
materials for, **11,** 570–575
measurement techniques for, **11,** 575–580
High-repetition-rate lasers, **29C,** 146–147, 150
High-resolution energy spectrometry, **17,** 183–184
High-resolution NMR, *see* Nuclear magnetic resonance spectroscopy
High-speed photography, **16C,** 478
High-spin state, **17,** 510, 515
High-T_c superconductors, **30,** 3
High-temperature chemistry, **3B,** 628
High-temperature flow, *see* Plastic deformation
High-velocity ion
charge equilibrium in matter, **17,** 73–148
energy loss in matter, **17,** 25–72
High-voltage ion accelerator, **17,** 466
High-voltage pulse, **22,** 404
Higher-order autocorrelators, **29C,** 213–215
Highest posterior density, **28,** 413
Hilger-Chance refractometer, **25,** 185
Hindered rotation and translation, **13B,** 44
Hindered rotations, **3B,** 984
Hinteregger discharge lamp, **11,** 121
Histogram(s), **28,** 157–158
grain images, **19,** 248–251, 271
grain size, **19,** 248–251, 271
History, **13A,** 3
HMT, *see* Hexamethylenetetramine
HNC, transition frequencies of, **12C,** 26
HO coefficient, **14,** 252, 255
HO, transition frequencies of, **12C,** 26
$HoCo_2$, **23C,** 216
Hoffman electrometer, **1,** 441–442
HoIG ferrimagnetic exchange resonance, **10,** 419
Hold time
definition, **18,** 530
diaphragm pressure gage, **18,** 568
free surface motion pressure gage, **18,** 606
pressure bar gage with end sensor, **18,** 603
stub pressure gage, **18,** 590
Holding pump, **14,** 462
Holdup process, **8,** 227

Hole burning experiment, NMR, **21,** 392, 393
Hole formation, **30,** 189
Hole-electron concentration profiles, **2A,** 53
Holey film, specimen support in electron microscopy, **16B,** 214
Hollow cathode discharge, **13A,** 218, 272
  use as source of
    metastable atoms, **29B,** 102
    free radicals, **29B,** 88–89
    neutral stoms and ions, **29B,** 221
  use in branching fractions measurements, **29B,** 244–248
Hollow cathode, **31,** 65–72
  low current, 66
Hologram embossing, **26,** 206
Hologram generation, **19,** 538–539
Hologram, Fourier transform, **26,** 178
  Faunhofer, **26,** 175
  Fresnel, **26,** 175
  Lippman-Denisyuk, **26,** 185
  binary, **26,** 189
  color, **26,** 180
  computer, **26,** 189
  half tone, **26,** 190
  image plane, **26,** 179
  phase, **26,** 190
  rainbow cylindrical, **26,** 189
  volume reflection, **26,** 174
Holograms, effects of thickness in, **26,** 172
  rainbow, **26,** 184
  reflection, **26,** 172
  transmission, **26,** 172
  volume, **26,** 172
Holographic bleaching, **26,** 204
Holographic grating, aspheric wavefront, **31,** 352–353
Holographic grating, spherical wavefront, **31,** 350–352
Holographic gratings, **13A,** 229
Holographic interferometer, **16C,** 246
Holographic interferometry, **18,** 381, 550, 746–750; **24A,** 148–152; **24B,** 403–404; **26,** 191
  plasma applications of, **9B,** 42–48
    diffuser in scene beam, **9B,** 42–43
    introduction of background fringes, **9B,** 43–44
    two-color type, **9B,** 45–46
Holographic materials, **26,** 201
Holographic microscopy, **26,** 195
Holographic optical element(s) (HOE), **26,** 197; **29C,** 361–362
Holographic processing, **26,** 204
Holographic recording, **26,** 204
Holographic scanners, **25,** 114
Holographic stereograms, **26,** 184
Holography, **19,** 533–562; **20,** 438, 439; **25,** 266; **26,** 167; **30,** 112–115
  aberrations, **19,** 544
  applications, **26,** 191
  applications of, **19,** 552–554
  area detector, **19,** 546–548
  backward wave reconstructing imaging, **19,** 559–562
  combined with interferometer method, **18,** 381
  combined with schlieren method, **18,** 368
  composite, **26,** 184
  computer reconstruction, **19,** 544–562
  fundamental concepts of, **19,** 538–546
  historical survey of, **19,** 536–538
  image parameters, **19,** 542–546
  image position, **19,** 542–543
  imaging applications, **19,** 552–553
  implementation of, **19,** 546–551
  interferometry, **19,** 554
  magnification, **19,** 544
  methods, **18,** 715, 743–750
  phenomenological foundation of, **19,** 533–536
  principle, **18,** 744–746
  reconstruction, **18,** 550, 743–746, 747, 748, 750
  resolution, **19,** 544–546
  sampling detectors, **19,** 549–551
  wave reconstruction, **19,** 540–542
Holstein trapping factor, **15A,** 88
Homocentric bundle, **1,** 335
Homocentric pencil of rays, **31,** 151–152
Homodiodes, **11,** 185
Homodyne detection, *see* Optical homodyne detection
Homogeneity, magnetic field, **3,** 412–415
Homogeneous electrostatic field, **8,** 172
Homogeneous nucleation, *see* Nucleation
Homogeneous vapor nucleation, **30,** 235
Homogenization, **30,** 198–200
Homojunction laser, **25,** 268
Homologous design principle, **12B,** 12
Homopolymer, **16A,** 3
Hook circuit relaxation oscillator, **2A,** 356

Hook collector transistor, **2A,** 98–99
Hook method, **7A,** 128–131; **18,** 351
Hooke's law, **1,** 62; **19,** 497; **28,** 502
Hopkinson pressure bar, **18,** 596
Hopper crystal face, **6A,** 334–335
Hopping gas model, **22,** 383
Horizontal deflection sensitivity, **8,** 151
Horn antennas, simple, **12B,** 11
Horn feeds, **12B,** 65–71
  corrugated, **12B,** 68, 78
  diagonal, **12B,** 67, 78
  Love, **12B,** 67
  multimode, **12B,** 70, 78
  rectangular, **12B,** 78
  ridge, **12B,** 70
  ridged square, **12B,** 78
  scalar, **12B,** 70
Horn, **1,** 205ff, 217
  conical, **1,** 206
  hyperbolic exponential, **1,** 206, 207
Horn-reflector antenna, **12B,** 13
Hornyak detector, **5B,** 466
Horology, **1,** 70
Horsepower, **1,** 93
  brake, **1,** 95
Host materials, in solid state lasers, **15A,** 169, 187–193
Hot filament discharge, **13A,** 219
Hot stage for optical microscope
  construction of, **16B,** 357–359
  crystallization of thin polymer films in, **16B,** 360
Hot wires, in shock-wave studies, **7B,** 180
Hot-body radiators, **10,** 13–16
Hot-cathode arc, **7A,** 196
Hot-electron plasma, ion acoustic wave observation in, **9A,** 16–18
Hot-film anemometer, *see also* Constant current anemometer; Constant temperature anemometer; hot-wire anemometer
  aspect ratio, **18,** 271, 304, 307, 309
  calibration, **18,** 258–289, 297–303
  compensating circuit, **18,** 283
  constant current, **18,** 276–289
  constant temperature, **18,** 289–303
  construction, **18,** 261; **18,** 266–268
  external coating, **18,** 282
  diameter, **18,** 261, 284
  directional dependence, **18,** 306–308
  effect of temperature variation along length, **18,** 303
  effective cooling velocity, **18,** 306
  flow interference effects, **18,** 308
  frequency response, **18,** 278, 282, 283, 297
  heat conduction, **18,** 272, 273, 382
  linearized energy balance equation, **18,** 275
  multiple probe arrays, **18,** 310–313
  physical characteristics, **18,** 260–268
  resistance, **18,** 268
    of wire, **18,** 268
  Reynolds number, **18,** 269
  sheath, **18,** 261, 265
  substrate, **18,** 267, 273
  supports, **18,** 265
  temperature sensitivity, **18,** 313–314, 461
  theory, **18,** 268–276
  thickness, **18,** 267
  time constant, **18,** 281–297
  X-probe, **18,** 310
$HoVO_4$, **23C,** 282
HS, transition frequencies of, **12C,** 27
Huang scattering, **21,** 166; **23B,** 94, 98
Hum signals, **8,** 135
Hum, amplifier, **2,** 219–220
Hund's cases (a) and (b), **10,** 332, 333
Huygen's principle, **8,** 153
Huygens
  eyepiece, **25,** 61, 132
  ocular, **1,** 339
  principle, **1,** 346
  wavelet, **26,** 49
Huygens sources, **8,** 156
Huygens-Fresnel
  principle, **26,** 49, 54
  theory, **26,** 51
Hybrid acoustic modes of superfluid helium, **19,** 442–444, 447
Hybrid integrated circuits, **2A,** 142–147
  assembly of, **2A,** 143–144
  ceramic insert in, **2A,** 145
  in shielded package, **2A,** 146
Hybrid-coil bridge, **6B,** 12–14
Hybrid-π common-emitter circuit, **2A,** 268
Hybrid-pi noise, **2B,** 472
Hydraulic dynamometer, **1,** 95–96
Hydraulic fracturing, **24B,** 393–401
Hydraulic jack, **1,** 67
Hydrides
  acoustic branches, **23B,** 137

first-row, correlation of low-lying electronic states, **3,** 180
transition metal, **23A,** 349
Hydrocarbons, adsorbed in graphite, **23B,** 29, 54, 62
Hydrodynamic flow source, **7A,** 237–252
for vapors, **7A,** 251–252
Hydrodynamic parameters, of space plasmas, **9B,** 77
Hydrodynamic regime, **23B,** 464
Hydrodynamic sputtering, ablation, **30,** 237
Hydrodynamic volume, **16A,** 3
in GPC calibration, **16A,** 56
Hydrodynamic waves, **1,** 113
Hydrodynamical ablation, **30,** 176
Hydrofrac, *see* Hydraulic fracturing
Hydrogen, **19,** 220; **22,** 154
adsorbed on graphite, **23B,** 27
atom fine structure, **3,** 630
atmospheric concentration, **12B,** 170
electrical breakdown, **14,** 421
from forevacuum pumps, **14,** 159–160
from glasses, ultraviolet irradiation, **14,** 329–333
from ionization gauges, **14,** 159
in metal vacuum systems, **14,** 485
outgassing from glass, **14,** 322
permeation in glass, **14,** 340, 342–343, 377, 385
permeability
glass, **14,** 394
palladium, **14,** 373, 409
physical properties, **14,** 420–421
physisorbed, **22,** 402
Raman spectrum, **3,** 141, 143
solubility, metals, **14,** 354
Hydrogen 1, transition frequencies of, **12C,** 26, 210
Hydrogen absorption
interferometer measures of, **12C,** 73
single antenna measurements of, **12C,** 72–74
spectrometer baseline determination with, **12C,** 74–77
Hydrogen bonds, **23A,** 358, 478
Hydrogen bromide
luminescence of, **15B,** 290–291
rotational states, **15B,** 116
Hydrogen chloride laser, **15B,** 79
purely chemical, *see* Purely chemical HCl and HCl-CO_2 lasers
Hydrogen chloride, *see* Hcl
rotational states of, **15B,** 116
Hydrogen cloud measurements
scale calibration in, **12C,** 69–72
spectrometer baseline determination for, **12C,** 66–69
Hydrogen clouds
emission surveys of, **12C,** 58–64
high velocity, **12C,** 61
measurements of, **12C,** 58–77
smaller region, **12C,** 62–64
standard regions in, **12C,** 36
Hydrogen continuum, **13A,** 206
Hydrogen cyanide, *see* HCN
Hydrogen discharge tube, **1,** 376
Hydrogen exchange, **20,** 376–378; **23C,** 355
Hydrogen fluoride
chemical laser, **15B,** 105–111, *see also* HF(DF) chemical lasers
molecular constants, **3,** 52–53
rotational spectrum, **3,** 48
rotational states of, **15B,** 116
Hydrogen fluoride lasers, **15B,** 95
in rotational relaxation experiments, **15B,** 118
Hydrogen halide reaction products, bimodal rotational distributions in, **15B,** 117
Hydrogen halides, chemiluminescence and, **15B,** 116
Hydrogen in amorphous metals, **23B,** 175
Hydrogen in metallic glasses, **23B,** 284
local vibrations, **23B,** 289
Hydrogen in metals, **23B,** 131; **23C,** 467
anharmonicity, **23B,** 161
dilute, **23B,** 132
isotope effects, **23B,** 165
nondilute, **23B,** 149
potentials, **23B,** 160
precipitation, **23B,** 174
self-trapping, **23B,** 168
shear constant anomalies, **23B,** 158
superconductivity, **23B,** 177
tunneling states, **23B,** 172
vibrations, **23B,** 153
acoustic, **23B,** 153
local, **23B,** 170
optic, **23B,** 156
table, **23B,** 166
Hydrogen lines, **8,** 161
Hydrogen plasma, continuum emission, **31,** 121–122

Hydrogen storage, **23B,** 147
Hydrogen sulfide
  absorption coefficient of, **12B,** 172
  atmospheric concentration, **12B,** 170
Hydrogen thyratrons, **2A,** 176–180
Hydrogen-bonded sheets, in polyamide crystals, **16B,** 88–90
Hydrogen-like *3d* functions, **3,** 225
Hydrogen-oxygen reaction, **3,** 573–574
Hydrogenic state, in beam-foil interactions, **17,** 549, 551, 564–566
Hydrogenous scintillators, **8,** 253
Hydrolysis, selective degradation by
  of cellulose, **16B,** 269–274
  of polyesters, **16B,** 274–277
Hydrophobic ions, transport of, **20,** 529, 530
Hydrophone, **1,** 217; **24B,** 15
  directional array, **1,** 225
Hydrostatic balances, **1,** 119
Hydrostatic correction, **18,** 244, 332
Hydrostatic density, **1,** 68
Hydrostatic force measurement, **1,** 67
Hydrostatic head correction, **1,** 242
Hydrostatic law, **18,** 505
Hydrostatic pressure, **18,** 501; *see also* Static high pressure
Hydrostatic weighing, **6A,** 281–291
Hydrothermal circulation, *see* Convection
Hydrothermal vents, detection, **24B,** 240
Hydroxyl lines, **10,** 334, 335
Hyperbolic flow equation, **18,** 760
Hyperfine field, **23C,** 262, 284
Hyperfine interaction, EPR, **21,** 496
  in foil excitation process, **17,** 592, 595
  NMR, *see* Knight shift
Hyperfine splitting, nuclear, **20,** 59
Hyperfine structure, **3,** 446; **3B,** 499, 837; **4B,** 101–102, 108–110, 112–113, 117, 119, 125, 150–191; **13A,** 342; **29B,** 151–152, 164
  anomoly, **4B,** 151
  Back-Goudsmidt limit, **4B,** 112, 116
  coupling constants, **4B,** 101, 103, 112–113, 151
  interval, **4B,** 103, 115, 151
    table of, **4B,** 156–161, 173, 174–175, 177
  interval rule, **4B,** 103, 151–152
  magnetic, anomalies, **5A,** 335
  optical shift, **5A,** 334
  splitting factor, **4B,** 102
  transitions, **4B,** 151–153
    weak field limit, **4B,** 112, 114
Hyperparameters, **28,** 426
Hyperprior distribution, **28,** 429
Hypersatellite Auger electron, **17,** 501–502
Hypersatellite lines, **13A,** 172
Hypersatellite production, **17,** 408–410
Hypersonic apparatus, **18,** 781–784, 791, 793
Hypersonic atmosphere entry, **18,** 664
Hypersonic relaxation and loss, **16A,** 194–196
Hypersonic viscosity, **16A,** 198–199
Hypothesis
  composite, **28,** 187–188
  simple, **28,** 187
    chi-squared, **28,** 189
Hysteresis
  as attenuation cause, **19,** 267
  curve, **1,** 550
  elastic, **1,** 108
  loss, **1,** 556
  not found in steel, **19,** 287–289
Hysteresis loops
  of ferromagnetic material, **6B,** 219
  shape of, **6B,** 209–210
  static, **1,** 555
  in stress-strain cycling, **16C,** 118

I

I on Pt(111), chemisorption, scanning tunneling microscopy of, **27,** 297
I^- photodetachment, **7A,** 197
I-V curve
  e/h constant and, **11,** 273–274
  for equivalent circuit model, **11,** 269–271
  external noise fn, **11,** 247–248
  Josephson oscillation and, **11,** 242
  microwave-induced steps in, **11,** 217–225
  for point contacts, **11,** 259
  quasi-particle steps in, **11,** 222–225
  subharmonic structure in, **11,** 239–243
  supercurrent steps on, **11,** 217–222
  for thin-film bridges, **11,** 257–258
  for tunnel junctions, **11,** 124, 193–194, 252–253
  for weak links, **11,** 209–212
IAES, *see* Ion-induced Auger electron spectroscopy
IBM 1670, **2B,** 166
IBM 701, **2B,** 166
IBM 7094, **2B,** 166
IBPBAC, **23B,** 559
IBS, *see* Ion-beam-sputtering deposition

IC (integrated-circuit) amplifier, **2A,** 287–288, *see also* Integrated circuits
IC testers, *see* Integrated circuit testors
ICANS, **23A,** 196
Ice point, **1,** 32, 237
ICN laser, **10,** 48, 50
Ideal detector, **2B,** 490, *see also* Detector
Ideal gas laws, **1,** 37, 119; **14,** 38–40, *see also* Kinetic theory of gases
Ideal gas, **1,** 238; **23A,** 37, 43
Ideal gas temperature scale, **1,** 238
Ideal solution, **19,** 196
Identification of materials by X-ray diffraction, **6A,** 237–238
IDSA method, *see* Inverted Doppler-shift attenuation method
IDT, *see* Surface acoustic wave transducers
IEL, *see* Inelastic energy loss
IETS, *see* Inelastic electron channeling spectroscopy
IGFEF (insulated fate field effect transistor), **2A,** 85
Ignitor, **2,** 107
Ignitrons, **2,** 105–108
  circuits, **2,** 305
Ignitrons, **2A,** 180–182
Iliovici permeammeter, **1,** 554, 555
Ill-conditioning, **28,** 274–276
Illuminance, **1,** 389, 390; **25,** 100, 104
Illuminated photovoltaic cell, equivalent circuit, **6B,** 369
Illumination, **12A,** 193; **25,** 106, 146, 170
Illumination, *see* Coherent illumination; Incoherent illuminations
Image, **1,** 330ff
  base width, spectrometer, **5A,** 351–352
  formation, latent, **5A,** 210–211
  photographic of droplet, **5A,** 447
  pre-, speck, **5A,** 211
  sub-, **5A,** 211
Image aberrations, **1,** 331
  mass spectrometer, **5B,** 7
Image charge, **22,** 151
Image compression factor, **22,** 357, 358
Image converter camera
  description, **18,** 732, 735–737
  use in LDV, **18,** 208–209
Image converter streak camera, **18,** 720
Image curvature, **8,** 184
Image dissection camera, **18,** 734–735
Image dissectors, **2B,** 351–352; **12A,** 278
Image distortion, geometric, **1,** 333
Image intensification, **22,** 378
Image intensifier, **29B,** 419
  camera, **18,** 735–737
Image intensifier-dissector, **12A,** 250, 279
Image isocon, **12A,** 281
Image orthicon, **12A,** 280
  camera tube, **2B,** 347, 350
Image parameters, **2,** 30
  design, **2,** 33–36
Image potential, **22,** 351, 379, 398
Image power, **25,** 155
Image rejection, **8,** 123
Image slicers, **12A,** 195–205
  Bowen's design, **12A,** 197
  Richardson's design, **12A,** 201
Image stabilization, **12A,** 206–220
  detectors for, **12A,** 213–216
  mirror drive systems, **12A,** 211
  wobbling secondary mirror, **12A,** 207
Image tubes, **12A,** 237–251, 252–276
  electrographic, **12A,** 252–276
  Phosphor-screen output, **12A,** 207
Image-force theory, **22,** 397, 398
Image-parameter design, in network filters, **2A,** 23–24
Image-plane hologram, **26,** 179
Imagery through diffusing media, **26,** 196
Imaging
  atom-probe, **22,** 415
  band edge states, theory of, **27,** 14
  dual polarity, **27,** 99, 104, 106, 163
  interpretation, **27,** 164, 167, 177
  radiometers, **26,** 252
  of single atom, **22,** 374
  of single molecule, **22,** 370
Imaging, NMR, **21,** 403, 404
  FONAR, **21,** 404
  spin imaging, **21,** 403–404
  zeugmatography, **21,** 404
Imidazole, **19,** 202; **23A,** 478
Immersion magnifier, **25,** 129
Immersion objectives, **25,** 168
Immersion refractometer, **1,** 345; **25,** 188
Immersion spectrophotometry, **11,** 670–671
Immunoassay, laser light scattering, **20,** 319
Impact avalanche and transit time (IMPATT)
  diodes, **2A,** 47, 67
  mode, **2B,** 275–279
Impact collision ISS (ICISS), **30,** 213
Impact parameter, **5A,** 5ff, 289; **21,** 289, 301

Impact pressure, **18,** 247, 503
Impact-parameter dependence, in ion-atom collisions, **17,** 303–344
  charge exchange processes, **17,** 79–80
  multiple ionization, **17,** 487
  projectile ionization, **17,** 113
  small, and electron loss, **17,** 117–118
  target ionization, **17,** 113
  vacancy production, **17,** 380–382
IMPATT, *see* Impact avalanche and transit time
Impedance, **1,** 461, 462, 466, 467; **2,** 21–22; **2A,** 6; **19,** 6–7, 42–50, 54; *see also* Compensated impedance; Source impedance
  acoustic, **1,** 201, 216, 232
  ammeter, **1,** 447
  bridge, transfer, **1,** 498
  bridges, **1,** 486–487
  characteristic, **19,** 6, 42–50, 54
    line, **1,** 478
  compensation, **8,** 115
  complex, **16C,** 386
  driving point, **2A,** 9
  level, **8,** 111
  locus, **8,** 110, *see also* Load impedance locus
  match, **8,** 119
  matching, **8,** 109, *see also* Balun impedance matching; Broad band; impedance matching; Internal impedance; Narrow band impedance matching
  measurements, **13B,** 142, 147
    microwave, **2,** 734–737
  mechanical, **1,** 234
  motional, **1,** 216
  radiation, **1,** 203ff
  shear, **19,** 140–165
  specific, **19,** 6
  surface, **11,** 544–550
  total, of parallel tuned circuit, **3,** 392
  transfer, **2A,** 9
  transformation equation, for a transmission line, **16C,** 412
  wave, **1,** 202, 477
Impedance matching, **2A,** 11; **2,** 28; **19,** 18–19, 31–34, 47–51, 59–65, *see also* Mechanical loading
  in de Haas-Van Alphen effect, **11,** 49–50
  electrical, **19,** 59–60, 62–65
    to liquid helium, **19,** 426–428
  Imperfection, **14,** 81, *see also* Leak
    levels, density of, **6B,** 353–354
  Imperfections, techniques for revealing, **6A,** 163–171, *see also* Dislocations, lattice defects and line imperfections
  Impermeability tensor, **26,** 146
  Impingement rate, molecular **14,** 7, 8
  Implosion hazard, **14,** 451
  Implosion, strong, **18,** 841–842
  Impulse approximation, **23A,** 38; **23B,** 322, 350
    for electron capture, **17,** 90–92
  Impulse voltages, in electric breakdown, **16C,** 465
    generators, **16C,** 466, 467
  Impulse, unit, **2,** 459
  Impulsive fields, **11,** 34
  Impurities, **16A,** 9–11
    detection of, **6A,** 29
    effect on colligative properties, **16A,** 10
    effect on electrical properties, **16A,** 10
  Impurities in polymers, **16B,** 353
    clean-up operation, **16B,** 354, 355
  Impurities in solids, control of, **6A,** 39–85
  Impurities in superconductors, **10,** 439–441
  Impurity absorption, **30,** 588
  Impurity induced absorption in alkali halides, **10,** 370–371
  Impurity induced lattice absorption, **10,** 368–372
  Impurity ions, **9A,** 128
  Impurity photoconductivity, **10,** 175–179
  Impurity photoconductors, **6B,** 341, 343
      in multielement arc plasmas, **7B,** 338–341
      relative line intensities, **7B,** 320–326
      stark widths and shifts, **7B,** 334–337
      with density known, **7B,** 327–334
      without density known, **7B,** 320–326
    well-stabilized types, **7B,** 312–314
In phase F1/F2, passing waves (amplitude difference), **8,** 44
In-hour equation, **8,** 250
In-shell transition, **17,** 579–580, 584, 596, 598
Inaccessibility postulate, **1,** 310, 322
Incandescent lamp, **8,** 27
Incandescent surface emittance, **1,** 376
Incident neutrons, **8,** 240, 264
Incident photon energy, **8,** 273
Incident power, **8,** 109, 139
Incident quanta, **8,** 216

Inclination factor, **26,** 52, 56
Inclined incidence method, for liquids, **19,** 148–149, 153–156
Inclined plane, **1,** 98–100
Inclusions, **30,** 421
Incoherent approximation, **23A,** 29
Incoherent detection, **10,** 227
Incoherent illuminations, **8,** 44–49
Incoherent light, **8,** 40
Incoherent light illuminated grating, **8,** 47–49
Incoherent optical sources, laser-produced plasmas and, **29C,** 1–18
Incoherent scattering, **21,** 148; **23A,** 20, 511
  isotope, **23A,** 528
    table, **23A,** 548
  from polymers, **23C,** 297, 329
  processes, **21,** 196, 198–206, 210, 211
  spin, **23A,** 528
Incoherent sources, **10,** 10, 11, 13–18
Incommensurate magnetic structures, **23C,** 131
Incommensurate overlayer, **22,** 256
Incommensurate phases, **23A,** 498; **23B,** 17, 42
Inconel, **14,** 378–380
Incrementing counter, **2A,** 417–418
Incubation effect, **30,** 579, 586–590
Indentation hardness, **6A,** 307–309
Independence, concept of, **28,** 12
Index matching, **23A,** 265
Index of air, **4B,** 94
Index of refraction, *see* Refractive index
  atmospheric, **12B,** 186
  molecular beam, **3,** 608
Indicatrix
  of biaxial crystal, **6B,** 266
  definition, **6B,** 264
  diagram, **6B,** 265
  equation for crystal, **6B,** 272–273
Indium, **22,** 154, 155
  chemical properties
    boiling point, **14,** 350
    melting point, **14,** 347, 350
    vapor pressure, **14,** 347, 350
  cyclotron mass measurements in, **11,** 60–61
  physical properties, **14,** 375, 377
Indium antimonide
  circular polarizer, **10,** 128, 129
  detectors, *see* Detectors, InSb
  junction laser, **10,** 67–69
  Landau level Raman scattering, **10,** 426–428
  spin-flip Raman scattering, **10,** 426–428
Indium arsenide
  junction laser, **10,** 67, 68
  Raman scattering, **10,** 427
Indium selenide, **22,** 167
Indium tin oxide (ITO), **30,** 551–552
Induced birefringence and rotation, **6B,** 271–278
Induced birefringence, **26,** 146
Induced charges, **16C,** 382
Induced emission, **13A,** 87
Induced polarization
  arrays, **24B,** 296, 301, 303–305
  data acquisition, **24B,** 300–302
  data processing, **24B,** 302–303
  mechanisms, **24B,** 285
  surveys, principles, **24B,** 292–299
  uses, **24B,** 292
Induced torque technique, in magnetoresistance measurements, **11,** 27–29
Inductance, **1,** 466
  bridges, **11,** 284
  kinetic, **11,** 267
  mutual, **2,** 18; **2A,** 2–3
  sensor, use with diaphragm gage, **18,** 570
Induction
  coefficient, **1,** 435
  decay, free, **3,** 394–395
  electromagnetic, **1,** 431
  magnetic, **1,** 518, 521–523, 532
  remanent, **1,** 557
  saturation, intrinsic, **1,** 550
  signal
    comparison with absorption, **3,** 391–392
    nuclear, **3,** 388–389
  transient nuclear, **3,** 394–400
  wind tunnel, **18,** 758
Induction (helicon) method, in magnetoresistance measurements, **11,** 26–27
Inductive energy, **8,** 133
Inductively coupled plasma (ICP), **30,** 384–393
Inductively coupled plasma (ICP), **31,** 57–59
Inductors, **2,** 44
  air-core, **2A,** 38
Industrial applications, **23C,** 405; **30,** 626
  approach to, **23C,** 475
  of PLD, **30,** 516–517
Inelastic collision, **17,** 22, 31

beam-foil interactions, **17,** 549–550
inner-shell cross sections, **17,** 279
Inelastic electron scattering, **3B,** 868–952; **22,** 200
collision cross sections, **3B,** 895, 945
diatomic molecules, **3B,** 910–912
experimental arrangements, **3B,** 872
intensities, **3B,** 886, 895
polyatomic molecules, **3B,** 912
resonances, **3B,** 934
Inelastic electron tunneling spectroscopy, **16A,** 149–169; **22,** 26, 28, 59, 95, 97
applications to polymers, **16A,** 167–169
experimental, **22,** 96
resolution, **22,** 30, 81
result tabulation, **22,** 121
sample preparation, **16A,** 154–156
sensitivity, **16A,** 161; **22,** 30
spectra, **16A,** 159–162; **22,** 97
spectrometer design, **16A,** 157–158
theoretical, **22,** 96
theory, **16A,** 162–164
tunnel junction fabrication, **22,** 98
tunneling current, **22,** 96
Inelastic energy loss (electronic energy loss), **17,** 26–27, 32–33, 38, 42–44, 150, 303–304
binary encounter, **17,** 159–173
broadening effects, **17,** 168–173
errors in analysis, **17,** 186–188
experimental determination of, **17,** 173–188
historical overview, **17,** 151–158
K-shell vacancy production, **17,** 332
at low velocities, **17,** 39–41
measurements of, in single collisions, **17,** 149–192
at medium velocities, **17,** 45–46
Inelastic energy loss in heavy particle collisions, **7A,** 442
Inelastic neutral-neutral collisons
experimental, **7A,** 325–335
state change method, **7A,** 330–334
theory, **7A,** 335–341
velocity change method, **7A,** 327–330
Inelastic neutron scattering, **3B,** 952–991
cross sections, **3B,** 956
experimental methods, **3B,** 969
spectroscopy, *see* Neutron scattering
Inelastic scattering, **23A,** 11
electrons, **22,** 193, 244
mean free path, **22,** 198, 244
of electrons, **7B,** 292–297
magnetic, **23A,** 77
single-crystal, spectrometers, **23A,** 245
spectrometers, **23A,** 237
Inelastic scattering, **8,** 257
Inelastic tunneling, **11,** 163–164
Inelasticity corrections, **23B,** 428, 472
Inertia
moment of, **1,** 82–84
terms, **23A,** 451
Inertial effects, in semiconductors, **6B,** 110–112
Inertial mass, **1,** 60, 62–63; **24B,** 4, 14,15, *see also* Pseudostationary point
Infinitely narrow slit source, **8,** 50–51
Infinity-corrected objectives, **25,** 165
Information
capacity
of a communication network, **2,** 695–696
of a noisy channel, **2,** 695
density
in continuous processes, **2,** 688–689
in discrete processes, **2,** 687–688
in several statistically connected processes, **2,** 689–690
efficiency of a system, **2,** 696
rate of a noisy channel, **2,** 692–695
theory
fundamental laws, **2,** 690–696
application to telemarketing, **2,** 696–697
unit, **2,** 687
Information quantity, **8,** 191
Information-storage gain, in nuclear-track plates, **12A,** 273
Infrared absorption, as measure of crystallization in thin polymer films, **16B,** 368
Infrared analysis
of irradiated polyethylene, **16B,** 286
of polyethylene, effect of halogenation, **16B,** 250–251
of unsaturation and branch location in polymers, **16B,** 267
Infrared catastrophe, **22,** 207
Infrared detector(s), **2B,** 331; **3,** 80–85; **3A,** 192–197; **9A,** 81
Infrared dichroism, **16C,** 173–175
Infrared emission lasers, **15A,** 232–234
Infrared instrumentation, **3,** 91–103
Infrared interferometer diagnostic method, **18,** 750–753

Infrared isotope effect, **3,** 43
Infrared laser(s), **22,** 77, *see also* Molecular infrared lasers
  frequency measurements for, **11,** 303–304
Infrared microscope, **25,** 174
Infrared optical materials, **3,** 77–80
Infrared photometry, **12A,** 415–465
Infrared pyrometer, **18,** 672
Infrared radiation, **1,** 398–402
  sources, **3,** 76–77
  detectors, **1,** 401–402
  dissociation or predissociation process in, **15B,** 295
  grating, **1,** 369
  monochromatization, **1,** 400–401
  spectrographs, **1,** 399–401
Infrared reflection
  from clean metal surface, **22,** 82
  from metal surface with adsorbate, **22,** 82
Infrared reflection-absorption spectroscopy, **22,** 27, 30, 31, 75
  detector, **22,** 79
  experimental, **22,** 76
  modulation systems, **22,** 79
  resolution, **22,** 30, 78, 81
  result tabulation, **22,** 117, 121
  sensitivity, **22,** 30
  theoretical considerations, **22,** 81
Infrared region, **13A,** 2, 8
  near, *see* Optical region
Infrared Region, **13B,** 1
  astronomical applications, **13B,** 48
  atomic spectra, **13B,** 2
  detectors, **13B,** 10, 11, 13
  emission spectra, **13B,** 3
  far-, *see* Far-infrared
  near-, *see* Optical region
  optical components, **13B,** 13
  sources, **13B,** 2
  spectra of gases, **13B,** 28
  spectra of liquids and solids, **13B,** 44
  windows, **13B,** 14
Infrared signal, upconversion of, **15B,** 144
Infrared source, **22,** 76
  diffraction grating, **22,** 77, 78
  Globar source, **22,** 77
  infrared laser, **22,** 77
  intensity, **22,** 77
  Nernst glower, **22,** 77
Infrared spectra, general theory, **3,** 42–73; **3A,** 129
Infrared spectrometer
  dispersion, **16A,** 79–80
  Fourier transform, **16A,** 80–82
  signal-to-noise ratio, **16A,** 81
Infrared spectroscopy, **3A,** 126–202
  attenuated total reflection, **16A,** 83
  end-group analysis by, **16A,** 71–72
  frequency range, **16A,** 77
  fundamentals, **16A,** 77–79
  instrumentation, **16A,** 79–82
  in polymer deformation studies, **16C,** 226–227
  sample preparation, **16A,** 83
Infrared techniques, **3A,** 172–201
Infrared transition, laser action and, **15B,** 9–10
Infrared transmission, **14,** 394–396
Infrared vibration spectrum, **20,** 127, 128
Infrared window, KRS–5, **3,** 50
Infrared, near, *see* Near infrared
InGaAsP laser, **29C,** 79
Inhomogeneities, **23B,** 93
Inhomogeneity, **8,** 183
Inhomogeneous distribution, **8,** 243
Inhomogeneous magnetic field, four, six-pole magnets, **4B,** 238–242
Inhomogeneous magnetic field, two-directional focusing property, **8,** 181
Inhomogeneous magnetic sector fields, **8,** 182
Initial density of states, **22,** 171
Injection efficiency
  of an emitter contact, **6B,** 81
  transistor, **2,** 144
Injection of carriers, **6B,** 80–82
Injection requirements for contact formation, **6B,** 87–91
Injection seeding, **29C,** 145–146
Inner coil radius, **8,** 89
Inner conductor, **8,** 113
Inner potential, **22,** 233
Inner-shell collision, **17,** 149, 163, 165
Inner-shell corrections to stopping power, **17,** 36–37
Inner-shell electron capture, **17,** 97
Inner-shell electron loss, **17,** 116–117
Inner-shell excitation, **17,** 22–23, 190
  excitation energy, **17,** 149
  heavy-ion-atom collision, **17,** 152
Inner-shell ionization, **17,** 384–385; *see also* K-shell ionization; L-shell ionization; M-shell ionization

binary encounter approximation, **17,** 200–202
electron loss processes, **17,** 75
experimental arrangements, **17,** 213–228
with heavy-ion projectiles, **17,** 203, 205–207, 221–228
molecular orbit excitation, **17,** 212–213
multiple ionization effects, **17,** 207–209
plane-wave Born approximation, **17,** 202–203
relativistic electron wavefunctions, **17,** 209–211
by relativistic projectiles, **17,** 211–212
semiclassical approximation, **17,** 202
and target thickness, **17,** 255–257
thick targets, **17,** 218–219
thin targets, **17,** 218–221
theoretical models, **17,** 197–213
and x-ray cross sections, **17,** 195–197
Inner-shell vacancy, **17,** 287, 393; *see also* K-shell vacancy; L-shell vacancy
Auger lines, **17,** 468–470, 516–517
collisional quenching quenching, **17,** 426–428
decay time, **17,** 311
doubly-excited state, **17,** 561
electron capture and, **17,** 139–142
electron promotion process, **17,** 484–485
fluorescence yield, **17,** 292–294, 297
heavy ion within solid, **17,** 257, 273–274, 277
impact-parameter dependence, **17,** 304–344
lifetime, **17,** 523
from multiple collisions, **17,** 224–227
multiple production, **17,** 374–386
two- and three-component models for target x-ray production, **17,** 257–262, 264–267, 270–272
Inner-shell vacancy cross section
charge dependence, **17,** 279–302
for heavy ions on thick targets, **17,** 224
Inner-shell vacancy fraction, beam Auger measurements, **17,** 519–524
Innershell excitation, **9A,** 137
Inorganic cluster compound, **22,** 108
InSb, *see* Indium antimonide
Inscattering
correction of neutron cross-section, **5B,** 378–382, 394
multiple, **5B,** 382–383
InSe, *see* Indium selenide
Insertion devices, **31,** 12–14, 20–21
Insertion devices, synchrotron radiation generation, **29C,** 28–30, 32–33
Insertion loss, **19,** 32–34, 61–65
Instabilities
microscopic method, use of, **21,** 142–143
postacquisition stabilization, **21,** 141, 142
reference spectra, use of, **21,** 140, 141
stabilization of, **21,** 139–143
Instantaneous field of view, **25,** 139
Instantaneous source, **6A,** 74
Institut Laue-Langevin, **23A,** 201, 222
Instrument broadening in diffractometers, **22,** 260
Instrument function, **25,** 195
Instrument profile, **25,** 195
Instrument resolution function, neutrons, **21,** 203
Instrument response, **22,** 263
function, **22,** 260
Instrumental errors, **28,** 30–31
Instrumental function of interferometer, **10,** 143
Instrumentation, **23A,** 197
lamp, **25,** 244
Instrumentation, radiometric characterization, **31,** 131–132
Instrumented heat gage models
thin wall, **18,** 670–671
thick wall, **18,** 671
surface temperature mapping, **18,** 671–672
Instruments, moving coil, **1,** 449
Insulation, **8,** 84
Insulator(s), **1,** 433; **14,** 307
ablation of, **30,** 92–106
band structure, **30,** 23, 139, 140, 144, 149, 584, 589
defects, **30,** 144, 148–149
desorption from, **30,** 71–75
experimental photoelectromagnetic measurements, **6B,** 363–364, *see also* Dielectric measurement
Integral counting, *see* Counting
Integral, convolution or superposition, **2,** 460
Integrated amplifiers, noise in, **2B,** 485
Integrated circuit testers, **2B,** 177–178
Integrated circuits, **2A,** 46, 103–146
beam-lead bipolar, **2A,** 141
biasing scheme for, **2A,** 284
capacitors in, **2A,** 119
complementary MOS (CMOS), **2A,** 124

defined, **2A,** 103, 312
design of, **2A,** 104–105
dielectric isolation in, **2A,** 108–110, 142
digital and linear, **2A,** 138–140
gate system in, **2A,** 135–136
hybrid, **2A,** 142–147
monolithic, **2A,** 105–142
MOSFETS in, **2A,** 121–126
noise figure and, **2A,** 330
NPN transistor in, **2A,** 115
resistors in, **2A,** 117
transfer efficiency in, **2A,** 131–132
2-phase clock in, **2A,** 134–137
Integrated injection logic (I^2L), **2A,** 138
Integrated optics, **29C,** 382
Integrated-circuit amplifier, **2A,** 287–288
differential amplifier and, **2A,** 292
noise level in, **2A,** 325
Integrated-circuit comparator, **2A,** 383
Integrated-circuit devices, 115–138
Integrated-circuit differential amplifier, **2A,** 303–305
Integrated-circuit diode, **2A,** 116–117
Integrated-circuit transistors, **2A,** 331, *see also* Transistor(s)
Integrated-intensity-versus-energy measurement, **22,** 268, 284, 291
Integrating chambers, **5A,** 109–110
Integrating spheres, **26,** 280
Integration, **28,** 490–491
electronic, **1,** 452
Integrator(s)
count, **2,** 354
fast, **5B,** 497
Miller, **2,** 370–371
use in multichannel spectrometers, **12B,** 272–274
Intensifying screen, for x-ray detector, **18,** 407
Intensities
of the diffracted rays, **6A,** 199–200
integrated, **3,** 324–329
of Raman bands, **3,** 124–125
relative, of Stark components of hyperfine lines, **3,** 661–662
of spectral lines, **3A,** 109, 161
Intensity, **13A,** 20, 73; **19,** 7; **30,** 24–25
electric field, **1,** 518
factor, molecular structure, **3,** 336, 338
gamma ray, **5A,** 637–641
level, **1,** 201
luminous, **1,** 389
magnetic field, **1,** 531–532
measurement, Raman, **3,** 139–140
neutron source, **5A,** 522–529
photon beam, **5B,** 508
relative internal conversion, **5A,** 585
sound, **1,** 227–228
of spectral lines, **1,** 391
standards, **3,** 178–179; **3A,** 221
ultrasonic, **1,** 227
of an unattenuated molecular beam, **3,** 593
of X-rays, **1,** 414, 415
Intensity (of light), **12A,** 193
Intensity autocorrelation function, **29C,** 205
Intensity autocorrelators, **29C,** 209–215
Intensity calculations for neutron scattering, **21,** 166
Intensity distribution measurements, **8,** 198–200
Intensity fluctuation spectroscopy, **20,** 306–310
Intensity-temperature measurement, **22,** 296
Intensity-versus-energy measurement, **22,** 269, 286
in LEED, **22,** 269
in RHEED, **22,** 270
Interaction between electrons and photons, **5B,** 233–235
Interaction of high energy electrons with atoms, **6A,** 247–248
Interaction of light and matter, **13A,** 79
Interaction parameter, **16C,** 277
determination, **16C,** 283–287
Interaction potential
inelastic energy loss, **17,** 163–167
nuclear energy loss, **17,** 72
Interaction potential, **8,** 277
Interactions
dipole-dipole, **3,** 678
dipole-quadrupole, **3,** 679
effects, **3,** 409, 411
electric quadrupole, **3,** 384–386
gamma ray photoelectric, **5A,** 621
multiple, **3,** 678–679
quadrupole perturbation, **3,** 660
of radiation with matter, **5A,** 1–89
spin-orbit, **3,** 205–221
vibronic, **3,** 190–200
X-rays with matter, **3,** 302–308
Interatomic potential, **7A,** 227–229
Interatomic spacing, **22,** 232, 234

Interband transitions, **23A,** 86
Intercalated compounds, **23B,** 1, 69
Intercavity generation, in nonlinear optical devices, **15B,** 166–167
Interception efficiency, **8,** 254
Intercombination transitions, **9A,** 138
Intercrystalline links, **16B,** 223, 231–233, 234, 346
  see also Tie molecules
Interface tension, **1,** 122
Interfaces
  BEEM, ballistic electron emission microscopy of, **27,** 307
  critical angle at
    for electrons, **27,** 312
    for holes, **27,** 321
  quantum mechanical reflection at, **27,** 317, 342
  transverse photoemission microscopy, **27,** 82
Interfacial impedance, **24B,** 286–288
Interference, **1,** 349–361; **26,** 1
  constructive, **1,** 351
  destructive, **1,** 352
  edge filters, **29C,** 350
  experiments, in superconducting rings, **11,** 286–294
  microscope, **1,** 343
  order of, **1,** 372
  Pendellösung, **23A,** 321
    by amplitude division, **23A,** 279
    interparticle, **23A,** 54
    by wavefront division, **23A,** 276
  refractometers, **1,** 359–361
  spectrum, **1,** 370–373
  X-ray, **1,** 421
Interference filter(s), **1,** 355; **7A,** 458; **9A,** 97, 99; **10,** 105–107; **26,** 31; **29B,** 377–379
  for optical pumping, **7B,** 17
Interference fringe(s), **26,** 1
  visibility, **18,** 711
Interference fringes, *see* Young's interference fringes
Interference function, **3,** 276–282; **22,** 241, 250
Interference pattern, **8,** 49
Interference polarizer, **31,** 242–243
Interferogram(s), **10,** 140–143, 282–284; **13A,** 333; **13B,** 61, 68, 70; **25,** 224
  function, **25,** 225
Interferometer(s); **1,** 57, 154, 155, 350–359; **8,** 162; **13A,** 304; *see also* Waveguide interferometer
  acoustic, **1,** 222, 223
  air-spaced interferometers, **29C,** 350
  Billet split-lens, **23A,** 278
  Bordè atom interferometer, **29C,** 295–298
  Bonze-Hart three-crystal, **23A,** 282
  cavity resonator, **4B,** 76
  connected-elements, **12C,** 158–173
  data inversion techniques for, **12C,** 151–152
  delay, **12C,** 147
  detection schemes for, **12C,** 157
  for electron plasma wave study, **9A,** 21–22
  Fabry-Perot, **13B,** 55, 84, 127
  fringes in, **12C,** 139
  gas flow, **1,** 171
  holographic, for craze displacement field, **16C,** 246
  lamellar grating, **13B,** 80
  Lummer-Gehrke, **23A,** 280
  Mach-Zender, **23A,** 280; **29C,** 389
  Marton, **23A,** 283
  Mock, **13B,** 86
  for measuring craze thickness profile, **16C,** 240–242
  Michelson, **24B,** 417–418, 432; **13B,** 62
  multielement, **12B,** 27
  noise analysis for, **12C,** 154–157
  observing techniques for, **12C,** 136
  perfect-crystal, **23A,** 280
  phase correction for, **12C,** 160
  phase measurements with, **12C,** 161
  plane-parallel interferometer, **29C,** 337–338
  for plasma measurements, **10,** 281–286
  Rayleigh, **23A,** 285
  receivers, **12B,** 221
  resolving power, **13B,** 62
  Sagnac interferometer, **29C,** 180
  scanning Michelson interferometer, **29C,** 315–331
  signal analysis for, **12C,** 141–147
  theory of two-element, **12C,** 139–157
  thermometer, **1,** 260
  two-crystal, **23A,** 283
  two-element, **12B,** 25
  white fringes in, **12C,** 142, 158
  use in measuring *g*, **24B,** 134
Interferometer, continuous wave, **19,** 118–120
  attenuation correction, **19,** 118–119

attenuation measurements from, **19,** 118
fixed length small volume, **19,** 120
parallelism, importance of, **19,** 120
velocity measurements from, **19,** 118
Interferometer, Fabry-Perot, **5B,** 53–54; **23A,** 280; **24B,** 417–419, 433
Interferometer,(s) Pierce variable path, **3,** 693–694;**29C,** 350, 362–366
Interferometer, superconducting, **11,** 294–296
IInterferometric autocorrelators, **29C,** 215–216
Interferometric principles, **8,** 147
Interferometric radiometer, **26,** 261
Interferometric spectrometer(s), **3A,** 172, 186; **10,** 146–149, 156
Interferometry, **13B,** 60, 69; **23A,** 275; **30,** 112–115, 354–355; *see also* Fourier transform spectroscopy
applications, **26,** 24
coherence in, **23A,** 303
diffraction grating, **18,** 380
evaluation procedures, **18,** 392–398
high speed recording, **18,** 739–743
holographic, **18,** 381, 550, 746–750
infrared, **18,** 750–753
measurement of electron density, **18,** 698
multiple beam resolving power, **18,** 196
principles, **18,** 196, 374–392, 739–743
reference beam, **18,** 375
shearing, **18,** 375
two color, **18,** 742–743, 794
Interlock, **14,** 455
with turbo-pump, **14,** 267
Intermediate-energy beam-foil experiments, **17,** 535–536
Intermediate-range fluctuations, **23B,** 270
Intermodulated fluorescence spectroscopy, **15B,** 235–236
Intermodulation, frequency translation and, **2B,** 236–237
Intermolecular electronic-vibrational excitation transfer, **15B,** 285–286
Intermolecular potentials at small internuclear separations, **7A,** 342–344
Internal circuit probing, in equipment testing, **2B,** 204
Internal conversion, **21,** 6
Internal electron distribution curve
L(*T*), **22,** 130
N(*T*), **22,** 132, 134, 135, 137, 144, 145
Internal energy distribution, **22,** 144
Internal friction, *see also* Attenuation
bar resonance, **24A,** 49–51
creep experiments, **24A,** 46–49
crystal imperfections, **24A,** 36
dislocation damping, **24A,** 37
errors in measurement, **24A,** 45–46
grain boundary relaxation, **24A,** 36–37
hysteresis loop data, **24A,** 39–41
pendulum free decay, **24A,** 47–49
phase transformations, **24A,** 37–38
pore fluids, effect on, **24A,** 35
pressure dependence, **24A,** 51
resonance bandwidth, **24A,** 50–51
sliding at crack surfaces, **24A,** 34
standard linear solid model, **24A,** 35
subresonance measurement, **24A,** 38–46
thermoelastic mechanisms, **24A,** 35–36
torsion measurement, **24A,** 41–45
ultrasonic measurement, **24A,** 52
Internal impedance, **8,** 11
Internal modes of vibration, **23A,** 450
Internal rotation, **3A,** 83
International Collaboration on Advanced Neutron Sources, *see* ICANS
International electrical units, **1,** 33, 34
International Gravity Formula, **24B,** 132
International Radio Consultative Committee, **12C,** 182
International temperature scale, **1,** 32, 255
Interphase, **16C,** 282
surface energy in superconductors, **6B,** 76–77
Interpolation
continuous frequency, **1,** 470
errors of, **1,** 21
Intersecting waveguides, **29C,** 382–383
Interstellar polarization, **12A,** 409
Interstitial, *see* Vacancy-interstitial
Interval estimation, **28,** 165–178
Bonferroni method, **28,** 178
comparing independent samples, **28,** 170–171
confidence interval
application, **28,** 169–170
interpretation, **28,** 167–168
confidence rectangles, **28,** 174–178
confidence regions, **28,** 173–174
prediction intervals, **28,** 171–173
single parameter confidence intervals, **28,** 165–167
transit time, **24B,** 476

Interval-censored data, **28,** 490–491
Intervals
  confidence, **5B,** 785–786
    for standard deviation of a normal distribution, **5B,** 791–792
  distribution, **5B,** 767–771
    s-fold, **5B,** 768–771
  involving *s* random events, **5B,** 768–771
  selection, "overlap" method, **5B,** 310–311
  between successive random events, **5B,** 767–768
  tolerance, **5B,** 812
Intra-atomic relaxation, **22,** 224
Intracavity frequency doubling, **29C,** 237–239
Intracavity spectral filters, **29C,** 164–169
Intramolecular coordinates, **23A,** 447
Intraresonator
  laser intensity, for steady-state laser, **15A,** 295
  short-pulse dye laser and, **15A,** 290–292
Intrinsic breakdown, **6B,** 24–30
Intrinsic curvature array, **28,** 260
Intrinsic efficiency, **8,** 219, *see also* Compton continuum; Photo-Peak spectrometer
Intrinsic electric strength, **16C,** 444
  specimen preparation, **16C,** 452
Intrinsic germanium detector, **17,** 358
Intrinsic magnetic-moment, **4B,** 164
Intrinsic peak efficiency, **8,** 225
Intrinsic photoconductors, **6B,** 341, 343
Intrinsic plasmon, **22,** 229
Intrinsic radiation, **6B,** 321
Intrinsic resolving power, **25,** 197
Intrinsic viscosity, *see* Limiting viscosity number
Invar, **14,** 379
Inverse Bremsstrahlung, **29C,** 9; **30,** 187, 391, 396
Inverse Brillouin effect, **3A,** 441
Inverse phase velocity surfaces, acoustooptic visualization, **19,** 470–473
Inverse problem, **28,** 414
Inversion, **3A,** 77
Inversion center, in crystal symmetry, **16B,** 8, 9Inversion doublet, **10,** 318, 319
Inverted Doppler-shift attenuation method, of energy-loss measurement, **17,** 65–68
Inverted magnetron gauge, **14,** 66, *see also* Ionization gauge
Inverted telephoto, **25,** 89
Inverter circuits, **2A,** 238
Inviscid flow, **1,** 128
Iodide cyanide laser, *see* ICN laser
Iodine ion laser, **15A,** 146–149
Iodine vapor Geiger counter, **7A,** 465
Iodine vapor vibrational levels, thermal distribution, **3,** 45
Iodine, atmospheric concentration, **12B,** 170
Iodine, in hollow-cathode discharge, **15A,** 148
Iodine-stabilized He-Ne lasers, optical wavelength standards, **29C,** 288–294, 307
Ioffe trap, **29B,** 136, 141
Ion accelerator, **17,** 464–467
Ion acoustic waves, **9A,** 6–7
  observation of, **9A,** 12–18
  wavelength and damping length, **9A,** 15
Ion beam
  accelerator life time measurements, **7A,** 136
  applications, **29A,** 69
  damage, **22,** 321
  positive ion sources, *see also specific source*
    acceptance, **29A,** 73–74
    brightness, **29A,** 73
    emittance, **29A,** 71–73
    selection, **29A,** 70
    thermal emitters, **29A,** 157–159
  pupil, **29A,** 192
  window, **29A,** 191
Ion beams, **8,** 169, 186
Ion bombardment, **2,** 95–97; **16C,** 439
  cleaning, **11,** 107–110
  and gas clean-up, **2,** 97
  method, **6A,** 34
Ion chambers, dosimeters, **21,** 34, 35
Ion clusters effect, in stopping cross sections, **17,** 71–72
Ion current measurement, **7A,** 104
Ion cyclotron resonance, **7A,** 383–390
  mass spectrometer, *see* Mass spectrometer
Ion desorption trajectory, **22,** 472
Ion detection, **7A,** 104
  two-photon transition monitoring and, **15B,** 242
Ion drift tube, **7A,** 373–383
Ion emission, **30,** 150, 155, 156
Ion engines, **9B,** 254–255
Ion exchange and chromatography, **6A,** 24
Ion exchange resins, ESR spin probe studies, **16A,** 473

Ion extraction slits, **8,** 186
Ion flux measurements, **9B,** 282–287
  Langmuir probe application, **9B,** 282–283
Ion formation mechanism, **22,** 467
Ion implantation, **21,** 303, 308–313, 317–320, 331; **24A,** 106
Ion laser excitation, **15A,** 90–93
  sudden perturbation excitation model of, **15A,** 92
Ion laser plasmas
  inductive or B-field coupling in, **15A,** 90
  radio-frequency excitation for production of, **15A,** 90
  self-compressed or Z-pinched discharges in, **15A,** 91
  excitation of by electron collision, **15A,** 59–97
Ion laser(s)
  continuous wave, **15A,** 71–74, 77, 89
  discharge, energy distributions of, **15A,** 81
  exciting of by atom or ion collision, **15A,** 120–166
  noble gas, *see* Noble-gas ion laser(s)
  oscillation, beam-generated plasmas for, **15A,** 92
  sputtered metal vapor, **15A,** 157–166
Ion mass analyzer, *see* Mass spectrometer
Ion microprobe, **22,** 338, 347
Ion mobility, **7A,** 201
  along linear vortex lines, **11,** 481
  liquid helium microscopic excitations and, **11,** 463–469
  temperature dependence in, **11,** 465
Ion neutralization spectrometry (INS), **21,** 257
Ion pairs, **19,** 193, 202
Ion propulsion, **7A,** 466
Ion pump(s), **1,** 180–182; **14,** 216–225, 227–230
  cleaning, **14,** 502
  rejuvenation of, **14,** 229
  titanium, **14,** 366
  in ultrahigh vacuum systems, **14,** 482, 488
Ion pumping, **4B,** 322
Ion reactions, application, *see* Heavy ion reactions
Ion scattering spectrometry (ISS), **21,** 224
Ion scattering spectroscopy (ISS), **30,** 293, 306–315
Ion separation, **7A,** 103
Ion source, **8,** 174, 177, 179, 185–186, *see also* Reactor fuel research
  shielding, **8,** 186
Ion sources, **7A,** 363–373; **22,** 313, 337, 403
  duoplasmatron, **22,** 347
  electron-impact, **22,** 313
  field-ion, **22,** 339
  field-ionization, **22,** 347
  liquid-gallium, **22,** 403
  liquid-metal, **22,** 402, 403
  liquid-metal field desorption, **22,** 403
  negative ions, **4A,** 280–282
  plasma, **22,** 314
  plasma-discharge, **22,** 337
  positive ions, **4A,** 256–279
    charge exchange, **4A,** 278–281
      multiply charged ions, **4A,** 278–279
    electron bombardment, **4A,** 260–264
    field emission, **4A,** 259
    photoionization, **4A,** 259–260
    plasma, **4A,** 264–278
      cold cathode, cyclotron type, **4A,** 267–268
      duoplasmatron, **4A,** 269–270
      extraction of ions, **4A,** 264–265
      high-current proton source, **4A,** 268–269
      history, **4A,** 265–266
      multiply charged ions, **4A,** 270–272
      occluded gas, **4A,** 276
      RF source, **4A,** 266–267
      sources for isotope separators, **4A,** 272–275
      for radioactive ions, **4A,** 274–275
      vacuum spark, **4A,** 275–276
      sputtering, **4A,** 276–278
    thermal (surface) ionization, **4A,** 257–259
  pulsed, **7A,** 370–372
Ion spectrometers, **4A,** 360–385
  double focusing, **4A,** 365
  electrostatic, **4A,** 363–365
    127° analyzer, **4A,** 364–365
  monopole, **4A,** 377–379
  quadrupole mass filter, **4A,** 366–377
  quadrupole trap, **4A,** 380–385
    electron containment, **4A,** 385
  resolving power, **4A,** 364–365, 374–376, 379, 382–384
  sector magnet, **4A,** 360–363
    focal length, **4A,** 361
    magnification, **4A,** 362

stability diagrams, **4A,** 368, 370–373, 375, 378, 381
Ion storage ring, *see also* Electron-beam ion trap; Penning trap mass spectrometer
applications, **29A,** 349
construction
electrodes, **29A,** 352–354
filament, **29A,** 355
oven, **29A,** 354
vacuum system requirements, **29A,** 355–356
cooling trapped ions, **29A,** 358
correlation techniques, **29A,** 357–358
detection of trapped ions
electronic detection, **29A,** 356
fluorescence detection, **29A,** 356–357
Paul-Penning trap, **29A,** 351
Paul-Straubel trap, **29A,** 350–351
Paul trap, **29A,** 350
Penning trap, **29A,** 351, 364
quadrupole, **29A,** 351
radiation sources for exciting trapped ions, **29A,** 359
temperature determination of trapped ions, **29A,** 358–359
Ion techniques, for studying microscopic quantum excitations, **11,** 469–481
Ion temperatures in plasmas
measurement by optical spectra, **4A,** 387–388
by probes, **4A,** 387
Ion temperatures, in noble-gas ion lasers, **15A,** 82
Ion threshold, **30,** 428
Ion transport mechanisms, **20,** 525–541
Ion traps, *see* Ion spectrometers
Ion wave echoes, **9A,** 34
observation of, **9A,** 36
Ion yield, **22,** 303, 334, 474, 482
curve, **22,** 483
negative-ion, **22,** 337
positive-ion, **22,** 337
Ion(s), **14,** 82, 83; *see also* Negative ion; Positive ion
angular distribution, **22,** 485
atomic collisions of, **7B,** 78–124
in flames, **7B,** 198
beam, **14,** 82
behavior in gases, **5A,** 91–94
bombardment, **14,** 220, 222
burial, **14,** 216
capture of by quantized vortex lines, **11,** 470–474
capture width for, **11,** 472–474
collector, **14,** 91
current, **14,** 83; **22,** 401
positive, produced by electron impact, **3,** 561
energy distribution, **22,** 492
escape of from quantized vortex lines, **11,** 474–476
fragment, kinetic energy, **3,** 537–539
free liquid surface and, **11,** 481–483
gauge, **14,** 60–75
magnetic, nuclear magnetic resonance of nuclei, **3,** 375
molecular, decomposition, **3,** 587–588
motions of in superfluid films, **11,** 483–484
heavy, response of semiconductor detector, **5A,** 273–274
negative
appearance potential, **3,** 536
drift velocity, **5A,** 92
electron impact methods, **3,** 535–539
neutralization, **5A,** 94
pair
energy loss per, **5A,** 103–105
in gas, energy to produce, **5A,** 43–44
parallel radial electrical field of, **15A,** 81
paramagnetic, in solid state, spin Hamiltonian for, **5B,** 178–179
path, **14,** 92
per second produced by electrons, **3,** 529
positive, **14,** 82
electron impact method, **3,** 527–535
pumping of noble gases, **14,** 217, 221
pressure, negative, emitted from hot filament, **3,** 561
in quantum liquid studies, **11,** 443–484
resonance, **14,** 91
retention on metal surfaces, **5B,** 77–78
source, **14,** 81, 82
structure of in liquid helium, **11,** 443–446
thresholds, **5A,** 196
trajectories, **14,** 94, 95
transitions, metastable, **3,** 587–588
transmission, **14,** 91
turbulent superfluid helium and, **11,** 479–481
vortex ring creation with, **11,** 476–479
Ion-atom collision, **17,** 1–23; *see also* Beam-foil experiments

Auger spectroscopic studies of, **17,** 468–527
charge equilibrium of high-velocity ions in matter, **17,** 73–148
direct Coulomb ionization or excitation, **17,** 486–488
electron capture, **17,** 89–110, 488–489
electron promotion, **17,** 484–486
energy loss of high-velocity ions, **17,** 25–72
heavy ions, **17,** 279
impact-parameter dependent processes, **17,** 303–344
mechanisms, **17,** 382–386, 482–503
multiple-collision phenomena, **17,** 426–431
single collision, **17,** 306–310, 401–416
inelastic energy loss in, **17,** 149–192
Ion-atom collisions, angular distribution, **7A,** 442–454
Ion-atom interchange, **7A,** 367–369
Ion-beam-sputtering (IBS) deposition, **29C,** 348
Ion-collection chambers, high resolution with, **5A,** 410
Ion-cutoff whistlers, **9B,** 134–135
Ion-induced Auger electron spectroscopy, **17,** 433–527
ion-atom collision phenomena, **17,** 468–527
emission-yield studies, **17,** 191–192
techniques, **17,** 436–467
Ion-induced x-ray spectroscopy, **17,** 355–431
chemical effects, **17,** 416–426
detectors, **17,** 356–371
fluorescence yields, **17,** 392–395
line energies, **17,** 386–392
multielectron transitions, **17,** 395–401
multiple-collision phenomena, **17,** 426–431
multiple inner-shell vacancy production, **17,** 374–386
single-collision phenomena, **17,** 401–416
targets, **17,** 371–374
Ion-ion interactions, in solid state lasers, **15A,** 179–181
Ion-ion mutual neutralization, by merging beams, **7A,** 478
Ion-molecule collision, Auger spectra following, **17,** 524–527
Ion-molecule reactions, **3,** 582–587; **3B,** 658; **7A,** 378–380, 473
with angular distribution, **7A,** 476
by crossed beams, **7A,** 472
at high pressure, **7A,** 373
using ion cyclotron resonance, **7A,** 389–390
Ion-neutral experiments, **7A,** 361ff
at low energies, **7A,** 361–390
Ion-neutralization process, **22,** 472
Ion-pair process, **3,** 536
Ion-solid interaction, **17,** 428–431
Ionic binding energies, **7A,** 380
Ionic clusters, **29B,** 41–43
Ionic conductivity, **23B,** 188
Ionic conductors, **6B,** 69–70
Ionic crystals, **23A,** 411
Ionic drift velocity, **7A,** 374, 376–378
Cunsolo method in, **11,** 456–458
measurement of in liquid helium, **11,** 454–463
signal averages in, **11,** 458–461
Ionic polarization, **10,** 347, 348
Ionic solutions, **23B,** 471
Ionization by electric and magnetic fields, **7A,** 208ff
Ionization by electron bombardment, **3B,** 713
Ionization by electron collisions, **7B,** 61–69
Ionization chambers, **8,** 249
Ionization cross section, **17,** 195, 383–385
heavy ion incident on solid target, **17,** 253, 255
K-shell ionization, **17,** 111–113, 195, 201, 210
L-subshell ionization, **17,** 196–197
Ionization cross sections, data on, **9A,** 167
Ionization cross-section (electron-atom), **7A,** 77
Ionization energy of impurity atoms in semiconductors, **10,** 177
loss of high-velocity ions, **17,** 26
Ionization energy, **17,** 35–37
Ionization gage, **1,** 182, 189–191
in shock-wave studies, **7B,** 170
Ionization gauge for atomic beam detection, **4A,** 400–401
Ionization gauge, **14,** 60–75, *see also* Pressure measurement
Bayard-Alpert, **14,** 61, 67–72
modulated, **14,** 70–72
cold cathode, **14,** 64–66
degassing, **14,** 491
errors, **14,** 161
filament, **14,** 62–63
gas sensitivity, **14,** 73–75
Groszkowski, **14,** 72
Helmer, **14,** 72, 73

inverted magnetron, **14,** 66
optical analysis, **14,** 510, 511
outgassing, **14,** 503
Penning gauge, **14,** 64–66
Phillips ionization gauge, **14,** 64–66
production of hydrogen, **14,** 159
pumping action, **14,** 65, 69–70
radioactive, **14,** 157
sensitivity, **14,** 61
X-ray limit, **14,** 61, 67, 70
Ionization of gases by heavy particles, **7A,** 428–454
energy and angular distribution of ions, **7A,** 441–454
energy and angular distribution of electrons, **7A,** 449–454
at low energies, **7A,** 435–438
mass and charge analysis of ions, **7A,** 438–440
Ionization of ions by electrons, **7A,** 100
Ionization potentials of atoms and ions, data on, **9A,** 166
Ionization potentials, **3B,** 576, 605, 606; **5A,** 231, 232–233
of free radicals, **3,** 543–546, 548
molecular, **3,** 526–527, 542–543
Ionization rate, study in shock tube, **18,** 792–794
Ionization spectroscopy, multiphoton absorption and, **15B,** 226–229
Ionization states, laser produced plasmas, **31,** 84
laser produced plasmas, **31,** 84
Ionization, **14,** 220; **17,** 383
amount of, **5A,** 103–105
associative, **15B,** 227
chamber(s), **2B,** 80; **5A,** 89–110; **5B,** 280–282; **18,** 407
argon in, **5A,** 92–93
energy measurement with, **5A,** 409–410
gas-filled, **5A,** 265
high energy, calibration, **5B,** 547–553
low energy free-air, **5B,** 539–541
parallel-plate, **5A,** 96ff
pulse type, **5A,** 89–91
quantitative operation, **5A,** 100–103
as relative beam monitor, **5B,** 499–501
of cloud chamber tracks, photometric measurement, **5A,** 449–451
coefficient, first Townsend, **2,** 72
efficiency, **5A,** 103–105; **22,** 410
curve, **3,** 527–528
element, impact, **2,** 124–125
loss
average for mu mesons, **5A,** 20ff
bubble chamber measurements, **5A,** 40–41
charged particles, **5A,** 1, 4–44
dependence on particle velocity, **5A,** 7–8
electron-positron pair, **5A,** 36–37
of electrons, **5A,** 18–19
evaluation, **5A,** 10–14
in helium, **5A,** 16
most probable, **5A,** 17–18
in oxygen, **5A,** 15
in photographic emulsions, **5A,** 34–37
in proportional counters, **5A,** 30–34
protons in solids, **5A,** 47–48
in a radiator, **5A,** 178
relativistic rise, **5A,** 7, 23–29
restricted, **5A,** 14–17
in scintillator, **5A,** 37–38
of lower atmosphere, **1,** 501
measurement
in cloud chamber, **5A,** 444–449
of electron pairs, **5A,** 678–681
in emulsions, **5A,** 240–245
multiphoton, **15B,** 264
optical avalanche, **15B,** 265
parameters, **5A,** 245–263
probabilities, for multiply ionized neon, **17,** 503–506
probability, **22,** 379
projectile, **17,** 113–114
of residual gas, **22,** 369
smallest measureable energy loss in, **5A,** 409
surface, partial pressures in, **3,** 554–558
solid targets and x-ray production, **17,** 193–278
of target, fur to light particles, **17,** 110–113
Ionization-range measurement of mass in cloud chamber, **5B,** 36–37
Ionized argon, laser transitions of, **15A,** 65, *see also* Argon
Ionized gas heat transfer measurement, **18,** 673–679
Ionized noble gases, laser lines in, **15A,** 62–64, *see also* Noble-gas laser(s)
Ionizer in atomic beam detection
electron bombardment, **4A,** 402–408
limiting sensitivity, **4A,** 407

probability of ionization, **4A,** 405–407
surface, **4A,** 397–399
Ionizers for atomic beams, **4B,** 248–250
efficiency, **4B,** 248
types, **4B,** 249–250
Ionophores, **19,** 234
Ionosphere, **12B,** 119–135
irregularities, large scale in, **12B,** 133–134
irregularities, small scale in, **12B,** 131–133
propagation in, **12B,** 120–125
Ionospheric measurements, **12B,** 134–135
Ionospheric radio wave scattering, **9B,** 139–186
from a diffuse medium, **9B,** 142–158
theory, **9B,** 142–146
electron density fluctuations, **9B,** 142
autocorrelation function, **9B,** 144
Fourier transform of, **9B,** 143
from equatorial electrojet, **9B,** 176–186
irregularities, **9B,** 183
VHF scatter observations from, **9B,** 178–179
frequency stepping, **9B,** 156
Gaussian random variable, **9B,** 145
higher-order spectral functions, **9B,** 155–156
bispectrum, **9B,** 155
incoherent type, **9B,** 158–176
plasma line, **9B,** 167–169
power spectrum, **9B,** 166–167
theory, **9B,** 160–164
total scattering cross section, **9B,** 165–166
measurement techniques, **9B,** 146–158
for polarization, **9B,** 149
for power, **9B,** 147–148
for spectrum and autocorrelation function, **9B,** 150–153
parameters of, **9B,** 140
power spectrum, **9B,** 144
pulse compression, **9B,** 156–158
ambiguity function, **9B,** 157
pulse repetition frequency (PRF), **9B,** 156
refractive index fluctuations, **9B,** 144
scattered power, **9B,** 143
true height profiles, **9B,** 141
virtual height profiles, **9B,** 141
IP, *see* Induced polarization
IPNS, **23A,** 132, 203, 226, 240
IR lasers, *see* Molecular infrared lasers
IRAS, *see* Infrared reflection-absorption spectroscopy
Iridium, as filament, **14,** 62, 63
alloys, **14,** 379
boiling point, **14,** 350
melting point, **14,** 347, 350
physical properties, **14,** 358–361
thermal conductivity, **14,** 356
vapor pressure, **14,** 347, 350
Young's modulus, **14,** 360
Iridium, reflectance spectra, **31,** 213–214
Iris coupling, **8,** 133
Iris diameters, **8,** 139
Iris size, **8,** 139–141
Iron fluoride, *see* FeF_2
Iron group ions, in solid state lasers, **15A,** 171–173
Iron transition metal ions, **15A,** 184–186
Iron, hyperfine Fe^{57} structure in, **11,** 343–344
Iron-group elements, solar abundances, **17,** 601–602
Iron-titanium, **22,** 150, 151, 152
Irradiance by light sources, **18,** 689
Irradiance, **30,** 24
Irradiated fuel samples, **8,** 177
Irradiated materials, **23C,** 441
Irradiation, **6A,** 359–361
Irradiation capsule, *see* Fuel irradiation capsule
Irradiation facilities, **21,** 19, 20; *see also* Sample preparation for irradiation
reactor, **5B,** 608–611
Irradiation test capsules, **8,** 245
Irreducible representation, **23C,** 114
Irreversible chemical reactions, **19,** 217
Irtran, **10,** 95, 96
Isentropic flow, **1,** 129
Ising model
two-dimensional, **23C,** 23
three-dimensional, **23C,** 27
Ising systems, random, **23C,** 41
ISIS, **23A,** 100, 137
Island growth, **30,** 308
Iso-butyl-phenyl-benzylidene-amino-cinnamate, *see* IBPBAC
Isochromat, **5B,** 468
Isochronal annealing, **6A,** 362–363
Isochronal focusing, **22,** 405
Isochronous reaction surfaces, **8,** 253–266
Isochronous sphere, **8,** 255
Isochronous spherical surface, **8,** 253

Isochronous surface, **8,** 258–260, 262, 265–266
detector, **8,** 258
Isochronous time-of-flight spectrometer, **8,** 253
Isocorrelation contours, **28,** 380
Isolation
collector diffusion, **2A,** 111
dialectric, **2A,** 108–110, 142
diffused, **2A,** 113–114
in monolithic integrated circuits, **2A,** 105–114
sound, **1,** 230
transformer, in magnetoresistance measurements, **11,** 22
Isolator(s), **10,** 129, 130; **29C,** 174–175, 358–361
Isomeric transition, **21,** 6
Isomorphous replacement, **20,** 252–255; **23C,** 341
Isonentropic compressibility, **19,** 183
Isoplanar technique, **2A,** 113
Isostasy, *see* Isostatic compensation
Isostatic compensation, **24B,** 148–150
Isosteric heat
of adsorption, **22,** 438, 440
of desorption, **22,** 439
Isotactic chain configuration, **16B,** 19, 340
Isothermal annealing, **6A,** 363
Isothermal crystallization
Avrami equation for, **16B,** 336
effect of molecular weight, **16B,** 337
Isothermal desorption, **22,** 436
Isothermal flames, **7B,** 218–220
Isotope abundances, **8,** 178
Isotope conversion efficiency, **21,** 20
Isotope dilution analysis, for species concentrations, **18,** 630
Isotope effects in diffusion, **21,** 79
half life separation technique, **21,** 71
spectrometry technique, **21,** 71–73
Isotope separator, **8,** 181–186, *see also* Beta-gamma-radioactive isotope separator
exercises, **8,** 186
Isotope separator, beam-foil experiment, **17,** 533
Isotope shift, **4B,** 104; **5A,** 334; **7A,** 140
Isotope substitution, **23A,** 340
Isotope(s)
availability, **23A,** 533
diffuse neutron scattering, **3,** 350
effects, **3,** 58
oxygen, **1,** 38ff
parity, **23A,** 533
production and preparation, **5B,** 80–85
shift, **5B,** 45–46
spins, **23A,** 533
table, **23B,** 476
Isotopic content, **8,** 177
Isotopic dilution method, **6A,** 34
Isotopic labeling, **20,** 42, 43
Isotopic replacement, **23C,** 397
Isotopic substitution, **23B,** 474
Isotropic angular distribution, **8,** 273
Isotropic crystals, **26,** 135
Isotropic media, **1,** 432; **26,** 113
Isotropic motion, effect on EPR spectra, **20,** 93–98
Isotropic solid, longitudinal waves in, **11,** 375–376
Isotropic yield of target x rays, **17,** 218
Isotropy, **28,** 380
Isotype heterojunctions, **15A,** 216–217
Itinerant magnetic systems, **23C,** 101

J

j - j Coupling limit, **4B,** 111
J coupling (spin-spin coupling), **16A,** 259–264
j-coupling, **3,** 369
Jacobian for particle reactions, **5B,** 825–828
Jacobian notation, **1,** 315ff
Jacquinot advantage, **13B,** 52; **25,** 198
Jacrat type chopper, **5A,** 503
Jahn-Teller effect, **3,** 193–197, 445; **3A,** 268
two-dimensional illustration, **3,** 195–196
Jamin interferometer, **26,** 12
Jamin refractometer, **1,** 361
Jansky, **12B,** 4
Jason effect, **22,** 382
Jason peak, **22,** 382
JDOS, *see* Joint density of states
JEDEC Electron Tube Council, **2B,** 339
Jeffries method of nuclear orientation, **5B,** 192
Jelley refractometer, **25,** 190
Jet
formation in cavitation, *see* Collapse of transient cavity
open, use in wind tunnel, **18,** 759, 764
shadowgraph visualization, **18,** 359
JFET, *see* Junction field effect transistor

Jicamarca Array, **12B,** 21
Jitter, **2,** 105; **2B,** 29–32
  in pulse amplitude measurement, **2B,** 112
  spark triggering, **18,** 702–703
  time of counter, **5A,** 534
Jog, **6A,** 330
Johann curved-crystal spectrometer, **17,** 365–369
Johannson curved-crystal spectrometer, **17,** 365–369
Johnson noise formula, **8,** 7
Johnson noise, **2B,** 72; **9A,** 105; **10,** 219–221; **11,** 288; **14,** 83; **22,** 202; *see also* Noise
Johnson-Matheson seismometer, **24B,** 54–55
Johnson-Nyquist relation, **11,** 245
Joining function, **23A,** 168
Joint density of states, **22,** 145, 146, 174
  energy distribution, **22,** 145, 146
Joint distribution, **28,** 14–15, 21
Joint, demountable, **14,** 294
Joly's steam calorimeter, **1,** 266, 274
Jones' matrices, **26,** 159
Josephson current
  alternating or oscillating, **11,** 233–238
  frequency dependence, **11,** 225–229
Josephson effect, **10,** 74–77, 201–210; **11,** 127, 199–200
  Dayem bridges and, **11,** 255–256
  e/*h* constant and, **11,** 273
  experiments on, **11,** 201–251
  first experimental evidence for, **11,** 212–213
  nature of, **11,** 201–208
  phonon generation by, **11,** 243–245
  supercurrent steps and, **11,** 217–22
  thermodynamic fluctuations and noise in, **11,** 245–251
  thin-film bridges and, **11,** 255–258
  volt and, **11,** 275
  voltage standards and, **11,** 274–275
Josephson frequency, **11,** 204, 301
  Riedel singularity and, **11,** 242
Josephson junction(s), **11,** 199; **13B,** 97; **24A,** 350–352
  detector, **10,** 201–210
  generator, **10,** 74–77
  mixing, **10,** 30
Josephson oscillation
  frequency modulation of, **11,** 246
  line width of, **11,** 247
  nature of, **11,** 265
  noise and, **11,** 247–249
  in thin-film bridge, **11,** 266
Josephson penetration depth, **11,** 205m238
Josephson radiation, self-coupling of, **11,** 231
Joule, **1,** 93, 272
  heat, **1,** 246, 292, 507
Joule-Thompson refrigerators, **11,** 356
Julius suspension, **1,** 448
Jump diffusion, **23B,** 230
  activated, **23B,** 160
Jump effect, **1,** 298
Jump frequency determination, **21,** 443–444, 456–457, 461–465
Jump sequences, **23B,** 135
Junction diodes, noise in, **2B,** 469–471
Junction field effect transistor (JFET), **2A,** 80–84, 89–90, 100, 121, 124; **2B,** 466, 479–481
  input resistance of, **2A,** 83
  output characteristics of, **2A,** 82, 363
Junction lasers, **10,** 67–69
Junction photoconductive cells, **2B,** 326
Junction transistors, as nonlinear devices, **2A,** 362, *see also* Transistor(s)
Junctions, **6B,** 87–88
  capacitance, **5A,** 270–271, 272
  detector, *p-n* diffused, **5A,** 265ff
  p-n, **2,** 118–121
    diode, **2,** 134–135
    rectifier cells, **2,** 127–132
  transmission line, **2,** 711–713
  width, **5A,** 270–271
J,lich, KFA, **23A,** 220

K

K shell
  corrections to stopping power, **17,** 37
  electron capture, **17,** 140, 350–354
  multielectron transitions, **17,** 396–401
  orbital velocity, **17,** 483
  proton capture from, **17,** 353–354
K-band, **28,** 197–198, 201
K-shell ionization, **17,** 195
  binary encounter approximation, **17,** 201
  cross section, **17,** 296–297, 380–381
  fluorescence yield, **17,** 478–479
    in multiple ionization, **17,** 207–209
  by heavy ions, **17,** 203, 250–272
  by light ions, **17,** 204
  low-velocity effects, **17,** 204–206

plane-wave Born approximation, **17,** 206–207, 323
projectile charge dependence, **17,** 290–292
relativistic electron wavefunctions, **17,** 209–211
by relativistic projectiles, **17,** 211
semiclassical approximation, **17,** 323–328
x-ray measurements, **17,** 229–240
K-shell vacancy, **17,** 305, 375–377, 417
Auger spectroscopy, **17,** 468–469, 476, 481, 516
charge state dependence, **17,** 292–297
Coulomb ionization, **17,** 492–499
cross section, **17,** 385
fluorescence yields, **17,** 392–395
heavy ions in solids, **17,** 273–277
hypersatellite production, **17,** 408–410
intermediate region, **17,** 338–344
ion passing through solid, **17,** 141–142
molecular orbit model, **17,** 331–337
multiple ionization, **17,** 379–382
$Ne^+ \rightarrow Ne$ collisions, **17,** 490–491
sharing ratios, **17,** 491, 493
in three-component model for target x-ray production, **17,** 270–272
in two-component model for target x-ray production, **17,** 257–267
K-shell x-ray yields, **17,** 140, 143, 287–292, 300–301, 386–392
K_2CuF_4, **23C,** 26
K_2MnF_4, **23C,** 24, 39, 151
K_2NiF_4, **23C,** 24
K_2SeO_4, **23C,** 60
Kanthal, **11,** 360
Kanzaki forces, **21,** 156, 170
Kaplan-Meier estimates, **28,** 205–206
Kapton films, **30,** 188
Kayser, **4B,** 50
Kbr transmission, **10,** 95
KCl, *see* Potassium chloride
$KCoF_3$, **23C,** 20
KDP, **9A,** 81; *see also* Potassium dihydrogen phosphate
Kellner eyepiece, **25,** 62
Kelvin
absolute voltmeter, **1,** 438, 439, 443
balance, **1,** 449
bridge, **6B,** 41–43
double bridge, **1,** 246, 457
dynamometer, **1,** 95, 96
probe, **22,** 22
relations, **6B,** 115
scale, **1,** 32, 237, 311
Kelvin-Helmholtz instability, **9A,** 338
KENS, **23A,** 132
Kepler ellipse, **24B,** 170
Keplerian telescope, **25,** 135
Kerr
cell, **1,** 386, 397
constant, **1,** 386–387
effect, **1,** 386
Kerr cell, **11,** 705; **15A,** 27; **16C,** 478; **18,** 717, 723, 729–731
Kerr effect, **6B,** 273–275, 276–278; **26,** 147, 153
optical, **20,** 194
Kerr effects, **11,** 705–706
Kerr method
longitudinal, **11,** 721–725
polar, **11,** 718–721
transverse, **11,** 725–726
Keystone distortion, **25,** 138
KF model, *see* Knotek and Feibelman model
Kiel probe, **18,** 249
Kikuchi lines, **6A,** 274–275
Kilogram, standard, **1,** 28
Kinematic calculations, **8,** 259
Kinematic conservation, **8,** 262
laws, **8,** 253
Kinematic equations, relativistic treatment, **8,** 270
Kinematic identification, **8,** 257
Kinematic recoil factor, **24A,** 119
Kinematic theory, **23C,** 84
of diffraction, **6A,** 258–261
Kinematic viscosity, **1,** 131; **18,** 797
Kinematics, *see* Q-value reactions
non-relativistic in neutron recoil experiments, **5A,** 486–488
point, **1,** 74–81
relativistic, two body problem, **5A,** 487–488
Kinematics of particle reactions, **5B,** 821–844
two-body decay, **5B,** 831–833
Kinescopes, **2,** 771–772
Kinetic coefficient, **19,** 184
Kinetic energy, **8,** 3, 7; **14,** 39–40; **22,** 467, 482; **23B,** 322
of ablation products, **30,** 492–493
of Compton electron, **5A,** 80
distribution, **22,** 491
of scatter particles, **17,** 159–161
window, **22,** 175

Kinetic inductance, **11,** 267
Kinetic parameters, **22,** 448
Kinetic properties, of ideal gases, **14,** 6
Kinetic theory, **29B,** 2–9
Kinetic theory of gases, **14,** 38, *see also* Ideal gas law hypotheses
Kinetics
chemical, mass spectrometric studies, **3,** 566–588
gas, **3,** 589, 597–602
of polymerization, DSC studies, **16B,** 332
King's furnace, **13A,** 260
King's law, **18,** 271, 297
Kingdon cage, **4A,** 401
Kinks and kink-step interactions, scanning tunneling microscopy of, **27,** 185
Kinoforms, **26,** 190
Kirchhoff
equation, **1,** 282; **3,** 561
law, **1,** 335, 373
radiation law, **18,** 465, 467, 471, 690, 707
relation, **9A,** 121
sound attenuation coefficient, **3,** 685–686
Kirchhoff's diffraction theory, **26,** 50, 51
Kirchhoff's law(s), **2,** 19; **2A,** 3, 8 n; **4A,** 291–292
Kirchhoff's radiation law, **10,** 290
Kirkendall effect, **6A,** 372–373
Kirkpatrick-Baez mirror system, **31,** 162–163, 179
Kirkwood-Riseman equation, **16A,** 435
Klein-Nishina formula, **5A,** 80–81
for Compton scattering, **5B,** 139
Kleinman's symmetry, for nonlinear optical devices, **15B,** 146
KLL-Auger transition, **17,** 469–470, 472, 474–475, 498, 519
KLM equivalent circuit, *see* Equivalent circuits
Klystron frequency instability, **3,** 491–492
Klystron tubes, **2B,** 265–269
in magnetic resonance methods, **2B,** 149–150
Klystron(s), **1,** 487; **2,** 718–724; **8,** 141, 149; **10,** 28–34; **13B,** 105; **16A,** 453 *see also* Reflex klystron
frequency multiplier, **2,** 723–724
reflex, **2,** 721–723; **3,** 20–24
range of oscillation frequency, **3,** 22
mode, **8,** 147
$KMnF_3$, **23C,** 20, 59
Knee whistlers, **9B,** 103–107, in plasmapause, **9B,** 102–103
$KNiF_3$ antiferromagnetic resonance, **10,** 411
Knife edge, **25,** 175, 178
acoustooptic technique, **19,** 484–487, 490–492
Knife-edge test, **26,** 39
Knight shift, **3,** 369–370; **21,** 398–400
pseudodipolar coupling, **21,** 400
pseudoexchange coupling, **21,** 400
Knoop hardness number, **6A,** 307
Knotek and Feibelman mechanism, **22,** 492
Knotek and Feibelman model, **22,** 469, 472, 505
Knotek-Feibel model, **30,** 60–61
Knudsen
cell, **14,** 236
effusion cell, **3,** 559–560
equation, **1,** 170
formulation, **14,** 15
gage, **1,** 191
geometry, *see* Orifice geometry
layer, **30,** 226–228, 239–242, 271–281, 492
number, **1,** 157, 167; **7A,** 231; **18,** 763, 831
Kodak spectroscopic plates, **12A,** 226–235
Koheler illumination, **25,** 99, 152
Kohn anomalies, **23A,** 386
Kohn effect, **8,** 197
Kolmogorov band, **28,** 197–198, 201
Kolmogorov length scale, **18,** 265
Kolmogorov test statistic, **28,** 201
two-sample, **28,** 208–209
Kolmogorov-Smirnov test, two-sample, **28,** 207–209
Kondo effect, **11,** 530
Korringa equation, **11,** 512
Korringa law, **23C,** 268
Kosterlitz-Thouless theory, **23C,** 26, 153
Kosterlitz-Thouless transition, **23B,** 45
Kovar, **14,** 315, 317, 319, 379
physical properties, **14,** 381
thermal expansion, **14,** 381
Kramer's degeneracy, **3,** 445
Kramers degeneracy, **10,** 391–393, 418
Kramers-Kronig relation, **31,** 268
Kramers-Kronig relations, **2B,** 136; **6B,** 210; **10,** 357–358; **13B,** 47, 179
Kramers-Kronig relationships, **15B,** 236
Kratky small-angle X-ray collimation system, **16B,** 157, 178–179

Kraus-type antenna, **12B,** 17
Kriging
application to sulfate deposition, **28,** 397–398
median-polish, **28,** 397
universal, **28,** 396–397
variance, **28,** 396–397, 400
Krishnan effect, **3A,** 420
Kron electrographic camera, **12A,** 261–265
Kronecker delta, **1,** 131
KRS transmission
Krypton fluoride laser, **15B,** 86–88
pulse energies for, **15B,** 126
tuning of, **15B,**93
Krypton, **14,** 82, 419, 420
atmospheric concentration, **12B,** 170
dense gas, **23B,** 406, 461
ion lasers, outputs for, **15A,** 93, 96
laser, **29C,** 68, 69
singly ionized, **15A,** 67–68
KS–36000 seismometer, **24B,** 67
K_sCoF_4, **23C,** 22
KTP, as OPO material, **29C,** 24, 25–259, 262
Kuizenga-Siegman equation, **29C,** 177–178
Kullback-Leibler distance, **28,** 412
Kundt constant, **6B,** 276
Kuster temperature recorder, **24B,** 192
Kuyatt-Simpson electron spectrometer, **22,** 35, 40
Kyoto University Research Reactor, **23A,** 214

L

L gasket, comparison to O ring, **14,** 479
L shell
corrections to stopping powers, **17,** 37
fluorescence yields, **17,** 393, 395
multielectron transitions, **17,** 397–398, 401
population probability, in ion-solid interactions, **17,** 428–430
λ-meter, *see* Scanning Michelson interferometer
L-shell excitation, **17,** 190
L-shell ionization, **17,** 195–197, 232–235
binary encounter approximation, **17,** 329–330
cross section, **17,** 380, 382
fluorescence yield, **17,** 479
with heavy ion projectiles, **17,** 203
multiple, **17,** 401–408
neon, **17,** 412–413
plane-wave Born approximation, **17,** 323
semiclassical approximation, **17,** 323, 329–330
x-ray measurements, **17,** 240–249, 271–272
L-shell vacancy, **17,** 375–376, 378–379, 416–418
Auger spectroscopy, **17,** 474, 476, 481, 516–518
charge-analyzed reaction to products related to production of, **17,** 345–347
Coulomb ionization, **17,** 493–495, 499
filling, **17,** 420, 422, 424–425
fluorescence yield, **17,** 292–293
fraction, **17,** 521–522
hypersatellite production, **17,** 409–410
impact-parameter dependence, **17,** 304–305
molecular orbital model, **17,** 330–331
multiple ionization, **17,** 379–382
L-subshell ionization, **17,** 195–197, 240–246
binary encounter approximation, **17,** 201
constrained binary encounter approximation, **17,** 202
with light ion projectiles, **17,** 204
low-velocity effects, **17,** 204–206
L-subshell vacancy, **17,** 380, 382
l-type doubling, **10,** 325, 328
LA branch, *see* Longitudinal acoustic branch
$La_2Mg_3(NO_3)_{12}$•$24H_2O$, *see* LMN
Label
fluorescent, **16A,** 438
paramagnetic (spin), **16A,** 457–465
Labyrinth seal, **14,** 292, 293
Lacquers, strain sensitive, **1,** 105
Ladder filters, low-pass, **2A,** 27
Ladder of powers, **28,** 268
Ladder sequencing, **30,** 432
Lag network, **2B,** 420
in series compensation, **2B,** 417
Lag window, **28,** 330–331
spectral estimators, **28,** 329–333
Lagging, **30,** 427
Lagrange equation, **25,** 22
Lagrangian and Eulerian mean square velocities, **18,** 38
Lagrangian strains, thermoelasticity and, **11,** 436
Lallemand electronic camera, **12A,** 256–260
Lamb dip, **15B,** 233–234
Lamb shielding field, **3,** 366–367
Lamb shift(s), **4B,** 131–133, 135, 138–139, 142, 144–145; **13B,** 256; **23A,** 308
in H, **4B,** 131, 133–139, 142–145

in heavy ions, **17,** 587–590
in ionized He, **4B,** 131, 139–142
n-dependence, **4B,** 142
in two-electron atoms, **4B,** 145–148
Z-dependence, **4B,** 139
Lamb wave, **19,** 501, 502
Lamb-Bohr criterion, **17,** 125–127, 142
Lamb-Dicke criterion, **29B,** 267–268
Lamb-Retherford experiment, **4B,** 131–137
Lamb-Retherford shift, **1,** 43
Lamber-Beer relation, *see* Beer's law
Lambert's cosine law, **6B,** 307
Lambert's law, **1,** 302, 335, 388; **8,** 29; **25,** 240
Lambert's law,
Lambert-Beer law, **7A,** 143, 174; **9A,** 152; **19,** 571
Lambertian source, **25,** 99
Lambertian sphere, **26,** 153
Lambertian surface, **25,** 99
Lamellar crystal(s), **16A,** 121–122, 505–506
core thickness from X-ray line shape, **16B,** 150
in melt-crystallized polymers, **16B,** 170, 244, 345
thickness from diffuse X-ray scattering, **16B,** 172–175
two-phase model, **16B,** 128
Lamellar grating, **13B,** 80
interferometer, **10,** 147–149
Lamellar thickening, **16B,** 230, 300
Lamellar thickness
effect of annealing on, **16B,** 299
effect on melting point, **16B,** 351–352
effect of molecular weight on, **16B,** 297
see also Fold period
Laminar boundary layer
heat transfer, **18,** 667
in hypersonic tunnel, **18,** 783–784
similarity solution, **18,** 832–834
study in wind tunnel, **18,** 766–767
Laminar flow, **1,** 132, 133
Laminar grating, **31,** 394–398
Lammbda doubling, **10,** 335, 336
Lamp(s)
gas-filled tungsten filament, **3,** 178
hydrogen discharge, **4A,** 311–312, 314
mercury, **4A,** 297–298, 302
microwave excited, **3,** 128
noble gas, **4A,** 314–315
quartz-iodine, **4A,** 296
sodium, **4A,** 300–301
Toronto, **3,** 127–128
tungsten, **4A,** 295–298
zirconium, **4A,** 298
Lanczos window weights, **28,** 293–295
Land microscope, **1,** 342
Landau damping, **9A,** 24, 346
of plasma waves, **9A,** 5, 6, 76
Landau effect, **5A,** 17
Landau fluctuations, **5A,** 117
Landau level, **10,** 80, 185, 186, 188, 234, 236, 260, 261, 426–428; **11,** 9
Raman scattering, **10,** 426–428
Landau tail, **28,** 445
Landau theory, **23B,** 377
Landau, L., **23B,** 326
Landau-Ginsburg theory, **11,** 201, 208
Landau-Khalatnikov theory, **23B,** 336
Landau-Placzek ration, **16A,** 178, 179, 188
Landau-Pomaranchuk model, **23B,** 386
Landé formula, **4B,** 114
Landé *g* factor, **4B,** 95
Lande splitting factor, **11,** 39
Landscape lens, **25,** 75
Langevin
function, **3,** 638–639
Langevin cross section, **7A,** 368–369
Langevin equation, **2,** 81
for static permeability, **3,** 491
Langevin function, **13B,** 171
Langevin method, **6B,** 172
Langmuir equation, **4A,** 87
Langmuir isotherm, **16C,** 319
Langmuir mode discharge operation, **2,** 92
Langmuir probe(s), **7B,** 199, 228–229; *see also* Flowing afterglow/Langmuir probe
applications, in ion flux measurements, **9B,** 282–283
measurements, **30,** 160–163
Langmuir space-charge-limited flow formula, planar diode, **29A,** 91–92
Langmuir-Childs relation, **4A,** 264
Langmuir-Saha equation, **4A,** 393, 399
Langmuir-Taylor detector, **7A,** 269–272
Lanolin, **14,** 415
LANSCE, *see* WNR
Lanthanide series ions, **15A,** 173
Lanthanides, **23C,** 195
Laplace equation, **1,** 289, 434, 443
Laplace transform, **1,** 199
in feedback control systems, **2B,** 381–385

inverse, **2B,** 382
pairs, **2B,** 384
Laplace transformation, **2,** 460–465
theorems for applying, **2,** 461
Laplace's equation, **22,** 350
Lapping in surface preparation,
hand methods, **6A,** 150
machine methods, **6A,** 150–151
Large molecules, **3A,** 299
Large-aperture seismic array, **24B,** 10, 36
Large-area thin layer photoconductor, **6B,** 342–343
Large-optical cavity laser, **15A,** 215
Large-signal conversion, in nonlinear optical devices, **15B,** 159–161
LARIS, *see* Laser ablation resonance ionization spectroscopy
Larmor
angular frequency, **3,** 361
precession, **3,** 360–363
angular frequency, **3,** 618, 619
of magnetic dipole, **3,** 617–618
Larmor equation, **20,** 5, 6
Larmor frequency, **4B,** 115, 176, 183–184; **5B,** 105, 143, 227; **23C,** 245
Larmor precession frequency, **16A,** 243
Larmor precession length, **23A,** 332
Larsen and McCleary plots, **28,** 269–270
LASA, *see* Large-aperture seismic array
Laser ablation transfer (LAT) printing, **30,** 634–636
Laser ablation, **29B,** 41, 91; **30,** 15, 78–106
resonance ionization spectroscopy, **29B,** 182–187
Laser action, infrared transition and, **15B,** 9–10
Laser anemometer, *see* Laser Doppler velocimeter
Laser applications, *see* Laser techniques and applications
Laser bands, Doppler broadening of, **15B,** 232
Laser beam
angle of incidence, **30,** 195–197
in beam-foil experiments, **17,** 530, 532
divergence, **25,** 260, 261
effects of optical elements, **30,** 197–198
focal spot properties, **30,** 191
Gaussian, **30,** 173, 179–180
homogenization, **30,** 198–200, 367, 426
polarization effects of, **30,** 195–197
spatial profile, **30,** 174
temporal profile, **30,** 174–175
waist, **25,** 260, 261
Laser breakdown, ion injection in, **11,** 453, 454
Laser centers, with highly concentrated modal beams, **15B,** 25–28, *see also* Color-center lasers
Laser cooling
Doppler cooling
lower temperature limit, **29B,** 131
principle, **29B,** 130
of atomic beams
beam deflection, **29B,** 135
longitudinal cooling
chirp slowing, **29B,** 132–133
Zeeman slowing, **29B,** 133–135
transverse cooling, **29B,** 135
sub-Doppler cooling techniques
optical pumping, **29B,** 131
stimulated Raman transitions, **29B,** 131
velocity-selective coherent population trapping, **29B,** 131
Laser degradation, **29C,** 84
Laser diodes, **29C,** 373; *see also* Semiconductor diode lasers
Laser Doppler anemometer, *see* Laser Doppler velocimeter
Laser Doppler velocimeter
characteristics of, **18,** 96
choice of technique, **18,** 232–235
combined with Raman scattering diagnostics, **18,** 431
combined with schlieren method, **18,** 369
compared with probe methods, **18,** 97
design calculation, illustration of, **18,** 235–240
illustrations of signal, **18,** 155
optical configurations, **18,** 108–110
optimization of performance, **18,** 235
photodetector output current, **18,** 110–115
principle, **18,** 97
rotating flow apparatus, **18,** 819
signal analysis, **18,** 229–232
signal processing methods, **18,** 227–228
Laser drilling, **30,** 627–628
Laser dye(s), **18,** 715–725; **25,** 263; **29C,** 66–71, 155, 159–160
list of, **15A,** 357, *see also* Dye laser(s)
Laser fireball, **7A,** 156
Laser frequencies, **3A,** 380–391
Laser fusion, **30,** 2

Laser goggles, **29C,** 157–158
Laser induced fluorescence
distinction from resonance Raman scattering, **29B,** 400–403
for metastable atom detection, **29B,** 204–206
from fast beams, **29B,** 306–308
in crossed heat pipes, **29B,** 71–72
laser sources, **29B,** 222–224
use in lifetime measurements, **29B,** 220–232, 236–239
Laser instabilities, reasons for, **15A,** 18–19
Laser intensity, **15A,** 1, 28; **30,** 24; *see also* Irradiance
Laser ions, *see* Ion laser; Aluminum ion laser; Beryllium ion laser; Copper ion laser; Lead ion laser
in solid state lasers, **15A,** 181–187
Laser levels, excitation mechanism models for, **15A,** 84–87
Laser line polarizing beam splitter cubes, **29C,** 354
Laser materials, in solid state lasers, **15A,** 181–194
Laser mixing
diodes, **10,** 21, 22
nonlinear crystals, **10,** 29
Laser operation of GaAs, **2,** 141
Laser oscillator, **15A,** 12–20
electron-excited, **15A,** 70
saturation characteristics of, **15A,** 12–14
threshold condition for, **15A,** 12
Laser output
mode-locked, **15A,** 28–30
spectral width of, **15A,** 16–17
Laser output characteristics, modification of, **15A,** 26
Laser photodetachment of I^-, **7A,** 197–200
Laser plasma, **31,** 83–90
Laser power, measurement of, **26,** 277
Laser rate equations, stability considerations and, **15A,** 18–20
Laser resonators, **10,** 50–53, 56, 64, 68, 71; **29C,** 56–58
Laser scattering, **20,** 299–336
Laser solution performance, improvements in, **15A,** 358–359
Laser sources, **25,** 258
Laser spectroscopy
Doppler-free, *see* Doppler-free laser spectroscopy
laser tuning and, **15B,** 210
two-photon absorption without Doppler broadening in, **15B,** 238–243
Laser surgery, **30,** 3, 461–465
Laser systems, **25,** 106
Laser techniques and applications, **15B,** 185–268
Laser temperature jump, **19,** 194
Laser threshold
defined, **15A,** 291
for dye lasers, **15A,** 258–263
Laser transitions
homogeneous broadening of, **15A,** 13–14, 17
inhomogeneous, **15A,** 17
inhomogeneous broadening of, **15A,** 14–16
Laser transitions, new developments in, **15B,** 83–85
Laser triggered spark gap, **18,** 703
Laser tuning, laser spectroscopy and, **15B,** 210–211
Laser, **22,** 406; *see also* individual lasers
active modulation, **18,** 718–719
argon ion, **18,** 716
for Brillouin scattering source, **16A,** 182
carbon dioxide, **18,** 714–716, 750–753, 794
chirped-pulse amplifier (CPA), **30,** 93–99
coherence properties, **18,** 710–712, 717
continuous emission, **18,** 715–716
damage testing, **29C,** 346
energy output, **18,** 718, 750–753
excimer, **30,** 334–335
femtosecond, **30,** 93–99, 190–191
free-electron (FEL), **30,** 17, 102, 190, 457–467
frequency stabilization, *see* Frequency stabilization
fundamental properties, **18,** 707–708
gas, **25,** 262
diode, **25,** 267
gasdynamic, **18,** 796
generation of harmonics, **18,** 722–725
giant pulses, **18,** 717–718
helium-neon, **18,** 714–716
infrared source, **16A,** 79
inversion of levels, **18,** 707–708
line shape, **18,** 708–709, 718
mode spectrum, **18,** 708–710
mode-locked pulses, **18,** 718–720
nanosecond, **30,** 188–190
Nd:YAG, **30,** 102, 334
Neodymium glass, **25,** 265
Neodymium YAG, **25,** 265, 267

neodymium-doped glass, **18,** 714–716
nitrogen, **18,** 721
nonlinear optical methods, **18,** 721–725
parametric methods, **18,** 724–725
picosecond, **30,** 93–94, 96–97, 190
population inversions, **29C,** 138
pumping by flash lamb, **18,** 692–693, 707
Q-switch, **18,** 717–718
Raman, **18,** 722–724
source, **16A,** 84–85
recording interferometry, **18,** 739–743
relaxation pulses, **18,** 716–717
Rubi, **25,** 265
ruby, **18,** 714–716, 717, 725, 746; **30,** 2, 72–73
saturable absorber, **18,** 717–719
semiconductor, **25,** 265, 267
solid state, **25,** 265
speckle, **18,** 707, 712–714
spectral ranges, **18,** 714–716
superradiant, **18,** 720–721
Ti:sapphire, **30,** 335
YAG, **18,** 714–716, 725

Laser-active centers, **15B,** 2–12, *see also* Color-center lasers
Laser-cooled absorbers, wavelength standards utilizing narrow resonances, **29C,** 394–303, 307
Laser-excited fluorescence, **22,** 398
Laser-induced cavitation, *see* Transient cavitation
Laser-induced desorption, **30,** 58–78, 139, 145, 154, 162
Laser-induced fluorescent line narrowing, **15B,** 245
Laser-induced line narrowing, in Doppler-free spectroscopy, **15B,** 243–248
Laser-induced periodic surface structures (LIPSS), **30,** 341–342
Laser-induced plasmas, **30,** 376–405
Laser-induced vaporization technique, **15B,**137
Laser-produced plasma, **29C,** 1–5
black-body radiators, **29C,** 6–7
debris, **29C,** 17–18
focusing, **29C,** 15
lasers used, **29C,** 4–5
physics of, **29C,** 7–13
short-pulse-length pumping, **29C,** 10, 15–17
targets, **29C,** 13–15
Laser-produced-plasma ion source
highly charged ion generation, **29A,** 185
laser types, **29A,** 185–186
Laser-selective chemistry, **15B,** 269–297
chemical kinetics in, **15B,** 278–283
dissociation with intense infrared radiation in, **15B,** 295–297
electronic excitation and, **15B,** 288–293
molecular spectroscopy and, **15B,** 277–278
principles for, **15B,** 270–271
relaxation processes in, **15B,** 283–287
Laser/matter interaction, **31,** 84–85
Lasers, **3A,** 213–216, 302–394; **13A,** 221; **20,** 164, 170, 171, 174–176, 315; *see also* Masers; Semiconductor diode lasers; Solid state lasers; Tunable lasers
application to absorption spectra, **13B,** 325
argon-chloride, **15A,** 116
argon ion, **15A,** 61
argon-oxygen, **15A,** 177
atomic and ionic gas, *see* Gas lasers
basic frequency selectivity and, **15A,** 3
chemical, **3A,** 379; *see also* Chemical lasers
chemically pumped, *see* Chemically pumped lasers
CO_2, **10,** 29, 52, 61–64
continuous wave, *see* Continuous wave dye laser(s)
copper vapor, **15A,** 44–45
crystalline solid state, **15A,** 16–17
cyclic, *see* Cyclic laser
density matrix formalism, **15A,** 2–3
depopulation of lower levels of, **15A,** 109
devices, **20,** 315
diode injection, **13B,** 290
distributed-feedback, *see* Distributed-feedback lasers
disturbed-feedback, **15A,** 247–249
dye, *see* Dye laser(s)
features common to, **15A,** 2
four-level system and, **15A,** 9
gain medium of, **15A,** 12
gas, **10,** 45–64
HCN, **10,** 48–61
helium-neon, *see* Helium-neon laser
high-pressure tunable, *see* High-pressure tunable lasers
high-pulsed output powers of, **15A,** 117
infrared, **15A,** 232–234
instabilities in, **15A,** 18
intensities of, **15A,** 1, 28

IR, *see* Molecular infrared lasers
as light amplification by stimulated emission, **15A,** 4–9
for light-scattering experiments, **9A,** 81–82
mercury ion, **15A,** 124–128
millimeter wave, **13B,** 54
mirror used with, **15A,** 10–11
mode locked, **20,** 164, 170, 171
molecular, **13B,** 280; *see also* Molecular lasers
narrow-band, **15B,** 232
near-infrared emission, **15A,** 230–231
neon-oxygen, **15A,** 116
neutral atom, **15A,** 33–59, 97–120
noble gas, **3A,** 378
noble-gas, *see* Noble gas ion laser(s)
as reactant in, **15A,** 120
optical cavity in, **15A,** 3–4
optical resonator as feedback mechanism of, **15A,** 9–12
optically pumped, **3A,** 375
p-n junction, **10,** 64–69
population inversion and, **15A,** 6–9, 20
properties of, **15A,** 1–3
pumping parameters and, **15A,** 19
purely chemical, *see* Purely chemical lasers
Q switched, **20,** 164
Raman, **10,** 70–74
red helium-neon, **15A,** 31
semiconductor diode, *see* Semiconductor diode lasers
simplest for of, **15A,** 3
single shot, **20,** 174–176
solid state, *see* Solid state lasers
spectral characteristics of, **15A,** 16–18
spectral width of, **15A,** 1–2, 11
spin-flip Raman, **13B,** 299
split-cylinder hollow-cathode, **15A,** 43
steady-state, *see* Steady-state laser
stimulated emission cross section of, **15A,** 5–6
tellurium ion, **15A,** 152–153
thallium ion, **15A,** 155–156
transient, **15A,** 35
tunable, **15A,** 205
types of, **15A,** 2
visible emission, **15A,** 231–232
ultrashort optical pulses from, **15B,** 185
Latent heat
of sublimination, **1,** 281
of transformation, **1,** 262, 281–283
of vaporization, **1,** 281ff
Latent interface, **30,** 182
Lateral diffusion of lipid spin labels, **20,** 108, 109
spin label haptens, **20,** 112, 113
Lateral resolution, **22,** 339
Lateral shearing interferometer, **26,** 17
Laterolog, **24B,** 511–516
Lattice
reciprocal, **3,** 275–276
sum factor, **3,** 278–279, 280
vector translation, **3,** 270
Lattice constants, **6A,** 285
Lattice defects, **6A,** 357–359
Lattice distortions, neutron scattering, **21,** 154, 169
Lattice dynamics, **23A,** 369
molecular, **23A,** 443
of perfect crystals, **21,** 173–174
under pressure, **23A,** 451
Lattice instabilities, **23A,** 423
Lattice parameters, **1,** 424–425
measurement, **6A,** 225–228
Lattice plane indices, **16B,** 7
Lattice planes in a crystal, **6A,** 194–197
Lattice relaxation, **30,** 45–49
Lattice site location, **21,** 277
angular scans, ion channeling, **21,** 278, 323, 331, 333, 352
displaced octahedral site, **21,** 338
flux peaking, **21,** 323, 327
interstitial sites, octahedral, **21,** 277, 324, 339
interstitial sites, tetrahedral, **21,** 324, 338, 343
triangulation technique, **21,** 277
Lattice vibrations, **1,** 293; **10,** 358–372
Lattice, crystallographic, various types, **16B,** 5
Laue
-Bragg diffraction, effect of thermal vibration, **3,** 282–283
method, **3,** 321–324
Laue atlas, **22,** 432
Laue back reflection technique, **22,** 432
Laue conditions, **22,** 241
Laue equations, **16B,** 32, 34
for oriented fiber, **16B,** 56
Laue group, in crystal symmetry, **16B,** 18
Laue method(s), **1,** 425; **6A,** 206
Laue technique, time-of-flight, **23A,** 341
Lauritsen electroscope, **4A,** 343

Lava, *see* Ceramics
Laval nozzle, **4B,** 235; **7A,** 245
  in wind tunnel, **18,** 772
Law of large numbers, **28,** 7, 25–26
Law of reflection, **31,** 147
Law of refraction, **25,** 10
Layer compound, **22,** 167, 168, 169, 180
Layer lines, **3,** 312; **16B,** 53–59
LBO, as OPO material, **29C,** 257
LC filters, passive, **2A,** 28
LC networks, **2A,** 4 n
LC-type oscillations, **2B,** 180
LDA, *see* Laser Doppler anemometer
LDS, *see* Lithium dodecyl sulfate
LDV, *see* Laser Doppler velocimeter
Le Boulengé ballistic chronograph, **1,** 75
Lead
  cleaning and plating of, **11,** 574
  dHvA studies in, **11,** 56–60
  for high-*Q* superconductivity resonators, **11,** 570, 573–575
  metaniobate, **19,** 30, 33, 54, 61–64
  total cross section for 10 Ú neutrons, **5A,** 504, 507
Lead network
  in series compensation, **2B,** 415–417
  stabilization with, **2B,** 428
Lead selenide infrared detector, **3,** 83
Lead sulfide detector, **3,** 81–82
Lead telluride infrared detector, **3,** 82–83
Lead zirconate-titanate (PZT), **19,** 30, 54
Lead-borosilicate glass, outgassing, **14,** 325
Lead-in wires, **25,** 241
Lead-ion laser, **15A,** 155
Lead-salt laser, **29C,** 78, 149
Leaderman technique, shear viscosity, **16C,** 5
Leak detection, **14,** 511–515
  acceptable rates, **14,** 510
  acetone, use of, **14,** 512
  bagging, **14,** 514
  bombing, **14,** 514–515
  bubble method, **14,** 512
  halogen leak detectors, **14,** 513
  in hermetically sealed devices, **14,** 514
  ion gauge leak detectors, **14,** 513
  ion-pump leak detectors, **14,** 513
  response time, **14,** 516
  Tesla coil, **14,** 511–512
  vacuum, **4B,** 316–318
Leak valve, coarse and fine, **14,** 297
Leak, *see also* Leak detection
  diffusion limited, **14,** 505, 506, 509, 510
  establishing presence, **14,** 506–509
  filamentary, **14,** 307, 431, 433
  high vacuum phase, **14,** 507–509
  metal porosity, **14,** 519
  outgassing and, **14,** 505–507, 509
  permeation, **14,** 505, 515
  pumpdown characteristics, **14,** 506, 508, 509
  pumpdown time, **14,** 106
  rate-of-rise method, **14,** 512
  repair techniques, **14,** 517–519
  in roughing phase, **14,** 506–507
  scratches in sealing surfaces, **14,** 518
  standard, **14,** 511
  virtual, **14,** 285, 505, 516–518
Leakage, **14,** 304, 309
  factor, magnetic, **1,** 559
  periodogram, **28,** 319–320, 324–325
  smoothing window, **28,** 330
  spectral analysis, **28,** 300–301
Leaky mode, **24B,** 585–586, *see also* Pseudo-Rayleigh wave; Reverberant wave; Stonely wave
Least count, **1,** 1, 3, 8; **6A,** 1
  of an instrument, **3,** 709
  instrument, **2,** 1; **5B,** 807
Least squares, **1,** 14–19; **2,** 10–13; **28,** 245–280
  analysis, **6A,** 10–12
  approximation function, **1,** 15
  compartmental analysis of humans, **28,** 248
  diagnostics, **28,** 266–277
  error process statistical view, **28,** 252–253
  errors in regressors, **28,** 277–280
  estimate, **1,** 4
  extreme points effect, **28,** 271–272
  fitting, **28,** 254–256
  growth model, **28,** 246–247
  method, **3,** 299–300, 718–721; **5B,** 16–18, 816–819
  nonlinear, **28,** 259–261
    weighted, **28,** 391
  Ohm's law, **28,** 246
  ordinary, properties, **28,** 256–257
  with and without outlier, **28,** 270
  n-pentane, catalytic isomerization, **28,** 247–248
  robust methods, **28,** 259
  S functions, **28,** 506–508
  solution freezing point, **28,** 248–250
  spray paint particle congealing, **28,** 250–252

statistical interference, **28,** 261–265
statistical modeling, **28,** 245–252
theory, **1,** 48
weighted, **28,** 257–259
Least-squares fit, **8,** 189
Lecher wires, **2B,** 16, 26–27
LED, *see* Light-emitting diodes
LEDs, **25,** 254
LEED, *see* Low-energy electron diffraction
Left truncation, **28,** 219–220
Left-censored observations, **28,** 218–219
fitting Weibull, **28,** 230–234
Legendre polynomials
first order, **8,** 72
recurrence formula, **8,** 73, 74
usage, **8,** 71–75
Lehner-Griffith short-period seismometer, **24B,** 52–54
Leiss grating monochromator, **7A,** 461–462
Lengevin function, **24A,** 313
Length, **1,** 27–28, 30, 53–60
distributions in scattering, **20,** 367
focal, **14,** 87
measurements, **26,** 41
standards, **25,** 253
thin down, **5A,** 300
Lens, *see* Electrostatic lens
bending, **1,** 337
cardinal points, **1,** 335
defocusing, matrix for, **5B,** 721–723
equation, magnet system, **5B,** 696–719
focusing, matrix for, **5B,** 719–720
high-current focusing, **5B,** 713–714
magnetic
defocusing action, **5B,** 699–701
focal length, **5B,** 696, 699
focusing action, **5B,** 696–699
power, **1,** 336
principal points, **1,** 335, 336
spectrometers, **5A,** 365–375
theory, **16B,** 196
Lens/eye, **8,** 43–44
Lenses, **1,** 335–337; **29C,** 343, 346, 351–353
electron, **2,** 767–768
Lenses for laser scanners, **25,** 114
Lenz-Jensen potential, **17,** 164, 166
Level crossing
in collision systems, **17,** 385, 409, 596
technique, **4B,** 148, 188–189; **7A,** 137, 138
Level-crossing spectroscopy, **29B,** 332–335
Level-crossing technique, **7B,** 4
Levitation, **19,** 314–319
Leybold R, **14,** 415
Li^+ ionization, **7A,** 114
Li-drifted semiconductors, **8,** 213
Li^7 (p,n) Be^7 reaction, **5B,** 369–370
back-angle total cross-section, **5B,** 385, 386
LIA, *see* Lock-in amplifier
Librational modes, **23A,** 442
LIF, *see* Laser induced fluorescence
Life, mean,
from counting rate, **5B,** 320
excited state, **5B,** 303
Life-period method, **7B,** 208
Lifetime (of positron), **21,** 79, 95, 97, 122–139
amplitude walk, **21,** 124, 126
coincidence fast-fast, slow-fast, **21,** 124
constant fraction discriminator, **21,** 124–125
data analysis, **21,** 132–138
deconvolution, **21,** 132, 133
detectors, **21,** 98, 123–131
electronics, **21,** 123–124
global fitting, **21,** 136, 137
mean lifetime by peakshift, **21,** 133
mistriggering of constant fraction discrimination, **21,** 131
phototube, quantum efficiency of, **21,** 126
phototubes, **21,** 124–130
positron sources, **21,** 138–139
pulseshaping, fast timing, **21,** 124, 127
scattering effects, **21,** 131
scintillator, PILOT-U, **21,** 97, 125, 126, 130
side channel selection, **21,** 130
source contribution, **21,** 134, 135
stabilization, **21,** 139–143
time-to-amplitude converter, **21,** 123–125
timing resolution, **21,** 125–130, 131, 133–134, 136
Lifetime broadening, **22,** 190, 192, 216
Lifetime measurement, **6B,** 99–109
on charge carriers, **6B,** 336, 350–351
determination by diffusion length, **6B,** 352–353
integrated detection methods, **6B,** 100–101
localized detection method, **6B,** 104–109
p-n junctions, **6B,** 107–109
steady state case, **6B,** 100–101, 104–107
transient case, **6B,** 101–104, 107–109
Lifetime measurements, **7A,** 131; **13A,** 141; *see also* Radiative lifetime; Time-of-flight measurements

beam-foil measurements, **17,** 529, 537, 570–590
and density effects, **17,** 133–134
Doppler-shift attenuation methods, **17,** 65
hydrogenic states, **17,** 566
measurements, **7A,** 131
photoelectric detection, **17,** 539
Lifetime of nuclear spin state, **3,** 378
Lifetime of the excited state(s), **3B,** 769; **22,** 171
definition, **4B,** 215
derived from Stark effect, **4B,** 223–224
direct measurement, **4B,** 218–222
relation to linewidth, **4B,** 218
Lifetime, **2,** 118
determination, **5B,** 275–338
long, **5B,** 275–302
short, **5B,** 302–338
excited state, **5B,** 353
Lifetimes, scintillator, **5A,** 129–131
Lifshitz point, **23C,** 39, 142
Lift, **1,** 147, 150
force, aerodynamic, **18,** 768–769
Ligands, **3,** 223
Light amplification, **6B,** 355–356
Light amplification by stimulated emission, **15A,** 4–9, *see also* Laser; Laser(s)
Light beating, *see* Optical mixing
Light beats, method of, **4B,** 207–208
Light curve, **28,** 351–353
Light diffraction, **19,** 76–85
Brag diffraction, **19,** 82–84, 461, 470
Raman-Nath diffraction, **19,** 76–81, 461, 470
visibility method, **19,** 84–85, 134
Light distribution function, in LDV measuring volume, **18,** 119–120
Light emitting diodes, **25,** 254
Light enhancement factor, **25,** 258
Light gas gun, **18,** 780
Light gathering power, **25,** 195
Light guides, in light-scattering optical systems, **9A,** 102
Light impurities, *see* Localized modes
Light integrating bar, **25,** 103
Light ion
Coulomb excitation, **17,** 384
ionization of inner shells, **17,** 383
projectiles, **17,** 402–403
target ionization due to, **17,** 110–113
Light monochromatization, *see* Monochromatization
Light path lengths, effects on LDV performance, **18,** 131
Light pen, in pulse amplitude measurement, **2B,** 132
Light pipes, **10,** 123, 124
in beam-foil experiments, **17,** 538
Light polarization, *see* Polarization
Light ray, vibrations of, **11,** 699
Light recording methods, **18,** 725–753
Light scattering mechanisms, **19,** 458–459
corrugation effect, **19,** 459
elastooptic effect, **19,** 458
molecular anisotropy in liquids, **19,** 147
Light scattering phenomena, **19,** 455–493
Light scattering theory of, **9A,** 73–78
autocorrelation function, **9A,** 65
correlation functions, **9A,** 66
density spectral function, **9A,** 65
dielectric coefficient, **9A,** 67
electron radius, **9A,** 65
probability density, **9A,** 66
scattered field, **9A,** 64
scattering angle, **9A,** 65
scattering cross section, **9A,** 65
spectral density, **9A,** 65
spectral function, **9A,** 66
Thomson cross section, **9A,** 67
Light scattering theory. review of, **19,** 458–467
bulk waves, scattering from, **19,** 459–464
characteristics of scattered light, **19,** 465–467
mechanisms for scattering, **19,** 458
surfaces waves, scattering by, **19,** 464
Light scattering, **3A,** 395–464; **23B,** 499; *see also* Raman scattering; Rayleigh scattering
Bessel function, **9A,** 73
coherent scattering, **9A,** 76
collective spectra, **9A,** 68–73
coherent scattering, **9A,** 73
critical mode, **9A,** 71
ion plasma oscillations, **9A,** 69
satellite lines, **9A,** 70
scattering cross section, **9A,** 70, 71
collisions, **9A,** 75
density measurements, **9A,** 106
design parameters for experiment, **9A,** 78–90
absorption coefficient, **9A,** 89
continuum radiation, **9A,** 84–85
emission coefficients, **9A,** 85

frequency spectrum measurement, **9A,** 79
light collection efficiency, **9A,** 82
light sources, **9A,** 81–82
noise, **9A,** 87–89, 104
quantum efficiency, **9A,** 83
scattered light intensity, **9A,** 82–84
scattering angle, **9A,** 81
stray light, **9A,** 81
energy and momentum transfer diagram, **16A,** 207
enhancement, **9A,** 77
by flow tracing particles, **18,** 52–60, 64, 410
from fluctuations, **9A,** 61
spectrum in a magnetic field, **9A,** 73–74
incoherent type, **9A,** 68, 77
lower hybrid frequency, **9A,** 74
optical systems, **9A,** 91–106
alignment, **9A,** 96–97
beam handling optics, **9A,** 91–94
detectors, **9A,** 102–103
grating spectrometer, **9A,** 100
light collection optics, **9A,** 94–96
multichannel devices, **9A,** 101–102
photomultipliers, **9A,** 103–105
solid-state detectors, **9A,** 105–106
spectral analysis of scattered light, **9A,** 97–101
plasma diagnostics by, **9A,** 61–113, 134–135
data interpretation, **9A,** 106–111
inhomogenous and nonstationary plasmas, **9A,** 74–75
Maxwellian plasmas, **9A,** 66–73
stable drifting plasmas, **9A,** 71
turbulent plasmas, **9A,** 77–78, 88
in polymer solutions
experimental, **16A,** 29
theory, **16A,** 27
scattering angle, **9A,** 62
scattering vector, **9A,** 62
scattering volume, **9A,** 74, 76, 111
spatial coherence in, **9A,** 62
spectral resolution observations, **9A,** 108
stimulated, **3A,** 429, 450
from thin polymer films, **16B,** 366–368
transit time broadening, **9A,** 77
upper hybrid frequency, **9A,** 73
Light sensor
image intensifier camera, **18,** 735–737
photographic material, **18,** 689, 715, 726–727
phototube, **18,** 689
spectral response, **18,** 689, 751
Light source(s), **8,** 24
absorptivity, **18,** 690
beam, **18,** 406
broad source, **18,** 406
chemical explosive, **18,** 705
coherence length, **18,** 130–132
continuous sources, **4B,** 55–57
"globars," **4B,** 56
hydrogen continuum, **4B,** 56
Lyman continuum, **4B,** 57
rare gas, **4B,** 56
synchrotron radiation, **4B,** 57
tungsten lamp, **4B,** 55
vacuum spark, **4B,** 57
diffraction-limited point source, **18,** 710
duration, **18,** 692–695, 700–702
effect of size on LDV performance, **18,** 124–126
energy, flash lamp, **18,** 693–695
exploding wire, **18,** 703–705
flash lamps, **18,** 692–695
flow visualization, **18,** 694, 695–703
general, **18,** 687–725
for holography, **26,** 199
infrared, **1,** 399; **18,** 750
for interferometers, **26,** 6
laser, **4B,** 62–63; **18,** 707–725
controlled wavelength shift, **4B,** 63
laser pump lamps, **18,** 693, 694
laser spectral ranges, **18,** 714–716
line sources, **4B,** 57–60
condensed spark, **4B,** 58
electrodeless, **4B,** 59
furnaces, flames, **4B,** 57
Geissler, **4B,** 59
"gliding spark," **4B,** 59
Pfund type, **4B,** 58
ring discharge, **4B,** 58–59
luminous efficiency, **18,** 691, 692, 698–703, 705
narrow line sources, **4B,** 60–62
atomic beam, **4B,** 62
hollow cathode tube, **4B,** 62
mercury 198 lamp, **4B,** 61
nonlinear optical methods, **18,** 721–725
physical and photometric aspects, **18,** 688–689

plasma focus, **18,** 705–706
point source, **18,** 356, 405, 710
polarization, **26,** 158
Raman scattering diagnostics, **18,** 419, 425
shadowgraph, **18,** 694, 695–703
short duration pulse, **18,** 710, 717–721, 727, 750
measurement, **18,** 720
spark, **18,** 695–703
triggering, **18,** 702–703
spatial coherence, **18,** 125, 406
spectograph, **1,** 373–376
spectral characteristics, **18,** 688, 691
spectral luminance, **18,** 690
spectral output
of lamps, **18,** 691–695
of spark, **18,** 699
temporal coherence, **18,** 125, 129
thermal, **18,** 688–707
units of output, **18,** 688–689
vacuum ultraviolet, **1,** 407
xenon flash lamp, **18,** 693–694
Zeeman, **4B,** 63–64
Light spectroscopy, application in LDV, **18,** 105–106
Light, **1,** 351; **13A,** 32
circularly polarized, **11,** 700
circulatory polarized, **1,** 104
collection uniformity, **5A,** 149
electromagnetic theory of, **1,** 376
as energy for chemical reactions, **15B,** 269
flop, **3,** 625
intensity, **13A,** 20, 73
interaction with matter, **13A,** 79
non-visible, **1,** 398–408
output
scintillation, **5A,** 147–148
saturation of CsI (Tl), **5A,** 415
plane of vibration, **1,** 377
plane polarized, **1,** 377
polarization of scattered, **3,** 122–123
propagation of in crystals, **11,** 700–702
pulse time, Čerenkov, **5A,** 454
refraction, **3,** 645–648
scintillation, collection of, **5B,** 307
sources, **3,** 161–169
atomic beam, **5B,** 52–53
to excite Raman spectra, **3,** 126–131
for hyperfine structure measurement, **5B,** 50–53
velocity of, **1,** 36–37
wavelength of gas scintillators, **5A,** 433
Light-emitting diodes (LEDs), **15A,** 195; **29C,** 373
Lightning discharge, **1,** 506
LiH, **23C,** 272
Likelihood displacement, **28,** 273
Likelihood function, **28,** 164, 213–214, 406–407
limited failure population model, **28,** 236
maximum, **28,** 222–224, 363
relative, **28,** 223–224
Weibull, **28,** 232–233
Likelihood ratio, **28,** 224–225
principle, **28,** 183
tests, **28,** 178
based on *F*-statistic, **28,** 262
Likelihood, log, **28,** 222
Lime-aluminum glass, outgassing, **14,** 325
Limited failure population model, **28,** 235–238
likelihood-based confidence, **28,** 236–238
likelihood function, **28,** 236
profile likelihood functions, **28,** 236–238
Limited Space-Charge Accumulation (LSA)
diodes, **2A,** 47
mode, **2B,** 278, 280
Limiter
circuits, **5B,** 317–318
defined, **2A,** 369
low-level, **2A,** 375
Limiters, **2,** 328–335
long-tailed pair, **2,** 333–335
Limiting in multielement tubes, **2,** 332–333
Limiting magnitude for a Fourier-transform spectrometer, **12A,** 527
Limiting viscosity number (intrinsic viscosity)
definition, **16A,** 16
effect of branching on, **16A,** 61, 62
in GPC calculations, **16A,** 58
measuring precision, **16A,** 20
relation to hydrodynamic volume, **16A,** 18
LINACS, *see* Radiofrequency linear accelerator
$LiNbO_3$ Raman laser, **10,** 72, 73, 429, 430
Linde cell, for permeability, **16C,** 365
Lindhardt potential, *see* Channeling, theory
Line
broadening, **3,** 460
spectral, **3,** 8–16
Line broadening, **4B,** 60–61, 120–123, 125, 172, 183, 224

diffraction, **23C,** 448
electrostatic broadening, **21,** 364
homogeneous broadening, **21,** 365
inhomogeneous broadening, **21,** 364, 365
neutrons, **21,** 195, 197
NMR, **21,** 364–366
Line currents, **8,** 63
Line currents, field, **8,** 55–59
examples, **8,** 58–59
Line energies
ion-induced Auger electron spectroscopy, **17,** 469–474
ion-induced x-ray spectroscopy, **17,** 386–392
Line imperfections, **6A,** 321–357
Line intensities, **3A,** 30, 109, 161
Line narrowing, laser-induced, *see* Laser-induced line narrowing
Line profile, X-ray, **1,** 426–427
Line reversal method of temperature, **18,** 466–470
Line shape(s), , **3A,** 353, **3B,** 762; **4B,** 119–125; **10,** 232–235, 239, 240, 343, 344
Ben-Reuven, **12B,** 148
ENDOR, **3,** 500
Lorentzian shape, **4B,** 125
resonance, **3,** 459–464, 618–620
Van Vleck-Weisskopf, **12B,** 148
Line sources for uv, **13A,** 215
Line spectrum, **8,** 188
Line splitting, spontaneous, **28,** 328
Line strength, **4B,** 215, 218; **7A,** 118
Line strengths, *see also* Line intensities, **3A,** 31
Line stretchers, **1,** 482
Line width(s), **3A,** 109; **5A,** 351–352; **7A,** 139
calculation, **12C,** 29–31
narrowing, motional, **3,** 381–384
resonance, **3,** 618–620, 624–625
spectral, impact theory, **3,** 677–678
X-ray, **1,** 426–427
Line-broadening parameters, electric-arc use in studies of, **7B,** 349–351
Line-shape function, **12B,** 146
Line-voltage stabilizers, **2A,** 192
Linear accelerator (linac), **10,** 27, 28, 44, 45; **17,** 14–17
beam-foil experiments, **17,** 533, 535
Linear active device, in network theory, **2A,** 22
Linear algebra, **28,** 490
Linear amplifier, **2A,** 245, *see also* Amplifier
Linear analog meter, **2B,** 9–10
Linear chain of atoms, *see* Diatomic chain
Linear chirp parameter, **29C,** 195
Linear flashlamp, **15A,** 302–305
ablating, **15A,** 305–307
reflection cavities, **15A,** 310–312
transverse dye flow in, **15A,** 311
Linear gates, **2A,** 377–382
Linear growth model, **28,** 14
Linear integrated circuits, **2A,** 138–140, *see also* Integrated circuits
applications of, **2A,** 140
Linear models, **28,** 184
Linear molecules, **3A,** 37; **10,** 303, 306–315
Linear momentum of radiation, **13A,** 39
Linear polarization analyzer, **31,** 245–246
Linear polarization, **26,** 110; **31,** 228–229
Linear polarizers, **26,** 126, 130
Linear pulse transformer, blocking oscillator and, **2A,** 408
Linear rotor, **3A,** 16, 57, 73
Linear source, **8,** 42
Linear sweep method, in energy distribution curves, **11,** 97
Linear undulator, **29C,** 33
Linear variable differential transducer, **24A,** 82; **24B,** 70, 417, 421
Linear variable differential transformer, in strain measurement, **16C,** 121
Linear velocity, **8,** 170
Linear-core devices, **2A,** 407–408
Linear-dependent transfer function, **8,** 235
Linear-sweep A/D converter, **2A,** 416–417
Linear-to-circular conversion (of polarization), **12A,** 397
Linearity, **14,** 84
of electrographic recording, **12A,** 269, 276
of helium-3 proportional counter, **5A,** 574–575
Linearization, **2,** 406–407, 411
in feedback control systems, **2B,** 425–428
Linearized rate equation, **19,** 208
Lines, **13A,** 20, 24, 254
conversion, **5A,** 582–599
Linewidth, **10,** 240, 343, 345
of crystal modes, **10,** 370, 371
of microwave gas spectra, **5B,** 23–24
natural, **4B,** 60, 215, 218, 224
parameter, **12B,** 148
spectral, *see* Spectral linewidth
Link, in network theory, **2A,** 7
Link, radio transmission, **2,** 680–681

Liouville's theorem, **29A,** 70–71
  brightness relationship, **29A,** 73
  emittance relationship, **29A,** 71–73
Lipid model membranes, **20,** 513–543
  Mueller-Rudin type, **20,** 520–525
  planar bilayers, **20,** 519–525
  solvent containing, **20,** 521, 522
  solvent depleted, **20,** 523
  solvent free, **20,** 524, 525
  spherical bilayers, **20,** 516–519
Lipid monolayers, **20,** 515, 516
Lipid/solvent systems, **20,** 520, 521
Lipoproteins, **20,** 387
Lippich polarimeter, **6B,** 271
Lippman-Denisyul hologram, **26,** 185
Liquid crystal
  infrared sensor, **18,** 751–753
  devices, **29C,** 355–357
  displays, **2B,** 8
Liquid crystalline order in polymers, **16A,** 350–352
Liquid crystals, **16A,** 430; **23B,** 545; **26,** 155; **29C,** 356, 357
  attractors, **23B,** 577
  cholesterics, **23B,** 549
  coherent scattering, **23B,** 567
  convective instabilities, **23B,** 575
  dielectric relaxation, **23B,** 551
  diffusion, **23B,** 563
  dynamic fluctuations, **23B,** 556, 558
  dynamics, **23B,** 550
  incoherent scattering, **23B,** 553
  nematic phase, **23B,** 545, 563
  order parameters, **23B,** 577
  phase transitions, **23B,** 556, 563
  pretransitional effect, **23B,** 570
  quasi-elastic scattering, **23B,** 553
  reorientations, **23B,** 550, 556
  rotational diffusion, **23B,** 554
  scattering functions, **23B,** 553
  self-diffusion, **23B,** 551
  smectic phases, **23B,** 546, 572
Liquid diffraction, **23A,** 226
Liquid droplet target sources, **31,** 88
Liquid ejector pump, *see* Ejector pump
Liquid helium
  ion production and structure in, **11,** 443–454
  negative ions in, **11,** 450
  phonon region of, **11,** 464, 467–468
  photoelectric injection of electrons in, **11,** 448–449
  roton region of, **11,** 464
  space-charge-limited diode and, **11,** 461–462
  superconductivity magnet and, **11,** 606–607
Liquid manometer, **14,** 40–47; **18,** 505; *see also* Manometer
Liquid metal, **22,** 402, 403
Liquid metals and semiconductors, **6B,** 66–67
Liquid piston pump, **14,** 112, 113, 133–136, *see also* Fore-vacuum pump
Liquid resonator, very high Q, **19,** 132
Liquid ring pump, *see* Liquid piston pump
Liquid scintillation detectors, **21,** 51, 52, 60, *see also* Pulse counting systems
Liquid seal, **14,** 278
Liquid target, charge state fraction measurements, **17,** 85
Liquid-crystal detectors, **26,** 302
Liquid-gallium source, **22,** 403
Liquid-liquid transitions, in amorphous polymers, **16B,** 331
Liquid-metal field desorption source, **22,** 403
Liquid-metal ion source
  brightness, **29A,** 142–143
  lifetime, **29A,** 139
  metals and alloys, **29A,** 146–147
  microfocused beam applications, **29A,** 140, 143–144
  needle-type source, **29A,** 144–145
  space-charge effects, **29A,** 144
Liquid-solid contact angle, **1,** 122ff
Liquids
  degassing of, **1,** 115
  drying, **1,** 115
  one-dimensional, **23C,** 60
  solvent shift in, **3,** 234
Lissajous figures, **2,** 550; **2B,** 38
  in phase measurement, **2B,** 39
$LiTbF_4$, **23C,** 33
Liter, **1,** 29
Lithium aluminum silicate, *see* β-Eucryptite
Lithium atomic form factor, **3,** 308
Lithium dodecyl sulfate, **23B,** 519
  LiCl solutions, **23B,** 527
Lithium drift germanium detector, **7A,** 408–409
Lithium fluoride, **14,** 395
  F_2^+ center lasers in, **15B,** 45–48
  window, **7A,** 458
Lithium iodide scintillator, **5A,** 417
Lithium ion conductors, **23B,** 228, 232
Lithium niobate ($LiNbO_3$), **12A,** 378; **30,** 100–101; *see also* $LiNbO_3$ Raman laser

acoustooptic devices, **29C,** 390
electrooptic effect, **29C,** 387–389
guided-wave electrooptic devices, **29C,** 389–390
as OPO material, **29C,** 257
photorefractive effect, **29C,** 391
properties, **29C,** 385–387
pyroelectric effect, **29C,** 391–392
quasi-phase effect, **29C,** 391–392
remote sensing, **29C,** 392
Lithium-drifted germanium detectors, **8,** 204
Lithium-drifted germanium diode, **17,** 358; *see also* Germanium (lithium) detector
Lithium-drifted silicon detector, for X rays, **16B,** 45
Lithium-drifted silicon diode, **17,** 358; *see also* Silicon (lithium) detector
Lithium-iodine trioxide, representative mixing experiments using, **15B,** 178–179
Lithium-like ions, **9A,** 128
Lithium-niobium trioxide
dye laser mixing in, **15B,** 160–165
temperature control for, **15B,** 176–177
Lithiumlike atom
anisotropies, **17,** 514–515
Auger decay, **17,** 586–587
excited-state populations, **17,** 506–514
K-Auger emission, **17,** 524–526
Lithography, **22,** 511
excimer laser, **30,** 367–369
Lithography, **31,** 353–354
Littman-Metcalf configuration, **29C,** 87, 260
Littrow arrangement, **25,** 201, 213
Littrow condition, **12A,** 470
Littrow configuration, **26,** 158
Littrow mounting, **1,** 363, 365, 399–400; **13A,** 286
Lloyd's mirror, **26,** 1
LMIS, *see* Liquid-metal ion source
LMM-Auger decay, **17,** 475–476
LMN, **23C,** 257
LN-trap, **14,** 300
LO branch, *see* Longitudinal optical branch
Load cell, **24A,** 74, 82
Load curve, **12A,** 421
Load deformation, solids, **6A,** 313–314
Load impedance locus, **8,** 111
Load reflection coefficient, **8,** 109
Load switching, **12C,** 33–34, 48–49
Loading method,
bending and torsion, **6A,** 321
characteristics of, **6A,** 319–321
compression, **6A,** 320
direct shear, **6A,** 320–321
tension, **6A,** 319–320
test apparatus and procedure, **6A,** 317–321
Lobe pump, **14,** 112, 113, 128–134, *see also* Forevacuum pump
compression ratio, **14,** 130, 132
with liquid piston pump, **14,** 136
power, **14,** 130
pressure limit, **14,** 131
ultimate pressure, **14,** 130
volumetric efficiency, **14,** 128–129, 132
Lobe rotation
VLBI use of, **12C,** 184
LOC, *see* Large optical cavity laser
Local density of states, **22,** 154; **27,** 7
Local field, **21,** 456, 462, 466, 471–472
Local mode absorption, **10,** 368–370
Local moments, **23B,** 120
relaxation, **23B,** 121
Local oscillators for VLBI
coherence function of, **12C,** 175
Local standard of rest, **12C,** 281
Local thermal equilibrium (LTE) in arc plasmas, **7B,** 341–347
Local thermodynamic equilbrium (LTE), **30,** 176
model, **9A,** 120–122
Local transient laser fluence, **30,** 174
Localized atomic moments, paramagnetic susceptibility of, **11,** 504–511
Localized fringes, **26,** 12
Localized modes, **21,** 190–193; **23B,** 124
of excitation, **30,** 38–42, 45–48
Localized positron, *see* Trapping of positrons
Locally enhance magnification, **22,** 370
Location measurements, **8,** 196–198
Lock-in (phase-sensitive) amplifier, **12A,** 432, 440, 461–462
Lock-in amplifier, **2A,** 331; **3B,** 569
noise reduction, **29A,** 449
principle, **29A,** 448
schematic, **29A,** 448
Lock-in amplifiers, in ac measurements, **2B,** 75
Lock-in detector, in nuclear magnetic resonance, **2B,** 142
Log-periodic array feed, **12B,** 79
Log-periodic feed, **12B,** 73
Logarithmic ammeter, *see* Ammeter
Logarithmic counting rate meter, **2B,** 10–11
Logarithmic decrement, **1,** 88, 465; **16C,** 41; **19,** 21

Logarithmic field gradient, **8,** 183
Logic circuitry, transistor, **2,** 446–449
Logic circuits, **2A,** 409–421
  release time in, **2A,** 415
  set-up time in, **2A,** 415
Logic design, **2A,** 409–416
  combinational, **2A,** 413–414
  sequential, **2A,** 414–416
Logic gates, characterization of, **2A,** 411–413
Logic, positive and negative, **2A,** 138, 411
Lond-pass filters, **25,** 284
London dispersion force, **23B,** 496
London energy, **3,** 236, 237
Londons' theory, **6B,** 75–76
Long pulse buffer rod method, **19,** 102–105
Long-chain branching, determination by GPC, **16A,** 61–63
Long-wavelength fluctuations, **23B,** 93, 109
  magnetic, **23B,** 113
Longitudinal acoustic branch, **10,** 361
Longitudinal acoustic mode, **16A,** 124–128
Longitudinal instabilities, **9A,** 279–299
Longitudinal modulus, **16A,** 178
Longitudinal optical branch, **10,** 361
Longitudinal plasma waves, **10,** 278, 279, 300, 301
Longitudinal vibrations, **8,** 3
Longitudinal waves, *see* Modes
  absorption of, **16C,** 70
  definition, **16C,** 59
Loop, **2,** 19
  analysis, **2,** 25–26
    in network theory, **2A,** 8–9
  feed, **12B,** 73, 79
  filter, laser frequency stabilization, **29C,** 124–126
Loop arrangement, **8,** 80
Loop experiments, **8,** 245
Loop system, *see* Second order loop system
Loran C stations, **12C,** 180–181
Loran C system, **12C,** 179
Lorentz (electron) microscopy, in magnetic domain studies, **11,** 726–734
Lorentz cavity, **3,** 364
Lorentz conductivity, derivation of, **7B,** 86
Lorentz equation, **16C,** 162
Lorentz factors, **3,** 281, 325; **23C,** 84
Lorentz ionization, **9A,** 154
Lorentz line shape, **13A,** 24
Lorentz polarization factor, **16B,** 66
  correction to X-ray diffraction line profile, **16B,** 139–140
Lorentz theory, **11,** 68
Lorentz-Sommerfield theory of thermoelectric power, **6B,** 116–117
Lorentzian line shape, **3,** 458; **7A,** 385
Lorentzian, width, **21,** 197, 201, 205, 218, 219
Lorenz-Lorentz equation, **3,** 646
Loss factor, **1,** 493
  dielectric, *see* Dielectric loss factor
Loss spectrum, **22,** 195
Loss tangent, **1,** 111; **16C,** 10
  dielectric, *see* Dissipation factor
Loss, *see also* Attenuation
  apparent, due to beam spreading, **19,** 239, 260–262, 276, 292
  apparent, due to side walls of specimens, **19,** 292–296
Loudness, **1,** 202, 218, 227
Love wave, **19,** 501, 502
Low density gas flows, *see* Rarefied gas flows
Low density wind tunnel, **18,** 784–785
Low EIS (LEIS), **30,** 306
Low energy ion scattering (LEIS), **21,** 224, 236–244, 263
  applications, **21,** 241–244
  calibration, **21,** 240
  equipment, **21,** 236
  spectra, **21,** 236–239
Low energy ion-neutral interactions, **7A,** 361–390
Low frequency loudspeaker design, **8,** 11–17
Low *Q*, *see Q*
Low temperature irradiation and absorption cell, **6B,** 330–331
Low temperature thermoelectric power apparatus, **6B,** 122–123
Low-density polyethylene, *see* Polyethylene
Low-dimensional magnetic systems, **23C,** 223
Low-energy beam-foil experiments, **17,** 533, 536–537
  lifetime measurements, **17,** 571, 576
Low-energy electron diffraction (LEED), **22,** 227, 241, 324, 326, 344, 487, 500, 502, 504, 505; **23B,** 1; **30,** 292
  detector, **22,** 283
  measurements, complementary to channeling studies, **21,** 354
  pattern, **11,** 635
    in photoemission sample preparation, **11,** 109–110
  theories, **22,** 271

Low-energy electron gun, **22,** 282
Low-energy electron loss spectra, **22,** 480
Low-energy electron microscopy (LEEM), **30,** 297
Low-energy ion scattering (LEIS), **30,** 310–312
Low-frequency amplifier, noise in, **2A,** 251, 329
Low-frequency instabilities, **9A,** 305–343
  ballooning and trapped particle modes, **9A,** 313–318, 332
    bounce period, **9A,** 317
    quadrupole configuration, **9A,** 315
  Bohm diffusion, **9A,** 323
  classification of, **9A,** 305–318
  collisionless type, **9A,** 329
  diffusion in, **9A,** 305
  dispersion relation, **9A,** 341–342
  drift frequency, **9A,** 308
  drift type, **9A,** 311–313, 318–320, 322–323
  experiments on, **9A,** 329–338
  floating potential, **9A,** 338
  fluctuations and particle loss, **9A,** 322–323
  flute type, **9A,** 309–311
    interchange mode, **9A,** 310
    residual mode, **9A,** 311
    resonant particle instability, **9A,** 310
  growth rates of, **9A,** 318–322
    calculation of, **9A,** 322
    finite Larmor radius stabilization, **9A,** 319
    shear stabilization, **9A,** 319
    trapped particle modes, **9A,** 321–322
  Kelvin-Helmholtz instability, **9A,** 338
  Langmuir probes, **9A,** 330, 338
  low-frequency oscillations, **9A,** 336–338
    correlation distance, **9A,** 337
    phase velocity, **9A,** 338
  multipoles, **9A,** 324, 332–336
    overstabilities, **9A,** 332
  orbit theory, **9A,** 306
  from resonant particles, **9A,** 337
  spherator or levitron, **9A,** 324–329
  stellarator, **9A,** 329
  toroidal equilibrium, **9A,** 328
  Vlasov equation, **9A,** 305
Low-noise amplifier, **2A,** 325–333
  defined, **2A,** 325
  design considerations in, **2A,** 326–332
Low-pass filter(s), **28,** 287; *see also* Filters
Low-spin state, **17,** 515
Low-temperature field-ion microscope, **22,** 384, 389
Low-temperature scanning tunneling microscopy
  experimental techniques, **27,** 355, 430
Low-velocity effects
  electron capture, **17,** 100–101, 104
  target ionization, **17,** 113, 204–205, 236
Low-velocity ion
  electronic energy loss, **17,** 39–41
  nuclear energy loss, **17,** 26, 28, 30
Lower laser levels, depopulation of, **15A,** 109
LOWESS, **28,** 500
  smoothing parameter, **28,** 505
Lowest order perturbation theory, **31,** 112
  disagreement in high-order harmonic generation, **31,** 110–111
LRC circuit, **8,** 7
LSA, *see* Limited Space-Charge Accumulation
LSI integrated circuits, **2A,** 139–140, *see also* Integrated circuits
LSS theory, of electronic energy loss, **17,** 39–44, 72
LST relation, *see* Lyddane-Sachs-Teller relation
LTE (local thermodynamic equilibrium), **7A,** 120
Lubricant, cutting, **14,** 428
Lubrication in vacuum, **4B,** 287–289
Lubrication, **1,** 98; **19,** 178
Ludwieg tube, **18,** 774
Lumen, **1,** 388, 389
Luminance, **1,** 389; **12A,** 463; **18,** 689; **25,** 100; *see also* Illumination
Luminescence, **6A,** 19; **6B,** 293–334
  processes, **5A,** 123
  rate of, **5A,** 128
Luminescent detectors, **26,** 302
Luminophors, *see* Phosphors
Luminosity, **5A,** 346, 352, 357; **12A,** 193; **13A,** 311; **22,** 220; **25,** 195
Luminosity, electron energy analyzer
  effect on figure of merit, **29A,** 218–220
  optimization, **29A,** 219–220
Luminous efficiency, **6B,** 306; **18,** 691, 692, 698–703, 705
Lummer-Brodhun cube, **25,** 123
Lummer-Brodhun photometer, **1,** 390–391
Lummer-Gehrcke plate, **1,** 370, 371
Lumped circuit measurements, **1,** 495–499
Lumped elements *R*, *L*, *C*, **2A,** 1–2
Lumped-constants resonant circuit, **8,** 138
Lunar diffraction, **12C,** 94
Lunar occultation
  bandwidth considerations in, **12C,** 109–111

finite beam effects in, **12C,** 104–105
measurements, **12C,** 91–117
methods of observations, **12C,** 92–93
noise effects in, **12C,** 106–109
occurrence of, **12C,** 92
point source in, **12C,** 95–96
refraction effects in, **12C,** 115–117
restoration technique for, **12C,** 101–104, 107–108, 111
source model fitting in, **12C,** 100–101
source-size effects in, **12C,** 98
surveys by, **12C,** 114–115
time scale of, **12C,** 96–97
Lund photometric method, **5B,** 41–43
Lutz-Edelman electrometer, **1,** 440
LVDT, *see* Linear variable differential transducer
Lyddane-Sachs-Teller relation, **10,** 375, 376, 378
Lyman capillary spark discharge, **7A,** 153–154
Lyman ghosts, **26,** 76
Lyman source, **1,** 407
Lyman-α, **7A,** 83, 458
from charge transfer of H^+, **7A,** 465
Lyot depolarizer, **12A,** 380, 386
Lyot filter, **29C,** 63, 178
Lysozyme, **20,** 386; **23C,** 348

M

M-shell excitation, **17,** 190
M-shell ionization
semiclassical approximation, **17,** 323
x-ray measurements, **17,** 249–250
M-shell vacancy, **17,** 375, 391, 416, 418, 424
Auger spectroscopy, **17,** 476, 517–518
beam-foil interaction, **17,** 524
M-shell x-ray production, **17,** 344
M/*n*, *see* Mass-to-charge ratio
Mach disc, **29B,** 28–29
Mach number
definition, **18,** 246, 830
measurement, **18,** 776–777
wind tunnel, **18,** 758, 759–761, 771–779, 844–847
Mach number, **1,** 129, 143, 158, 170, 195; **7B,** 144; **9A,** 171, 214; **29B,** 24, 33–34
free stream, **1,** 153
Mach-Zehnder interferometer, **1,** 357; **7B,** 171; **9B,** 13–20; **18,** 377, 379, 740; **26,** 12, 17; **29C,** 389
Mach-Zehnder modulator, **29C,** 178
Machining, **19,** 350–352
MacNeille polarizing prisms, **26,** 130
MaCor, **14,** 481
Macrocrack, **14,**425
Macromolecular aggregates, scattering from, **20,** 381–385
Macromolecule(s), **23C,** 287, 369
definition, **16A,** 3
shape determination, **23C,** 443
Madelung potential, **22,** 223
Madistors, **2,** 141
Magetotellurics
data acquisition, **24B,** 327–332
exploration method, **24B,** 313–314
sensor deployment, **24B,** 328–330
sources of fields, **24B,** 314–315
sources of noise in exploration, **24B,** 339–340
TE and TM modes, **24B,** 317
Magic angle, **7A,** 46; **21,** 467
Magic eye tubes, **2B,** 335
Magic tee, **6B,** 237; **8,** 143, 145, 147, 148
Magic-angle spinning, **16A,** 325
experimental implementation, **16A,** 340–343
in semicrystalline polyethylene, **16A,** 356
Magnesium
electrical conductivity, **14,** 363
thermal conductivity, **14,** 356
Young's modulus, **14,** 360
Magnesium fluoride, **12A,** 370, 374
Magnesium ion laser, **15A,** 154
Magnesium oxide (MgO), **30,** 144, 150
diffuse reflector, **3,** 129, 164
Magnesium sulfate, **19,** 193
Magnet parameter, **8,** 95
Magnet(s)
alternating gradient synchrotron, **5B,** 665
Cambridge electron accelerator, **5B,** 677
configuration, **11,** 362
design
electron synchrotron, **5B,** 654
proton synchrotron, **5B,** 657–659
electro-, **1,** 528–530
geometry, AVF cyclotron, **5B,** 679ff
hazard, **14,** 451
in Mössbauer spectroscopy, **11,** 361–363
for nuclear and quadrupole resonance, **5B,** 112
permanent, **1,** 530
pole faces for molecular beam apparatus, **3,** 609–611

pole shaping, betatron, **5B,** 628
quadrupole, **5B,** 692–694
comparison with wedge-shaped magnets, **5B,** 746–747
focal properties, **5B,** 756
with rectangular aperture, **5B,** 712–713
shimming, **1,** 529
superconducting, *see* Superconducting magnets
system
focal length, **5B,** 726–731
lens equation, **5B,** 696–719
position of focal planes, **5B,** 726–731
"strong focusing," matrix method of calculation, **5B,** 719–731
three-, matrices for, **5B,** 724–726
two-, matrices for, **5B,** 723–724
temperature requirements, **3,** 417–418
wedge-shaped
focusing equations, **5B,** 731–747
focusing with uniform field, **5B,** 740–742
horizontal focusing, **5B,** 734–735
for $n < 1$, **5B,** 736–738
for $n > 1$, **5B,** 738
vertical defocusing for $n < 0$, **5B,** 738–740
vertical focusing
by fringing field, **5B,** 732–734
for $n > 0$, **5B,** 735–736
Magnet-free instruments, **8,** 169
Magnet-free mass spectrometer, **8,** 169
Magnetic amplifiers, **2B,** 359–273, *see also* Amplifiers
gate voltage in, **2B,** 365
gross half-cycle average output voltage of, **2B,** 366
multicore circuits in, **2B,** 366–368
saturable reactors and, **2B,** 369–370
single-core circuits of, **2B,** 362–360
transient behavior and feedback circuits for, **2B,** 368–369
voltage reset circuit of, **2B,** 361–364
Magnetic analog memories, **2B,** 372–374
Magnetic analyzer, in beam-foil experiments, **17,** 536
Magnetic anisotropy, **1,** 547; **6B,** 243–247
magnetic resonance method, **6B,** 245–246
magnetization curve, **6B,** 244–245
susceptibility method, **6B,** 246–247
torque method, **6B,** 245, 247
Magnetic breakdown, galvanometric properties and, **11,** 9
Magnetic circular dichroism (MCD), **31,** 7–10, 253
Magnetic deflection measurements, in photoelectricity, **6B,** 379–380
Magnetic devices
B-H relations in, **2B,** 359–360
gating half-cycle in, **2B,** 361
hysteresis loop in, **2B,** 360–361
inductance of, **2B,** 360
operation and terminology of, **2B,** 360–362
Magnetic dipole, **8,** 63–64; **24A,** 309–310
decay process, **17,** 583–585
moment, nuclear, **4B,** 102–103
Magnetic dipole-dipole interactions, **20,** 17–19
Magnetic domain
Bitter patterns, **24A,** 457–461
Faraday effect, **24A,** 463–464
Kerr effect, **24A,** 464
multidomain minerals, **24A,** 326
scanning electron microscopy, **24A,** 462–463
transmission electron microscopy, **24A,** 461–462
Magnetic domains, **11,** 683–687
antiferromagnetic domains and, **11,** 740–743
chemical polishing and, **11,** 686
defined, **11,** 675
Faraday rotation and, **11,** 703–704
formation of, **11,** 676–683
Hall probes in, **11,** 683
Lorentz (electron) microscopy in, **11,** 726–734
magnetostatic effects in, **11,** 684
observation of, **11,** 675–743
optical and magnetooptical properties in, **11,** 697–726
phase position in, **11,** 715
powder pattern (colloid) technique in, **11,** 687–692
scanning electron microscopy in, **11,** 734–735
specialized techniques and pattern formation in, **11,** 692–697
specimen preparation in , **11,** 684–687
strain-free surfaces and, **11,** 685
X-ray studies in, **11,** 693–694
Magnetic energy selection, **8,** 188
Magnetic excitations, **23C,** 159
experimental techniques, **23C,** 165
theory, **23C,** 162
Magnetic field intensity, **8,** 59
calculation, *see* Current loop

Magnetic field(s)
  critical, **6B,** 72–73
  gradient, force on neutron, **23A,** 330
  Helmholtz pair, **7B,** 15
    for optical pumping, **7B,** 14–16
  linearity of, **2B,** 426–427
  low temperatures and, **11,** 20
  in Mössbauer spectroscopy, **11,** 361–363
  modulation in ESR, **16A,** 454
  near photomultiplier tubes, **2B,** 321
  for nuclear resonance, **6B,** 239–240
  production and measurement, **6B,** 164–165
  shift in neutron wavelength, **23A,** 322
  strength measurement, **16A,** 456
Magnetic field, **8,** 14, 24, 55, 59, 61, 63, 70, 88, 182–185, *see also* Current loop; Electron collimating magnetic field; Inhomogeneous magnetic field
  calculation, *see* Axial symmetric systems; Circular loop; Rectangular
  cross sectional area
  lines, **8,** 132
  uniformity, **8,** 99
Magnetic films
  Faraday rotation in, **11,** 703–704
  powder patterns in, **11,** 695–697
  vapor condensation method for, **11,** 693–694
Magnetic flux density, **8,** 61, 68; **24A,** 305–307
Magnetic force microscopy, **27,** 90
Magnetic hysteresis
  saturation, **24A,** 325
  remanent magnetization, **24A,** 325
Magnetic induction
  equations for, **11,** 45
  Faraday law, **24A,** 310
Magnetic interaction operator, **23A,** 59
Magnetic levitation bearing, **14,** 265
Magnetic losses
  A. C. and dynamic measurements, **6B,** 221–224
  measurements, **6B,** 222–223
  pulse measurements, **6B,** 223
  small excitation, **6B,** 222
Magnetic materials, relationships between B, H, and M vectors, **6B,** 209
Magnetic measurements at high pressures, **6A,** 429–430
Magnetic minor, **7A,** 209
Magnetic moment, **1,** 520, 523, 530ff
  electron, **1,** 43, 45–46
  of electron
    effect of uncertainty principle, **4B,** 23
    Feynman diagrams, **4B,** 24
    measurement of, **4B,** 25–30
    nuclear effects, **4B,** 26
    radiative corrections, **4B,** 23–25, 28
    theoretical value, **4B,** 24, 41
  nuclear, **1,** 46ff
  of positron, **4B,** 42
  proton, **1,** 44–45
Magnetic moment, **8,** 6
Magnetic momentum analysis, **17,** 464–465
Magnetic octupole moment, **4B,** 100
Magnetic order
  diamagnetic, **24A,** 313
  ferromagnetic, **24A,** 314
  paramagnetic, **24A,** 313, 330
Magnetic permeability, **24A,** 306, 311
Magnetic phases in rock, **24A,** 315–325
Magnetic potential
  scalar, **24A,** 307
  vector, **24A,** 305
Magnetic precession of electron, **4B,** 35
Magnetic properties of solids, **6B,** 171–247
Magnetic quadrupole decay process, **17,** 583–584, 586
Magnetic quantum number, **4B,** 64
Magnetic relaxation, **3B,** 468
Magnetic resonance in anisotropic crystals, **6B,** 245–246, *see also* Resonance phenomena
Magnetic resonance(s), **2B,** 133–151; **3B,** 465–573, 753; **10,** 385–419; *see also* EPR, ESR, NMR; Nuclear magnetic resonance
  in atomic beams, **2B,** 134
  in bulk matter, **2B,** 134–137
  diatomic molecules, **3B,** 791
Magnetic scattering, **23A,** 6, 58
  elastic, **23C,** 71
  of neutrons, **16A,** 206
  single-atom, **23C,** 71
Magnetic sector fields, *see* Inhomogeneous magnetic sector fields
Magnetic sector mass analyzer, *see* Mass spectrometer
Magnetic sector mass spectrometer, **22,** 444, 473, 493
Magnetic shield, **24A,** 359
Magnetic shielding constant, **16A,** 257
  probe for molecular structure, **16A,** 259
Magnetic shielding, **12A,** 62, 256

anisotropy, **3B,** 789
of image tubes, **12A,** 256
of photomultipliers, **12A,** 62
Magnetic spectrometers, **22,** 219
Magnetic structures, **23C,** 69
Magnetic susceptibility, *see also* Ferromagnetism and Ferrimagnetism
A. C. bridge, calibration, **6B,** 187–188
A. C. method, **6B,** 184–188
analytical balance, **6B,** 179
ballistic methods, **6B,** 188–190
corrections, **6B,** 191–193
couple method, **6B,** 183–184
Curie formula for, **2B,** 134
experimental methods, **6B,** 178–193
Faraday method, **6B,** 178–179, 181, 184
force methods, **6B,** 178–184
Gouy method, **6B,** 178–179
Hartshorn bridge, **6B,** 184–187, 188
induction methods, **6B,** 184–190
Langevin method, **6B,** 172
measurement, **6B,** 171–193
measurement in strong fields, **6B,** 190–191
metals and semiconductors, **6B,** 176–178
solids with exchange interactions, **6B,** 174–176
torsion balance, **6B,** 170–182
vibration methods, **6B,** 190
Magnetic tape recording, in telemetering, **2B,** 228, 242–245
Magnetic tape storage, in telemetering, **2B,** 220
Magnetic top, nuclear, **3,** 360–362
azimuthal component, **8,** 82
Magnetic vector potential, **8,** 55, 59, 67
azimuthal component, **8,** 82
Magnetic volume dipoles, **8,** 63
Magnetic x-ray scattering, **29C,** 40
Magnetic-core materials, **2B,** 359–360
Magnetic-susceptibility anisotropy, **3B,** 788
Magnetically confined thermonuclear fusion plasma, **17,** 599–600
Magnetically focused gun, **22,** 282
Magnetism, **1,** 517–559
Magnetization, **1,** 517, 528, 530–537; **23A,** 60; **24A,** 305–311
curve(s), **1,** 550; **6B,** 219, 244–245
computation of, **11,** 681
diffusion, **3,** 398
direction of easy, **1,** 547
induced nuclear, **3,** 376
nuclear, of system of nuclei, **3,** 386ff
nonlinear, **6B,** 218–221
orbital, **23C,** 74
oscillations in, **11,** 33
paramagnetic, Langevin expression, **3,** 376
ripple, **11,** 731–733
spin, **23C,** 74
spontaneous, **1,** 546–549
technical, **1,** 549–559
time variation, **3,** 396, 400
total, **23C,** 74
transverse, decay of, **3,** 383
Magnetization-field relationship, **6B,** 200
Magnetized dipoles, influence, *see* **8,** Fields
Magnetizing field, **24A,** 305
Magneto-acoustic waves, **9A,** 226–227
Magneto-kerr effect, **10,** 259
Magneto-mechanical ratio, **3,** 442
Magneto-optic effect, **26,** 153
Magneto-optic trap, **29B,** 137–140
Magneto-optical effect, **6B,** 276–278
Magneto-optics, **1,** 385–388
Magneto-Raman laser spectrometer, **10,** 430
Magneto-thermoelectromotive force, **6B,** 120
Magnetoacoustic measurements, **11,** 430–431
Magnetochemistry, **1,** 543
Magnetocrystalline anisotropy, **11,** 40–41
Magnetodiodes, **2A,** 61
Magnetogyric ratio, **20,** 3, 67
Magnetohydrodynamic flow studies, in shock tube, **18,** 795
Magnetohydrodynamic instability, **9A,** 395
Magnetohydrodynamics (MHD),
electromagnetic propulsion as a problem in, **9B,** 255–269
distributed currents, **9B,** 266
electrical efficiency, **9B,** 259
energy inventory in a coaxial gun, **9B,** 260
linear channel, **9B,** 257–260
magnetic probes, **9B,** 258–260
particle analyzers, **9B,** 268
pulse-forming network, **9B,** 261
transducers, **9B,** 268
Magnetometer
astatic, **24A,** 343–345
circuit, superconducting rings and, **11,** 279–285
computer interface, **24A,** 377–379
controlled magnetic field environment, **24A,** 379–381
flux-gate, **2B,** 371
as magnetic flux meter, **11,** 283, 286
based on optical pumping, **4B,** 185

signal enhancement, **24A,** 374–377
spinner, **24A,** 345–347
superconducting
gradiometer, **24A,** 366–373
principles of operation, **24A,** 348–361
sample handling, **24A,** 363–366
SQUID, *see* Superconducting quantum interference device
superconducting quantum interference device, **24A,** 352–354
vibration sample, **24A,** 427–433
Magnetomotive force (MMF), **8,** 57
Magnetooptic isolators, **29C,** 359–360
Magnetooptical effects
plasmas, **10,** 286, 287
semiconductors, **10,** 257–259
Magnetopause
definition of, **9B,** 50
plasma parameters of, **9B,** 51
Magnetoplasma effects, **10,** 249–259, 273–287
Magnetoresistance measurements
ac method in, **11,** 23–25
attaching leads in, **11,** 12
boundary-value problem in, **11,** 26–27
components in, **11,** 20
data collection and processing in, **11,** 29
de Haas-Van Alphen effect in, **11,** 214
high-pressure, **11,** 29–30
induced torque method in, **11,** 27–29
inductive (helium) method in, **11,** 26–27
isolation transformer in, **11,** 22
low-field effects in, **11,** 30
measurement techniques in, **11,** 21
noise in, **11,** 22
pulsed field method in, **11,** 25–26
pulsed magnets in, **11,** 20
sample evaluation in, **11,** 15
sample holders in, **11,** 15–20
sample preparation and evaluation in, **11,** 10–15
spark clutter in, **11,** 11
special topics in , **11,** 29–31
Magnetoresistance, **6B,** 160–165
analysis of data, **6B,** 165
anomalous longitudinal, **11,** 30–31
background, **11,** 6
of bismuth, **1,** 524
electric measuring technique, **6B,** 165
field dependence of, **11,** 4
high-field, **11,** 1–31
magnetic breakdown and, **11,** 8–9
preparation of materials, **6B,** 162–164
quadratic, **11,** 9
ratio, **6B,** 160–161
rotation curve, **11,** 5
temperature control, **6B,** 165
Magnetorotation, **7A,** 140
Magnetosheath, definition of, **9B,** 50
plasma parameters of, **9B,** 51
Magnetosonic waves, **9A,** 226
Magnetospherically-reflected whistlers, **9B,** 115–117
Magnetostatic effects, **11,** 684
Magnetostatics
compass, **24A,** 308
magnetic pole strength, **24A,** 308
potential energy, **24A,** 309
Magnetotail, definition of, **9B,** 50
Magnetovibrational scattering, **23A,** 78
Magnetron(s), **1,** 487, 518; **2,** 726–727; **10,** 38–41; **14,** 226
harmonics, **10,** 29, 154
in microwave circuits, **2B,** 272–273
voltage-tuned, **2B,** 272–273
Magnification, **1,** 336; **22,** 354
of cloud chamber pictures, **5A,** 447
of cylindrical microscope, **22,** 354
of field-electron emission technique, **22,** 357
of field-ion image, **22,** 388
of field-ion microscope, **22,** 385, 389
lateral, double focusing magnet system, **5B,** 709, 729–730
longitudinal, magnet system, **5B,** 730, 740
of point projection microscope, **22,** 357
telescope, **1,** 338
wedge-shaped magnet, **5B,** 739
Magnifier, **25,** 67, 127
Magnifying power, **1,** 340–341
Magnitude scale, infrared, **12A,** 451, 453, 456–460
zero point, **12A,** 456–460
Magnitude squared coherence, **28,** 342–346
Magnon, **10,** 401
scattering, **23A,** 80
Magnus law, **1,** 510–511
Mahan cones, **22,** 163, 164
primary, **22,** 163, 164
secondary, **22,** 163, 164
Majorana transition(s), **3B,** 763; **4B,** 250; **29B,** 136
Maksutov corrector, **1,** 340
Malter effect, **4A,** 80
Manely-Rowe equation, **12B,** 226

Manganese ammonium Totton salt, **11,** 357
Manganese fluoride, **10,** 410–412, 414, 415
Manganese nodules, **24B,** 100
Manganin resistance gage, **6A,** 415–417
Manley-Rove equation, **15B,** 149
Mann monochromator, **7A,** 185, 186
Manometer(s), **7A,** 394–396; **14,** 38–51
  accelerometer, **1,** 78, 79
  aneroid capsule, **14,** 48–49
  Bourdon, **14,** 47–48
  capacitance, **14,** 49–51
  dead-weight, **1,** 117
  diaphragm, **14,** 49–51
  differential, **1,** 166
  fluids, **14,** 417–418
  liquid, **14,** 35–36, 40–47
  McLeod gauge, **14,** 43–47
  mechanical, **1,** 117; **14,** 47–51
  spiral tube, **14,** 47–48
  U-tube, **1,** 166; **14,** 40–43
Manoxol OT, in magnetic domain studies, **11,** 688–692
Map statistics, *see* Field statistics
Mapping of band structure, see Band structure mapping
Mapping sources
  interferometer techniques for, **12C,** 149, 166–173
  VLBI techniques for, **12C,** 248–256, 270–271
Marginal distribution, **28,** 15
Marginal oscillator method, **6B,** 233–234
Marginal oscillator techniques, continuous wave, **19,** 124–129
  amplitude stability, TOUS, **19,** 127, 128
  applications, ROUS, **19,** 129–130
  automatic gain control, TOUS, **19,** 128
  calibration, electrical, TOUS, **19,** 128
  calibration, true acoustic, with TOUS, **19,** 128
  electromagnetic leakage effects, ROUS, **19,** 130
  frequency tracking, FM TOUS technique, **19,** 129
  frequency tracking, TOUS, **19,** 126
  marginal oscillator ultrasonic spectrometer, MOUS, **19,** 126
  nuclear resonance, use with, **19,** 126
  particulates in flowing fluids, measurement of, **19,** 128
  reflection oscillator ultrasonic spectrometer, ROUS, **19,** 129–130
  transducer geometry, ROUS, **19,** 129–130
  transmission oscillator ultrasonic spectrometer, TOUS, **19,** 124–129
Marginal oscillators, in nuclear magnetic resonance, **2B,** 139
Marginal testing, of equipment, **2B,** 171, *see also* Equipment Testing
Marine sediments
  characterization by acoustic backscatter, **24B,** 97
  sound transmission
    attenuation, **24B,** 82
    penetration, **24B,** 94
    velocity, **24B,** 85
Mark-Houwink constants, **16A,** 17, 58
Marking, laser, **30,** 636
Markoffian random process, **2B,** 497
Markov chain, **16A,** 289
Markov model, first order, **28,** 463
Marx circuit, **6B,** 30
Marx surge generator, for nitrogen laser, **18,** 721
Maser amplifiers, **2B,** 485–486
  basic properties of, **12B,** 246–265
  cryogenic cooling of, **12B,** 260
  millimeter wave, **12B,** 262–265
Masers, **2,** 817–818; **3,** 626; *see also* Lasers
  action
    spectral-line sources, **12C,** 30
  ammonia, **3,** 37
    beam, **15A,** 6
  electron cyclotron, **10,** 77–81
  HCN, *see* HCN maser
  molecular beam, **3A,** 303
  optical, **3,** 154–155
  systems, **12B,** 257–260
Mason equivalent circuit, *see* Equivalent circuits
Mass ablation rate, **30,** 387–389
Mass analysis, **22,** 473
Mass defects, **21,** 183–186
Mass discrimination, **8,** 170, 172
Mass dispersion coefficient, **8,** 183
Mass exchange reactions, dynamics, **8,** 277–281
Mass flow, *see* Throughput
Mass radiator, **10,** 11, 16–18
Mass resolution, **22,** 303, 344, 405
Mass separation field, **8,** 182
Mass separation, **22,** 314, 338
Mass spectra, interpretation of, **14,** 99
Mass spectrometer(s), **7A,** 362–363; **7A,** 381–383; **14,** 81, 85–87; **22,** 404, 444; *see also* Penning trap mass spectrometer

cyclotron motion, **29A,** 339–340
double-focusing, **22,** 328, 340
for electron-positron beams, **4B,** 14–16
Fourier transform ion cyclotron resonance mass spectrometer
apparatus, **29A,** 341, 344
ion cyclotron excitation, **29A,** 346
mass range, **29A,** 346
mass resolution, **29A,** 345–346
principle, **29A,** 343–345
ion cyclotron resonance mass spectrometer
apparatus, **29A,** 341–343
drift velocity of ions, **29A,** 343
operation, **29A,** 342
magnetic sector mass analyzer
double-focusing instrument, **29A,** 337–339
geometry, **29A,** 339
resolution, **29A,** 336–337, 339
magnetic-sector, **22,** 444, 473, 493
mass resolution, **29A,** 321–322, 325–356
mass-to-charge ratio, **29A,** 322–323, 326
monopole mass filter
apparatus, **29A,** 331–332
resolution, **29A,** 332–333
Omegatron, **22,** 44
apparatus, **29A,** 341–342
resolution, **29A,** 342
peaks in turbo-pump, **14,** 258, 259
photo detachment, **7A,** 200
quadrupole, **22,** 444, 487, 493
quadrupole ion storage trap
apparatus, **29A,** 333
operation, **29A,** 333–334
quadrupole mass filter
apparatus, **29A,** 327–328
mass range, **29A,** 331
operation, **29A,** 328–329
resolution, **29A,** 329–331
radial electrical field
particle trajectory, **29A,** 334
velocity focusing, **29A,** 334–335
time-of-flight, **22,** 404, 444
time-of-flight instrument
apparatus, **29A,** 324–325
axial field, **29A,** 323
ion flight time, **29A,** 324–325
kinetic energy of ions, **29A,** 325
principle, **29A,** 323
reflectron, **29A,** 326
resolution, **29A,** 325–326
sensitivity, **29A,** 326–327
transverse particle motion in uniform magnetic field, **29A,** 335–336
use at high pressure, **7A,** 370
use in ion-molecule reactions, **7A,** 365–366, 370
vacuum requirements, **29A,** 21
Wien filter
apparatus, **29A,** 340–341
mass resolution, **29A,** 340–341
focusing properties, **29A,** 340–341
Mass spectrometer, **8,** 175, *see also* Magnet-free mass spectrometer; Time-of-flight mass spectrometer
usage, *see* Reactor fuel research; Upper atmosphere research
Mass spectrometry, **3,** 525–588; **3B,** 573–667; **6A,** 32–36; **22,** 340, 398
advantages in composition diagnostics, **18,** 645–646
of afterglows, **7B,** 111–113, 124–139
in combustion studies, **7B,** 190, 198, 221–227
chemical kinetic studies, **3,** 566–588
composition measurement, **18,** 645–661
detectors, **18,** 652–653
fast reaction rates by, **3,** 579–582
flame composition measurement, **18,** 659–660
fragmentation, **18,** 646–647
free radicals, **18,** 648–649
ion sources, **18,** 646–649
Kantrowitz-Grey molecular beam inlet, **18,** 654–656
molecular structure applications, **3,** 542–565; **3B,** 605
photoionization methods in, **3,** 539–542
in polymer fracture studies, **16C,** 227–228
quadrupole (QMS), **18,** 650; **30,** 210–211
resonance ionization (RIMS), **30,** 210–217
sampling systems, **18,** 653–656
time-of-flight (TOF-MS), **18,** 649–650; **30,** 2, 143, 208, 429–430, 577
use in shock tube, **18,** 656–659
Mass spectrometry, **8,** 180
Mass spectrum, **22,** 408
Mass(es)
atomic
from microwave spectra, **5B,** 20–35
from nuclear reaction and disintegration energies, **5B,** 9–20
comparison, doublet method, **5B,** 5–7

determination of nuclei and particles, **5B,** 1–44
difference ration, **5B,** 33
excess, **5B,** 11
atomic, **5B,** 342
number, **5B,** 1–2
reduced, **5B,** 45
unit
atomic, **5B,** 1, 9
relative nuclidic, **5B,** 1, 9
Mass, **1,** 28–29, 31, 60–69
absorption coefficient for gamma rays, **5A,** 88
analyzer, **14,** 81, 82
residual gas, **14,** 85
atomic, **1,** 38–39
attenuation coefficient, total, for photons, **5A,** 673–674
center of, **1,** 82
correction factor, **5A,** 51
discrimination, **14,** 85, 87
dispersion, **14,** 87
electromagnetic measurement of, **1,** 64–66
filter, quadrupole, **14,** 92–94
flow rate, **1,** 170
fluid, **1,** 115
gravitational, **1,** 60, 61–62
inertial, **1,** 60, 62–63
moment of, **1,** 81
range, **14,** 82
resolving capability, **14,** 82
spectograph, **1,** 64
units, **14,** 86
Mass-to-charge ratio, **22,** 405
Mass/frequency, relation, **8,** 172
Mössbauer effect, **13A,** 25, 146
experimental configurations in, **11,** 308
susceptibility thermometer and, **11,** 532
Mössbauer experiments, radioactive sources of, **11,** 350–351
Mössbauer gamma-ray transition energy, **11,** 311
Mössbauer spectra
Auger processes and, **11,** 310
scattering geometry in, **11,** 309
Mössbauer spectrometers, **11,** 316–351
accuracy of, **11,** 339
calibration and standards for, **11,** 342–346
constant-velocity systems and, **11,** 318–319
drives and data collection in, **11,** 316–346
electromagnetic transducers and, **11,** 331–337
error signal in, **11,** 326–327
fast flyback operation in, **11,** 240
feedback velocity control in, **11,** 326–331
gamma-ray detection in, **11,** 346–350
modulator mode in, **11,** 320–323
nuclear counter instrumentation for, **11,** 348–350
preamplifiers and, **11,** 349
resonance detectors in, **11,** 347
scale factor for, **11,** 344
scintillation counter and, **11,** 346–347
sinusoidal wave form and, **11,** 339
time mode in, **11,** 323–326
transducers for, **11,** 331–337
"tuning" of, **11,** 317
zero-velocity point for, **11,** 344–345
Mössbauer spectroscopy
auxiliary equipment for, **11,** 352–366
constant acceleration in, **11,** 337–338
cooled sources in, **11,** 356–357
data analysis in, **11,** 366–367
data handling in, **11,** 365–366
defined, **11,** 307
Dewars techniques in, **11,** 352–356
electric field gradient in, **11,** 314
experimental methods in, **11,** 307–368
furnaces for, **11,** 359–361
graphs in, **11,** 367–368
high-pressure experiment in, **11,** 363–365
introduction to, **11,** 307–316
low-temperature techniques in, **11,** 352–359
magnets and magnetic fields in, **11,** 361–363
proportional feedback temperature control circuit for, **11,** 358
quadrupole coupling in, **11,** 314
refrigerators and flow coolers in, **11,** 356
soruces for, **11,** 350–351
techniques used in, **11,** 309–310
temperatrure control in, **11,** 358
thermal leaks in, **11,** 354–355
triangle generators for, **11,** 339
ultralow temperatures in, **11,** 357
wave form generators and, **11,** 337–342
Massive support principle, **24A,** 247–248
Master formula, **23A,** 15
Matched filtering, **26,** 197
Matched load, **8,** 139
Matched source, **8,** 117
Matching network, **1,** 468
Matching problem, **8,** 200–201
Material balance equation, **22,** 430
Materials for optical windows, **26,** 331

Materials science studies, EPR, **21,** 425–428
Mathieu's equation, **4A,** 367, 380, 382
Matrix cathodes, **4A,** 8–19
impregnated, **4A,** 10–11
"L" cathode, **4A,** 8–10
molded, **4A,** 12–15
pressed, **4A,** 11
special matrix, **4A,** 15–18
Matrix crystallization, **30,** 417–425
Matrix effect, **22,** 339, 346; **30,** 316
Matrix element(s), **22,** 147, 150, 151, 156, 157, 169, 180, 230
effect, **22,** 147, 149, 169, 179
electric dipole transitions, **10,** 304, 341
harmonic oscillator transitions, **10,** 235
Matrix isolation technique, **3,** 230
Matrix-assisted laser desorption/ionization (MALDI), **30,** 3, 413–439
Matter, molecular theory of, **3,** 1–4
Matter, optical properties of, **1,** 329–335
Mattis-Bardeen theory, **11,** 550–553, 560
Maximum entropy method, **28,** 415–416
interferometer use of, **12C,** 150–151
Maximum likelihood
estimators, **28,** 164–165
fitting semivariogram models, **28,** 392
Maximum reading thermometer, **24B,** 192–194
Maxwell bridge, **1,** 463, 464
Maxwell equations, **10,** 2, 114, 267; **24B,** 266–267, 524; **25,** 274; **26,** 108, 134
in electromagnetic phenomena, **15B,** 250
Maxwell relations, **1,** 315ff, 431, 476, 521, 525
Maxwell relaxation time, **19,** 138
Maxwell-Boltzmann distribution, **10,** 46; *see also* Boltzmann distribution
Maxwell-Boltzmann velocity distribution, **14,** 6
Maxwellian distribution, **29B,** 8, 256
Maxwellian energy distribution, **8,** 242
Maxwellian plasmas, light scattering diagnostics of, **9A,** 66–73
Maxwellian velocity distribution, in plasma analysis, **9A,** 131–134
Mayer, Johann Tobias, **28,** 2–3
McBain balance, in sorption measurements, **16C,** 350
McGee spectracon, **12A,** 265–268
McLeod gauge, **1,** 164–166, 191; **7A,** 28, 57, 291, 390–395; **14,** 43; **16C,** 340; **18,** 509; *see also* Manometer errors
MCP, *see* Microchannel plate; Microchannel-plate detector
McPherson monochromator, **29C,** 35
Mean, **1,** 4; **28,** 19
grand, **1,** 10
weighted, **1,** 10
Mean charge state, **17,** 189–190
Mean coil diameter, **8,** 97
Mean coil spacing, **8,** 97
Mean excitation energy, **17,** 35–37
Mean field theory, **23C,** 3, 61
Mean free path, **4A,** 158–160; **14,** 8, 9; **22,** 176, 245, 505; **29B,** 2
of electrons in gases, **5A,** 113
gamma rays in toluene, **5A,** 144
high-energy electron, **22,** 245
inelastic-electron, **22,** 198, 244
inelastic scattering, **22,** 192
neutron, proton collision in toluene, **5A,** 145
for scattering plus absorption in scintillation liquid, **5A,** 148
x-ray, **22,** 245
Mean harmonic thermal conductivity, **24B,** 216
Mean inelastic energy loss, **17,** 153, 155, 189–190
Mean life of atomic state, **4B,** 217
Mean residence time, **14,** 232
Mean response, **28,** 264
Mean square circuit, **2B,** 66
Mean square displacement, **8,** 6
Mean square speed, **14,** 39
Mean squared error, **28,** 160–161
Mean squared prediction error, **28,** 396
Mean value(s), **1,** 4–5; **2,** 3–4; **3,** 711–712; **5B,** 772–773, 803, 809–810; **6A,** 3–4
difference of two, **5B,** 784–785
Mean-free-path,
Measure, quantitative, **2,** 687
Measured amplitude with random phase method, **28,** 469–470
Measurement of diffracted beam intensity
channel electron multiplier, **22,** 286
Faraday cup detector, **22,** 283, 287
fluorescent screen detector, **22,** 283, 284, 287
parallel detection, **22,** 285
photography, **22,** 482
synchronous detection, **22,** 285
vidicon camera, **22,** 285, 488
Measurement of intensity of X-ray beam, counter tube methods, **6A,** 225

measurement of nucleon mass, **5B,** 37–44
measurement of scattered particle flux, **5B,** 505–506
Measurement of phase transitions, **22,** 279
Measurement(s), **2B,** 1–249
absolute, **2B,** 48
ac, **2B,** 63–70
counting, **2B,** 1–14
digital displays and recordings in, **2B,** 6–9
direct, **1,** 6–12; **2,** 5–8; **3,** 713–716; **6A,** 5–8
of directional reflectance, **26,** 274
of emissivity, **26,** 269
errors of computation, **1,** 11–12
evaluation of, **1,** 1–24; **2,** 1–15; **3,** 709–710; **5B,** 807–821
statistical control of, **5B,** 810–811
frequency, **2B,** 15–28
indirect, **1,** 12–19; **2,** 8–13; **3,** 716–721; **6A,** 8–12
of laser power, **26,** 277
length, **1,** 53–55
of microwaves, **2B,** 251–300
phase, **2B,** 38–47
pulse amplitude, **2B,** 77–132
of reflectivity, **26,** 272
relativistic effects on, **1,** 53
significance of results, **1,** 11–12
statistical control of, **1,** 5–6; **2,** 4–5; **3,** 712–713; **6A,** 4–5
telemetering, **2B,** 220–249
of time, **2B,** 29–39
of voltage, current, and charge, **2B,** 48–76
weighted, **1,** 9–11
Measuring flask, **1,** 116
Measuring spin relaxation (T_1)
inversion recovery technique, **21,** 452–453
saturating-comb technique, **21,** 453
saturation recovery technique, **21,** 452–453
Measuring spin-spin relaxation time T_2, **21,** 444–447
Carr-Purcell sequence, **21,** 444–445
Meiboom-Gill modification, **21,** 446
shape of FID, **21,** 444–445
spin echo technique, **21,** 444–445
Measuring T_{1D}, **21,** 464–465
ADRF method applied to, **21,** 463–464
phase shifted pulse pair method applied to, **21,** 464–465
Measuring $T_{1D\rho}$, **21,** 468–470
Measuring $T_{1\rho}$, **21,** 457–460
ADRF method applied to, **21,** 459
spin locking method applied to, **21,** 459–460
Measuring volume, *see also* Spatial resolution
dimensions in LDV, **18,** 64, 83–84, 96, 126, 131–134
distribution of light in, in LDV, **18,** 119–120
Measuring wave plates, **26,** 145
Mechanical booster pump, *see* Lobe pump
Mechanical dispersions, origin of, **16C,** 13–16
Mechanical equivalent of heat, **1,** 128, 261, 269–272
Mechanical loading, **19,** 32–33,44–45, 49–51, 58–59, 61–65
Mechanical matching, *see* Impedance matching
Mechanical oscillator, **8,** 7
Mechanical properties
effect of pressure on, **16C,** 114–116
of fields, **13A,** 38
at high pressure, **6A,** 432–434
Mechanical pump, *see* Lobe pump
Mechanical resonance, **19,** 31–34, 40, 46–47, 56–60
Mechanical resonator, *see* Resonators
Mechanical shear impedance, *see* Impedance
Mechanical stability in holography, **26,** 200
Mechanical strain in optical measurements, **6B,** 271–273
Mechanical tube length, **25,** 165
Mechanical vacuum pump, *see* Rotary oil sealed pump; Turbomolecular pump; Lobe pump; Liquid piston pump
Mechanical zoom, **25,** 96
Mechanics of solids, **1,** 53–114
deformation, **1,** 100–114
force, **1,** 60–69
frequency, **1,** 69–73
friction, **1,** 97–100
length, **1,** 53–60
mass, **1,** 60–69
point kinematics, **1,** 74–81
rigid-body, **1,** 81–97
time, **1,** 69–73
Median value, **5B,** 773
Median-polish algorithm, **28,** 384–385
Median-polish kriging, **28,** 397
Medium-energy ion scattering (MEIS), **30,** 291, 306
Medium-velocity region, electronic energy loss, **17,** 45–46
Meiboom-Gill modification, *see* Measuring spin-spin relaxation time T_2

Meissner effect, **11,** 214, 555, 557, 609, 614
Meissner-Ochsenfeld effect, **6B,** 71
Melinex, *see* Mylar (polyethylene terephthalate)
Melt conduction, **24B,** 291
Melt fracture, **16C,** 48
Melt interface, **30,** 182, 183
Melt-crystallized polymers, morphology, **16B,** 163, 230–233, 238, 341–346
Melt-quench process, **6A,** 68–69
Meltese cross extinction pattern in spherulites, **16B,** 341
Melting, **23A,** 503; **30,** 180
  of semiconductors, **30,** 120
  pulsed laser, **30,** 180–183, 347–350
  two-dimensional, **23B,** 42
Melting behavior of polymers, **16B,** 294–316
  dependence on crystallization conditions, **16B,** 297
  effect of heating rate, **16B,** 300–302, 313–314
  effect of irradiation, **16B,** 305
  effect of molecular weight, **16B,** 298
  effect of sample mass, **16B,** 308–312
Melting point, **14,** 187–189
  definition on DSC curve, **16B,** 295
  effect of crystal morphology, **16B,** 298
  effect of pressure on, **16C,** 107, 115
  of elements, **14,** 347–349, 352
  equilibrium, **16B,** 296, 350
    determination, **16B,** 362, 375–381
  of lamellar crystals, **16B,** 296, 352
  multiple peaks, **16B,** 303–308
Melting threshold, **30,** 176
Melting transition, DSC/DTA trace, **16B,** 289
Membrane calorimeter, **18,** 664
Membrane polarization, **24B,** 288–290, *see also* Induced polarization, mechanisms
Membrane pressure gage, *see* Diaphragm pressure gage
Membranes, **20,** 47–49, 112–122, 137–142, 197–228, 445–543
  current measurement, **20,** 459–461
  fluidity, **20,** 142, 156
  model, **20,** 513–543
  NMR in, **20,** 47–49
  Raman scattering in, **20,** 137–142
  spin labeling in, **20,** 112–122
  viscosity, **20,** 202, 204, 216–222
Memories, dynamic, pulse spectrometer, **2,** 619–621
Memories, for integrated circuits, **2A,** 140
Memory effects, **8,** 179
Memory, **2B,** 1, 372–374
Menzel-Gomer-Redhead model, **22,** 468; **30,** 60–63
Mercury, **22,** 403
  boiling point, **14,** 350
  corrosion, *see specific materials, physical properties*
  density, **14,** 41
  diffusion pumps, **14,** 181
  ion laser, **15A,** 60, 124–128
  lamps, hydrogen production in glass vacuum systems, **14,** 329, 332
  melting point, **14,** 347, 350
  physical properties, **14,** 382–383
  vapor pressure, **14,** 40, 320, 347, 350
Mercury arc, **1,** 374
  discharge lamp, *see* Arc discharge in mercury
  quartz, **3,** 76
Merged transistor logic (MTL), **2A,** 138
Merging beams, **7A,** 115, 477ff
  limitations of, **7A,** 448, 484
  theory of experiment, **7A,** 478–480
Meridian, of X-ray fiber pattern, **16B,** 55
Meridional plane, **25,** 7
Meridional ray tracing, **25,** 14
Merzbacher-Lewis relationship, **17,** 218–219
Mesa diode, fabrication of, **2A,** 56
Mesa transistor, **2A,** 69, 71
Mesh, **7A,** 23
Meson
  mu
    average ionization loss, **5A,** 20ff
    decay electron spectrum, **5A,** 155
    decay time spectrum, **5A,** 156
  π^0
    energy, **5A,** 681
    gamma rays from decay, **5A,** 679–682
Mesoscale properties, **30,** 78–79
Metabolism, **20,** 49–52
Metal(s)
  magnetoresistance of, **11,** 1–31
  phonon spectrum of, **11,** 146–148
  state of compensation of, **11,** 3
  ultrasonic attentuation in, **11,** 428–430
Metal beam splitters, **29C,** 349
Metal gasket, **14,** 287–291, 293
Metal grid polarizers, **26,** 157
Metal hose, **14,** 301

Metal inert gas (MIG) welding, **14,** 429, 434
Metal insulator semiconductor (MIS), **2A,** 85
Metal overlayers on Ge(111)
  scanning tunneling microscopy of of, **27,** 241
Metal oxides, **29C,** 375
Metal seal, **14,** 284, 286, 287
  design, **14,** 290, 292, 296
Metal vacuum system
  design aspects, **14,** 460–464
  ultrahigh vacuum, **14,** 478–489
Metal vapor cyclic laser, **15A,** 35–48
Metal, *see specific metals*
  glasses for sealing, **14,** 318–319
  mirror protection, **26,** 33
  nuclear relaxation of, **3,** 430
  vapor composition, **3,** 564–565
Metal-insulator-metal (MIM) diode, **29C,** 251
Metal-insulator-metal (MIM) junctions, **11,** 147, 178–181
Metal-insulator-semiconductor junction, **11,** 130
Metal-matrix ceramics, **30,** 188
Metal-metal diode, **10,** 22
Metal-organic chemical plate deposition, film growth for negative electron affinity, **29A,** 7
Metal-oxide semiconductor (MOS) bucket brigade, **2A,** 137
Metal-oxide semiconductor (MOS) capacitor, **2A,** 120
  on P^+ substrate, **2A,** 134
  surface charge transfer devices and, **2A,** 129–130
Metal-oxide semiconductor (MOS) devices, **2A,** 46, 85–90
  devices, in pulse almplitude measurement, **2B,** 125
Metal-oxide semiconductor (MOS) digital integrated circuits, **2A,** 90
Metal-oxide semiconductor (MOS) integrated circuits, **2A,** 114, *see also* Integrated circuits
Metal-oxide semiconductor (MOS) logic circuits, **2A,** 139–140
Metal-oxide semiconductor (MOS) transistors **2A,** 85–90, *see also* MOS, MOSFET
  amplifier and, **2A,** 265
  drain characteristics of, **2A,** 89
  drain potential of, **2A,** 88
  dual-gate, **2A,** 90
  in electrometer amplifier design, **2A,** 299
  P-channel, **2A,** 89
Metal-oxide-semiconductor field effect transistor (MOSFET), **2A,** 85–90, 121–126; **2B,** 143–144, 480
  basic structure of, **2A,** 87
  depletion-mode, **2A,** 88
  hole packet generated by, **2A,** 131
Metal-semiconductor barrier diode, **2A,** 57
Metal-oxide-semiconductor field effect transistor (MOSFET) feedback biasing, **2A,** 267
Metal-semiconductor barrier rectifier, **2A,** 59
Metal-semiconductor diodes, **10,** 18–23, 158–166
Metal-semiconductor interfaces, **22,** 155
Metal-semiconductor tunnel junctions, **11,** 181–184
Metal-to-ceramic joint, **14,** 277
Metal-to-glass seals, **14,** 316–319
Metallic glasses, phonon density of states, **23B,** 293
Metallic mirrors, **29C,** 347
Metallic reflection, **10,** 110–112
Metallicity, **30,** 38–40, 42–44
Metallization, during ablation, **30,** 87–89
Metals
  ablation, **30,** 79–86
  bulk measurement of resistance, **6B,** 53–57
  desorption, **30,** 62–67
  lattice dynamics, **23A,** 419
  law of intermediate, **1,** 511
  photoemission data, **6B,** 380–381
  surface modification of, **30,** 363–364
  wave propagation for, **26,** 120
Metals in vacuum systems, **4B,** 272–274
  corrosion effects, **4B,** 272
  outgassing, **4B,** 272–274
  permeability for gases, **4B,** 273
Metals on GaAs(111), spectroscopy of, **27,** 265
Metascope, **25,** 9
Metastability effects, **23C,** 48
Metastable atom detection, **7A,** 85–86, 268–269
Metastable atom diffusion, afterglow studies of, **7B,** 78
Metastable atom quenching, *see* Optical quenching
Metastable atomic hydrogen, **7A,** 83, 465
Metastable atomic states, optical pumping of, **7B,** 3–4

Metastable atoms, **4A,** 183–187; **4B,** 133, 136–137, 140, 145, 166, 198
  detection, **4B,** 136, 139, 141, 145, 166, 198
  metastable beam, **4B,** 134, 136, 165, 198, 254–255
Metastable atoms/molecules
  absolute detection
    gas cell methods, **29B,** 207–210
    laser photoionization, **29B,** 210–211
  beam considerations
    focusing, **29B,** 109–110
    purity, **29B,** 108
  detection methods
    collision-induced photon emission, **29B,** 203–204
    electron emission from surfaces
      ejection mechanisms, **29B,** 194–195
      practical considerations, **29B,** 195–197
      secondary electron yield, **29B,** 197–201
    Penning ionization, **29B,** 201–202
    photexcitation/photoionization, **29B,** 204–206
    radiative decay, **29B,** 202–203
  optical quenching, **29B,** 109
  properties, **29B,** 192
  sources
    charge transfer, **29B,** 105–106
    discharges
      gas dynamic, **29B,** 102–104
      hollow cathode, **29B,** 102
      hot filament, **29B,** 102
      overview, **29B,** 101–102
      supersonic, **29B,** 104–105
    electron beam bombardment
      coaxial geometry, **29B,** 99–100
      overview, **29B,** 98–99
      transverse geometry, **29B,** 100–101
    optical pumping, **29B,** 106–107
    summary, **29B,** 95–97
    thermal, **29B,** 107–108
  velocity selection, **29B,** 110
Metastable density curve, **15A,** 107
Metastable helium, **7A,** 91, 426
Metastable hygrogen, **3B,** 812
Metastable ions (of He^+), **7A,** 106, 107
Metastable mercury, **7A,** 86
Metastable molecules, **3B,** 710, 768
  detection, **3B,** 721
Metastable oxygen, **7A,** 141
Metastable species, **22,** 494, 495
Metastable state, **17,** 427–428, 583–584
  detection of, **17,** 89
  forbidden Auger transition, **17,** 586
  K-Auger emissions, **17,** 524–526
  lifetime of, **4B,** 225
  lithiumlike atoms, **17,** 511–512
Metastable states, **3B,** 810–828
  in charge transfer, **7A,** 426–428
Meter(s), **1,** 27
  definition, **26,** 41
  flow, **1,** 117
  flux, **1,** 522, 523
  friction, **1,** 100
  positive displacement utility, **1,** 155
  Q, **1,** 465–466
  rectifier, **1,** 459
  sound level, **1,** 218, 227, 228
  standing wave, **2,** 734
  thermocouple, **1,** 460
  velocity, **1,** 78, 93
  Venturi, **1,** 118, 155, 156
  watt-hour, **1,** 469
Metering nozzle, **18,** 324–331
Methane
  absorbed on graphite, **23B,** 17, 37, 43
  atmospheric concentration, **12B,** 170
  solid, **23B,** 108
Methanol, **14,** 499–500; **19,** 189, 190
Method of differences, **23B,** 474
Method of least squares, point estimation, **28,** 162–164
Method of maximum likelihood, **28,** 211–242
  applications, **28,** 211
    to exponential distribution, **28,** 221–230
      approximate confidence regions, **28,** 228–230
      approximate likelihood-based confidence regions and intervals, **28,** 225–226
      asymptotic distribution of estimators, **28,** 226–228
      asymptotic normality of estimators, **28,** 228–230
      data and model, **28,** 221–222
      Fisher information, **28,** 227
      likelihood-based confidence intervals, **28,** 225
      likelihood function, **28,** 222–224
      normal-theory method, **28,** 228–230
      profile likelihood, **28,** 224–225

density approximation, **28,** 216
potential problems, **28,** 216–218
distributions with threshold parameters, **28,** 239
estimators
asymptotic behavior, **28,** 241
asymptotic distribution, **28,** 226–228
asymptotic normality, **28,** 228–230
explanatory variables, **28,** 220–221
fitting limited failure population model, **28,** 235–238
generalized gamma distribution, **28,** 239
interval-censored observations, **28,** 214–216
left-censored observations, **28,** 218–219
fitting Weibull, **28,** 230–234
left truncation, **28,** 219–220
modeling variability, with parametric distribution, **28,** 212
nonregular models, **28,** 241–242
numerical methods, **28,** 240
point estimation, **28,** 164–165
point processes, **28,** 240
random truncation, **28,** 239
regression, **28,** 220–221
regularity conditions, **28,** 241–242
right-censored observations, **28,** 219
right truncation, **28,** 220
simulation-based confidence intervals, **28,** 242
software packages, **28,** 240
Method of mixtures, **1,** 263ff, 273, 283
Method of moments
point estimation, **28,** 161–162
references for, **12B,** 61
Methods of spectroscopy, **13A,** 19
Methyl alcohol laser emission, **10,** 49, 50
Methyl amine, **19,** 202
Methyl fluoride laser emission, **10,** 50
Methyl groups, **23C,** 355
Methyl halide molecular fundamental frequency, **3,** 58
Methyl methacrylate, bulk polymerization, DSC study of, **16B,** 332–333
Methyl radical-oxygen reaction, **3,** 573–574
Metmyoglobin, **23C,** 347, 359
Meyer formula, **1,** 140
Meyer hardness, **6A,** 307
Mgal, *see* Gal
MgZn glasses, correlated motions, **23B,** 296
Mica, **14,** 395–396
Micellar solutions, ionic, **23B,** 524
Micelle formation, **19,** 203
Micelle theory, **16A,** 2
Micelles, **23B,** 490
Michelson fringes, **26,** 12
Michelson interferometer, **1,** 350, 355ff; **10,** 146–148; **12A,** 495–498, 516; **13B,** 62, 79; **16A,** 80; **25,** 223; **26,** 4, 9; **29B,** 387–390
diagnostic uses, **18,** 606–610, 740
Fox-Smith adaptations of, **15A,** 322
infrared, **18,** 752–753
measure coherence time, **18,** 711
Michelson stellar interferometer, **26,** 22
Michelson-Gale tiltmeter, **24B,** 425, 528
Michelson-Morley experiment, **23A,** 333; **26,** 38
Mickey Mouse ears, **29C,** 173
Microanalyzer, X-ray, **1,** 419
Microbar, **1,** 201
Microbeam methods, **6A,** 221
Microblister, vacuum system, **14,** 321
Microcalorimetric methods, **1,** 285
Microchannel plate (MCP), **29C,** 199, 200
composition, **29A,** 253
configuration and output, **29A,** 255–256
detection efficiency
degradation, **29A,** 258
determination, **29A,** 255
uniformity, **29A,** 258
electron detection, **29A,** 210
gating, **29A,** 259
output pulse amplitude and count rate, **29A,** 257–258
positron-sensitive particle detection
interpolative anode arrays, **29A,** 261, 263
optical systems, **29A,** 260–261
resistive anode
accuracy, **29A,** 265
noise sources, **29A,** 266–268
resistance parameters, **29A,** 263, 265
resolution, **29A,** 266
resolution, **29A,** 253, 260
wedge-and-strip anode
configuration, **29A,** 268
error sources, **29A,** 269–270
resolution, **29A,** 268–269
principle, **29A,** 263–254
response time, **29A,** 211–212, 258–259
stacking, **29A,** 254–255

Microchannel-plate detector, **22,** 390, 391, 487, 492
Microclimatic effect on geothermal gradient, **24B,** 191
Microcracks, **24A,** 132
Microcropy, **26,** 42
Microdensitometer, **6A,** 223–224; **25,** 174
Microdensitometry, **12A,** 222
Microearthquake network
basic elements, **24B,** 36–37
short-period telemetered, **24B,** 11
U. S. Geological Survey, **24B,** 36
Microelectrode (voltage clamp), **20,** 473–482
Microemulsions, **23B,** 534
Microfabrication, **30,** 636–638
Microhardness test, **6A,** 308
Micromachining, **30,** 194
Micromanometer, **18,** 506
Micrometer electrodes, **6B,** 4
Micron, **1,** 172
Microphone, **1,** 217, 225
power output, **1,** 231
Microphotography, **25,** 159; **26,** 29
Microphotometer, **1,** 391
Microphotometry, **12A,** 222
Microprobe sampling, **7B,** 203
in flame studies, **7B,** 221–227
Microradiography, **1,** 420–421
Microscale properties, **30,** 54–62
Microscope(s), **1,** 340–344; **22,** 21, 353; **25,** 67, 159
cathode-emission, **22,** 21
electron-emission, **22,** 21
eyepieces, **25,** 71
field-electron-emission, **22,** 349, 353, 359
field-emission, **22,** 21
field-ion, **22,** 376, 384, 387, 392, 412
infrared, **3,** 96–97
low-temperature field-ion, **22,** 384, 389
objectives, **25,** 164
physical parameters, **25,** 162
point projection, **22,** 357
reflected electron, **22,** 22
Microscopy
synchrotron radiation studies, **29C,** 41
X-ray, **1,** 428
Microstructure
description of, *see* Polycrystalline medium, description of
determination, **23C,** 426
in Fe–30% Ni alloy, **19,** 279–281
within grains, **19,** 242, 253–256, 267, 277–290
due to phase transformations, **19,** 253–256, 267, 277–290
in steel, **19,** 253–256, 267, 278, 284–290
titanium, **19,** 281–284
Microtoming (sectioning) polymer specimens for electron microscopy, **16B,** 215
Microtron, **5B,** 648–651
Microtubules, (structure), **20,** 293–296
Microturbulence, relaxation effects of, **9A,** 32–33
Microwave "noise" power, determination of, **7B,** 101
Microwave absorption, **4B,** 170–173; **6B,** 48
anomalous dispersion, **4B,** 171
clocks, **1,** 71–72
detection in ESR, **16A,** 454
detector, **4B,** 171–172
Microwave bridge, **7B,** 98
networks, **3,** 471–473
Microwave cavity, **16A,** 453
Microwave cavity resonator, **8,** 129–134
exercises, **8,** 134
Microwave cavity/waveguide, coupling, **8,** 135–141
exercises, **8,** 140–141
Microwave circuits, **2B,** 251–263
antennas in, **2B,** 263
attenuation measurements in, **2B,** 295–296
attenuators in, **2B,** 260–261
circuit junctions and obstacles in, **2B,** 259
directional couplers in, **2B,** 260
duplexers and diplexers in, **2B,** 285–286
ferrite devices in, **2B,** 261–263
frequency conversion in, **2B,** 284–285
hybrid ring or rat-race circuit in, **2B,** 258–259, 286
impedance measurements in, **2B,** 289–293
phenomenological devices in, **2B,** 275–278
rat-race circuit in, **2B,** 258–259, 286
relative power measurements in, **2B,** 294–295
resistive wave attenuators in, **2B,** 261
resonant window in, **2B,** 259
resonators and filters in, **2B,** 256–257
transmission lines in, **2B,** 252–256
waveguide attenuation in, **2B,** 255
waveguides in, **2B,** 258
Microwave crystal video detector, **8,** 123

Microwave electron spin resonance, **6B,** 235–239
components, **6B,** 236–239
sources, **6B,** 235–236
Microwave Faraday rotation, **6B,** 243
Microwave frequencies, **8,** 133, 153
Microwave gas tubes, **2A,** 186
Microwave measurements, **2B,** 287–300
antenna patterns and gain in, **2B,** 298–300
attenuation measurements and, **2B,** 295–296
crystal detectors in, **2B,** 295
field strength measurements in, **2B,** 298–300
frequency measurements and, **2B,** 296–297
hybrid tee in, **2B,** 292
impedance, **2B,** 289–293
phase measurement in, **2B,** 297–298
power measurements in, **2B,** 293–296
probe indication in, **2B,** 291
reflectometer in, **2B,** 291–293
relative power in, **2B,** 294–295
standing wavemeter in, **2B,** 289
VSWR in, **2B,** 290–291
Microwave optics, **8,** 153–159, 158–159
Microwave plasma interferometer, **10,** 281–283
Microwave power sources, **2B,** 264–282
solid-state, **2B,** 273–282
Microwave power, **14,** 454, *see* Pulsed microwave power
Microwave radiation measurements, in afterglow studies, **7B,** 100–102
Microwave receiver, **8,** 123
Microwave region, **13B,** 102
absorption cells, **13B,** 116, 127
cavity resonators, **13B,** 161
detectors, **13B,** 161
harmonic generators, **13B,** 128
modulation techniques, **13B,** 114
radio astronomy, **13B,** 132
sources, **13B,** 104, 105, 108
spectrometers, **13B,** 121
Microwave resonators, superconducting, *see* Superconducting microwave resonators
Microwave scattering from plasmas, **9A,** 37–59
advantages of, **9A,** 40–41
Alfvén waves and, **9A,** 52
apparatus for, **9A,** 54–57
"boxcar" integrator, **9A,** 59
phase-lock detector, **9A,** 57
signal processing, **9A,** 57–59
Born approximation and, **9A,** 40
coherent type, **9A,** 37
forward scattering from, **9A,** 48
conversion process in, **9A,** 37
definition of, **9A,** 37
electromagnetic problem in, **9A,** 41–48
boundary effects, **9A,** 46–48
combination scattering, **9A,** 42
conductivity tensor, **9A,** 42
dielectric tensor, **9A,** 42
dynamic form factor, **9A,** 44
scattering cross section, **9A,** 44
turbulence, **9A,** 46
experimental problem in, **9A,** 52–59
Bremsstrahlung and, **9A,** 53
resonance heating, **9A,** 54
superheterodyne methods, **9A,** 52
fluctuation problem in, **9A,** 48–52
fluctuation spectrum in, **9A,** 49
incoherent reflection process in, **9A,** 37
incoherent type, **9A,** 37
theory of, **9A,** 46
instability studies by, **9A,** 40
from optical mixing, **9A,** 52
plasma-wave investigation by, **9A,** 40–44
restrictions upon methods for, **9A,** 38–40
Tonks-Dattner resonances in, **9A,** 46
study by combination scattering, **9A,** 46
turbulence studies by, **9A,** 40
as wave-mixing interaction, **9A,** 52
Microwave semiconductor devices, **2B,** 273–282
Microwave source, **8,** 143; **16A,** 452
Microwave spectrographs, **3A,** 119
Microwave spectrometers, **1,** 66; **10,** 150–155
Microwave spectroscopy, **3A,** 11–125; **5B,** 23–29, 85–103
conventional spectrometers
cell design, **29B,** 279–281
description, **29B,** 274–276
detectors, **29B,** 281–282
frequency sources, **29B,** 276–277
modulation techniques, **29B,** 277–278
electric-resonance optothermal spectrometer
applications, **29B,** 284–287
description, **29B,** 282–284
Fourier transform spectrometers
applications, **29B,** 292–295
instrumentation, **29B,** 287–292

in fast beams, **29B,** 309–311
in gases, **5B,** 93–100
high resolution, **5B,** 92–93
paramagnetic resonance, **5B,** 95–96, 100–103
of Rydberg atoms, **29B,** 118
slit-jet spectrometers, **29B,** 295–296
Microwave techniques, **3A,** 119–125
Microwave transmission line, impedance in, **2B,** 288–293
Microwave tubes, **2B,** 264–273
amplifying, **2B,** 266
avalanche diodes in, **2B,** 275–277
grid-control tubes, **2B,** 264–265
klystrons, **2B,** 265–269
magnetrons, **2B,** 272–273
reflex oscillators, **2B,** 267–268
TR and ATR, **2B,** 285–286
traveling-wave, **2B,** 269–271
wideband traveling-wave oscillator, **2B,** 271
Microwave(s), **2,** 705–742; **2B,** 251–300; **13A,** 10
bolometers and thermistors in, **2B,** 284
circuits, **2,** 705–716
crystal detectors in, **2B,** 283–284
defined, **2B,** 251
detectors, **2,** 728–732
detectors and receivers for, **2B,** 283–286
devices, ferrite, **2,** 714–716
discharge, **13A,** 220
excitation, **13A,** 261
frequency conversion in, **2B,** 284–285
frequency determination in, **2B,** 296–297
measurements, **2,** 732–742
mixer system, balanced, **3,** 481
radar and, **2B,** 251
resonant cavity impedance, **3,** 467
receivers, **2,** 728–732
spectra, atomic mass determinations from, **5B,** 20–35
TEM mode in, **2B,** 253–254
use in afterglow studies, **7B,** 78–79, 85–87
Microwavecomponents, **3,** 20–29
electron spin resonance, **3,** 469–480
for EPR, **3B,** 564
Mid-infrared, sum frequency generation in, **15B,** 181–183
Mie scattering theory, **12B,** 177
Mie theory of light particle scattering, **18,** 54
Mie-Grüneisen equation of state, **6A,** 376; **24A,** 228, 264
MIG, *see* Metal inert gas
Migration, surface of pumping fluids, **14,** 167–169
Miller indices, **1,** 423; **3,** 275; **16B,** 7; **16C,** 152; **22,** 355
Millikan hot vacuum spark, **7A,** 153–154
Millimeter, waves, **13B,** 54, 128, *see also* Far-infrared and microwave regions
Millman effect, **3,** 621
MIM diode, *see* Metal-insulator-metal diode
Minerals, cleavage planes of, **11,** 103
Minicomputers, **2B,** 168
Minimum angle of resolution, **22,** 263, 264
Minimum detectable signal, **12B,** 208–210
Minimum number of spins detectable, EPR, **21,** 407–409, dependence on FM noise of microwave source, **21,** 408–409; *see also* NMR, double resonance techniques
Minimum variance, **28,** 161
unbiased estimator, **28,** 161
Minority carrier depletion, **6B,** 83–84
Minority carrier drift, **6B,** 359
Minority carrier lifetime from photoelectromagnetic measurement, **6B,** 359
Mirror fabrication, **31,** 175–179
Mirror of circular cross section, aberration, **31,** 164
Mirror optics, **31,** 145
Mirror plane, in crystal symmetry, **16B,** 9, 10
Mirror plane, *see* Symmetry axis
Mirror symmetry, **22,** 168
Mirror systems, **25,** 21, 97
Mirror transmission, CW dye lasers, **29C,** 60–61
Mirrors, **1,** 337; **29C,** 343, 345–348
cold, **25,** 296
Fabry-Perot cavities and, **15A,** 10–11
lasers and, **15A,** 10–11
MIS, *see* Metal insulator semiconductor
Mixed circuits, testing of, **2B,** 210
Mixer tubes, **2,** 65
Mixer, crystal, **3,** 29
Mixer, in nonlinear circuit, **2A,** 402–403
Mixing, *see* Nonlinear effects
diodes, **10,** 21, 22, 163–165
InSb, **10,** 29, 184
Josephson junction, **10,** 30
Mixtures of polymers, **16C,** 276–278

Miyamoto-Wolf theory, **26,** 53
ML relationship, *see* Merzbacher-Lewis relationship
MMF, *see* Magnetomotive force
$MnAu_2$, **23C,** 70
$MnCl_2 \cdot 4H_2O$, **23C,** 151
$MnCO_3$ antiferromagnetic resonance, **10,** 411
MnF_2, **23C,** 20, 28
  antiferromagnetic resonance, **10,** 411
MnO, **23C,** 69
  antiferromagnetic resonance, **10,** 411
MnO_2, **23C,** 70
MNOSFET (silicon nitride MOSFET), **2A,** 127, *see also* MOSFET
MnP, **23C,** 39
$MnTiO_3$ antiferromagnetic resonance, **10,** 411
MO model, of ion-atom collision, *see* Molecular-orbital model
Mo(001)
  scanning tunneling microscopy of, **27,** 296
Mobile-precursor model, **22,** 454
Mobility, **2,** 115–116; **7A,** 376–378
  electrophoretic, **20,** 324–326
  Hall, **2,** 123
  measurement, of a carrier pulse, **6B,** 91–93
Mobility, ambipolar, **6B,** 93–94
  determination from photoconductivity measurements, **6B,** 353
  drift, **6B,** 91–95
Mobility, of carriers in polymers, **16C,** 441
  determination by "time and flight" methods, **16C,** 442
MOCVD, *see* Metal-organic chemical vapor deposition
Modal beam, laser cavities with, **15B,** 25–28
Modal value, **5B,** 773
Mode locked lasers, **15A,** 28–30
  short pulses from , **15A,** 29
Mode(s), **25,** 259
  acoustical, **10,** 361
  area, for cw dye lasers, **15A,** 347–351
  cavity, **3,** 27
  conversion, **19,** 241–242
    during scattering, **19,** 241–242
  dominant in waveguides, **10,** 113, 117, 118
  of hot-cathode arcs, **2,** 93–94
  longitudinal, **19,** 241–245, 268, 272, 274–278, 280, 284–290
  multiple, of narrow specimens, **19,** 294–296
  natural, **2,** 22
  optical, **10,** 361
  scattering for, **19,** 242–245
  shear, **19,** 241–245, 273
  transverse, **10,** 362
  waveguide, **2,** 707ff; **3,** 24
  waves in solids, **19,** 241–245
Mode-locked cw dye laser, amplified pulses from, **15B,** 203, *see also* cw dye laser; Mode-locked lasers
Mode-locked laser
  pumping of dye laser by, **15B,** 175
  ultrashort pulses from, **15B,** 186
Mode-locked laser, **18,** 718–720
Model atmosphere
  exponential (CRPL), **12B,** 189
  U.S. Standard, **12B,** 139, 188
Model building programs, **20,** 264
Model testing principles, **18,** 821–848
Modelocking
  active modelocking, **29C,** 175–180
  amplitude modulation, **29C,** 176–179
  nonlinear loop mirror modelocking, **29C,** 180–182
  nonlinear polarization rotation modelocking, **29C,** 182–183
  passive modelocking, **29C,** 180–182
  phase modulation, **29C,** 179
  semiconductor saturable absorbers, **29C,** 180
  "soliton" modelocking, **29C,** 180
  stretched pulse modelocking, **29C,** 183–184
Models of viscoelasticity
  Burgers, **19,** 143
  Maxwell, **19,** 141, 143, 145, 146
  Voigt, **19,** 143
Moderation
  of thermal neutrons, **5A,** 515–517
  time for fast neutrons, **5A,** 519–521
Moderators, pulsed source, **23A,** 157, 199
  beam current, **23A,** 157
  cold, **23A,** 164
  emission time, **23A,** 168
Moderators, steady source, **23A,** 143, 145, 199
  cold, **23A,** 146
  hot, **23A,** 148
Modulated Bayard-Alpert gauge, **14,** 70–72, *see also* Bayard-Alpert Gauge; Ionization gauge
Modulated excitation sources, **6B,** 302–304
Modulated receivers, **12B,** 214–217
Modulated structures, excitations of, **23C,** 206
Modulated-potential Faraday cup

angular response, **9B,** 58–59
in deep-space plasma measurements, **9B,** 53–62
diagram of, **9B,** 56
energy window of, **9B,** 57
extraneous effects, **9B,** 60–62
high-order corrections for, **9B,** 57–58
transmission function, **9B,** 58
"Venetian blind" collimeter of, **9B,** 59
Modulating signal, amplifier, and, **2A,** 248
Modulation, **1,** 451; **2,** 263–265
high-frequency, cavities for, **3,** 479–480
lamp, **25,** 244
magnetic, **3,** 488–490, 507–508
in microwave spectroscopy of gases, **5B,** 90–92
noise, **2B,** 462
sideband, **3,** 405–407
Stark, and the phase sensitive detector, **3,** 666–668
techniques, **7A,** 55, 74, 76, 91, 93, 109–113, 188, 189
Modulation transfer function (MTF), **12A,** 242, 278, 286–291; **25,** 76; **26,** 85
Modulator, **8,** 169, 173, *see also* Single modulator
electrodes, **8,** 174
system, *see* Multimodulator
Modulator-demodulator circuit, **2A,** 338
Modulators, **1,** 217; **2A,** 336–338; **12A,** 437, 440, 445–447, 450, 460
defined, **2A,** 336
even harmonic, **2,** 793–794
Modulus, compression and shear, **16A,** 174
Modulus, tangent and secant, **16C,** 118
Moho, **24B,** 3
Moiré pattern, **6A,** 348
Molar heat, **1,** 261
Molding, vacuum, **16C,** 52–56
in electric breakdown specimen preparation, **16C,** 452
Molecular beam epitaxy (MBE), **30,** 291
laser, **30,** 480
Molecular beam(s)
absorption spectrometers, **3B,** 672
deflection and focusing, **3B,** 723
deflection experiments, **3B,** 670
detectors, **3B,** 711; **7A,** 80, 85–86, 267–274
diatomic molecules, **3B,** 799
electric deflection, **3B,** 766
electric resonance (MBER), **3B,** 750
focusing, **7A,** 261–267
formation, **3B,** 685–691
light sources, **1,** 375
magnetic deflection, **3B,** 765
magnetic resonance MBMR, **3B,** 753
maser spectrometers, **3B,** 675
maser spectroscopy, **3B,** 839
resonance spectrometers, **3B,** 673
separated oscillating fields, **3B,** 761
in shock wave studies, **7B,** 183
single oscillating field, **3B,** 757
sources, **7A,** 71, 74–75, 230–255
state selector, **7A,** 259–267
supersonic, **3B,** 692–710
transition process, **3B,** 757–765
translational spectroscopy, **3B,** 676
velocity selector, **7A,** 255–259
Molecular biology, **23C,** 367
Molecular continua, light sources, **4A,** 314–315
Molecular correlation, **13B,** 175
Molecular density, **14,** 40
Molecular dissociation, **4B,** 234–235
at high energies, **7A,** 466–472
at low energies, **7A,** 472
Molecular drag pump, **14,** 247
Molecular dynamics, **3B,** 956; **23A,** 441, 502
of an associated liquid, **3B,** 989
Beeman algorithm, **23A,** 503
computer simulation, **23B,** 238
fluorescent probe studies, **16A,** 432–436
by high resolution NMR, **16A,** 302–320
of the solid state, **16A,** 361–365
internal coordinates, **23A,** 505
simulations, **30,** 120–121
by spin probe techniques, **16A,** 471
unstressed, **23A,** 503
Molecular effusion source, **7A,** 71, 74–75, 232
multichannel array, **7A,** 84, 235–236
Molecular effusion, **3B,** 678; **4A,** 156–167
from ideal aperture, **4A,** 156–161
angular distribution from, **4A,** 156
commonly used shapes, **4A,** 160
cloud formation, **4A,** 158
directivity, **4A,** 157
from long canals, **4A,** 161–167
directivity pattern, **4A,** 163
flow rate, **4A,** 161–162
use of hypodermic needle, **4A,** 167, 169

from multiple-channels, **4A,** 162–163, 166–169
use of klystron grid, **4A,** 169
Molecular energy levels, populating and depopulating of, **15B,** 75
Molecular entanglement, effect on creep, **16C,** 15
Molecular imaging, **22,** 370, 416
Molecular infrared laser(s), **15B,** 55–85
collisional processes and, **15B,** 63–75
continuous tunability in, **15B,** 80
device configurations in, **15B,** 75–85
electron impact excitation and, **15B,** 78–85
high-pressure tunable, **15B,** 79–82
new developments in, **15B,** 79
new transitions in, **15B,** 83–85
optically pumped, **15B,** 76–78
simple harmonic oscillator rules for, **15B,** 60
tunability of, **15B,** 59
Molecular infrared physics, **15B,** 58–75
Molecular interaction with static fields, **3B,** 723
Molecular lasers, **15B,** 55–93, *see also* Gas lasers; Molecular infrared lasers
finite rotational relaxation rate and, **15B,** 67
Molecular light scattering, advantages of diagnostics, **18,** 409, 418–421
Molecular motion, **20,** 197–200, 208–216
macroscopic, **20,** 198, 199, 208–216
methods of detection, **16A,** 7
microscopic, **20,** 197–199
rotational, **20,** 199, 200
Molecular motions in solids and liquids, **3B,** 979
Molecular network, creep of, **16C,** 6
Molecular orbitals, **15B,** 271
Molecular photoionization, **7A,** 162
Molecular physics, **3A,** 1–9
Molecular pump, **14,**180, *see also* Diffusion pump; Turbomolecular pump
Molecular recombination spectra, **1,** 407
Molecular replacement, **20,** 260–263
Molecular rotation spectra, **10,** 306–321
Molecular sieve, **14,** 232, 238
Molecular spectra, **13B,** 1
Molecular spectroscopy, **3A,** 11–301; **15B,** 58–63; **23A,** 441
in laser-selective chemistry, **15B,** 277–278
Molecular states, **13A,** 106
Molecular structure
by high resolution NMR, **16A,** 281–301
from spin-spin coupling constant, **16A,** 263
Molecular systems, **23A,** 441
Molecular theory, **3A,** 2
Molecular vapors, sum mixing in, **15B,** 220
Molecular vibrations, **10,** 321, 328
Molecular weight
changes during mechanical fracture, **16C,** 229
distribution, effect on creep compliance, **16C,** 15
ESR spin probe sensitivity, **16A,** 472
number-average
definition, **16A,** 14
measurement by end-group analysis, **16A,** 71
measurement by osmometry, **16A,** 24
standards for GPC calibration, **16A,** 54
viscosity-average
calculation from GPC data, **16A,** 59
definition, **16A,** 16–17
weight-average
definition, **16A,** 14–15
measurement by light scattering, **16A,** 28
measurement by ultracentrifugation, **16A,** 67
z-average
definition, **16A,** 15
measurement by ultracentrifugation, **16A,** 67
Molecular x rays, **13A,** 182
Molecular-beam scattering, **22,** 497
Molecular-beam spectroscopy, **3B,** 669–846
diatomic molecules, **3B,** 799
line shapes, **3B,** 762
methods, **3B,** 678–685
non-resonance experiments, **3B,** 765–772
polyatomic molecules, **3B,** 828–837
resonance experiments, **3B,** 772–846
spectrometer design and operation, **3B,** 745–757
Molecular-orbital excitation, **17,** 190, 212–213, 330–337, 384–386
Molecular-orbital model, of ion-atom collision, **17,** 112, 152, 155, 482
K-shell ionization, **17,** 483
L-shell vacancy production, **17,** 345–347
Molecular-structure determination, **3A,** 90
Molecule(s)

asymmetric top, **3,** 13–15, 654–655, 657–659
centrifugal distortion due to rotation, **5B,** 30
diamotic, **3,** 42–53
energy levels, **3,** 12
nuclear-molecular interactions, **3,** 633
rotational problem, **3,** 46–48
transitions, **3,** 12–13
vibrational problem, **3,** 42–46
ionization potentials, **3,** 526–527
linear, **3,** 656
Stark effect for, **3,** 661
triatomic, **3,** 13
in a magnetic field, **3,** 605–606
magnetic interactions, **3,** 200–221
moments of inertia, **3,** 11–12
multiphoton dissociation of, **15B,** 265–268
per cell, **3,** 289
polar, **3,** 637
polarization, electric, **3,** 637–645
polyatomic, **3,** 53–73
linear, **3,** 54–56
non-linear, **3,** 56–57
rotational problem, **3,** 59–73
spectra of, **3,** 208–221
vibrational modes, **3,** 54ff
vibrational problems, **3,** 54–59
properties, electric, **3,** 637–680
quadrupole coupling, **3,** 19
resonance spectra,
electric, **3,** 634–635
magnetic, **3,** 630–634
simple free electronic spectra, **3,** 179–222
symmetric top, **3,** 13, 655–656, 659–661
symmetric top, **5B,** 20–21
triatomic, **3,** 186, 187
bent, fundamental frequency, **3,** 56
triplet states, **3,** 207
Molière approximation, **22,** 311
Molière potential, *see* Channeling, theory
Molière theory of multiple scattering, **5A,** 392ff
Møller scattering, **5B,** 242
Molybdenum
boiling point, **14,** 351
disulfide, **14,** 416
for high-temperature furnaces, **11,** 359
hydrogen solubility, **14,** 354
melting point, **14,** 348, 351
oxide, **14,** 369
physical properties, **14,** 369–370
vapor pressure, **14,** 348, 351
Molybdenum, reflectance spectra, **31,** 218
Moment
anomalous magnetic, of electron, **3,** 629–630
atomic magnetic, **3,** 604, 605
electric quadrupole, **3,** 384
evaluation of, **5B,** 49–50
of force, **1,** 84–86
of inertia, **1,** 82–84
of a molecule, **3,** 11–12
molecule, contribution of molecular electrons to, **5B,** 31–32
magnetic, **1,** 520, 523, 530ff
of mass, **1,** 81
neutron magnetic, **3,** 635–636
nuclear
magnetic, **3,** 632
quadrupole, **3,** 651
nuclear, determination, **5B,** 44–214
indirect methods, **5B,** 129–147
microwave method, **5B,** 85–103
spectroscopic methods, **5B,** 44–129
nuclear quadrupole, **5B,** 99
quadrupole, **3,** 384, 503, 523, 651
determination, **5B,** 120–121, 147–148
ratios, nuclear quadrupole, **3,** 523
Moments, **28,** 19
Momentum distribution, **8,** 220
Momentum transfer, **8,** 279
Momentum, **14,** 39
acceptance
focusing system wtih, **5B,** 742–743
image width due to, **5B,** 716–719
angular, **3,** 18
orbital, quenching of, **3,** 455–456
determination, **5A,** 341–438
dispersion
at image, **5B,** 714–716
wedge-shaped magnet, **5B,** 739
gamma ray
with curved crystal spectrometer, **5A,** 599–616
with pair spectrometer, **5A,** 641–652
linear, measured by recoil techniques, **5A,** 461–494
maximum detectable, **5A,** 198, 384
measurement
with cloud or bubble chamber, **5A,** 375–388

in nuclear emulsions, **5A,** 388–408
neutrino, **5A,** 692–698
neutron
crystal diffraction measurement, **5A,** 566–570
determination with He^3 spectrometer, **5A,** 570–581
resolution relation to energy resolution, **5A,** 345
of singly charged particle, **5A,** 396
spin angular measured by recoil techniques, **5A,** 461–494
transfer, **17,** 198
and energy loss, **17,** 33, 40
Momentum-ionization measurement in cloud chamber, **5B,** 35–36
Momentum-range measurement in cloud chamber, **5B,** 36
Monel, **14,** 378, 380
bellows, **4B,** 286
thermal conductivity, **14,** 356
Monitoring, induced radioactivity, **5B,** 501
Monitors, charged particle,
primary, **5B,** 486–498
secondary, **5B,** 499–507
Monoatomic layer, in vacuum tubes, **2A,** 149
Monochromatic electrons, **22,** 368
Monochromatic plane waves, **6B,** 249
Monochromatic radiation, **8,** 39
Monochromatization
neutrons, **20,** 353–355
of X rays, **16B,** 46–50, 136
x rays, **20,** 346, 347
Monochromatization (light), synchrotron radiation, **29C,** 33–39
Monochromator layout, **31,** 30
Monochromator, **6B,** 285; **8,** 30, 35–36; **10,** 130–140; **17,** 539–544; **21,** 213; **22,** 32, 33, 76, 78, 144, 150, 176, 178, 179, 184, 481; **25,** 191
bandpass, *see* Trapezoidal monochromator bandpass
Brown-Lien grazing-incidence "grasshopper," **22,** 144
crystals (neutron scattering), **21,** 178–179
electron, **7A,** 4, 54, 65
entrance slit, **8,** 31, 35, 36
exit slit, **8,** 37
extended-range "grasshopper," **22,** 144
neutron crystal, **5A,** 569–570; **5B,** 392
photon, **7A,** 185–186
toroidal-grating, **22,** 144
transmission function, **22,** 483
ultraviolet, **1,** 365, 406
Monochromators and light sources, in photoemission studies, **11,** 120–121
Monochromators, **6A,** 221–223; **9A,** 97; **13A,** 234; **23A,** 208, 216; **29C,** 34–35, 38–39, 350
Monochrometer
optical, design, **16A,** 85–86
thermal neutron, design, **16A,** 218–219, 488
Monoclinic crystal system
definition, **16B,** 4
interplanar spacing formulae, **16B,** 26
reciprocal lattice relationships, **16B,** 24
Monoclinic lattice, definition, **16B,** 5
Monoenergetic ions, **22,** 401
Monohydride molecular fragments, **3,** 185
Monolayer, **14,** 25, 285, 190, 299
removal, **14,** 190
time, **14,** 7, 190
Monolithic cavity design, **29C,** 242–243
Monolithic integrated circuits, **2A,** 105–142, *see also* Integrated circuits
examples of, **2A,** 140–142
isolation of, **2A,** 105–114
types of, **2A,** 138–140
Monomer, heat of polymerization by DSC, **16B,** 332
Monometer, **16A,** 3
Monopole mass filter, *see* Mass spectrometer
Monopole mass spectrometer, **14,** 96
Monoxymyoglobin, **23C,** 351
Monte Carlo experiment, **28,** 163–164
Monte Carlo integration technique, **28,** 429
Monte Carlo simulation
field statistics assessment, **28,** 462–466
critical area estimation, **28,** 465
field significance testing, **28,** 465–466
local critical correlation estimate, **28,** 463–464
random time series, **28,** 462–463
simulation of field statistics, **28,** 463
global ozone data fields, **28,** 473–475
Monte Carlo technique, estimating map's statistical significance, **28,** 467–470
Monte-Carlo calculation, **8,** 215
Moon observations
VLBI techniques for, **12C,** 273
Mooney rhomb, **26,** 133

MOPA laser (monolithic master-oscillator/power-amplifier laser), **29C,** 80, 82, 90, 162
Mori formalism, **23B,** 381
Morin transition, hematite, **24A,** 363
Morphology
of block copolymers, **16B,** 234–235
of a craze, **16C,** 245of block copolymers, **16C,** 294–299
effect on electrical conduction, **16C,** 495
effect on electric breakdown, **16C,** 471–473
effect on gas transport properties, **16C,** 328–331
of melt-crystallized polymers, **16B,** 163, 230–233, 238, 341–346
by solvent-etching, **16B,** 239–241
interior, by plasma-etching, **16B,** 241–245
modification of surface, **30,** 353–359
one-dimensional model in small-angle X-ray scattering, **16B,** 167–171
of polyethylene, effect on melting point, **16B,** 298
of polymer single crystals, **16B,** 128, 174, 229–230, 230, 237, 342, 343–344
control by self-seeding, **16B,** 382
deductions from halogenation, **16B,** 248–251
deductions from oxidation, **16B,** 261–267
positron lifetime sensitivity to, **16A,** 397
MOS (FAMOS), **2A,** 127–128
MOS, *see* Metal-oxide semiconductor
Mosaic block model, of paracrystallinity in single crystals, **16B,** 151
Mosaic structure, **22,** 248
Moseley's law, **1,** 417; **5A,** 294–295
MOSFET, *see* Metal-oxide-semiconductor field effect transistor
Mössbauer effect, **5B,** 273–274
MOST (MOS transistor), *see* MOS transistor
Most probable velocity, **29B,** 9
MOT, *see* Magneto-optic trap
Motes, *see* Cavitation nuclei
Motility, **20,** 326–330
Motion
control in vacuum, **4B,** 286–289
critically damped, **1,** 89
forced harmonic, **1,** 89–90
nuclear rotational, **3,** 11–15
overdamped, **1,** 89
underdamped, **1,** 88
Motional averaging, EPR, **20,** 94
Motional mass, **8,** 11
Motional narrowing of NMR peak, **16A,** 249
Motional narrowing, **21,** 441–444, 473–477
Motor
acceleration requirements of, **2B,** 440
for servomechanisms, **2B,** 437–440
speed control circuit, **2A,** 235–238
Motorboating, **2,** 218
Mott
asymmetry function, **5A,** 68ff
scattering, **5B,** 249–259; **7A,** 40
cross section, **5A,** 55
of electron from heavy nuclei, **5A,** 68
Mott polarimeter
atomic-target polarimeter, **29A,** 239–240
calibration
calculation of Mott asymmetry, **29A,** 232, 235, 243–244
double-scattering measurements, **29A,** 244
measurement of analyzing power, **29A,** 232, 235, 244
use of electrons with known polarization, **29A,** 244–245
concentric-electrode configuration
advantages of design, **29A,** 237
configuration, **29A,** 236
geometry optimization, **29A,** 237–239
retarding-field analyzer, **29A,** 239
diffuse scattering polarimeter, **29A,** 240
Mott scattering error sources
background, **29A,** 247
extrapolation error, **29A,** 245
instrumental asymmetry, **29A,** 245–247
performance, **29A,** 248–250
standard configuration polarimeter
configuration, **29A,** 234
energy range, **29A,** 236
operation, **29A,** 234–235
Mountain peak, **3A,** 417
Mountings, *see* Gratings
antenna, **12B,** 14–16
Moving-coil instruments, **2B,** 49–50
Moving-film methods, **6A,** 208–214
MT transfer function, **24B,** 333–335
MT, *see* Magnetotellurics
MTR element, **8,** 208
MTR fuel element, **8,** 207

Mu-mesic atoms, formation of transient, **5A,** 329–330
Mueller bridge, **1,** 246
Mueller matrix ellipsometers, **25,** 201
Mueller's matrices, **26,** 159
Multi-gamma-ray spectrum, **8,** 217, 218
Multi-*k* structures, **23C,** 80, 122
Multiaccelerator, **17,** 17–19
Multialkali cathodes, **6B,** 389
Multianvil press
  calibration, **24A,** 259–265
  cubic anvil press, **24A,** 240–243
  design and construction, **24A,** 238–249
  internal heater, **24A,** 249–252
  multiple anvil sliding system press, **24A,** 244–246
  pressure cell-sample assembly, **24A,** 249–259
  spilt sphere press, **24A,** 243–244
  tetrahedral anvil press, **24A,** 239–240
Multibeam sonar, *see also* Sea Beam; Sonar
  artifacts, **24B,** 92–93
  navigation, **24B,** 93
  performance parameters, **24B,** 92
Multichannel analyzers, **7A,** 77; **11,** 319–322; **21,** 56, 57; *see also* Angular correlation (of annihilation radiation), electronics; Doppler broadening (of annihilation radiation), electronics; Lifetime (of positron), electronics; Mössbauer spectrometers
  data handling in, **11,** 365–366
  "fast flyback" in, **11,** 324
  thermometry and, **11,** 522–523
  time mode in, **11,** 323
  triangle generator in, **11,** 339–340
  up-down multiscale mode in, **11,** 325, 330
Multichannel filter spectrometers
  applications in millimeter range, **12C,** 44
  filter spacing in, **12C,** 42
  measurements with, **12C,** 38–45
  use in survey work, **12C,** 44
Multichannel scaling, **11,** 323–324
  sinusoidal waveform in, **11,** 339
Multichannel-filter spectrometers, **12B,** 266–279
Multicollinearity, **28,** 274–277
Multicritical points, **23C,** 37, 53
Multielectron transitions, **17,** 395–401
Multilayer coatings, **31,** 271–272
Multilayer films, **26,** 35, 128
Multilayer filters, **31,** 282–283, 330–338
Multilayer gratings, **31,** 379
  design, **31,** 380–385
Multilayer images, **22,** 412
Multilayer polarizer, **31,** 244–245, 284–285
Multilayer structures, epitaxial, **30,** 480
Multilayer thin films, **25,** 274
Multilayer zone plate, **31,** 296
Multilayers, **23A,** 279; **23B,** 51
  construction, **31,** 285–286
  performance, **31,** 285–286
Multimeters, **11,** 279–284
Multimode fiber, **29C,** 377, 380
Multimode laser, **25,** 260
Multimodulator system
  energy gain, **8,** 173–175
  phase condition, **8,** 173–175
Multiparticle coherent emission, **31,** 2
Multipass systems, **13A,** 296
Multiphase analysis, **23C,** 444
Multiphase systems, **23C,** 310
Multiphonon scattering, **23A,** 29
Multiphoton absorption, **20,** 190; **30,** 25–27
  in excitation processes, **15B,** 279
  ionization spectroscopy and, **15B,** 226–229
Multiphoton axcitation, **30,** 585
Multiphoton dissociation
  collisionless, **15B,** 268
  of molecules, **15B,** 265–268
Multiphoton ionization, **15B,** 264; **22,** 498; **29B,** 173–178; **30,** 69, 186
  spectroscopy, technique of, **15B,** 227–229
Multiphoton processes, resonant enhancement and, **15B,** 211–216
Multiphoton transitions, **30,** 24–26
Multiplate resonant reflector (MRR), **31,** 340–343
Multiple circuits, in nonlinear circuits, **2A,** 403
Multiple electron capture, **17,** 104–107
  cross section, **17,** 98, 104–105
Multiple excitation
  and charge-state density effect, **17,** 135–139
  shell effect, **17,** 107
Multiple inner-shell vacancy production, **17,** 135–139
Multiple ionization, **17,** 379–382, 417; **22,** 216
  Auger decay, **17,** 468–482
  Coulomb ionization, **17,** 487, 496–500
  fluorescence yields, **17,** 207–209, 393–395, 503

L shells, 247–248, 401–408
molecular orbital collisions, **17,** 484
neon, **17,** 411
Multiple melting peaks, **16B,** 303–308
Multiple pass spectrometers, **25,** 201
Multiple reflection interferometer, **26,** 30
Multiple reflections in a thin film, **26,** 126
Multiple relaxation, **13B,** 180
Multiple scattering, **7A,** 39; **12B,** 183–184; **23A,** 405; *see also* Scattering ions
of charged particles, **5A,** 73–76
corrections, **23B,** 419, 422
electron, **22,** 234
of electrons in solids, **7B,** 299–301
ion, **22,** 307, 308, 310
neutrons, **21,** 212, 217
Multiple scattering, in ion-atom collisions, **17,** 309–310
beam-foil experiments, **17,** 538, 577–578
Multiple spark camera, **18,** 737–739
Multiple spectrometer, **12B,** 220
Multiple-beam interferometer (Fabry-Perot), **18,** 195–227, 708, 717, 740; **26,** 25
Multiple-collision effects, **17,** 224–227
beam-foil experiments, **17,** 549
charge-changing probabilities, **17,** 79
collisional quenching, **17,** 426–428
Multiple-electron loss, **17,** 116–118
cross section, **17,** 98, 100, 105–106
Multiple-ion plasmas, wave propagation in, **9B,** 135–137
Multiple-order waveplates, **29C,** 353
Multiple-pass
Fizeau interferometer, **26,** 37
Twyman-Green interferometer, **26,** 37
interferometer, **26,** 37
Multiple-pass laser, **29C,** 162–164
Multiple-quantum transitions, **3B,** 762
Multiple-reflection
Fizeau interferometer, **26,** 28
interferometer, **26,** 25
Multiplet tables, **9A,** 165
Multiplex advantage, **13B,** 53; **25,** 198; **29B,** 366–369, 416
Multiplex factor, **25,** 198
Multiplex gain (or Fellgett advantage), **10,** 141
Multiplex spectrometry, **10,** 140–149
Multiplexing, **12A,** 404; **18,** 555; *see also* Telemetry, modes
frequency division, **2,** 678
in telemetering, **2B,** 247
time division, **2,** 675–678
Multiplication
factor, proportional counters, **5A,** 111
gas, **5A,** 110–118
neutron, **5B,** 405–407
Multiplicative law, **28,** 13
Multiplicity factors, in X-ray peak intensity, **16B,** 67
Multiplier
detector, **22,** 474
frequency, **2,** 793–794
noise, **7A,** 276
secondary electron, **14,** 84
servodriven, **2,** 641
Multiply excited states, **13B,** 224
Multipole expansion, **23C,** 382, 393
Multipole fields, **3B,** 732
Multipole focusing fields, **7A,** 261–267
Multipole moments, **4B,** 107, 151; **23A,** 454
Multipoles, **13A,** 93, 115
Multiradix index, **12C,** 287
Multireflector antennas, **12B,** 48–60
Multiregion devices, **2A,** 92–99
Multishot autocorrelation, **29C,** 206
Multishot emission spectroscopy, **20,** 177–179
Multistage turbo-pump, **14,** 253
Multitaper estimator, cross spectrum, **28,** 344–346
Multitaper spectral estimators, **28,** 335–338
Multivariate distribution, covariance, **28,** 22
Multivariate nonlinear model, **28,** 248
Multivariate normal distribution, **28,** 256–257
Multivibrator(s) **2,** 363–371, 538
asymmetrical, **2,** 366–368
bistable, **2A,** 398–400
in equipment testing, **2B,** 188
in frequency measurement, **2B,** 18
monostable, **2A,** 397–398
square-loop core driver, **2A,** 405–408
symmetrical, **2,** 363–366
transistor, **2,** 422–433
bistable, **2,** 422–427
free-running, **2,** 427–433
monostable, **2,** 427–433
square-loop, **2,** 435–437
Muonic x rays, **13A,** 189
Muonium, **4B,** 189–191
hyperfine structure, **4B,** 190
Muscle fiber (voltage clamp), **20,** 497–499
Mutual coherence function, **18,** 712–714

Mutual inductance, **1,** 542; **2A,** 2–3
Myelinated nerve (voltage clamp), **20,** 492, 497
Mylar (polyethylene terephthalate), **10,** 97, 146
beam splitter, **10,** 146–148
Mylar, **14,** 414
Myoglobin, **19,** 202; **20,** 386

N

n-p scattering process, **8,** 254
n-Paraffins
crystallization of polyethylene from dotriacontane, **16B,** 346
electric field effect on postronium yield, **16A,** 402
infrared active rocking-twisting modes, **16A,** 112
infrared spectra of C_4H_{10}, C_5H_{12}, and $C_{17}H_{36}$, **16A,** 135
infrared spectrum of $C_{24}H_{50}$, **16A,** 111, 114
isothermal compressibility of $C_{16}H_{34}$, **16A,** 190
long-chain, preparation from degraded polyethylene crystals, **16B,** 268–269
Raman spectrum of $C_{24}H_{50}$, **16A,** 112, 113
Rayleigh-Brillouin spectrum of $C_{16}H_{34}$, **16A,** 171, 181, 189
N-particle system, **8,** 277
n-Pentane, catalytic isomerization, **28,** 247–248
N-th order desorption kinetics, **22,** 460
n-threshold studies, **8,** 269
N-type epi layer, **2A,** 83–84
n.d.f., *see* Normal distribution function
N_2 dissociative photoionization, **7A,** 164
N_2 fluorescence, **7A,** 170
N_2^+ dissociation, **7A,** 115
N_2^+ fluorescence, **7A,** 167, 170–171
N_2^+ Frank-Condon curves, **7A,** 168
N_2^+, excitation by collisions with N_2, **7A,** 464
N_2O laser, **10,** 49, 51, 52
NaCl, **10,** 88, 90–95, 359, 370, 371
NaI crystal scintillator, **7A,** 406
NALM, *see* Nonlinear loop mirror modelocking
NAND function, **2A,** 410
Nanolight, **18,** 700, 702
Nanosecond OPO systems, **29C,** 261–262
Naphthalene, **23A,** 445
Narrow band impedance matching, **8,** 117–122
exercises, **8,** 121–122
Narrow specimens, errors due to, **19,** 294–297
Narrow-band interference filter, in beam-foil experiments, **17,** 538–539
Narrow-band photoconductive detector, **10,** 185–188
Narrowband transducers, **19,** 55–60
electrical characteristics, **19,** 56–57
electrical loading effects, **19,** 59–60
mechanical loading effects, **19,** 58–59
National Bureau of Standards (NBS)), **2B,** 17; **11,** 343
nonlinear test fluid No. 1, creep properties of, **16C,** 18–24
National Bureau of Standards, **8,** 28
Natural broadening, **13A,** 23
Natural line breadth, **4B,** 60, 120, 138, 141
Natural line broadening, **8,** 161
Navier-Stokes equation, **1,** 127, 131, 133, 160, 161; **9A,** 225
NBS, *see* National Bureau of Standards
$NbSe_2$
2H phase
scanning tunneling microscopy of, **27,** 430, 432, 441
scanning tunneling microscopy of charge density wave in, **27,** 386
spectroscopy of, **27,** 386–389
superconducting properties of
scanning tunneling microscopy of, **27,** 430, 432
$NbSe_3$
scanning tunneling microscopy of, **27,** 402, 405
spectroscopy of, **27,** 418
Nd (neodymium) lasers, laser-produced plasmas, **29C,** 4, 5
Nd (neodymium):glass lasers, **15B,** 186–188
passively mode-locked, **15B,** 198
Nd, **23C,** 207
Nd-glass laser(s), **9A,** 81; **15A,** 168; **29C,** 144
properties of, **15A,** 199–202
Nd:glass oscillator, destructive effects in, **15B,** 188
ND:YLF lasers, *see* Neodymium/yttrium lithium fluoride lasers
NdYAG, *see* Neodymium-yttrium aluminum garnet
Near infrared, *see also* Infrared radiation; Molecular infrared lasers
frequency generation in, **15B,** 174–181
pulse generation in, **15B,**191

Near point of the eye, **25,** 120
Near-edge x-ray absorption fine structure (NEXAFS), **29C,** 40
Near-field scanning optical microscopy, **30,** 110–111
Near-infrared emission lasers, **15A,** 230–231
Nearsightedness, **25,** 68
Neck region, deformation in, **16C,** 139
Néel temperatures of antiferromagnetic crystals, **10,** 411
Needle arc, **14,** 427
Néel theory, **6B,** 195–196
Néel's equation, **6B,** 176
Negative absorption, **18,** 406, 707, 720–725
  by electrons, **10,** 77–81, 259–263
Negative common-mode feedback, **2A,** 295
Negative feedback systems, testing of, **2B,** 208–209, *see also* Feedback control systems
Negative immitance converter (NIC), **2A,** 335–336
Negative ion, bubble for, **11,** 443–445
  beam sources, **7A,** 181
  resonance scattering, **22,** 52, 53, 54
Negative ion-neutral reactions, **7B,** 139
Negative ions, **7A,** 179ff; **22,** 466, 493
Negative mass instability, **9A,** 291–295
  bunching in, **9A,** 291
Negative resistance devices, in relaxation oscillators, **2A,** 351
Negative-electron-affinity semiconductor photocathode
  comparison of polarized electron souces, **29A,** 30–32
  electron gun
    brightness values, **29A,** 25–26
    configuration, **29A,** 27
    electron-optical considerations, **29A,** 24–25
    low-energy electron diffraction gun, **29A,** 27–28
    spin rotation, **29A,** 28, 30
    stability, **29A,** 26
  gallium arsenide
    applications, **29A,** 32
    development, **29A,** 1–2
    diffusion length, **29A,** 4, 6
    electron escape, **29A,** 4–5
    optical spin orientation, **29A,** 3
    polarization
      film growth, **29A,** 7, 9, 12
      formula, **29A,** 2–3, 6
      optimization, **29A,** 6–7, 9–10
    steps in photoemission, **29A,** 3
  incident radiation, **29A,** 22–23
  photocathode
    aniodizaiton, **29A,** 12–13, 34
    heat cleaning, **29A,** 13–14
    lifetime, **29A,** 21–22
    limitations of response, **29A,** 20–21
    material, **29A,** 11–12
    photocurrent monitoring, **29A,** 15
    surface activation, **29A,** 15–18
    surface cleaning, **29A,** 12–15, 33–34
  principle of negative electron affinity, **29A,** 4–5
  quantum efficiency
    carbon contamination and quantum efficiency, **29A,** 15
    factors affecting, **29A,** 11
    formula, **29A,** 18
    yield curve, **29A,** 19
Negative-U interactions, **30,** 48–49
Nelsonian illumination, **25,** 152
Nematic liquid crystals, **29C,** 356, 357
Neodymium glass lasers, *see* ND:glass lasers
Neodymium-yttrium aluminum garnet (NdYAG) crystals, **15A,** 198; **15B,** 38–39
Neodymium-yttrium aluminum garnet (NdYAG) lasers, **15A,** 168, 195; **25,** 265; **29C,** 172, 259
  energy spectrum, **29C,** 142
  fourth harmonic of, **15B,** 215
  intercavity frequency doubling, **29C,** 237–239
  properties of, **15A,** 196–198
  pulsed, **15B,** 46
  as pump source, **15B,** 45
  pumping of, **29C,** 150
  Q-switched, **15B,** 192
  Q-switching, **29C,** 143–144, 146–147, 260
  synchronous pumping with, **15B,** 175
  as two-step ionization source, **15B,** 42
  wavelength, **29C,** 151
Neodymium/yttrium lithium fluoride (ND:YLF) lasers, **15A,** 204–205; **29C,** 262
Neodynium glass laser, **25,** 265, 267
Neon laser emission, **10,** 48–49
Neon, **14,** 419–420
  in air, **14,** 419
  atmospheric concentration, **12B,** 170
  permeation in glasses, **14,** 340–343, 477

target, **17**, 410–416
Neon-oxygen laser, **15A**, 116
Neoprene, **14**, 407–410, *see also* Elastomer
NEP, *see* Noise equivalent power
Nernst calorimeter, **1**, 266
Nernst effect, **6B**, 166, 168–169
Nernst filament, **1**, 399; **3**, 76
Nernst glower, **10**, 15, 16; **13B**, 5; **22**, 77
Nernst relation, **1**, 276
Nernst-Eucken method, **6A**, 377–380
Nernst-Lindemann relation, **1**, 269
Network analysis
complex notation in, **2A**, 3–4
unit function in, **2A**, 5
Network function, **2A**, 4, potential analogy for, **2A**, 5
Network parameters, **2A**, 17–20
Network structures
active devices in, **2A**, 22–23
transformers and, **2A**, 21–22
T structures in, **2A**, 21
Network theory, **2A**, 7–17
filters in, **2A**, 23–28
lattice structure in, **2A**, 21
linear active devices in, **2A**, 22–23
loop analysis in, **2A**, 8–9
node analysis in, **2A**, 9–10
Norton's theorems in, **2A**, 10–11
power transfer and impedance matching in, **2A**, 11
reciprocity in, **2A**, 10
simple resonant circuits and, **2A**, 11–14
simple transients in, **2A**, 14–15
specific structures in, **2A**, 21–23
superposition principle in, **2A**, 10
Thevenin's theorem in, **2A**, 10–11
Network topology, **2A**, 7–9
Network(s), *see also* Circuits; Electronic circuits; Integrated circuits
all-pass, **2B**, 40
branches and nodes in, **2A**, 7
bridged T, **2**, 31,32
distributed constant, **2**, 37–41; **2A**, 29–33
electromechanical devices in, **2A**, 32–33
function, **2**, 20
image-matched, **2A**, 19
impedance and admittance in, **2A**, 6
input impedance in, **2A**, 19
lattice, **2**, 31, 32
natural frequencies of, **2A**, 9
passive, **2**, 17–46
linear, **2A**, 1–44
π, T, **2**, 31, 32
series and parallel components of, **2A**, 6
theory, **2**, 24–28
transmission lines and, **2A**, 29–32
two-terminal-pair, **2**, 28–37; **2A**, 17–23
voltage and current sources in, **2A**, 3
Neurons, thermal, **20**, 338
Neutral atom density, in noble-gas lasers, **15A**, 83
Neutral atom lasers, excited by collision with atoms, **15A**, 97–120, *see also* Argon-chlorine laser; Helium-neon lasers
Neutral beams, as probes, **9A**, 130
Neutral beams, high velocity, **4A**, 279–280
Neutral particle, energy analysis, **17**, 180
Neutral species, **22**, 496
Neutral-neutral beams, center-of-mass transformation, **7A**, 306–310, 327–330
Neutral-neutral collisions, theoretical survey, **7A**, 297–303
Neutral-neutral interactions, **7A**, 227ff
Neutral-particle detection, **22**, 320
Neutralization, **22**, 303, 304, 307, 310, 326, 473
Auger neutralization, **22**, 304
heat of, **1**, 262
of ions, **5A**, 94
quasiresonance neutralization, **22**, 304–305
resonance neutralization, **22**, 304
Neutrals, sputtering of, **21**, 237–244, 257, 267
Neutrino, **5A**, 682–698
energy determination, **5A**, 683–692
free, experiment, **5A**, 685–692
momentum, **5A**, 692–698
reactions, **5A**, 682–683
rest mass, **5A**, 692
spin, **5A**, 698
Neutron absorption, **23A**, 340, 343
for density measurement, **18**, 705
Neutron beam, **8**, 241, 243, 244
Neutron bombardment, glass, **14**, 322
Neutron charge, **4B**, 8
Neutron coherent scattering length, **16C**, 305
Neutron detection efficiency, **8**, 254
Neutron detection, **21**, 176, 177
Neutron detectors, **3B**, 976; **8**, 268; **16A**, 219, 495–497
Neutron detectors,
Neutron diffraction, **3**, 348–358; **16A**, 209–210

analysis of results, **3,** 354–358
apparatus and procedures, **3,** 352–354
applications, **6A,** 281–282
methods, **6A,** 278–282
spectrometer, **6A,** 279–281
Neutron emission, **8,** 267
Neutron filters, **23A,** 153, 154, 204, 215, 216
Neutron filters, *see* Crystal neutron filters
Neutron flux, **21,** 19
measurement, **5A,** 466–481
with nuclear emulsions, **5A,** 475–478
by recoil techniques, **5A,** 461–494
tandem counter, **5A,** 466–471
nonexponential attenuation, **5B,** 383–385, 394–395
Neutron generating time, **8,** 233
Neutron guides, **23A,** 150, 152, 204, 270
Neutron leakage, **23A,** 163
Neutron lifetime, **8,** 251
Neutron optics, **23A,** 259
in magnetic scattering, **23A,** 75
Neutron polarimeter, **8,** 255
Neutron polarization studies, **8,** 264–265
application, **8,** 255–257
Neutron reflection, **23A,** 265
total, **23A,** 267
Neutron refraction, **23A,** 265
Neutron resonances, **23C,** 267
Neutron scattering (thermal neutron scattering)
coherent, **16A,** 208
comparison with other types, **16A,** 206–207
cross section of selected nuclei, **16A,** 209
energy and momentum transfer diagram, **16A,** 207
elastic, **16A,** 208
experimental techniques, **16A,** 218–229
incoherent, **16A,** 208
incoherent inelastic, **16A,** 212
inelastic, **16A,** 208
multiphonon contributions, **16A,** 215
quasi-elastic, **16A,** 210
small-angle, **16A,** 210
differential scattering cross section, **16A,** 481–485
polymer studies, **16A,** 497–506
source intensity, **16A,** 490–495
spectrometer design, **16A,** 485–497
study of low-frequency vibrations in crystalline polyethylene, **16A,** 229–235
study of segmental diffusion in rubbers, **16A,** 236–239
study of torsional dynamics in methyl side groups, **16A,** 235–236
Neutron scattering principle, **6A,** 278–279
techniques, **6A,** 279–281
Neutron scattering, **5A,** 317–319, **8,** 256; **24A,** 28; **24B,** 531–534; *see also* Inelastic
differential cross section, **5A,** 466
free atom recoil from, **3,** 351–352
by the nucleus, **3,** 350–351
wavelength, average, in reactor core, **3,** 348
Neutron scattering, Bayesian methods, **28,** 414–47
Neutron scattering, experimental
analyzer and monochromator crystals, **21,** 178, 179
Bragg, **21,** 158
diffuse, **21,** 161
Huang, **21,** 166
small angle neutron scattering (SANS), **21,** 159
small Q values, **21,** 152, 156, 165–167
Neutron source(s), **3B,** 971; **16A,** 218, 485; **23A,** 99
comparison, **23A,** 100, 110
energy spectrum, **16A,** 491
history, **23A,** 99
intensity, **5A,** 522–529
monoenergetic, **5B,** 369–373
portable, **23C,** 472
practical, **23A,** 109
radioactive, **5B,** 555–567
slow, **5B,** 390–392
types, **23A,** 99
Neutron spectra, **23A,** 159, 168
effect of cooling, **23A,** 171
effect of material, **23A,** 173, 176
Neutron spectrometers, *see* Time-of-flight neutron spectrometer
backscattering type, **16A,** 228–229
modified triple axis, **16A,** 226–228
small-angle, **16A,** 485–497
resolution, **16A,** 487–490
time-of-flight, **16A,** 220–224
triple axis, **16A,** 224–226
Neutron spectrometry, **8,** 197
Neutron spectroscopy, application, *see* Fast neutron spectroscopy
Neutron threshold measurements, **8,** 267–270

Neutron time-of-flight spectrometer, **8,** 201
Neutron transmission measurements, **8,** 239–244
  basic principle, **8,** 239–242
  exercises, **8,** 244
  final remarks, **8,** 243
  numerical results, **8,** 242–243
Neutron velocity, **8,** 241, 263
Neutron(s)
  background
    extraneous, from beam direction, **5B,** 377–378
    room scattered, **5B,** 376–377
  beam
    cold, intensity, **5A,** 511–513
    counting rate, **5A,** 517ff
    magnetic resonance, **3,** 635–636
    polarized, **5A,** 465
  capture by boron, probability, **3,** 350
  capture in cadmium, **5A,** 157
  cold, studies at Brookhaven reactor, **5A,** 510–511
  confinement, **23A,** 270
  counting, **5A,** 107–109
  cross-section
    differential elastic, **5B,** 472–475
    nonelastic, **5B,** 397–411
    total, transmission measurement, **5B,** 367ff
  detection, **24B,** 538–542
  distribution, liquid nitrogen cooled Be filtered, **5A,** 506
  emission, **5B,** 440–442
  energy measurement, **5A,** 466–481
    with bubble chamber, **5A,** 478–481
  epithermal, detectors for, **5A,** 533–538
  filtered, energy spread, **5A,** 509–510
  gauge interactions, **23A,** 312
  generated per nonelastic collision, **5B,** 405ff
  helium scattering, **5A,** 481–482, 493–494
  induced reactions, leading to emission, **5B,** 448–460
    charged particle, **5B,** 449–452
    γ-ray, **5B,** 459–460
    neutron, **5B,** 452–459
    proton, **5B,** 450–451
  interactions, **5A,** 496–497
    recoil energy, **5A,** 463–464
    time-of-flight measurement, **5A,** 495–566
  magnetic moment, **3,** 635–636
  moderated, energy spectrum shape, **5A,** 521–522
  moderating material, **5B,** 02
  moderation, **24B,** 463–466
  outgassing from glass, **14,** 333–337
  polarization, **3,** 635–636
  production, **5A,** 495
    in beryllium, **5A,** 525–526
  prompt fission, energy distribution, **5B,** 594–595
  properties, **5A,** 496–497
  proton
    collision mean free path in toluene, **5A,** 145
    cross section, **5A,** 488–493
  response of semiconductor detector, **5A,** 273
  sensing instrumentation, **5B,** 604–605
  slow, detector, **5A,** 139, 140
  slowing down, **5A,** 145
    length, **24B,** 464–466, 537–538
  solid state detector, **5A,** 535–537
  spectroscopy, time analyzers, **5A,** 554–566
  spin-angular momentum, **5A,** 464–466
  thermal, **5A,** 497–517
    moderation, **5A,** 515–517
    sources, **5A,** 497–499
  threshold measurement, **5B,** 360–362
  transport and diffusion, **24B,** 534–538
  users, **23A,** ix
  wavelength selection by diffraction, **5A,** 568
  yield
    from gas-containing targets, **5A,** 524ff
    from heavy ice target, **5A,** 524
    per incident electron, **5A,** 528, 530
    from uranium target, **5A,** 527–528, 529
  zirconium, cross section, **14,** 365
Neutron, angular momentum about center of Earth, **23A,** 295
  basic properties, **23A,** 1
  effective mass, **23A,** 329
  interactions with matter, **23A,** 4
  wavelength, **23A,** 1
Neutron-absorbing materials, **8,** 249
Neutron-nucleus scattering potential, **5A,** 318
Neutron-proton scattering total cross-section, **5B,** 385, 386
Neutrons, *see* Collimated neutrons; Fast neutrons; Incident neutrons; Spin-down; Well-thermalized neutrons

transmission, *see* Thermal neutrons
well-collimated beam, **8,** 240
New chemical lasers
chemical pumping rate requirements in, **15B,** 122
efficient visible, **15B,** 139–140
ideal reaction mechanism in, **15B,** 139–140
laser-induced vaporization technique and, **15B,** 137
metal atom or free radical production in, **15B,** 135–136
phantom yield for, **15B,** 127–132
problems in search for, **15B,** 120–140
quenching of electronic chemiluminescence in, **15B,** 133
radiation life of, **15B,** 123
rare gas monohalide excimer lasers as, **15B,** 125–126
reaction systems for, **15B,** 126–135
special initiation techniques for, **15B,** 135–139
NEW, *see* Negative-electron-affinity semiconductor photocathode
Newton relation, **25,** 28
Newton's law, **1,** 30, 35
of cooling, **18,** 665
Newton's method, **8,** 90
Newton, **1,** 30, 432
interferometer, **26,** 7
rings, **26,** 4, 5
Newtonian field, **18,** 507, 797–800
Newtonian fluid, **1,** 123, 130
Newtonian liquid, **16C,** 2; **19,** 140, 141
Neymann-Pearson hypothesis testing, **28,** 182–183
NH, transition frequencies of, **12C,** 26
NH2-adsorbed
Si(111) 7 × 7
scanning tunneling microscopy of, **27,** 192
NH_3
laser emission, **10,** 49, 50
spectrum, **10,** 318, 319
NHHCO, transition frequencies of, **12C,** 27
Ni, **23C,** 31, 181
Ni-Cu alloys, *see* Cu-Ni alloys
NiAu(111)
epitaxy scanning tunneling microscopy of, **27,** 296
NiB glasses, **23B,** 259
phonon density of states, **23B,** 294
$NiBr_2$, **23C,** 141
NIC, *see* Negative immitance converter
Nicholas chart, **2,** 481–483
Nichols chart, **2B,** 411–413
Nichrome, **11,** 360; **14,** 378
Nickel
alloys, **14,** 377–379
boiling point, **14,** 351
de Haas-Van Alphen effect in, **11,** 63–64
Fe–30% Ni alloy, attenuation in, **19,** 268–274
hydrogen solubility, **14,** 354
melting point, **14,** 348
oxide (Nio), **30,** 125
physical properties, **14,** 357–360
polycrystalline, attenuation in, **19,** 268–275
spin-dependent tunneling in, **11,** 154
thermal conductivity, **14,** 356
vapor pressure, **14,** 348, 351
Nickel, reflectance spectra, **31,** 217–218
Nicol prism, **1,** 379, 380; **26,** 139
Nicotonic acid, **30,** 417, 418
NiF_2 antiferromagnetic resonance, **10,** 411
Night-sky background, **12A,** 338, 393
NIM, *see* Normal incidence monochromator; Nuclear instrument module
NiO antiferromagnetic resonance, **10,** 411
Niobium
boiling point, **14,** 351
cavities, **11,** 559, 563–566
as cavity material, **11,** 559, 563–566
energy gaps in, **11,** 142
for high-*Q* superconducting resonators, **11,** 570–573
hydrogen solubility, **14,** 354
melting point, **14,** 348, 351
physical properties, **14,** 370–371
in point contacts, **11,** 258–259, 301–302
selenide, **11,** 258
Siemens process for, **11,** 571–573
Stanford process for, **11,** 571–572
in superconducting ring, **11,** 280
tunneling in, **11,** 142
vapor pressure, **14,** 348, 351
NiTi metallic glasses, **23B,** 261
Nitric acid degradation of polymers, **16B,** 256–257
Nitrogen, **22,** 184
adsorbed on graphite, **23B,** 22
adsorption, **22,** 360
afterglow, **3,** 168

electron induced dissociation, **3,** 551
laser, **9A,** 81
physical properties, **14,** 421–422
quadrupole resonance, **3,** 513–514
rotational lines, pressure broadening, **3,** 146, 147
Nitrogen-laser-pumped dye laser, **15B,** 217
Nitrogen13, half-life, **8,** 187–190
Nitromethane, **23A,** 511
Nitrous oxide, **22,** 66, 75
absorption coefficient of, **12B,** 169–171
atmospheric concentration, **12B,** 170
microwave lines of, **12B,** 171
on Ru(001), **22,** 66, 75
Nitrous oxide laser, **15B,** 83
Nitrous oxide spectrum, **10,** 334, 337
Nitroxide free radical, as spin probe or label, **16A,** 458–465
Nixie tube, **2B,** 7
NiZr metallic glasses, **23B,** 264
NMR, *see* Nuclear magnetic resonance; Nuclear magnetic resonance spectroscopy
NNDR, *see* Nuclear magnetic resonance, Double resonance techniques
Noble-as ion laser(s)
cascade contribution to, **15A,** 86
characteristics of, **15A,** 71–81
mechanisms of, **15A,** 81–93
continuous neutral, **15A,** 55–59
electric field of, **15A,** 83–84
electron temperature in, **15A,** 82–83
exciting by atomic collision, **15A,** 121–124
exciting mechanism models of, **15A,** 84–87
ion density in, **15A,** 83
ion temperatures in, **15A,** 82
neutral atom density in, **15A,** 83
neutral-atom temperature in, **15A,** 83
parallel or perpendicular fields in, **15A,** 81–82
performance of, **15A,** 92–97
plasma properties of, **15A,** 81–82
pulsed, **15A,** 74–76, 95
radiation trapping in, **15A,** 87–90
single-step excitation model of, **15A,** 87
spatial development in output of, **15A,** 75–76
spectroscopy of, **15A,** 62–71
transient, **15A,** 53–55
ultraviolet, **15A,** 98–99
Noble-gas atom, outer electron shell of, **15A,** 62
Noble-gas energy levels, transient laser lines in, **15A,** 51
Noble-gas mixtures, high-pressure, **15A,** 117–118
Nodal planes, **25,** 25
Node, **2,** 19
analysis, **2,** 26–27; **2A,** 9–10
in network theory, **2A,** 7
of Ranvier, **20,** 487, 492–497
Noise analysis, amplifier and, **2A,** 326–327
Noise bandwidth, in low-noise amplifier, **2A,** 328
Noise conductance, **2B,** 498–500
Noise equivalent power (NEP), **2B,** 489; **10,** 158; **12A,** 422–424, 515
bolometers, **10,** 169–171
diode superheterodyne detectors, **10,** 164
diode video detectors, **10,** 163
germanium bolometer, **10,** 171
Golay cell, **10,** 167
InSb detector, **10,** 187
Josephson junction, **10,** 208
in photocells, **2B,** 331
pyroelectric detector, **10,** 173
superconducting bolometer, **10,** 170
Noise figure, **12B,** 208
in amplifier, **2A,** 77–78
of superheterodyne receiver, **10,** 164, 165
Noise filtering, in pulse measurement, **2B,** 79
Noise fluctuation
effects of, **12C,** 22
Noise immunity, of logic circuits, **2A,** 412
Noise in detectors, **26,** 328
Noise in infrared systems, **12A,** 422–426, 439–441, 447–451
detector, **12A,** 422–426, 444, 450, 461
generation and recombination, **12A,** 419
Johnson, **12A,** 419
modulator, **12A,** 447, 450, 461
phonon, **12A,** 422
photon, **12A,**425, 435, 439–441, 443
preamplifier, **12A,** 432, 461
sky, **12A,** 435, 444, 447–451, 461
Noise in interferograms, **12A,** 511–515
Noise limitations in photon detectors
detection sensitivity limits, **4A,** 421–429
detectivity-D*, **4A,** 419–420, 423–428
kinds of noise, **4A,** 417–419
noise-equivalent power, **4A,** 419

radiation noise, basic formula, **4A,** 421–422
signal-to-noise ratio, **4A,** 417, 419, 427
Nitrogen dioxide
absorption coefficient of, **12B,** 172
atmospheric concentration, **12B,** 170
Noise measure, **2B,** 500
Noise models, **2B,** 469–481
Noise power
crystal, **5B,** 89
output, crystal rectifier, **3,** 29
Noise reduction, **14,** 268
Noise resistance, **2B,** 498–499
Noise signals, calculation of, **2B,** 498
Noise sources in tube amplifiers
effect of bandwidth, **4A,** 321, 325, 327, 329
electron spin resonance, **3,** 491–493
flicker noise, **4A,** 328–329
grid current noise, **4A,** 326–328
induced grid noise, **4A,** 329–330
Johnson noise, **4A,** 319–321
noise referred to grid, **4A,** 321, 325, 329
partition noise, **4A,** 324–326
space charge smoothing, **4A,** 324–325
shot noise, **4A,** 321–325
space charge limited, **4A,** 323–324
temperature limited, **4A,** 321–323
Noise specification, for amplifiers, **2A,** 250–252
Noise temperature, **2A,** 252; **2B,** 499; **10,** 211, 217–219
cascaded stage, **12B,** 208
definition of, **12B,** 5
equivalent, **12B,** 203
Noise, **7A,** 189–191, 274–278; *see also* Signal-to-noise ratio
1/*f* and low frequency modulation, **21,** 416–417, 434
1/*f* noise, **29A,** 440
in ac measurements, **2B,** 71–76
amplifier, **2,** 808–818
in amplifiers, **2B,** 482–486
avalanche multiplication, **2B,** 470
in bipolar transistors, **2B,** 471–479
burst, **2B,** 467, 470
chamber track, **5A,** 381–382, 396, 398, 399
channel, **20,** 472, 473, 497
circuit representation of, **2B,** 498
comparison of in devices, **2B,** 483
current, **10,** 211, 221, 222
current, in semiconducting materials, **2,** 807–808
in electronic devices, **2,** 795–818; **2B,** 461–500
elimination, **5A,** 400–401
equivalent power, **1,** 402; *see also* NEP
experimental problems in scanning tunneling microscope operation, **27,** 71
fabrication technology and, **2B,** 466
figure, **2,** 811–813
amplification, **2,** 158
flicker, **2B,** 465–467; **10,** 211, 224; **22,** 202, 259, 402
in tubes, **2,** 806–807
in four-terminal networks, **2B,** 499–500
generation-recombination, **2,** 805–806; **2B,** 464–465
ground loop pickup, **29A,** 441
hybrid-pi, **2B,** 472
inductive noise pickup, **29A,** 441
at input of photodetector, **2B,** 489–492
ionization chamber amplifier, **5A,** 105–107
Johnson, **10,** 211, 219–221; **22,** 202; **18,** 143
noise, **29A,** 439–440
in junction diodes, **2B,** 469–471
at low frequencies, **2B,** 477–479
measurement, **2,** 808–810
mechanical vibration sources, **29A,** 441–442
minimization, **29A,** 438–440
modulation, **2,** 806–808; **2B,** 462
optical, **18,** 141
partition in pentodes, **2,** 803
in photoconductive detectors, **2B,** 494–496
in photodetector(s), **2B,** 486–496
in LDV, **18,** 142–143
in photodiodes and photomultipliers, **2B,** 693
in photoelectric detectors, **17,** 545
in photojunction devices, **2B,** 494
photon, *see* Noise radiation
in pulse amplifiers, **2,** 813–815
radiation, **10,** 211, 217
recombination, **10,** 211, 221
resistor, **1,** 458
Schottky, **2,** 701
shielding and guarding of, **2B,** 73
shot, **2,** 700; **2B,** 463–464, 499; **10,** 211, 222–224; **18,** 143, 185; **22,** 259
in diodes and triodes, **2,** 801–803

noise, **29A,** 438
solid state detector, **5A,** 271
solid state photocells, **2,** 761
source of, **2B,** 461–468
subjective evaluation of, **1,** 229
temperature and, **2B,** 478
thermal, **2,** 700, 798–801; **2B,** 462–463
in two-terminal networks, **2,** 799–801
thermocouple junction noise, **29A,** 442
transition, **2B,** 461
transistor, **2,** 158–159
transport, **2B,** 461
of two-terminal networks, **2B,** 498–499
vacuum tube oscillator, **2,** 273–274
voltage, **3,** 32, 490
white, **1,** 220
Zener breakdown, **2B,** 470
Noise, **8,** 3, 8–10, *see also* rms noise
currents, **8,** 7
fluctuations, **8,** 4
force, **8,** 5–6
formula, *see* Johnson noise formula
influence, *see* Background noise
spectrum, *see* White-noise spectrum
voltage generator, **8,** 7
Noise-adding receivers, **12B,** 216–217
Noisethermometer, **11,** 286–289
Nomarski microscope, **30,** 581
Non resonant s-wave production, **8,** 267
Non-Bravais lattices, magnetic, **23C,** 127
Non-conjugate losses, **1,** 316ff
Non-linear effects, **1,** 498
Non-Newtonian shear flow, **16C,** 47
Nonblackbody modifications, **8,** 30
Noncoincidence method of inelastic energy-loss measurement, **17,** 173–184, 188
Noncolinear, structures, **23C,** 127
Nonconductive material, **8,** 157
Nonconservative freezing methods, **6A,** 55–57
Nonconstant variance, **28,** 267–268
Noncontract transducers, **19,** 69–71, 134
capacitive transducers, **19,** 69
electromagnetic acoustic transducers, **19,** 69
phase-insensitive transducers, **19,** 70
Nondestructive burnup determination, **8,** 203
Nondestructive evaluation, *see* Surface acoustic wave applications
Nondimensional parameters, **8,** 83, 90
Nondispersive detector, **17,** 358, 372
Nondissipative regulators, **2B,** 456–458
Nonequilibrium carrier concentrations, generation by light, **6B,** 79–80
methods of production, **6B,** 79–84
minority carrier depletion, **6B,** 83–84
production by injection, **6B,** 80–82
Nonequilibrium charge state distributions, **17,** 80–83
Nonequilibrium semiconductor phenomena, **6B,** 78–109
Nonequilibrium system
composition measurements, **18,** 631–634, 637–643, 645–649
Nonlinear affects, **19,** 507, 517, 529
Nonlinear capacitance diode, **2A,** 366–367
Nonlinear circuits, **2A,** 361–421
automatic gain control in, **2A,** 404
balanced diode linear gate in, **2A,** 381
clamping circuits and, **2A,** 385–389
comparators in, **2A,** 382–385
flip-flops in, **2A,** 398–400
gating circuits in, **2A,** 378–380
limiters in, **2A,** 369–371
mixers in, **2A,** 402–403
multiple circuits and, **2A,** 403
multivibrators and, **2A,** 397–400
nonregenerative, **2A,** 369–395
pulse stretchers in, **2A,** 389–395
regenerative, **2A,** 395–402
square-law device for, **2A,** 403
Nonlinear devices, **2A,** 361–368
Nonlinear effect in holography, **26,** 172
Nonlinear index of refraction, **30,** 37–38
Nonlinear loop mirror modelocking (NALM), **29C,** 180
Nonlinear optical applications, large non-linearity, **15B,**257
Nonlinear optical devices, **15B,** 143–183
coupled amplitude equations in, **15B,** 148–151
efficiency reduction in, **15B,** 167
focused beams in, **15B,** 161–164
generation of by mixing of two inputs in, **15B,** 174–183
and interactions with short pulses, **15B,** 164–165
intracavity generation in, **15B,** 166–167
Kleinman's symmetry for, **15B,** 146
large-signal conversion in, **15B,** 159–161
nonlinear polarization in, **15B,** 144–147
output angle and, **15B,** 165–166
parametric oscillations as, **15B,** 168–174

phase matching in, **15B,** 151–159
recent parametric oscillator experiments and, **15B,** 155
subharmonic generation and, **15B,** 148–150
tuning and linewidth in, **15B,** 172–174
Nonlinear optical effects
coherent transient effects and, **15B,** 262–263
multiphoton dissociation of molecules in, **15B,** 265–268
multiphoton ionization and, **15B,** 264
nonlinear susceptibility tensors in, **15B,** 252–255
optical breakdown in, **15B,** 264–265
optical-field-induced birefringence in, **15B,** 260–261
self-focusing as, **15B,** 261–262
stimulated light scattering in, **15B,** 259–260
Nonlinear optical frequency conversion, **29C,** 231–233
difference-frequency generation, **29C,** 248–252, 254
four-wave mixing, **29C,** 253–254
optical parametric amplifiers and oscillators, **29C,** 255–266
Raman shifter, **29C,** 233, 266–268
second-harmonic generation, **29C,** 233–237
external enhancement cavity, **29C,** 239–243
intracavity frequency doubling, **29C,** 237–239
quasi-phase-matching, **29C,** 233, 234, 243–247
synchronously pumped frequency doubling, **29C,** 243
sum-frequency generation, **29C,** 247–248, 254
third-harmonic generation, **29C,** 252
up-conversion laser, **29C,** 233, 268–270
Nonlinear optical media, **3A,** 369
Nonlinear optical processes, **15B,** 255–268, *see also* Nonlinear optical devices; Nonlinear optical effects
optical mixing and harmonic generation in, **15B,** 255–258
Nonlinear optics
beginnings of, **15B,** 144
fundamentals of, **15B,** 250–252
Nonlinear phenomena in picosecond spectroscopy, **20,** 190–194
Nonlinear polarization, **15B,**144–145
Nonlinear polarization rotation modelocking, **29C,** 182–183
Nonlinear susceptibility
resource enhancement of, **15B,** 257
tensors, in non-linear optics, **15B,** 252–255
Nonlinear wave mechanics, **23A,** 308
Nonlinearity, **8,** 234
Nonlocal diffusion, *see* NMR, Applications of, bound diffusion
Nonoconfocal cavities, **29C,** 110, 364–366
Nonpolarizing beam splitter cubes, **29C,** 349
Nonpolarizing electrodes, **24B,** 328
Nonradiative electron capture, **17,** 89
Nonrelativistic stopping formula, **17,** 34
Nonscalar reconstruction, **19,** 573
Nonsilica optical fiber, **29C,** 375–376
Nonthermal distribution, **12C,** 28–29
Nonuniform impurity distributions, **6A,** 61–65
Nonvanishing terms, **8,** 97
Nonvolatile memory element, **2A,** 128
NOR function, **2A,** 410
Normal bivariate, **1,** 14
Normal dispersion, **25,** 201
Normal distribution, **1,** 4; **3,** 712
function, **3,** 711; **6A,** 4
Normal error model, **28,** 164
Normal freezing, **6A,** 48–49
Normal incidence method, **19,** 149–153, 156–159
Normal incidence monochromator (NIM), **29C,** 34, 35–36
Normal magnifying power, **25,** 70
Normal metals, surface impedance of, **11,** 544–550
Normal modes, **16A,** 90; **21,** 174, 183
of vibration, **23A,** 23, 372
Normal reactions, **19,** 210
Normal stresses, **16C,** 49
Normal vibrations, **10,** 322, 325
Normal-metal tunneling
barrier excitations in, **11,** 167–169
elastic, **11,** 157–161
electrode excitations in, **11,** 169–170
electron energy-momentum relationships in, **11,** 161
high-voltage tunneling in, **11,** 161–163
phenomena in, **11,** 159–160
phonon interaction in, **11,** 170
trapezoidal barrier in, **11,** 160–161
tunneling conductance and, **11,** 159

zero-bias anomalies in, **11,** 170–177
Normal-theory method, **28,** 228
Normality, tests for, **28,** 203
Normalization of cross section measurements, **7A,** 39, 73
Norton's theorem, **2,** 27–28; **2A,** 10–11
Nose whistlers
frequency and time delay of, **9B,** 93
propagation delay of, **9B,** 93
NOT function, **2A,** 410
Notation
complex, **2,** 19–20
in geometrical optics, **25,** 13
HCN laser states, **10,** 328
molecular states with electronic angular momentum, **10,** 330–339
Notch-noise method, in telemetering, **2B,** 246
Nozzle
beam, **7A,** 237–252, 342
wind tunnel, **18,** 758, 760, 764, 772
NPN transistor, **2A,** 94, 98–99, 106, 120, 352
characteristic curves for, **2A,** 256–257
fabrication of, **2A,** 107–108, 115
in integrated circuits, **2A,** 115–116
NPNP device, thyristor as, **2A,** 228
NRZ-L (nonreturn to zero level), **2B,** 228
NRZ-PCM/FM telemetering, **2B,** 233
Nuclear absorption, **3B,** 520; **23A,** 7
Nuclear collision, energy loss, **17,** 26, 28–29
Nuclear counters, electronic instrumentation for, **11,** 348–350
Nuclear counting experiments, **8,** 191–201
exercise, **8,** 201
Nuclear cross section, **21,** 15, 16
barns, definition, **21,** 15
charged particles, **21,** 21
excitation function, **21,** 16
fast neutrons, **21,** 21
thermal neutrons, **21,** 19, 20
Nuclear effects in atomic spectra, **13A,** 341
Nuclear energy levels, electronic perturbation of, **11,** 307
Nuclear energy loss, **17,** 26–28, 32, 41–44, 72
Nuclear explosion, **7A,** 466
Nuclear fuel technology, **8,** 178
Nuclear fuels, **23C,** 464
Nuclear *g* factor, **4B,** 100, 112, 114, 153
Nuclear hexadecapole interactions, **3B,** 785
Nuclear hyperfine interaction, **16A,** 448
Nuclear hyperfine splitting, *see* Hyperfine splitting, nuclear
Nuclear hyperfine structure in EPR, **20,** 77–79
Nuclear induction, **3B,** 517; **6B,** 232–233
Nuclear instrument module (NIM); **2A,** 43
standard, **2A,** 43
system, **11,** 349
Nuclear magnetic dipole interactions, **3B,** 783
Nuclear magnetic moment, **20,** 2, 5
Nuclear magnetic resonance (NMR) spectroscopy
broadline, **16A,** 302, 320
carbon-13, *see* Carbon-13 NMR
compared with ESR spin probe techniques, **16A,** 479
continuous wave, **16A,** 265
experimental, **16A,** 265–284
high resolution, as molecular structure probe, **16A,** 284–301
peak narrowing, **16A,** 249
principle, **16A,** 242–245
as probe for accurate magnetic field strength measurement, **16A,** 456
probe for molecular dynamics, **16A,** 302–320
pulsed Fourier transform
computer requirements, **16A,** 274–278
free induction decay, **16A,** 267
measurement of relaxation parameters, **16A,** 278–284
principles, **16A,** 266–270
spectrometer, **16A,** 271–274
rapid scan method, **16A,** 278
resonance line width, **16A,** 254
of the solid state, **16A,** 320–368
experimental, **16A,** 338–343
measurement of orientation, **16A,** 350–356
molecular dynamics, **16A,** 361–365
sensitivity, **16A,** 331–338
sources of line broadening, **16A,** 323–331
for sorption studies, **16C,** 356
stochastic excitation method, **16A,** 278
Nuclear magnetic resonance (NMR), **1,** 523–524; **2B,** 133, 137–147; **3B,** 472–497; **20,** 1–52, 198
absorption, **21,** 363
application to borehole fluid properties, **24B,** 494–497
basic phenomena, **21,** 361–364
Bayesian methods, **28,** 417–422

continous time averaging in, **2B,** 142–143
continuous wave, **20,** 7–9
display or visualization, of, **2B,** 141
double resonance, **3B,** 495
vs electron spin resonance, **2B,** 147
electric field perturbation, **21,** 430
Fourier transformed, **20,** 8, 10–12
free-induction decay, **20,** 7–9
free induction decay in, **2B,** 146
free precession technique in, **11,** 516
gyromagnetic ratio, **21,** 362
high resolution, **3B,** 480
Larmor frequency, **21,** 362
light perturbation, **21,** 428
limitations of magnetic resonance, **21,** 423–424
line shapes, **3B,** 525, 534
lineshift, **21,** 372
linewidth, **21,** 372
lock-in detector in, **2B,** 142
magic-angle spinning, **20,** 30–32
marginal oscillators and, **2B,** 138–139
Mössbauer spectra and, **11,** 312–314
measuremetns of, **11,** 513–518
motional narrowing, **21,** 372
multiple-pulse methods, **20,** 28–30
non-selective relaxation methods, **3B,** 538
position of resonance line, **21,** 371
principal, **24B,** 494
pulse spectrometer and, **2B,** 145–146
pulsed, **3B,** 487
spectrometers, **3B,** 535
resonance condition, **21,** 362
resonance frequency in, **11,** 530
rotating reference frame, **21,** 362–363, 368
selective relaxation methods, **3B,** 541
specimens of, in de Haas-Van Alphen effect, **11,** 51–52
spectra, **3B,** 484
spin relaxation time, **24B,** 494–495
steady-state methods in, **2B,** 137–147
stress perturbation, **21,** 429
transient methods in, **2B,** 144–147, 150
Nuclear magnetic resonance (NMR), applications
bound diffusion, **21,** 473
CaF_2, **21,** 473–477
chemical shifts, **21,** 400–402
domain structure determination, **21,** 397–398
F^- self-diffusion, **21,** 447
hydrogen NMR in amorphous-Si:H film, **21,** 392, 393
impurity diffusion, **21,** 469
KCN(^{13}C), **21,** 482
Knight shifts, **21,** 398–400
K^+ self-diffusion, **21,** 469
lineshape studies, **21,** 391–394
liquid crystals, nematic (PAA) (MBBA), **21,** 477–478
localized diffusion, **21,** 473
molecular diffusion, **21,** 477–480
molecular reorientations, **21,** 479–482
NMR imaging, **21,** 403, 404
platinum NMR in small particles, **21,** 400
between potential wells of unequal depth, **21,** 479–482
quadrupole resonance, **21,** 394–398
sodium in Na beta alumina, **21,** 393, 394
SrF^2, **21,** 473–477
translational diffusion, **21,** 473
Nuclear magnetic resonance (NMR), artificial line narrowing, **21,** 402–403
double resonance, **21,** 403
magic angle, **21,** 373
multipulse sequence, **21,** 403
Nuclear magnetic resonance (NMR), calibration techniques
exact resonance, **21,** 469–470
of local field (H_D), **21,** 471–472
RF pulse monitor (H_1), **21,** 470–471
Nuclear magnetic resonance (NMR), double resonance techniques, **21,** 403, 430–438
dynamic nuclear polarization, **21,** 435–436
electron-electron double resonance (ELDOR), **21,** 438
electron-nuclear double resonance (ENDOR), **21,** 432–435
Hahn condition, **21,** 431
nuclear-nuclear double resonance (NNDR), **21,** 430–432
optically detected magnetic resonance (ODMR), **21,** 427, 428, 436–438
optically detected nuclear magnetic resonance (ODNMR), **21,** 436–438
Nuclear magnetic resonance (NMR), measurement of diffusion coefficient, **21,** 446–447
ADRF applied to the measurement of *D*, **21,** 456–457
divacancy, **21,** 447
fluorine, self-diffusion in PbF_2, **21,** 447

Nuclear magnetic resonance (NMR), spectrometer design, **21,** 373–391
  block diagram, **21,** 375
  computers and interfaces, **21,** 387
  detector, **21,** 385–387
  diode detector, **21,** 385
  frequency source, **21,** 381–382
  gate, **21,** 382
  magnet, **21,** 376–378
  magnetic field stabilization, **21,** 378
  phase sensitive detection, **21,** 385–387
  power amplifier, **21,** 383–384
  preamp, **21,** 384–385
  pulse generator, **21,** 382–383
  pulsed NMR design, **21,** 375–391
  pulsed versus CW spectrometers, **21,** 374–375
  receiver, **21,** 384–387
  recovery time, **21,** 383–384, 389
  sample probe, **21,** 378–381
  signal averaging devices, **21,** 387
  signal-to-noise considerations, **21,** 379, 381, 386, 387–391
Nuclear magnetic resonance (NMR), ultraslow motion techniques
  adiabatic demagnetization in a rotating frame (ADRF), **21,** 456, 459, 463–464
  Ag^+ diffusion in RbCl, **21,** 449
  frequency dependence, **21,** 448, 451, 477–478
  lab frame (T_1), **21,** 447–449
  metallic Li, **21,** 473
  NaCN, **21,** 449
  off-resonance pulsing, **21,** 459
  rotating frame ($T_{1\rho}$), **21,** 454, 456
  spin lattice relaxation in a rotating frame ($T_{1\rho}$), **21,** 455
  strong collision theory, **21,** 456–457, 461–467
    for homonuclear systems, **21,** 456–457
  weak collision theory, **21,** 455–456
  weakly magnetic spins' motions, **21,** 461–467
Nuclear magnetism, **23C,** 241
Nuclear magneton, **4B,** 100
Nuclear moments, paramagnetic susceptibility of, **11,** 511–513
Nuclear octupole interactions, **3B,** 785
Nuclear orientation measurements, **11,** 518–533
Nuclear Overhauser effect, **20,** 33, 34
Nuclear Overhauser enhancement, **16A,** 264, 283
Nuclear Q values, **8,** 268
Nuclear quadrupole interactions, **3B,** 779
Nuclear quadrupole resonance (NQR), **3B,** 510
  spectrometers, **3B,** 544
Nuclear radiation, outgassing from glass, **14,** 333–337
Nuclear reaction analysis (NRA), **21,** 244–256
  channeling, **21,** 301, 333, 342
  charged particle activation analysis (CPAA), **21,** 245–246
  cross-section, **21,** 247
  neutron activation analysis (NAA), **21,** 245–246
  particle induced x-ray emission (PIXE), **21,** 250–251, 256
  prompt reaction analysis (PRA), **21,** 246–250
  resonant reaction analysis (RRA), **21,** 248–250
Nuclear reaction technique, channeling, **21,** 301, 333, 342
Nuclear reaction(s), **13A,** 137; **22,** 347
  (α,α), **21,** 248
  (d,α), **21,** 248
  (d,n), **21,** 247
  (d,p), **21,** 247
  (n,α), (n,p), **21,** 19, 20
  (n,γ), **21,** 14, 19, 20
  (p,α), **21,** 248
  (p,$\alpha\gamma$), **21,** 248, 250
  (p,γ), **21,** 247, 249
  (^{15}m,$\alpha\gamma$), **21,** 250
  charged particle, **21,** 15, 21
  energy loss, **17,** 26
  fission, **21,** 21
  spallation, **21,** 21
  yield, **21,** 17
Nuclear reaction, **8,** 253
Nuclear reactor, **8,** 233, **21,** 19–21, *see also* Water-cooled nuclear reactor
  technology, **8,** 233
Nuclear relaxation, enhancement by spin labels, **20,** 109, 110
Nuclear resonance bridge circuits, **6B,** 230–232
Nuclear resonance techniques, **6B,** 229–234
Nuclear scattering, **23A,** 5, 17
Nuclear scattering methods, for impurity analysis, **6A,** 37

scattering of electrons, **5A,** 63–71
cross section, **5A,** 67, 69
Nuclear spin, **2B,** 133
correlations, **23A,** 93
ordering, **23A,** 90
Nuclear tracks, *see* Autoradiography
Nuclear transitions, **13A,** 100
Nuclear, *see also* Gamma rays
binding energy, **5A,** 335–336
emulsions, **5A,** 216–224
microscope procedures, **5A,** 224–226
range-energy relation, **5A,** 229–236
interaction, inelastic, in C?erenkov medium, **5A,** 178
model of atom, **5A,** 289
periphery, **5A,** 307
radiation, response of semiconductor detector, **5A,** 273
size measurement, **5A,** 307–341
electromagnetic, **5A,** 322–325
nuclear force method, **5A,** 309–322
Nuclear-quadrupole hyperfine structure, **3A,** 105–109
Nuclear-spin statistical weights, **3A,** 36
Nuclear-track emulsion, **12A,** 270–276
Nucleation, **22,** 366
of crazes, **16C,** 251
in crystal growth from solutions, **6A,** 99
effect on melting point, **16B,** 303
effect of mode of crystallization, **16B,** 364
heterogeneous
in Avrami theory, **16B,** 348
measurement of, **16B,** 396–397
homogeneous
in Avrami theory, **16B,** 348
measurement by droplet technique, **16B,** 385–396
rate constant, **16B,** 349, 390
theory of crystallization, **16B,** 349
Nuclei
angular shapes, **5A,** 338–341
atomic, determination of charge, **5A,** 289–307
charge distribution, **5A,** 325, 327
chart of, **21,** 15
condensation, **5A,** 195
dynamic methods, **5B,** 171, 190–194
mirror, **5A,** 330–333; **5B,** 365
oriented, production of, **5B,** 168–184
positron emitters, **21,** 81
static methods, **5B,** 168ff
transient methods, **5B,** 172, 190–194
Nucleic acids, **20,** 44, 45
analysis, **30,** 436, 437
Nucleon mass measurement in emulsions, **5B,** 37–44
constant sagitta method, **5B,** 37–41
photometric method, **5B,** 41–43
Nucleus
compound, energy level spacing, **5A,** 538, 541–544
in constant magnetic field, torque on, **3,** 360, 362
ground state configuration, **5B,** 352
initial recoil velocity, **5B,** 333
magnetic field at, **3,** 363–375
optical model, **5A,** 30ff
potential at surface, **5A,** 312–313
Nucleus-nucleus scattering, **5A,** 321–322
Nuisance parameter, **28,** 166
planning value, **28,** 169–170
Null coil pendulum, **6B,** 206–207
Null detection, voltage comparison by, **2,** 567–569
Null detectors
in dc measurements, **2B,** 59–67
in phase measurements, **2B,** 43
Null hypothesis, **28,** 179
Null pendulum magnetometer, **6B,** 206–207
Null-balance principle, **6B,** 9
Nullpoint calorimeter, **18,** 676–677
Number of spins, counting experiments, **21,** 405
Numerical aperture, **25,** 24, 29, 166
Numerical mathematics, **28,** 489
areas important in statistical computing, **28,** 490–491
Numerical methods, **28,** 488–489
method of maximum likelihood, **28,** 240
Numerical techniques, use in antenna design, **12B,** 62
Nusselt number, **1,** 138, 157, 300; **18,** 460, 461
Nutation
VLBI techniques for, **12C,** 274
Nuvistors, **2A,** 164
NYAB, **29C,** 238
Nylon 12, radicals generated during fracture, **16C,** 204, 207
Nylon 6, 10
effect of crystallinity on gas diffusion, **16C,** 329

dielectric constant and loss, **16C,** 380
yield stress and Young's modulus, **16C,** 131
Nylon 6, 6
α form, unit cell and crystal structure, **16B,** 88–90
contact electrification in, **16C,** 430
etching with argon and nitrogen plasma, **16B,** 88, 244
radicals generated during fracture, **16C,** 204, 207
selective degradation by hydrazinolysis, **16B,** 277
selective degradation by hydrolysis, **16B,** 276
work function, **16C,** 431
yield stress and Young's modulus, **16C,** 131
Nylon 6, *see* Polycaprolactam
radicals generated during fracture, **16C,** 197, 200, 202, 204, 205, 207
stress relaxation, effect of moisture, **16C,** 55
drawn, true melting point by methoxymethylation, **16B,** 253–254
α form, repeat distance and unit cell, **16B,** 81–82
γ form
crystalline thickness from X-ray line broadening, **16B,** 151
repeat distance and unit cell, **16B,** 82
positron lifetime in, **16A,** 395
selective degradation by hydrolysis, **16B,** 276
solvent-etching, **16B,** 240
spherulitic growth rate, **16B,** 363
typical crystallinities, **16B,** 126
Nylon, **14,** 414
Nyquist equation, **1,** 458
Nyquist frequency, **28,** 309–310, 313, 316, 319
Nyquist plot, **2B,** 403
Nyquist sampling, **29B,** 365
theorem, **12A,** 403, 505
Nyquist theorem, **9B,** 162
Nyquist's formula, **2,** 798–799
Nyquist's theorem, noise and, **2B,** 462–464

O

O adsorbed
GaAs(110)
spectroscopy of, **27,** 110
Ge(111)
scanning tunneling microscopy of, **27,** 244
Si(111) 7 × 7
scanning tunneling microscopy of, **27,** 195
O on Cu(100)
chemisorption, scanning tunneling microscopy of, **27,** 303
O on Cu(110)
chemisorption, scanning tunneling microscopy of, **27,** 300
surface diffusion, **27,** 302
O ray, **1,** 379
o ray, interaction with e ray, **15B,** 146
O ring
Buna-N, **14,** 282–286
comparison to L gasket, **14,** 479
diameter, **14,** 281–283
elastomer, **14,** 281–283
gas load, **14,** 285
groove, **14,** 281–283
lubrication, **14,** 285
roll, **14,** 304
tolerance, **14,** 281–285
Viton, **14,** 282
O, radiative attachment, **7A,** 207
O, *see* Oxygen
O^- detachment, **7A,** 115
O^-, **7A,** 187
O_2 spectrum, **10,** 336
O_2^+ dissociation, **7A,** 115
O_2, photoabsorption by, **7A,** 458
Object slit, **8,** 183
Objective lens, in an electron microscope, **16B,** 208–211
Objectives
microscope, **1,** 340; **5A,** 224
telescope, **1,** 339ff
OBK approximation, *see* Oppenheimer-Brinkman-Kramers approximation
Oblique shock waves, **9A,** 240–241
Obliquity of field
circular mirror, **31,** 170
non-symmetrical system, **31,** 164–165
Observation point, **8,** 56, 57, 61, 62, 71–73, 78
Observing techniques, *see also source types*
Occultation times
duration of, **12C,** 98–99
OCS
lines, **10,** 313, 338
molecule, **10,** 310, 313, 314
transition frequencies of, **12C,** 26
Octupoles, *see* Multipoles
Ocular, **1,** 339

ODMR, *see* Optically detected magnetic resonance
ODNMR, *see* Nuclear magnetic resonance, double resonance techniques
Oersted, **1,** 518
Off-axis ablation in PLD, **30,** 513
Off-axis holography, **26,** 168
Off-center inteferograms, **12A,** 506–508
Off-the-shelf components, **25,** 3
Offset, in infrared photometer, **12A,** 446–447, 461
OFHC, *see* Copper
OH lines, **10,** 334, 335, 337
OH masers
  VLBI measurement of, **12C,** 251
OH, transition frequencies of, **12C,** 26
Ohm's law, **1,** 432, 446, 456
  superconducting resonators and, **11,** 546
Ohm, **1,** 33, 34
Ohmmeter measurement, **6B,** 33, 53–54
Ohmmeter method, **6B,** 57
Ohmmeter, in equipment testing, **2B,** 176
Oil
  contamination, **14,** 124–126, 234
    in turbo-pump, **14,** 257
  diffusion pump, **14,** 417–418
  immersion microscope, **1,** 341
  lubricated bearing, **14,** 263
  removal of, **14,** 443
  saturation by C/O ratio, **24B,** 484–486
Oil-drop experiment, **1,** 39
Olefins, addition of bromine atom to, **15B,** 288–289
Oligonucleotides, **30,** 436
Oligosaccharide, **30,** 437–438
Omegatron, **1,** 65; **4B,** 307; **14,** 86, 90; *see also* Mass spectrometer
  spectrometer, **22,** 444
OMS (ovonic memory switch), **2A,** 128
On-off measurements, **12C,** 15, 48–50, 52
On-resonance two-photon spectrum, in Doppler-free laser spectroscopy, **15B,** 247–248
One-dimensional conductors, **23A,** 360
One-dimensional magnets, **23C,** 223
One-electron system
  capture cross section, **17,** 299
  electron capture, **17,** 96
  Lamb shift, **17,** 587–588
  line energies, **17,** 387–388
  target x-ray cross section, **17,** 286–287
  x-ray decay, **17,** 398
One-mile telescope, **12C,** 162
One-over-v ($1/v$) law, **23A,** 525
One-phonon scattering, **23A,** 26
One-step method for rainbow holograms, **26,** 186
Onsager equation, **13B,** 172
Onsager relations, **6B,** 115
Opacity, **24A,** 21; *see also* Absorption
  measurement of emulsion tracks, **5A,** 303–305
OPAs, *see* Optical parametric amplifiers
Open channel liquid metering, **18,** 332–336
Open systems, **1,** 325
Open-loop control, **2B,** 375–378
Open-loop transfer function, **2B,** 422, 424, 427
Opening angle, **22,** 143
Operating systems, **28,** 482–483
Operation of turbo-pump, **14,** 267–273
Operation of vacuum systems
  back filling, **14,** 465
  general consideration, **14,** 465–467
  valve control, **14,** 466–467
Operational amplifier, **2B,** 446–452
  in analog computer, **2B,** 155
  closed-loop characteristics of, **2A,** 309–310
  comparators and, **2A,** 382–383
  design considerations in, **2A,** 311–314
  feedback impedance in, **2A,** 311
  feedback network in, **2A,** 311
  input or source impedance in, **2A,** 311
  nonlinear devices and, **2A,** 395
  open-loop characteristics of, **2A,** 309
OPO, *see* Optical parametric oscillator
OPOs, *see* Optical parametric oscillators
Oppenheimer, J. R., **23A,** 371
Oppenheimer-Brinkman-Kramers approximation, **17,** 90–92, 97, 286, 341–342, 353–354
Optic axis of a crystal, **26,** 135
Optic branch, **16A,** 101–103
Optical absorption, **22,** 174
  coefficient, **22,** 145, 179
  experiments, **7A,** 124, 143ff
  microscopy, **27,** 86
  measurement of, **7B,** 104–106
    in spectroscopy, **7B,** 102
Optical accelerometer, **1,** 80–81

Optical activity, **1,** 383–384; **6B,** 250, 270–271; **12A,** 374; **26,** 138
Optical avalanche ionization, **15B,** 265
Optical axis, **25,** 7
Optical cavity, feedback from, **15A,** 3–4, 210
Optical characteristics of flow tracing particles, **18,** 51–60
Optical components, **25,** 59
  antireflective coatings, **29C,** 346, 348
  beam splitters, **29C,** 348–349
  lenses, **29C,** 343, 346, 351–353
  mirrors, **29C,** 343, 345–348
  optical filters, **29C,** 350
  polarization-controlling components, **29C,** 353–358
Optical constants, **10,** 96, 101, 102, 104, 365, 366; **13B,** 43, 134, *see also* Dielectric constants
  calculation, **31,** 260
  determination, **31,** 267–269
  metals, **13B,** 94
  semiconductors, **13B,** 94
  solids and liquids, **13B,** 92
Optical damage, **30,** 581–612
Optical delay line, **18,** 738–739
Optical density of states, **11,** 76
Optical depth, **7A,** 49
  definition of, **9A,** 142
Optical design, **25,** 4
  by computer, **25,** 51
Optical determination of crystal orientation, **6A,** 160–163
Optical diode, **29C,** 63
Optical elements, holographic, **26,** 197
Optical emission experiments, **7A,** 120–123
Optical excitation, **11,** 75–77; **22,** 148, 163
  cross sections by electrons, absolute determinations, **7A,** 59
  functions, **7A,** 43, 82 *see also* Excitation functions
Optical extent, *see* Ötendue
Optical feedback, diode lasers, **29C,** 83–84, 85
Optical fibers
  properties, **29C,** 371–373
  types, **29C,** 373–376
Optical filters, **25,** 273; **29C,** 350
  bandpass filters, **29C,** 350
  birefringent filter, **29C,** 63
  edge filters, **29C,** 63
  interference edge filters, **29C,** 350
  intracavity spectral filters, **29C,** 164–169
  loop filter, **29C,** 124–126
  Lyot filter, **29C,** 63, 178
  multiple-beam interferometer, **18,** 105, 201, 195–227, 708, 717, 740
Optical frequency measurement, **29C,** 303–306, 311
Optical frequency mixing, with tunable dye lasers, **15B,** 174
Optical frequency references, **29C,** 106–121, 325
  atomic transistors, **29C,** 120
  Fabry-Perot cavity, **29C,** 107–120
  optical phase locking, **29C,** 121
Optical gain
  in color center laser, **15B,** 22–25
  self-absorption in, **15B,** 23–25
Optical heterodyne detection
  in LDV, **18,** 109, *see also* Heterodyne efficiency
Optical homodyne detection, **18,** in LDV, **18,** 116–118
Optical image, **1,** 335
Optical instruments, **25,** 119
Optical intensity calibration, **7A,** 53–54, 458–460
Optical interferometers, *see* Interferometry
  Fabry-Perot, **4B,** 79–84
    adjustment, **4B,** 83
    construction, **4B,** 82–83
    efficiency, **4B,** 80
    finesse, **4B,** 80
    resolving power, **4B,** 81–82
    use of secondary dispersion, **4B,** 81
  Michelson, **4B,** 76–79
    Fourier transform method, **4B,** 78
    resolution, **4B,** 77
    variations, **4B,** 79
    visibility curve, **4B,** 78
Optical interferometry, **9B,** 13–20
  Mach-Zehnder interferometer for, **9B,** 13–20
    effect of bending of the rays, **9B,** 18–20
    fringe localization, **9B,** 15
    light sources, **9B,** 15–18
  Schlieren methods, **9B,** 20–22
    derivation of angular deflection, **9B,** 22–27
    double-inclined-slit type, **9B,** 24–25
    grid projection technique, **9B,** 27
    phase contrast methods, **9B,** 27
    sensitivity limits, **9B,** 27

shadowgrams, **9B,** 25–27
Optical Kerr shutter, **29C,** 214
Optical lever, **16C,** 474
Optical locking, **29C,** 98–99, 121
Optical materials, **3A,** 241–245
glasses, **29C,** 343–346
for the infrared, **3A,** 189
OPAs and OPOs, **29C,** 257–259
Optical measurements,
general experimental equipment, **6B,** 252–253
at high pressures, **6A,** 430–432
on thin films, **6B,** 278–281
Optical media, **4B,** 53–54
Optical methods, **25,** 1
Optical mixing, **3A,** 450; **18,** 106–115; **27,** 82
application of, **15B,** 256
harmonic generation and, **15B,** 255–258
microwave signal from, **9A,** 52
Optical modes, **23A,** 374
Optical molasses, **29B,** 131
Optical monochromaters, **4B,** 69–72
Czerney-Turner type, **4B,** 70
design considerations, **4B,** 70–72
Ebert-Fastie type, **4B,** 70
Optical multichannel detector, **18,** 429
Optical parametric amplifiers (OPAs), **29C,** 255–266
Optical parametric oscillator (OPO), **15B,** 144
output frequency of, **15B,** 172
Optical parametric oscillators (OPOs), **29C,** 233, 255–266
Optical path (OP), **25,** 11
difference, **25,** 11, 44
lengths, **1,** 330; **4B,** 82, 98
Optical plumbing, **29C,** 382–383
Optical polarization, **26,** 107
Optical properties of solids, **6B,** 249–292
refractive indices, **6B,** 250–251
Optical pumping, **7B,** 1–28
of alkali stoms, **29B,** 163–166
of alkaline earths, **29B,** 166–167
apparatus for, **7B,** 13–18
containment, **7B,** 13–14
diagram, **7B,** 18
light sources, **7B,** 16
magnetic fields, **7B,** 14–16
photodetectors, **7B,** 17–18
basic concepts, **29B,** 146–147
beam diagnostics, **29B,** 161–163
definition of, **7B,** 1–2
experimental factors
beam deflection, **29B,** 159
Doppler shifts, **29B,** 158
laser beam profile, **29B,** 158
laser polarization, **29B,** 159–160
radiation trapping, **29B,** 161
stray magnetic fields, **29B,** 160–161
FIR wavelengths and, **15B,** 77–78
hyperfine frequency shifts, **7B,** 26–28
interaction of the alkali spin with the nuclear moment, **7B,** 26
level-crossing technique, **7B,** 4
light sources for, **7B,** 16
limitations of, **7B,** 4
matrix, **7B,** 7
of metastable atomic states, **7B,** 3–4
of metastable atoms, **29B,** 109, 167–168
methods, **4B,** 179–180, 182–187, 205–206
relaxation processes, **7B,** 21–28
crossed beam technique, **7B,** 24
hyperfine pumping, **7B,** 23
magnetic resonance line-shape, **7B,** 22
measurement of relaxation times, **7B,** 22–24
relaxation interactions, **7B,** 24–26
transient methods, **7B,** 22–24
spin exchange, **7B,** 18–21, 23
in sub-Doppler cooling, **29B,** 131
theoretical considerations
coherence effects, **29B,** 155–158
laser frequency dependence, **29B,** 154–155
multilevel atoms, **29B,** 150–154
two-level atoms, **29B,** 148–149
theoretical foundations, **7B,** 4–12
Optical pyrometers, **1,** 254, 257
Optical quenching of metastable atoms, **29B,** 109
Optical radiation
absorbed, **18,** 405–408, 621–634
emitted, **18,** 341–345, 641–645, 687–725
scattered, **18,** 408–433, 643–645
Optical refractivity of plasmas, **9B,** 1–48
cine-interferometry, **9B,** 46–48
experimental methods requiring laser sources, **9B,** 31–48
gas laser interferometry, **9B,** 31–39
Faraday rotation, **9B,** 27–31
heterodyne interferometry, **9B,** 37–38

holographic phase measurements, **9B,** 39–48
noninterferometric methods for, **9B,** 47–48
standard experimental methods, **9B,** 13–31
optical interferometry, **9B,** 13–20
theory of, **9B,** 3–13
nonelectronic refractivity contributions, **9B,** 11–13
wave equations, **9B,** 3–9
Optical region, **13A,** 253
detectors, **13A,** 314
grating spectrographs, **13A,** 287
prism spectrographs, **13A,** 279
Optical resonator, laser and, **15A,** 9
Optical rotation, **6B,** 270–278
Optical selection rules, **29B,** 152–154
Optical sensor for surface displacement or velocity, **18,** 549, 606
Optical sine condition, **25,** 24
Optical spectra in SCANIIR, **21,** 268
Optical spectroscopy
basic principles, **29B,** 363–369
for Raman spectroscopy, **29B,** 413–417
instrumentation
diffraction gratings, **29B,** 371–377
multiple beam interferometers
acousto-optic, **29B,** 383–386
Fabry-Perot, **29B,** 379–383
interference filters, **29B,** 378–380
two-beam interferometers, Michelson, **29B,** 387–390
prisms, **29B,** 369–371
Optical spectrum analyzer, **29C,** 188, 364
Optical studies on flames, **7B,** 209–220
Optical surface, **1,** 329
Optical system, transmissivity, **8,** 30–31
Optical systems, **1,** 335–338
for light scattering experiments, **9A,** 92
Optical theorem, **23C,** 244
Optical transfer function, **26,** 82, 84
Optical transmission, **14,** 394–396
Optical tube length, **1,** 341; **25,** 165
Optical waveguide
energy density and, **15B,** 158
propagation velocity for given velocity in, **15B,** 158
Optical wavelength measurement, **29C,** 311–312, 338
Fizeau wavemeter, **29C,** 331–337
plane-parallel interferometer, **29C,** 337–338
scanning Michelson interferometer, **29C,** 312–331
Optical wavelength range, **4B,** 52–54
Optical wavelength standards, **29C,** 279–280, 307
calcium-stabilized laser, **29C,** 294–295, 307
iodine-stabilized He-Ne lasers, **29C,** 288–294, 307
laser-cooled absorbers, **29C,** 294–303, 307
optical frequency measurement, **29C,** 303–306
theory, **29C,** 280–281
electronic frequency servo control, **29C,** 286–288
error signal generation, **29C,** 281–286
reproducibility, **29C,** 288
stability, **29C,** 288
uncertainty, **29C,** 288
Optical window, **24A,** 150–152
material, **26,** 31
Optical zoom, **25,** 96
Optically detected magnetic resonance (ODMR), **21,** 427–428, 436–438
Optically isotropic materials, **6B,** 254–263
Optically pumped devices, **15B,** 76–78
Optics, **1,** 329–429
atmospheric, **1,** 395–396
of molecular beams, **3,** 607–615
Optics, *see* Microwave optics
Optimal telemetry system, **24B,** 41
Optimization, **28,** 490
Optoacoustic effect, **13B,** 338
Orbital angular momenta, **8,** 274, 279
Orbital angular momentum, **21,** 398, 401, *see also* Chemical shift, NMR
Orbital configuration
for homonuclear first-row diatomic molecule, **3,** 184
for linear symmetrical AB^2 molecule, **3,** 186
Orbital degeneracy, effect of crystalline field in removing, **3,** 448
Orbital energies, **3,** 183
Orbital shrinkage, in beam-foil experiments, **17,** 598
Orbital symmetry properties, **3,** 183
Orbiting cross section, **7A,** 368–369
Orbitron ion pump, **14,** 225–227
Orbits, particle, in FFAG accelerator, **5B,** 681–682
Order contamination, **23A,** 403

Order of interference, **25,** 191, 210
Order parameter, **20,** 48, 49; **23C,** 5
  nematic, **23B,** 549
  temperature dependence, **23C,** 12
Order sorting, **25,** 213
Order-disorder phase transitions, **23A,** 434
Order-disorder transitions, **22,** 279
Ordinary beam, **26,** 136
Ordinary differential equation model, **28,** 501, 503, 506, 508–509
Ordinary ray, **12A,** 372
Ordinary wave in plasma, **10,** 255, 256, 273–275, 279
Organic superconductors, **23A,** 361
Organic/metal film filters, **31,** 324–326
Orientation
  of atoms
    by electron bombardment, **4B,** 145, 208
    by optical methods, **4B,** 179–180, 186–187
    transfer of, **4B,** 186
  by cold drawing, **16C,** 138–140
  in craze fibrils, **16C,** 265
  determination using fluorescent probe, **16A,** 428–432
  during polymerization, **16C,** 138
  effect on electric strength, **16C,** 473
  effect on EPR spectra (single crystal), **20,** 85–90
  effect on inelastic electron tunneling, **16A,** 163
  effect on infrared spectrum, **16A,** 83–84
  effect on neutron scattering, **16A,** 216, 231–232
  of films, **16C,** 141
  high resolution carbon–13 NMR measurements, **16A,** 350–356
  measurement by
    birefringence, **16C,** 165–167
    infrared dichroism, **16C,** 173–175
    small-angle X-ray diffraction, **16C,** 175–184
    sonic modulus, **16C,** 169–172
    wide-angle X-ray diffraction, **16C,** 158–161
  by melt flow, **16C,** 143–145
  nuclear, **5B,** 162–214
    degree of, **5B,** 194–201
    dynamic, **5B,** 124–127
    ferromagnetics and antiferromagnetics in, **5B,** 184
    Overhauser method, **5B,** 191–192
    paramagnetic salts for, **5B,** 181–182, 184–186
  in partially crystalline polymers, **16C,** 146, 148
  planar, **16C,** 156
  production of, **16A,** 8–9; **16C,** 138–146
  ultimate, in polymers, **16C,** 140
Orientation etching, sample preparation, **6A,** 162
Orientation function, **16C,** 148–150, 166
  expressed in Legendre polynomials, **16A,** 416–418
Orientational polarization, **10,** 347–351
Oriented crystalline polymers
  morphological model, **16C,** 219
  radicals generated during deformation, **16C,** 203–213
Oriented crystallization, **16C,** 143–145
Oriented polymers, electron microscopic studies, **16B,** 233–235
Orifice flowmeter, **18,** 324–331
Orifice geometry, **14,** 103–104, *see also* Pumping speed, conductance
Ornstein-Zernike equation, **23C,** 5
Orphée reactor, **23A,** 205
Orthicon
  image, **2,** 773–778
  image-intensifier, **2,** 778
Orthogonality, **1,** 19
Orthomode transducers, **12B,** 67
Orthorhombic crystal system
  definition, **16B,** 5
  interplanar spacing formulae, **16B,** 26
  reciprocal lattice relationships, **16B,** 24
Orthorhombic lattice
  definition, **16B,** 5
  reciprocal, **16B,** 23
Orthorhombic unit cell, direction indices in, **16B,** 6
Orthoscopic eyepiece, **25,** 62
Orthoscopic mode, **25,** 173
Orthoscopic recording in holography, **26,** 186
Oscillating mechanical systems, **1,** 70
Oscillation(s), **1,** 87–93
  basic threshold condition for, **15A,** 12
  betatron, **5B,** 628, 648, 735
  center of, **1,** 84
  frequency of, **1,** 199
  parasitic, **2,** 274–275
  period of, **1,** 83, 84, 87, 110

Oscillator stability, defined, **2A,** 341
Oscillator strength(s), **3B,** 901–910, 941, 944; **4B,** 215–217; **7A,** 35–38, 117–118, 144–145; **13B,** 239; **22,** 233
  data on, **9A,** 166
  definition, **29B,** 218
  derived from absorption measurement, **4B,** 222
  measurement
    in heat pipes, **29B,** 72–74
    through radiative transition probabilities, **29B,** 219–248
    by absorption spectroscopy, **29B,** 248–250
  relative measurement, **4B,** 223
Oscillator systems, continuous-wave composite, **11,** 397–398
Oscillator, **8,** 236, *see also* Mechanical oscillator; Reactivity oscillator; Transfer oscillator
  frequency, **8,** 235
Oscillator-detector
  autodyne, **3,** 401–402
  regenerative, **3,** 509–515
  superregenerative, **3,** 402–404, 515–520
Oscillators, **1,** 474–475; **2A,** 341–359; **3,** 469–471
  amplitude of oscillation in, **2A,** 346
  Astable circuits and, **2A,** 355–359
  backward-wave, **2,** 725–726
  blocking, **2,** 382–383
  Clapp
    transistor, **2,** 324
    vacuum tube, **2,** 280
  Colpitts
    transistor, **2,** 324
    vacuum tube, **2,** 276–277, 280
  coupling of to load, **2A,** 342
  crystal
    transistor, **2,** 325
    vacuum tube, **2,** 284–289
  damped harmonic, **1,** 90
  feedback, amplitude stabilization, **2,** 508–509
  general-purpose, **2B,** 180–181
  harmonic, **2A,** 341
  Hartley
    transistor, **2,** 324
    vacuum tube, **2,** 276–277, 279
  heterodyne, **2,** 540–541; **2B,** 22–23
  hydrogen fluoride, **3,** 44
  klystron reflex, **2,** 721–723
  laboratory, calibration of, **2,** 537–538
  linear, **1,** 87–89
  marginal, **2B,** 139; **3,** 401–402
  mechanical energy, **1,** 90
  microwave, **1,** 487–488
  negative resistance devices for, **2A,** 351–352
  periodic sawtooth, **2,** 401
  phase-shift, **2A,** 345
    transistor, **2,** 323
    vacuum tube, **2,** 293–296
  Pierce
    transistor, **2,** 325
    vacuum tube, **2,** 287
  piezoelectric, **1,** 469
  Pound-Watkins, **3,** 513
  programmable unijunction transistor and, **2A,** 353
  Q, **1,** 90
  quench, **3,** 515
  regenerative, *see* Regenerative oscillators
  relaxation, **2A,** 341
  strength, **3,** 242, 243
    first order vibrationally induced, **3,** 251
    temperature dependence, **3,** 253
    theoretical, **3,** 245–249
    vibrationally induced, **3,** 254–255
  superregenerative, **2B,** 140
  transistor, **2,** 323–326
  types of, **2A,** 341
  vacuum tube, **2,** 266–298
    beat-frequency, **2,** 291–293
    bridge, **2,** 296–298
    butterfly, **2,** 281
    LC, **2,** 278–281
    magnetostriction, **2,** 290–291
    Meacham bridge, **2,** 288–289
    Miller, **2,** 286–287
    resonant-line, **2,** 281–284
    Seiler, **2,** 280
    tuned-circuit, **2,** 275–291
    tuning fork, **2,** 290–291
    tuning range limitations, **2,** 267–269
    Wien bridge, **2,** 297
  voltage control of, **2A,** 347–348
  wide-band traveling-wave, **2,** 725–726
Oscilloscope techniques
  in equipment testing, **2B,** 191–202
  finite-bandwith corrections in, **2B,** 191–193
  low-frequency cutoff in, **2B,** 193

sag corrections in, **2B,** 193
Oscilloscope, **1,** 471–472; **2B,** 71
ac, **2,** 581
accuracy and resolution in, **2B,** 194–195
amplitude calibration, **2,** 663–664
choice of, **2B,** 201–202
current probes in, **2B,** 196
dc, **2,** 561
diagnostics with, **29C,** 189–190
in equipment testing, **2B,** 184
input impedance in, **2B,** 195
jitter in, **2B,** 198
multiple signal input, **2,** 663
operation of, **2B,** 341–342
recorder output of, **2B,** 201
risetime specifications and measurements in, **2B,** 191–192
sampling, **2B,** 199–200
signal delay *vs* sweep delay, **2B,** 198
single-event sampling and, **2B,** 201
techniques, **2,** 661–664
time calibration, **2,** 664
time deflection, **2,** 664
time domain reflectometry in, **2B,** 199–200
voltage probes in, **2B,** 195–196
Oscilloscope, **8,** 124, 126, 127, 129, 151, *see also* Dual channel oscilloscope
Osmium, reflectance spectra, **31,** 214–215
Osmometry, **16A,** 24–26
Ostwald viscosimeter, **1,** 124
OTS, *see* Optimal telemetry system
Out-of-shell transition, **17,** 579–580
Outer conductor, **8,** 113
Outer-shell collision, **17,** 149, 163, 165
Outer-shell electron capture, **17,** 299
Outer-shell electron loss, **17,** 116
Outer-shell excitation, **17,** 190, 376
Auger lines, **17,** 470
excitation cross section, **17,** 284, 287
excitation energy, **17,** 149
inelastic energy-loss measurements, **17,** 157
Outer-shell ionization, **17,** 384–385
Outer-shell rearrangement, in gaseous molecules, **17,** 424–425
Outer-shell relaxation, effect on x-ray spectra, **17,** 416–418
Outer-shell vacancy
Auger spectroscopy, **17,** 473–474
production at higher energies, **17,** 490
Outflow
from reservoir, **30,** 243–245
into ambient gas, **30,** 245–247
Outgassing, **14,** 292, 228, 257, 279, 304, 442
elastomer seals, **14,** 479
by electrons
common glasses, **14,** 328
of glass, **14,** 327–328
of fore-pump oil, **14,** 116–117
of glass, **14,** 332–337, 396
due to nuclear radiation, **14,** 333–337
by ultraviolet radiation, **14,** 329–333
porous material, **14,** 353
from pump, **14,** 493
pumpdown characteristics, **14,** 506–509
pumpdown time, **14,** 106
species, **14,** 448
thermal, of glass, **14,** 322–327
Outliers, **28,** 184, 259, 267
Output devices, use in multichannel spectrometers, **12B,** 276–277
Output power, dye lasers, **29C,** 54–56
Outrigged probe, **24B,** 232–233
Ovens for atomic beams
alkali, **4A,** 178
creepage in, **4A,** 171, 173
electron bombardment type, **4A,** 174, 176
for bismuth, **4A,** 180
for radioactive beams, **4A,** 215–228
channeled ovens, **4A,** 221–224
mounts, **4A,** 224–228
high temperature, **4A,** 172–178
moderate temperature, **4A,** 169–172
solid gold, **4A,** 180–181
tubular type, **4A,** 176–179
two-stage, **4A,** 172
Ovens, *see* Baking
Overcoring, **24B,** 383–389
Overhauser effect, **3B,** 506; *see also* Nuclear Overhauser effect
Overheat ratio, hot-wire and hot-film probes, **18,** 269
Overlapping orders, **10,** 134, 135
Overlayer phase diagram, **22,** 297
Oversampling, weak periodic signals, **28,** 366–367
Oversize waveguide, **10,** 122, 123
Overtone(s), **1,** 209
in near infrared, **3,** 108–109
in Raman spectroscopy, **29B,** 394–399
Overvoltage, **2,** 73
Ovonic memory switch, **2A,** 128
Owen bridge, **1,** 463, 464

Oxalic acid, **19,** 202
Oxazine IR dyes, **29C,** 155
Oxidation and reduction, **6A,** 24
Oxidation of polyethylene, selective, with ozone, **16B,** 260–261
Oxidation of polymers, studied by thermal analysis, **16B,** 335
Oxidation state, **22,** 223
Oxide cathodes, **4A,** 3–7, 35
  effect of dc field, **4A,** 5–6
  emission characteristics, **4A,** 4–7
  fabrication and processing, **4A,** 3
  life, **4A,** 7
Oxide emitters, **2A,** 149–150
Oxide fibers, **29C,** 375
Oxide film replica technique, **6A,** 179–180
Oxide glasses, **23B,** 267
  composition and batch calculations, **6A,** 143–144
  melting, **6A,** 144–146
  procedures for making in laboratory, **6A,** 143–147
  raw materials, **6A,** 143
  sample forming and annealing, **6A,** 146–147
  sample inspection, **6A,** 147
  supplementary procedure, **6A,** 147
  viscosity-temperature data, **6A,** 145
Oxide layer, in glass-to-metal seals, **14,** 316–317
Oxide thin films, **30,** 611–616
Oxide, surface layers, **14,** 354
Oxygen ($^1\Delta$) molecules, in purely chemical lasers, **15B,** 111–112
Oxygen ($^3\Sigma$) molecules, in chemical lasers, **15B,** 111–112
Oxygen molecule
  absorption coefficient of, **12B,** 58
  microwave lines of, **12B,** 166–168
Oxygen partial pressure, **24A,** 60
Oxygen point, **1,** 247
Oxygen poisoning of cathodes, **4A,** 35–36
Oxygen, **22,** 154, 155, 184
  activation energy for diffusion, **22,** 365
  adsorbed on graphite, **23B,** 24
  on Ag(100), **22,** 326
  on Cu(110), **22,** 328
  degradation of hot-cathode filaments, **29A,** 126–127
  elastic electron scattering, **5A,** 325–326
  electron induced dissociation, **3,** 550–551
  filter, using silver, **14,** 375
  in flashlamp-excited laser systems, **15A,** 359
  from glass, by electron bombardment, **14,** 327–328
  Moffit's "atoms-in-molecules" treatment, **3,** 182
  outgassing from glass, **14,** 322
  physical properties, **14,** 421–422
  thermal emission by, **12B,** 172
  on W(110), **22,** 365
Oxymyoglobin, **23C,** 349
Ozone
  absorption coefficient of, **12B,** 164–169
  atmospheric concentration, **12B,** 170
  atmospheric lines, measured, **12B,** 176
  correlation with solar flux, **28,** 475–477
  degradation with, **16B,** 257–261
  generation of, **16B,** 258–259
  properties of, **16B,** 257–258
  stress cracking, **16C,** 214–215
Ozone production, **31,** 41

P

P branch of vibration-rotation spectrum, **10,** 327, 328
P-I curve, CW diode laser, **29C,** 80
p-I-n Junction diode, **4A,** 464
P-K-W system, **3,** 401–402
P-N boundary, **2A,** 51
P-n homodiode junction, **11,** 185–186
P-N junction diode, **2A,** 51–57
  equivalent circuit of, **2A,** 62
  as nonlinear device, **2A,** 365
  silicon rectifier as, **2A,** 216
p-n junction laser, **10,** 64–69
p-n junction photodiodes, **7B,** 17
P-N junction rectifiers, **2A,** 57–59
  temperature in, **2A,** 58
P-N junction, **2A,** 51
  capacitance of, **2A,** 108
  concentration profile of holes and, **2A,** 70
  depletion region in, **2A,** 70
  forward bias across, **2A,** 53
  isolation of, **2A,** 108
  in monolithic integrated circuits, **2A,** 106
  as nonlinear devices, **2A,** 363–367
  reverse bias in, **2A,** 54, 72, 129, 136
  self-saturation in, **2A,** 72
  temperature in, **2A,** 46, 54
  Zener diodes and, **2A,** 60

zero bias in, **2A,** 86, 130
zero-bias depletion at, **2A,** 134
P-*n* junctions, method of measuring lifetime, **6B,** 107–109
P-N transistor(s), **2A,** 94, 98–99, 120–121, 352
in differential input amplifier, **2A,** 297
in integrated circuits, **2A,** 115–116
lateral, **2A,** 288
P-n tunnel junctions, **11,** 177, 184–187
P-polarization, **26,** 114
p-test, **28,** 181–182
p.d., *see* Parent distribution
PAA, **23B,** 545, 568, 577
Packer
impression, **24B,** 397, 399
inflatable, **24B,** 396, 397
straddle, **24B,** 397–398
Packing fraction, **5B,** 2
Packing of chains, in polymer crystals, **16B,** 87–90
Paint, temperature indicating, **18,** 671–672, 771
Painted contacts, **6B,** 52
Pair correlations
anion-water, **23B,** 478
cation-water, **23B,** 477
ion-ion, **23B,** 479
Pair density function, **23B,** 246
partial, **23B,** 252
Pair distribution function, **23B,** 471
Pair production cross section, **8,** 227–231
exercises, **8,** 231
measurements, **8,** 228–230
Pair production, **5A,** 83–87, 622, 641ff, 671, 672; **8,** 227–228
effect, *see* Gamma rays
energy distribution, **5A,** 84, 85
events, **8,** 228
polarization in, **5B,** 237–238
total cross section, **5A,** 86
Pair spectrometer
axially focusing, **5A,** 647
determination of momentum and energy of gamma rays, **5A,** 641–652
intermediate-image focusing, **5A,** 648–649
measurement of photon beam energy, **5B,** 546
multichannel homogeneous field, **5A,** 651
three crystal, **5A,** 629
uniform field, **5A,** 643–647
Pair-density function, **23A,** 31
Pair-peak fractions, **8,** 223
Pakadakis velocity-measurement system, **11,** 401
Paleomagnetism
data analysis
field tests of stability, **24A,** 410–411
Fisher statistics, **24A,** 384–388
projections, **24A,** 382–384
signal-to-noise discrimination, **24A,** 393–411
transformation of coordinates, **24A,** 388–393
paleointensity
Koenigsberger-Thellier-Thellier method, **24A,** 412–417
Shaw analysis, **24A,** 415–416
remanent magnetization, measurement, **24A,** 343–417
sampling
collection, **24A,** 331–342
drilling, **24A,** 335–340
orientation, **24A,** 331–335, 338
sediments, unconsolidated, **24A,** 341–342
Palladium, **22,** 378
leak, **4A,** 168
thimble, **22,** 389
Palladium, chemical properties
boiling point, **14,** 351
hydrogen solubility, **14,** 354
melting point, **14,** 347, 351
physical properties, **14,** 373, 374
vapor pressure, **14,** 347, 351
Palladium, hydrogen purification, **4B,** 327
Pancharatnam's half-wave plate, **12A,** 376, 390
Paper tape punchers, in pulse amplitude measurements, **2B,** 123
Pappus' theorem, **1,** 60
Parabolic cylinder antenna, **12B,** 19
Parabolic reflector, **8,** 158
Parabolic-path-type instrument, **8,** 169
Paraboloid antennas, **12B,** 12–14
Paracrystalline lattice factor, **16B,** 106
for one-dimensional lattice, **16B,** 110
Paracrystallinity, **16B,** 128; **23C,** 448
effect on X-ray diffraction pattern, **16B,** 105–111, 144–149
one-dimensional lattice distribution functions, **16B,** 106–108
Parafocusing principle, X-ray, **1,** 423–424
Parallel arrangement, **8,** 194

Parallel bands, **10,** 327
Parallel detection, **22,** 285
Parallel projection operator, **19,** 566
Parallel radiation, **8,** 153
Parallel wire line, **8,** 121, 122
Parallel-plate analyzer, **22,** 220
Parallel-plate electrostatic deflection analyzer, **17,** 178, 180, 439–454, 467–468
Parallel-T bridge, **1,** 464
Parallel-T network, **6B,** 14–15
Paramagentic phases, **23C,** 188
Paramagnetic anisotropy, **6B,** 246–247
Paramagnetic ion(s), **15A,** 181–182
  transitions, **10,** 392–395
Paramagnetic probes, **20,** 43, 44
Paramagnetic relaxation, **20,** 68–72
Paramagnetic resonance, **3A,** 360; **4B,** 198–203; **10,** 385–394; *see also* Electron spin resonance
  EPR spectrometer, **4B,** 198–200
  field sampling error, **4B,** 197, 202
  measurement of *g* factor
    of H, **4B,** 201–202
    of other atoms, **4B,** 202–203
Paramagnetic susceptibility
  of localized atomic moments, **11,** 504–511
  of nuclear moments, **11,** 511–533
Paramagnetism, **1,** 537ff; **6A,** 18; **6B,** 171, 176, *see also* Magnetic susceptibility
  parasitic, **1,** 548
Paramagnets, **23A,** 62, 66
Parameter, **28,** 245
  values, statistical inference, **28,** 264
Parameter-effects array, **28,** 260
Parameter-effects curvature, **28,** 264
Parameters, computational methods for refinement, **3,** 295–301
Parametric amplification, **15B,** 258
  weak links for, **11,** 304
Parametric amplifier(s), **2A,** 332; **12B,** 225–241
  cryogenic cooling of, **12B,** 238–239
  design considerations for, **12B,** 233–241
  equivalent circuits for, **12B,** 228
  for millimeter wavelengths, **12B,** 239
  noise in, **12B,** 231–233
  wide-band tunable, **12B,** 237
Parametric calculations, **13A,** 347
Parametric conversion, **13B,** 315
Parametric emission, in picosecond spectroscopy, **15B,** 195–196
Parametric families, tests for, **28,** 202–203
Parametric oscillations, laser, **18,** 724–725
Parametric oscillators, **13B,** 322; **15B,** 168–174, 258, *see also* Doubly resonant oscillators; Singly resonant oscillators
  conversion efficiency of, **15B,** 170–172
  experiments, nonlinear optical devices and, **15B,** 155
  pump power and damages threshold for, **15B,** 170
  risetime for, **15B,** 169–170
  threshold pump power density in, **15B,** 168–170
  tuning and linewidth for, **15B,** 172–174
Parametric superfluorescence, **15B,** 258
Parasitic
  paramagnetism, **1,** 548
  resistor, **1,** 458
  voltage, **1,** 454–455
Parasitic signal, **8,** 40
Paraxial computation system, **25,** 19
Paraxial optics, **31,** 148–150
Paraxial ray, **25,** 17, 39
Parent distribution, **1,** 3; **3,** 711; **6A,** 3
Parent isotopes, **8,** 181
Parfocal distance, **25,** 165
Parity conservation, **5A,** 3
Parity determination, **5B,** 44–214
  angular correlation, **5B,** 137–142
  indirect methods, **5B,** 129–147
  spectroscopic methods, **5B,** 44–129
Parity, **22,** 160, 183
Parker's electrooptic modulation light scattering technique, **19,** 488–490
Parseval's theorem, **26,** 69
Partial coherence, **25,** 109, 146; **26,** 89
Partial crystallization, in DTA technique, **16B,** 375, 376, 379
Partial differential equation models, **28,** 501
Partial ionization cross section, **17,** 480
Partial melting, **24A,** 37–38
Partial pressure
  analyzer, **14,** 81, 82, 510
  measurement, **14,** 81
    in diffusion pump systems, **14,** 161–163
  use in description of composition, **18,** 611
Partial residual plot, **28,** 270
Partial structure factor, **23B,** 472
Partial yield, **22,** 179

Partial-yield spectroscopy, **22,** 173, 179, 183
Partially transparent materials, **26,** 231
Particle acceleration, **19,** 6–7
Particle continuity equations, **7B,** 79–82
Particle detection, **22,** 320
Particle detectors, **2,** 141; *see also* Detector
simulation, **28,** 446–447
Particle diagnostics, **9A,** 130–131
charge transfer in, **9A,** 130
neutral beams in, **9A,** 130
particle detectors, **9A,** 133
relaxation time, **9A,** 132
stripping, **9A,** 130
suprathermal electrons, **9A,** 132
Particle displacement, **19,** 4–6, 498
Particle energy loss, simulation, **28,** 445
Particle momentum, **14,** 87
Particle multiple scattering, simulation, **28,** 445
Particle size, **6A,** 241–244
X-ray measurement of, **1,** 428–429
Particle tracking, *see* Flow tracing particles
Particle tunneling, concept of, **11,** 123, *see also* Electron tunneling; Tunneling
Particle velocity, **19,** 5–7, 505
Particle-solid interactions, **7B,** 231–305
"atomistic" description of solids, **7B,** 289–304
in crystalline solids, **7B,** 304–305
elastic scattering, **7B,** 290–291
electron-solid interactions, *see* Electron-solid interactions
group velocity of reflected beams, **7B,** 273
heavy particles, **7B,** 231–275
penetration experiments, *see* Penetration experiments . . .
at high-electron energies, **7B,** 290–297
inelastic scattering, **7B,** 292–297
at low electron energies, **7B,** 297–299
multiple scattering of electrons, **7B,** 299–301
secondary electron processes, **7B,** 301–304
Auger electron emission, **7B,** 302
electron slowing-down spectrum, **7B,** 302–303
radiation, **7B,** 303
secondary electron emission, **7B,** 302
structural and chemical change, **7B,** 303
"stagnation detector," **7B,** 273
"sticking probability," **7B,** 271
surface atom interactions, **7B,** 264–274
surfaces free of adsorbed gases, **7B,** 271–272
thermal energy collisions at surfaces, **7B,** 268, 275
time of flight techniques, **7B,** 274
Particles, **30,** 376, 381–393; *see also* α-particles
charge determination, **5A,** 289–307
charged
distribution of lateral displacement, **5A,** 74
heavy
response of NaI (Tl), **5A,** 412, 413
spectrum, **5A,** 411
ionization loss, **5A,** 1, 4–44
multiple scattering, **5A,** 73–76
scattering mean square angle, **5A,** 73–74
scintillation spectrometry, **5A,** 411–435
detection, **5A,** 1–288
elementary, intrinsic properties, **5B,** 845
energy loss, **5A,** 444
Particulate generation, in PLD, **30,** 501–502
Particulates, acceleration by flow, **18,** 8–15, 795
Particulates, in PLD, **30,** 481
Partition function, **7A,** 120; **18,** 427
linear molecule, **12B,** 145–146
nonlinear molecule, **12B,** 145–146
Parzen lag, window, **28,** 332–333
Pascal (pressure unit), **18,** 508
Pascal, unit of stress, **16C,** 122
Paschen's law, **2,** 75–76
Paschen-Back effect, **4B,** 119, 133
Paschen-Rung grating mount, **1,** 369
Paschen-Runge mounting, **13A,** 293
Passivation of stainless steel, **14,** 355
Passive linear circuit elements and networks, **2A,** 1–44
basic principles and terminology in, **2A,** 1–7
lumped elements *R*, *L*, and *C* in, **2A,** 1–2
Passive memory, as pulse spectrometer, **2B,** 122
Passive modelocking, **29C,** 180–182
Passive optical devices
Fabry-Perot interferometer, **29C,** 350, 362–366
holographic optical elements, **29C,** 361–362
isolators, **29C,** 174–175, 358–361
Passively mode-locked cw dye lasers, **15B,** 190–191, *see also* cw dye lasers

pulses form, **15B,** 200
Patch voltage clamp, **20,** 482–484
Path difference, optical, **12A,** 372
Path length
atmospheric, **12B,** 196–200
effects in, **12C,** 264
correction, **12B,** 197
of electrons, **22,** 244
ionospheric effects in, **12C,** 263
radiance, **26,** 236
Patterning, **30,** 354–355
Patterson function, **20,** 240–244; **23A,** 344
distribution of maxima, **3,** 292, 293
Patterson inversion, **20,** 367
Patterson method in X-ray diffraction, **6A,** 232–233
Patterson series, **3,** 287–289
Paul trap, *see* Ion trap
Pauli spin paramagnetism, **6B,** 176
$PbNb_2O_6$, *see* Lead metaniobate
PBSCCO plume, **30,** 258–261, 285
PbSe junction laser, **10,** 67–69
PC connectors, **29C,** 172
PCM, *see* Pulse code modulation
Pd_2MnSn, **23C,** 176
Pd_3Fe, **23C,** 166, 172
PdGe glasses, **23B,** 259
PdSi glasses, **23B,** 255
Peak efficiencies, **8,** 213, *see also* Intrinsic peak efficiencies
Peak flux, **23A,** 100, 110
Peak fractions, *see* Full energy peak fractions; Reciprocal peak fraction; Spectrometers

Peak fractions, *see* Full energy peak fractions; Reciprocal peak fraction; Spectrometers

Peak height, **14,** 83
Peak intrinsic efficiency, **8,** 215
Peak matching technique, **5B,** 6
Peak power capacity, **8,** 14
Peak separation, **14,** 89
Peak shapes, *see* Gaussian peak shapes
Peak surface temperature, **30,** 179
Peak tails, **8,** 216, 223
Peak transmission, **14,** 83
Peak-tail contributions, **8,** 225
Peak-to-valley, **25,** 106
Peaking, cathode or screen-grid, **2,** 231
Pearlite, anisotropy of, **19,** 267
Pebble-bed storage heater, **18,** 782
Pechan prism, **25,** 66
Pechan-Schmidt prism, **25,** 64
Pedersen mobility, **9B,** 177
Pedestal, in LDV signal, **18,** 114, 121, 156, 159, 184
Peirels barriers, Bordoni peaks and, **11,** 422
Peliln-Broca prism, **25,** 208
Pellicle beam splitter, **29C,** 349
Pellicles, **5A,** 221
Peltier cooling, **8,** 205
Peltier effect, **1,** 506–507
definition, **6B,** 115
measurement, **6B,** 125–126
Pencil angle
aperture radius relationship, **29A,** 191–192
considerations in combining lenses, **29A,** 202–203
determination, **29A,** 192, 202
Pencil tubes, **2A,** 164
Pencil-beam antennas, **12B,** 10–24
Pendellösung fringes, **23A,** 313
Pendulum, *see also* Pseudostationary point; Inertial mass
absolute measurement of *g*, **24B,** 133, 140
accelerometer, **1,** 78, 79
ballistic, **1,** 75–76
compound, **1,** 83–84
damped
forced oscillation, **24B,** 27–28
motion, **24B,** 25–27
equation of motion, **1,** 110
galvanometer coupling, **24B,** 32–35
garden gate, **24B,** 6, 23
horizontal
equations of motion, **24B,** 22–23
period, **24B,** 22, 24
period lengthening, **24B,** 24
inverted, **24B,** 6, 24
natural period, **24B,** 5
torsion, **1,** 82–83
use in seismic system, **24B,** 2, 4–6
vertical
angular frequency, **24B,** 16, 17
equation of motion, **24B,** 15
equilibrium condition, **24B,** 15
period, **24B,** 16, 17, 19
period lengthening, **24B,** 16–21
Penetrating radiation, **22,** 241
Penetration depth in superconductors, **6B,** 75–76
Penetration depth, optical, **30,** 18, 20, 24–25, 176

Penetration experiments on particle-solid interactions, **7B,** 232–264
backscattering from thick targets, **7B,** 257–258
"channeling" in, **7B,** 245–264
of fast heavy ions, **7B,** 254–257
ranges in axial types, **7B,** 248–251
crystal lattice effects, **7B,** 245–264
emergence patterns, **7B,** 260–262
emission from lattice sites, **7B,** 258–259
energy loss in thin layers of material, **7B,** 237–240, 252–257
interactions with surface atoms, **7B,** 264–268
ionic charges of heavy particles, **7B,** 236–237
ranges in, **7B,** 240–245
secondary particle yields, **7B,** 258–260
surface barrier detectors for, **7B,** 262–264
surfaces free of adsorbed gases, **7B,** 271–272
thin-surface-layer removal, **7B,** 243–245
transmission and reflection studies, **7B,** 251–252
Penetration length, **23C,** 407
Penetration neutrons, **23C,** 406
Penetration of gamma rays, **5A,** 76–89
Penetration weld, **14,** 432
Penning collisions, **15A,** 151–152
Penning discharge ion pump, **14,** 217–221, 223, *see also* Cold cathode discharge gauge; Ionization gauge
Penning discharge source
cold-cathode sources
cathode potential, **29A,** 108
discharge types, **29A,** 107
geometry, **29A,** 106–107
hot-cathode sources
applications, **29A,** 109, 111
high-intensity sources, **29A,** 111
Penning discharge, 72–76
Penning effect, **2,** 76–77; **2A,** 178
Penning electron, **15A,** 147
Penning ionization, **29B,** 172, 201, 209
source, **17,** 280
Penning lines, **15A,** 159
Penning process, **15A,** 133, 144–145, 159–160
Penning trap mass spectrometer
accuracy, **29A,** 363, 387
anharmonic detection, **29A,** 378–380
axial frequency resolution, **29A,** 368
cyclotron frequency, perturbation sources
electrostatic perturbation, **29A,** 386
ion position shift, **29A,** 385
misalignment, **29A,** 384
mode energy change, **29A,** 386–387
number of ions, **29A,** 384–385
ejection-detection method, **29A,** 380, 382
experimental parameters for various instruments, **29A,** 366–367
frequency-shift detector, **29A,** 376–378
input detection circuit, **29A,** 367–370
ion sample preparation, **29A,** 371–372
magnetic field stability, **29A,** 374–375
pulse-and-phase method, **29A,** 382–383
sideband cooling resonances, **29A,** 372–374
signal-to-noise ratio, **29A,** 370–371
trap
construction, **29A,** 366, 368
electrode voltage, **29A,** 365
performance, **29A,** 364–365
voltage sources, **29A,** 368–369
Pentagrid tube, **2A,** 160–161
Pentaprism, **25,** 66
Pentodes, **2,** 65–66
as load resister, **4A,** 343
Peptide antibiotics, **19,** 234
Peptide mapping, **30,** 436
Peptide sequencing, **30,** 432
Percolation, **23C,** 42
Percussion, center of, **1,** 84
Perfect crystals, **23A,** 312
Performance, **14,** 84, 184
factor, **14,** 85
Perfusion, internal, **20,** 456, 457
Period counting, LDV signal processing, **18,** 174–180
Periodic boundary conditions, **23A,** 374
Periodic heat flow, **6A,** 399–402
Periodic lattice distortion (PLD), **27,**51
Periodic signals, weak, **28,** 34–371
dc excess *vs* periodic strength, **28,** 361–364
H-test, **28,** 360
power of statistical tests, **28,** 360–361
Protheroe test, **28,** 358–359
Rayleigh test, **28,** 355–358
Z^2m-test, **28,** 359–360
independent Fourier frequencies, **28,** 366–368
light curve, **28,** 351–353
multiple data sets, **28,** 367–371

coherence, **28,** 367–368
use of all the data, **28,** 369–371
oversampling, **28,** 366–367
tests for unifomity of phase, **28,** 352, 354–361
epoch folding, **28,** 354–355
white noise, **28,** 351–352
Periodic strength, versus dc excess, **28,** 361–364
Periodic structure, **10,** 35, 37
Periodogram, **28,** 285, 316–320
bias sources, **28,** 318–319
cross, **28,** 343
data taper, **28,** 324
leakage, **28,** 319–320, 324–325
spectral window, **28,** 317
statistical properties, **28,** 317
Permanickel, **14,** 378
Permeability coefficient
definition, **16C,** 318
of gases in polymers, empirical relation, **16C,** 326
units, **16C,** 324–325
Permeability, **1,** 542; **6B,** 209–227
constant, definition, **14,** 338
various glasses, **14,** 338–343
Darcy's law, **24A,** 173
definition, **6B,** 209–214
effective, **1,** 550
of gases in glass, **14,** 337–343
of helium, in glass, **14,** 409
of hydrogen, in Pd, **14,** 409
in quartz, **14,** 409
initial, **1,** 550
measurement, **6B,** 211–214
of N_2, in iron, **14,** 409
non-constant, **6B,** 218–224
radioactive tracers, **24A,** 175–181
of rubber, **14,** 409
static, **3,** 491
steady-state techniques, **24A,** 173
tensor, **6B,** 214–216
transient technique, **24A,** 174
Permeability, to gases and vapors
closed receiving volume methods
apparatus, **16C,** 362–367
data analysis, **16C,** 356–362
continuous flow methods
apparatus, **16C,** 366–367
data analysis, **16C,** 369–371
measurement of penetrance concentration, **16C,** 372
cup method, **16C,** 366–367, 369
membrane-type sample installation, **16C,** 373
sources of measurement error, **16C,** 372–377
historic perspective, **16C,** 316–324
utility, **16C,** 339–340
Permeammeters, **1,** 552ff
method, **6B,** 201
Permittivity, **1,** 488–489, 494; *see also* Dielectric constant
Permittivity of free space, **16C,** 381
Perovskite crystal structure, **10,** 373
Perpendicular bands, **10,** 327, 328
Perturbation calculations, **13A,** 79; **31,** 112
of transition probability, **10,** 339–342
Perturbation Hamiltonian, **4B,** 112–114
Perturbation method of energy-loss calculation, **17,** 33–34
Perturbation theory
electron capture, **17,** 91, 95
inner-shell vacancy production, **17,** 321, 338, 340–341
optical excitation and, **11,** 75
semiclassical approximation calculations, **17,** 322
Perturbations, external, in electronic spectra, **3,** 222–241
Perturbations, externally applied to defects, **21,** 428–430
electric fields, **21,** 430
light, **21,** 428
stress, **21,** 429
Perturbed correlations, **13A,** 140
Perturbed-stationary-state approximation, **17,** 112
Perveance, **2,** 52
Perveance (microperveance), **4A,** 86, 99, 106–107, 111–112
Petzval
coefficient, **25,** 42
curvature, **25,** 42
lens, **25,** 90
Petzval surface, **12B,** 46
Pf-adsorbed
Si(111) 7×7
scanning tunneling microscopy of, **27,** 200
Pfund arrangement, **3,** 99–101

Pfund spectrometer, **3A,** 183
PGM, *see* Plane grating monochromator
PHA, *see* Pulse height analyzer
Phase, **1,** 306; *see also* Acoustoopic measurement of sound wave, phase
  acceptance, **5B,** 637
  coherence, **3,** 435
  curves for reaction intermediates, **3,** 570–571
  difference in gratings, **26,** 72
  holograms, **26,** 190
  meters, direct reading electronic, **2,** 552–557
  plate, **26,** 91
  retarders, **26,** 126, 130
Phase angle, **1,** 473–474, 492; **8,** 173, *see also* Travel
  boundary, **1,** 122
  comparison, **1,** 223
  contrast microscope, **1,** 342, 343
  difference, **1,** 348
  plate, **1,** 342
  response, **1,** 231
  response of seismic unit, **1,** 92
  shifters, **1,** 482, 483
  transformation, **1,** 281
  transition lines, **1,** 325
Phase cancellation effects, **19,** 70–71
  acoustoelectric transducers, **19,** 71
  phase-insensitive transducers, **19,** 70–71
Phase charges under high pressure, **6A,** 421–423
Phase condition, *see* Multimodulator system; Single modulator
Phase constant, **25,** 173
  in electron microscope image, **16B,** 219
Phase contrast, **20,** 401–403
  microscope, **26,** 90
  as visualization method, **18,** 389
Phase correction of interferograms, **12A,** 509
Phase delay, ionospheric, **12B,** 127–128
Phase detector
  circuits, **2B,** 43–47
  diode, **2B,** 44
  output waveforms of, **2B,** 46
Phase determination, in XRDA, **20,** 252–255
Phase diagram(s), **22,** 278
  magnetic, **23C,** 146
Phase difference, **31,** 248–250
Phase explosion, **30,** 226, 235, 245, 275–278, 271, 281
Phase factor, **21,** 149, 155, 174
Phase factors for two- and three- particle final states, **5B,** 838–844
Phase gratings, **13A,** 231
Phase interference, **19,** 587
Phase margin, of feedback control system, **2B,** 414, 418
Phase matching
  in birefringent crystals, **15B,** 154
  difference frequency generation and, **15B,** 160
  in nonlinear optical devices, **15B,** 151–156
  two-parameter change in, **15B,** 153
Phase measurements, **2,** 549–558; **2B,** 38–47
  commercial phase meters in, **2B,** 41–43
  comparison methods using calibrated phase shifters, **2,** 551–552
  with dual-channel recorder or oscilloscope, **2,** 549–550
  electronic counter method in, **2B,** 39–40
  in microwave circuits, **2B,** 297
  oscilloscope methods in, **2B,** 38–39
  phase shifting networks and devices in, **2B,** 40–41
  pictorial vector display method, **2,** 557
  substraction method in, **2B,** 42–43
  time delay method, **2,** 554–557
  time-delay method in, **2B,** 41–43
Phase meter, with digital readout, **2B,** 43
Phase modulation, **2,** 265; **2A,** 249–250; **13B,** 70; **26,** 146
  modelocking, **29C,** 179
Phase null detector, **2B,** 43
Phase object, **18,** 345
Phase position, in magnetic domain studies, **11,** 715
Phase prediction
  in frequency standards, **12C,** 207–208
Phase problem, **23A,** 344
Phase problems in X-ray diffraction, **6A,** 231–234
Phase relations, **8,** 155
Phase retarder, **31,** 284–285
Phase shift, **14,** 90; **22,** 233
  control, thyratron, **2,** 301–303
  networks and devices, **2,** 550–551
  oscillator, **2,** 293–296; **2A,** 345–346
  in plasmas, **10,** 280–286
  on reflection, **4B,** 82, 98
  upon reflection, **26,** 115
Phase shifter(s), **8,** 143–145
  calibrated, **2B,** 43

continous, **2B,** 40–41
exercises, **8,** 145
Phase shifting networks, **2B,** 40–41
Phase smearing, **8,** 174
Phase spectrum, **28,** 341
Phase splitters, **2B,** 45
Phase stability
alternating gradient synchrotron, **5B,** 666–668
electron cyclotron, **5B,** 652–653
linear accelerator, **5B,** 636–638
synchrocyclotron, **5B,** 646–647
Phase transformation, **30,** 103, 462, 463
Phase transformations and grain anisotropy, **19,** 253–256, 267
Phase transition(s), **23C,** 1; **30,** 335, 346–350
displacive, **23C,** 56
first-order, fluctuation-driven, **23C,** 35
incommensurate, **23C,** 60
incommensurate-commensurate, **23C,** 136
magnetic, **23C,** 3, 110
order-disorder, **23C,** 10, 56
random systems, **23C,** 40
structural, **23C,** 2, 56
Phase transitions, fast ion conductors, **23B,** 191
Phase velocities in a crystal, **26,** 136
Phase velocity matching, in VUV spectroscopy, **15B,** 214
Phase-lock loop (PLL), in telemetering, **2B,** 230
Phase-lock method, in ac measurements, **2B,** 75
Phase-matched interactions, **15B,** 143, 156–159
Phase-modulated lasers, **29C,** 179
Phase-sensitive (lock-in) detection, **12A,** 432, 440, 461–462
Phase-sensitive (or synchronous) detection, **10,** 226, 227
Phase-sensitive autocorrelators, **29C,** 215–220
Phase-sensitive demodulator, **2B,** 441
Phase-sensitive detector, **3B,** 569
Phase-shift lifetime measurement, **7A,** 136
Phase-space area, **31,** 21, 23–24
Phases contrast microscopy, of polymeric alloys, **16C,** 288–291
Phasography, **9A,** 207
Phasons, **23A,** 502
Phasor
diagram, **1,** 233
peak current, **1,** 467
peak voltage, **1,** 467
Phenol-formaldehyde resin, historic reference, **16A,** 4
Phenolic resin
density, **16C,** 72
elastic constants, **16C,** 81
ultrasonic properties, **16C,** 72, 73, 74,
Phenolic, **14,** 414
Philips gage, **1,** 190
Phillips ionization gauge, *see* Cold cathode discharge gauge; Ionization gauge
Phon, **1,** 202
Phonograph pickup, **1,** 74, 77
Phonon(s), **10,** 367, 368; **12A,** 10; **30,** 27, 28, 30
generation and detection of, **11,** 150–151
longitudinal and transverse, **16A,** 179
in polymers, **23C,** 321
recombination of, **11,** 150–151
region of, **11,** 467–468
spectrum of, with Van Hove singularities in, **11,** 146–147
thermal acoustic, **16A,** 170
Phonon branch, tracing using triple-axis neutron spectrometer, **16A,** 225
Phonon dispersion, **24A,** 28
relation, **23A,** 28, 373
Phonon kicking, **30,** 53, 54
Phonon scattering, **11,** 79–80; *see also* One-phonon scattering
with neutrons, **16A,** 210–211
Phonon-Raman scattering, **15B,** 260
Phosphor decay time, **8,** 254
Phosphor screen, **7B,** 282
Phosphor, cathode ray tube, **2,** 769
Phosphorescence, **1,** 396–398; **6B,** 294; **16A,** 406; **20,** 224, 225
spectra, lifetime, **3,** 257
Phosphoroscope, **1,** 397; **6B,** 309–310
Phosphorous NMR, **20,** 50–52
Phosphors, **6B,** 293–294, 296–301
classes of, **6B,** 300
electroluminescent, **6B,** 299–300
films, **6B,** 298–299
preparation of, **6B,** 296–301
single crystal, **6B,** 297–298
symbols, **6B,** 301
Phospor-screen image tubes, **12A,** 237–251
Photo dissociation, in glass, **14,** 329
Photo emission, **27,** 82
Photo-electro-magnetic detector, **2B,** 327
Photo-ionization, **3B,** 596, 847

Photo-Peak spectrometer, **8,** 213–218, 220, 225
intrinsic efficiency, **8,** 214–216
performances, discussion, **8,** 217–218
questions, **8,** 216–217
response function, **8,** 214–216
Photoabsorption spectroscopy, **29C,** 40
Photoacoustic deflection (PAD), **30,** 217–218
Photoacoustic spectroscopy, **30,** 463, 578–580
Photocathode, **29C,** 200–201; *see also* Negative-electron-affinity semiconductor photocathode
spectral response, **2,** 743–744
Photocathode surfaces, **8,** 24
Photocathodes, **6B,** 384–391; **12A,** 19–27
antimony-cesium type, **6B,** 385–388
bismuth-silver-oxygen-cesium, **6B,** 390–391
characteristics, **6B,** 391
composition of, **2B,** 302–303
formation, **6B,** 384
of image dissector, **2B,** 351
multialkali type, **6B,** 389
for photocathode tubes, **2B,** 310
silver-oxygen-cesium type, **6B,** 389–390
spectral response characteristics of, **2B,** 305
spectral response curves, **6B,** 387
spectral response of, **26,** 307
Photocells
applications of, **2B,** 333–334
characteristics of, **2B,** 326–333
detectivity in, **2B,** 331
frequency response of, **2B,** 330
materials used in, **2B,** 328–329
noise in, **2B,** 330–331
noise equivalent power of, **2B,** 331
stability of, **2B,** 332
temperature and, **2B,** 330
types and ratings in, **2B,** 334
Photocells, solid-state, **2,** 757–765; **2B,** 323–334
noise, **2,** 761
spectral response, **2,** 759–760
speed of response, **2,** 760
Photochemical ablation, **30,** 189
Photochemical effects, **30,** 455
Photochemical process, **30,** 176, 339
Photochemical reactions, in nature, **15B,** 269
Photoconductance method of lifetime measurement, **6B,** 100–101
Photoconductive cells, **13A,** 321
Photoconductive detectors, **10,** 174–188; **26,** 311
Photoconductive response, **26,** 319
Photoconductivity, **4A,** 451–453; **6B,** 335–337
band gap determination, **6B,** 350
basic concepts, **6B,** 335–340
characteristic phenomena, **6B,** 345–349
charge carrier, sign, **6B,** 350
color centers, **6B,** 333
decay, **6B,** 347–348
density of imperfection levels, **6B,** 353–354
excitation transitions, **6B,** 350
free-carrier, **10,** 179–188
impurity (or extrinsic), **10,** 175–179
level location, **6B,** 350
infrared quenching, **6B,** 349
lifetime of charge carriers, **6B,** 336, 350–351
primary, **6B,** 338–339
secondary, **6B,** 338–339
speed of response, **6B,** 347–348
surface effects, **6B,** 349
use of in experimental measurements, **6B,** 350–354
variation with light intensity and temperature, **6B,** 348–349
Photoconductors
defined, **2A,** 302–303
definition, **6B,** 335
electrode properties, **6B,** 345–346
energy bands, **6B,** 337–340
examples of, **6B,** 341–342
extrinsic, **26,** 317
impurity, **6B,** 341, 343
intrinsic, **6B,** 341, 343
preparation of, **6B,** 342–345
sensitization, **6B,** 346
spectral response, **6B,** 346–347
spectral response curves, **6B,** 347
types, **6B,** 340–342
use in devices, **6B,** 354–357
Photocurrent, **22,** 483
Photodesoprtion, *see* Desorption
Photodetachment, **7A,** 179ff, 181; **29B,** 50–54, 312–314, 317–318
using drift tube, **7A,** 200
Photodetector(s), **13B,** 6, 52; *see also* Light sensor output, LDV statistical character
noise in, **2B,** 486–496
responsivity of, **2B,** 492

signal response in, **2B,** 487
Photodiode(s), **2,** 758; **22,** 482, 485
avalanche, **2B,** 326–328
as short-pulse detector, **29C,** 199
in solid-state detectors, **9A,** 105–106
Photodisintegration, **5B,** 461–472
using bremsstrahlung, **5B,** 468–472
Photodissociation, **29B,** 63, 314–317, 396–400
Photodissociative selective chemistry, **15B,** 291–293
Photoelastic measurement, **6B,** 272
Photoelasticity, **1,** 103–105; **6B,** 271–273
Photoelectric cell, **8,** 40, 41
Photoelectric cells, **13A,** 321
Photoelectric detection, **8,** 162
Photoelectric detectors, **22,** 208
in beam-foil experiments, **17,** 538–539, 545
Photoelectric devices, **2,** 743–765; **2B,** 301–334
gas-filled phototubes, **2B,** 307–309
photomultiplier tubes, **2B,** 309–323
solid-state phototubes, **2B,** 323–334
vacuum phototubes, **2B,** 301–307
Photoelectric effect, **5A,** 77–79, 671, 672; **18,** 407; *see also* Gamma rays
cross section from K shell, **5A,** 77, 78
external, **5A,** 586–594
polarization in, **5B,** 238
Photoelectric emission, **4A,** 53–69; **6B,** 376–391; **16C,** 134; **22,** 7
basic process, **4A,** 53–54
cathode materials, preparation of, **6B,** 378–379
definitions, **6B,** 376–378
energy distribution measurements, **6B,** 379–380
experimental tube, **4A,** 67–69
Fowler analysis, **22,** 7
Fowler theory and, **11,** 72–73
interpretation of data, **6B,** 380–384
from metals, **4A,** 54–60, 62–69
effect of accelerating fields, **4A,** 60
energy distribution, **4A,** 57–60
yield, spectral distribution, **4A,** 54–57
in metals, **6B,** 380–381
methods of measurement, **6B,** 378–380
measurements of, **4A,** 58, 65–69
patchy surface, **22,** 7
photoelectric yield, **22,** 7
from semiconductors, **4A,** 60–61
in semiconductors, **6B,** 381–384
spectral response curve, **6B,** 379
surface effects, **4A,** 53, 57–65
threshold frequency, **22,** 7
vacuum requirements, **4A,** 63–65
Photoelectric method, **22,** 8
Photoelectric phenomena, **6A,** 20; **6B,** 335–391
Photoelectric photometry, **12A,** 1–92
atmospheric effects, **12A,** 97–101
diffraction effects, **12A,** 95–97
mechanical effects, **12A,** 102–104
polarization effects, **12A,** 104–105, 110
thermal effects on components, **12A,** 105–108
Photoelectric quantum yield, in photoemission measurements, **11,** 81–84
Photoelectric spectroscopy, **3B,** 602, 847–867
high energy, **3B,** 866
Photoelectric threshold, **6B,** 131, 378
Photoelectromagnetic effect, **4A,** 456; **6B,** 357–364
description of, **6B,** 357–358
limitations, **6B,** 359–360
measurements on insulators, **6B,** 363–364
measurements on semiconductors, **6B,** 360–362
results on semiconductors, **6B,** 362–363
schematic diagram, **6B,** 360–361
utility and limitations, **6B,** 358–359
Photoelectromagnetic method of lifetime measurement, **6B,** 101
Photoelectron angular distribution, **3B,** 859
Photoelectron parameters, **22,** 133
Photoelectron spectroscopy (PES), **11,** 68, 99; **13A,** 162, 250; **22,** 133, 508; **30,** 292
angle-resolved, **22,** 133, 135
Photoelectrons, **30,** 2, 140
energy analysis, **7A,** 161, 172
escape depth of, **11,** 104, *see also* Electron(s)
Photoemission, **12A,** 3–14; **22,** 20, 21; **29C,** 40; *see also* Photoelectrons
angle-resolved, **22,** 160, 161
cross section, **22,** 484
current-voltage curve in, **11,** 90
direct transitions in, **11,** 76
electron kinetic energy distribution in, **11,** 89–90
from metals, **11,** 67; **12A,** 5–7
nondirect model of, **11,** 77

from non-metals, **12A,** 7–14
quantum yield in, **11,** 81–84
reverse, **11,** 88
spectra, **22,** 170
temperature-energy relationship in, **11,** 101–102
three-step model of, **11,** 75–81
transport and scattering events in, **11,** 77–80
vacuum-chamber monochromator system and, **11,** 89
Photoemission analyzer, **11,** 84–89
resolution in, **11,** 85–88
zero potential in, **11,** 87
Photoemission chambers
design in, **11,** 116–120
high vacuum, **11,** 114–120
Photoemission measurements
cesium channels or ampoules for, **11,** 113, 121
cleavage in, **11,** 103–104
contaminants in, **11,** 106–107
current bridge circuit in, **11,** 93–94
electron multipliers in, **11,** 99
energy distribution measurements in, **11,** 84–102
evaporation in, **11,** 104–105
experimental techniques for, **11,** 67–122
film thickness in, **11,** 105
future research in, **11,** 121–122
heat cleaning in, **11,** 105–107
ion bombardment cleaning in, **11,** 107–110
LEED pattern in, **11,** 109–110
lowered electron affinity in, **11,** 110–114
monochromator and light sources in, **11,** 120–121
photoelectric quantum yield and, **11,** 81–84
resistance heating in, **11,** 104
retarding field analyzer in, **11,** 84–89
sample preparation in, **11,** 102–114
vacuum pumps in, **11,** 114–116
Photoemission process, physics of, **11,** 74–75
Photoemission studies, at high photon energies, **11,** 121, *see also* Photoemission measurements
Photoemitters, **6B,** 385
as camera tubes, **2B,** 350
Photoexcitation, **9A,** 141–142
application in
metastable stom detection, **29B,** 204–206
resonance ionization spectroscopy, **29B,** 173–176
Rydberg atom production, **29B,** 121–125
cross section, **29B,** 122
Photoflash trigger circuit, **2A,** 239
Photoflash, lamp, **25,** 247
Photographic detection, **8,** 162
Photographic emulsion, **13A,** 316; *see also* Autoradiography
for holography, **26,** 201
Photographic film, **1,** 411
Photographic photometry, **12A,** 222
Photographic plate, **4B,** 84–87; **7B,** 282
advantages and disadvantages, **4B,** 86–87
Schumann plate, **4B,** 86
sensitized emulsion, **4B,** 85
SWR plate, **4B,** 86
Photographic recording, of X rays, **16B,** 46–53, 154
Photographic speed, classical, **12A,** 228
increased by baking, **12A,** 233
Photographic techniques in optical microscopy, **16B,** 360
Photography
high speed, **18,** 725–753
recording material, **18,** 726–727
short duration light sources, **18,** 727
Photoionization detectors, **7A,** 458
counters, **4A,** 488–489
effect of electronegative gases, **4A,** 489
ion chambers, **4A,** 490–491
Photoionization, **3,** 526, methods in mass spectrometry, **3,** 539–542
Photoionization, **4B,** 257; **7A,** 150; **9A,** 152, 252; **17,** 481; **30,** 166–167
application in
metastable atom detection, **29B,** 204–206, 207–210
resonance ionization spectroscopy, **29B,** 173–176
chamber, **7A,** 460
dissociative, *see* Dissociative photoionization
molecular, **7A,** 162
process, two-step, *see* Two-step photoionization process
rare gases, **7A,** 157–161
source, highly charged ion generation, **29A,** 187
theory, **7A,** 172–178
Photojunction devices, noise in, **2B,** 494
Photolithography, **25,** 138

Photoluminescence, **1,** 396; **6B,** 294–295, 301–313
Photoluminescent efficiency, **1,** 397–398
Photolysis, **29B,** 39–40, 89–90; **30,** 352–353
Photomechanical effects, **30,** 456
Photometers, **26,** 246
  Lummer-Brodhun, **1,** 390–391
Photometric and radiometric
  nomenclature, **26,** 214
  symbols, **26,** 214
  units, **26,** 214
Photometric aspects of light sources, **18,** 688–689
Photometric bands, infrared, **12A,** 441, 447–460
Photometric measurements, **26,** 263
Photometric standards, **26,** 264
Photometric units, **26,** 286
Photometry, **1,** 388–393; **26,** 213
  heterochromatic, **1,** 398
  methods, **1,** 390–393
Photomicrography, **25,** 159
Photomixing of radiation signals, **4A,** 468–469
Photomultiplier tubes, **2B,** 309–323
  amplification factor, **4A,** 477
  applications of, **2B,** 322
  background pulse spectrum for, **2B,** 317–318
  channel type, **2B,** 310
  characteristics of, **2B,** 305–306, 310–322
  chopper for, **2B,** 315
  dark current, **4A,** 480–482
  dark current in, **2B,** 312–313
  dynode secondary emission coefficient, **4A,** 477
  focusing, **4A,** 486–487
  grid control, **4A,** 487
  light detection in, **2B,** 312–313
  light response in, **2B,** 319
  linearity, **4A,** 479
  at low temperature, **5B,** 203–204
  magnetic fields and, **2B,** 321
  manufacturers of, **2B,** 324–325
  materials for, **2B,** 311–312
  noise, **4A,** 483
  noise in, **2B,** 314–315, 493
  operating principle of, **2B,** 309
  output design of, **2B,** 320
  pulse rise time, **4A,** 478
  scintillation counting with, **2B,** 316–317
  types and ratings of, **2B,** 322–323
  unsealed, **4A,** 474–475
Photomultiplier(s), **1,** 391; **5A,** 179; **7A,** 55, 145, 457–458; **12A,** 3–94; **17,** 545; **18,** 104–107, 407; **26,** 304, 309; **29B,** 417–418
  application to angular correlation, **21,** 107, 109–111
  application to positron lifetime, **21,** 123–130
  cathode spectral response, **5A,** 133
  circuitry, **5A,** 151–154
  crystallization studies, **16B,** 366
  current, **5A,** 129
  current mode pulse counting, **7B,** 108
    noise in, **7B,** 109
  photocathode, **21,** 51
  pulse counting, **21,** 50, 51
  quantum efficiency, **21,** 126
  RC time constant, **21,** 51
  selection, **5A,** 151–154
  as short-pulse detector, **29C,** 199
Photomultiplier, **8,** 204, 255, 266
  tubes, **8,** 265
Photomultiplier, in vacuum ultraviolet generation, **15B,** 218
  basic physics of, **12A,** 3–15
  dark current, **12A,** 37–46
  dc *vs* pulse-counting, **12A,** 76–94
  fatigue, **12A,** 49–51
  gain, **12A,** 33–36
  materials and construction, **12A,** 16–45
  non-linearity, **12A,** 51–60
  polarization effects, **12A,** 63–65
  pulse-height distribution, **12A,** 66–72
  pulses, time distribution of, **12A,** 72–74
  shielding, **12A,** 60
  stability, **12A,** 46–65
  temperature effects on stability, **12A,** 46–49
  weighting functions for detection, **12A,** 89–94
Photon burst method, **29B,** 307–308
Photon counting correlation, LDV signal processing, **18,** 180–186
Photon detection efficiency for polarized light, **7A,** 459
Photon detectors, **7A,** 53–54, 107, 145, 457–458; **17,** 538; **26,** 303
Photon detectors, ideal types
  narrow bandwidth detector, **4A,** 425–427
  quantum detector (BLIP), **4A,** 424–425

signal-radiation noise limited detector, **4A,** 427–429
thermal detector, **4A,** 422–424
Photon emission rate, **8,** 30
Photon flux, **8,** 31, 34, 35, *see also* Spectroscopic system
Photon flux intensity measurements, **7A,** 53, 54, 458–460
Photon flux, synchrotron radiation, **29C,** 25–26
Photon momentum, **29B,** 130
Photon monochromator, **22,** 481, *see also* Monochromator
Photon noise, *see* Noise
Photon optics, monochromatization, **29C,** 33
Photon parameters, **22,** 133
Photon polarization, **22,** 158
in angle-resolved photoemission, **22,** 159
selection rules, **22,** 156
Photon shot noise, **7B,** 17
Photon source(s), **13A,** 205; **22,** 133
conventional, **22,** 141, 147
Photon statistics, **4A,** 429
Photon yields, for new chemical lasers, **15B,** 27–132
Photon(s), **1,** 408, 409; **13A,** 67
beam energy
absolute high-energy, **5B,** 541–546
relative, **5B,** 546–547
beam sources, **5B,** 509–515
bremsstrahlung, **5B,** 509–511
monoenergetic of fixed energy, **5B,** 514–515
positron annihilation-in-flight, **5B,** 511–514
Čerenkov, scattering, **5A,** 177
emission, **17,** 191
flux, differential, **5B,** 508–553
generation of, **17,** 26
initiated showers in lead, **5A,** 661–662
and particle detectors, **6B,** 354
polarization, **5B,** 196,197, 214–274
detection, **5B,** 264–274
quantum yield in, **11,** 72
relative intensities for calibrated sources, **17,** 216–218
response of semiconductor detector, **5A,** 277
total absorption counter, **5A,** 189–191
total mass-attenuation coefficients, **5A,** 673–674
Photon-assisted tunneling, **11,** 223
Photon-difference method, **5B,** 469–472
Photon-drag detect, **26,** 327
Photon-excited fluorescence, **7A,** 166
Photon-induced x rays, **17,** 375
Photon-stimulated desorption
angle-resolved, **22,** 185, 491, 493
cross section, **22,** 466
Photoneutron sources, **5B,** 562–565
Photons, **8,** 23, **8,** 168, *see also* Scattered photons
Photopeak area, **5B,** 156
Photoproduction, coherent neutral pion, **5A,** 336–338
Photorefractive effect, lithium niobate, **29C,** 391
Photoresists for holography, **26,** 204
Photosensitized reactions, mercury, **3,** 577–578
Photosynthesis, **20,** 181–185
antenna pigments, **20,** 181
reaction centers, **20,** 181
Photothermal ablation, **30,** 175, 189, 462–463
Photothermal effects, **30,** 105–106, 343–347, 455–456
Photothreshold voltage, **16C,** 436, 437
Phototransistor, **2,** 758; **2B,** 327; **4A,** 453, 457–459
Phototube(s), **1,** 391, 392; *see also* Photomultiplier tubes, Vacuum phototubes
calibration, in photoemission measurements, **11,** 82–84
gas-filled, **2,** 747–748
multiplier, **2,** 748–757
time resolution, **2,** 753
vacuum, **2,** 743–747
Phototubes (diodes)
dark current, **4A,** 479
equivalent noise input, **4A,** 485
gas filled, **4A,** 477
high speed biplanar, **4A,** 485–486
noise in, **4A,** 481–484
signal-to-noise ratio, **4A,** 483
spectral response, **4A,** 471–473
vacuum, **4A,** 475–476
fatigue, **4A,** 476
frequency response, **4A,** 477
Photovoltage microscopy, **27,** 91
Photovoltage surface, **6B,** 140–141
Photovoltaic cells, **2B,** 327

i-V characteristics, **6B,** 370
spectral characteristics, **6B,** 374
Photovoltaic devices, **26,** 322
Photovoltaic effect, **4A,** 453–456; **6B,** 365–376
basic concepts and theory, **6B,** 367–370
current versus voltage characteristics, **6B,** 370–372
definition, **6B,** 365
energy band configuration, **6B,** 367–368
measurement of effects of high energy bombardment, **6B,** 375–376
in *p-n* junction, **6B,** 365–366
spectral response, **6B,** 373–375
Physical constants, **19,** 10–14
of biological media, **19,** 12–14
of various materials, **19,** 10–11
Physical contact connectors, *see* PC connectors
Physical optics, references for, **12B,** 61
Physical properties, ultrasonic measurement of, **11,** 410–433
Physical system, **1,** 305
Physics parameter, estimator, **28,** 452
Physics, molecular, **3,** 4–6
Physisorbed layer, **22,** 364
Physisorbed systems, **23B,** 2
phase diagrams, **23B,** 16
Physisorption, **14,** 25, 26, 299
process, **22,** 131
PI isolators, **29C,** 174
Pi theorem, **18,** 826–828
π, value, estimation, **28,** 436–437
Picein, **14,** 414, 415
Pickling, **14,** 442
solution, **14,** 443
Pickup coil
crossed, **21,** 378–380
single, **21,** 378–381
voltage from, **21,** 369
Pickup, biaxial, **1,** 225
Picosecond continuum for monitoring absorption spectra, **15B,** 203
Picosecond interrogating pulse, **15B,** 202, *see also* Short optical pulses
in biological studies, **15B,** 206–207
continuum generation of, **15B,** 196–197
generation of by spontaneous anti-Stokes scattering, **15B,** 195
sources of, **15B,** 185–197
Picosecond lasers, **30,** 93–94, 96–97, 190
Picosecond spectroscopy, **15B,** 185–209
frequency conversion in , **15B,** 193–197
molecular dynamics and, **15B,** 204–206
parametric emission in, **15B,** 195–196
pulse measurement techniques in, **15B,** 198–201
repetitive-pulse measurements in, **15B,** 198
second harmonic generation and, **15B,** 194, 199–200
shorter wavelengths in, **15B,** 194–195
subpicosecond amplification in, **15B,** 198–199
two-photon fluorescence in, **15B,** 198–199
Picosecond streak camera techniques, **15B,** 198
Picture reproduction, **6B,** 335–356
Picture tubes, TV, **2B,** 342–343
Piecewise interferometric generation, of gratings, **15B,** 51–54
Pierce method for electron beams, **4A,** 94, 100–102, 116, 121
Pierce variable path interferometer, **3,** 693–694
Piezo optical effect, **31,** 252
Piezo-optical birefringence modulator, **12A,** 378, 396
Piezoelectric accelerometer, **1,** 78, 80, 219
Piezoelectric ceramics, *see* Ceramics
Piezoelectric crystals, **1,** 80
Piezoelectric disk, **1,** 222
Piezoelectric effect, as transducer, **11,** 392
Piezoelectric equations of state, **19,** 498
Piezoelectric excitation of oscillation, **6A,** 305–306
Piezoelectric head, **18,** 333
Piezoelectric materials, **1,** 387
Piezoelectric scanning interferometer, **18,** 201
Piezoelectric sensor, **18,** 542
Piezoelectric stiffening, **19,** 37, 502
Piezoelectric transducer (PZT), **16C,** 61, 62, 168; **29C,** 113, 122
Piezoelectricity, **1,** 70–71, 74; **19,** 30, 34–40, 51–55
constitutive relations, **19,** 34–36
definition of, **19,** 30, 35
material properties, **19,** 51–55
physical principles, **19,**36–38
Piezoresistors, **2A,** 51
PIG, *see* Penning discharge source
Pile-of-plates polarizer, **31,** 229–230
Pile-up, in pulse amplitude measurement, **2B,** 90, 95

PIN structure, **2A,** 56–58
Pin-cushion distortion, **16B,** 203, 204
Pinch integrated circuit resistor, **2A,** 118
Pinger, **24B,** 110–111
Pinhole camera, **26,** 61
Pinhole collimator, **6A,** 220
Pinhole transmission method, **6A,** 221
Pinholes in thin metal films, **31,** 323–324
Pion
  gap length distribution, **5A,** 256
  photoproduction, coherent neutral, **5A,** 336–338
  production, cross section, **5A,** 339
  response of semiconductor detectors, **5A,** 278–279
  scattering, **5A,** 319–321
Pipe flow geometry, **14,** 103–104, *see also* Pumping speed; Conductance
Pipette, **1,** 116
Pippard-Buckingham-Fairbank relations, **19,** 437
Pirani detector, **7A,** 80, 268–269
Pirani gage, **1,** 167–168, 191, 299
Pirani gauge for atom detection, **4A,** 390–393
  construction, **4A,** 391–393
  sensitivity of, **4A,** 391, 393
Pirani gauge, **14,** 54–57, *see also* Thermal conductivity gauge
Piston, *see* Rayleigh piston
  peak velocity/displacement, **8,** 15
Piston, *see* Rayleigh piston
  peak velocity/displacement, **8,** 15
Pitch, **1,** 147
  aerodynamic, **18,** 768–769
Pitot probe
  angular sensitivity, **18,** 248
  calibration, **18,** 243, 248, 253, 515–524
  corrections for turbulence, **18,** 252
  corrections for viscous effects, **18,** 251
  general, **18,** 240, 242–254, 515–524
  principle, **18,** 243
  tube construction, **18,** 248
  use in low density wind tunnel, **18,** 784
  use in velocity gradient, **18,** 250
  use near wall, **18,** 250
  wind tunnel, **18,** 776–777
Pitot tube, **1,** 130, 153, 154; *see also* Pitot probe
Pitot-static probe, **18,** 249
Pitting, laser damage to coatings, **29C,** 346
Pivots, **28,** 166, 178
Placzek corrections, *see* Inelasticity corrections
Placzek expansion, **23B,** 433
Plamons, surface, **11,** 184
Planar channeling, **21,** 286
Planar diode, fabrication of, **2A,** 57
Planar Hall effect, **6B,** 147–148
Planar waveguide, **29C,** 369, 370
Planck function, **9A,** 121
Planck law, **1,** 253, 301, 376
Planck's constant, **1,** 46–48, **8,** 6, 28, **10,** 2
Planck's formula, **4A,** 289
Planck's law, **8,** 29, **29C,** 6
Planck's radiation law, **8,** 27, **10,** 288, **18,** 465–466, 690, **31,** 49
Planck-Einstein equation, **3,** 689
Plane crystal spectrometer, **17,** 387
Plane diffraction grating, **31,** 29–30
Plane grating monochromator (PGM), **29C,**37
Plane gratings in convergent light, **12A,** 481
Plane hologram techniques, **26,** 181
Plane mirror
  extreme ultraviolet beam deflection, **31,** 161
  x-ray beam deflection, **31,** 161
Plane polarized light, **1,** 377
Plane states, **13A,** 73
Plane wave propagation, **19,** 3–9, *see also* Wave propagation
  physical parameters for water, **19,** 8
  relation among physical parameters, **19,** 6–7
Plane, focal, **14,** 88
Plane-holograms, **26,** 169
Plane-mirror analyzer, **22,** 138
Plane-mirror electrostatic deflection analyzer, *see* Parallel-plate electrostatic deflection analyzer
Plane-parallel interferometers, with CCD readout, **29C,** 337–338
Plane-wave Born approximation, **17,** 111–112, 199–200, 202–206, 210–211, 229, 236–239, 241–252, 321, 323, 383, 482–483, 486
Planes waves, **13A,** 41
Planets, size and mean disk temperature of, **12B,** 94
Planimeter, **1,** 58ff
Plano objectives, **25,** 165
Plano-convex lens, **29C,** 351

Plasma
  conductivity, **10,** 268–271
  in deep space, *see* Deep Space plasma measurements
  interferometry, **10,** 280–286
  optical refractivity of, **9B,** 1–48
  temperature versus particle density, **29A,** 169–170
Plasma absorption, **30,** 186, 187
Plasma characteristics, **2,** 80–97
Plasma column, **8,** 118, 119
Plasma column, equations for, **9A,** 7–8
Plasma desorption, **30,** 414, 415
Plasma detectors, for spacecraft, calibration, **9B,** 73
Plasma discharge ion source, **22,** 337
Plasma electron temperature, **29C,** 11
Plasma emission, **30,** 376–379
Plasma formation, **30,** 165
  in PLD, **30,** 482, 483
Plasma frequency, **9A,** 3; **10,** 110
  definition, in afterglow studies, **7B,** 86–87
  definition of, **9B,** 6
Plasma heating by strong shock waves, **9A,** 169–212
  shock-wave theory, **9A,** 169–173
Plasma heating, **30,** 163
Plasma jet, light sources, **4A,** 308–309
Plasma light source, **17,** 550
Plasma radiating system, **8,** 121
Plasma radiator, **8,** 120
Plasma refractive index, definition of, **9B,** 6
Plasma sheet
  definition of, **9B,** 50
  plasma parameters of, **9B,** 51
Plasma shielding, **30,** 49, 100, 390, 393, 396
Plasma studies, and beam-foil experiments, **17,** 599–605
Plasma temperature, **30,** 380, 381
Plasma wave echoes, **9A,** 27–36
  collision effects on, **9A,** 32–34
  observations of, **9A,** 34–36
Plasma waves
  electron type, **9A,** 4–6
  finite geometries of, **9A,** 7–10
  finite plasma dispersion relation for, **9A,** 9–10
  ion acoustic type, **9A,** 6–7
  Landau damping of, **9A,** 5, 6, 24
  of large amplitude, **9A,** 5–6
    resonant electrons of, **9A,** 5
  observation of, **9A,** 11–27
    experimental techniques, **9A,** 13–18
    general requirements for, **9A,** 12
    ion waves, **9A,** 12–13
  space and time damping of, **9A,** 10–11
  studies by microwave scattering, **9A,** 40
Plasma(s), **13A,** 191, 208; **13B,** 97
  analysis by experimental methods, **9A,** 119–134
    excitation processes, **9A,** 125–128
    Maxwellian velocity distribution, **9A,** 131–134
    partical diagnostics, **9A,** 130–131
    radiative energy losses, **9A,** 128
    rate equations, **9A,** 124–130
    statistical models, **9A,** 120–124
  atomic number dependence of spectra, **29C,** 13
  atomic processes in, *see* Atomic processes
  diagnostics by light scattering, **9A,** 61–113
  dialectric function of, **9A,** 3
  equations for, **9A,** 1–3
  laser-produced, *see* Laser-produced plasma
  microwave scattering from, **9A,** 37–59
  rate coefficients in, **9A,** 119–120
Plasma-arc harmonic generator, *see* Arc harmonic generator
Plasma-etching, **16B,** 241–245
Plasmapause, knee whistlers in, **9B,** 102–103
Plasmaphere, definition of, **9B,** 50
Plasmas, detection of charged particles in
  by microwave interferometers, **4A,** 386
  by Langmuir probes, **4A,** 387
  by radio-frequency probes, **4A,** 387
  by resonant cavities, **4A,** 386
Plasmon excitation, **7B,** 292
Plasmon polariton, **30,** 33
Plasmon resonance spectrometry, **30,** 107–108
Plasmon satellites, **13A,** 180
Plasmon(s), **22,** 195, 196, 229; **30,** 177
  decay, **22,** 197
  gain satellites, **22,** 229
  losses, **22,** 150
Plastic crystalline phase, **23A,** 506
Plastic deformation
  gas-medium apparatus, **24A,** 68–83
  load relaxation, **24A,** 60–68
  solid-medium apparatus, **24A,** 83–85
  unconfined creep, **24A,** 58–62
Plastic detector, **8,** 259
Plastic film, as dielectric, **2A,** 37
Plastic flow, **1,** 109–110

Plastic packaged components, **2A,** 46
Plastic replica technique, **6A,** 177–179
Plastic, *see* Plastomer
Plasticity of solids, **6A,** 309–321
Plasticizer, effect on gas transport, **16C,** 330
Plasticizers, **16A,** 477
Plastomer
acrylic, **14,** 414
nylon, **14,** 413
phenolics, **14,** 414
polyethylene, **14,** 413–414
polypropylene, **14,** 413–414
Teflon, **14,** 411–413
Tygon, **14,** 413
Plate
half-wave, **1,** 381
Lummer-Gehrcke, **1,** 370, 371
phase, **1,** 342
quarter-wave, **1,** 380ff
refracting, **1,** 340
Wilson test, **1,** 502–503, 504
Plate efficiency of tube, **2,** 67
Plate factor, **12A,** 467, 471
Plate tectonics, VLBI measurement of, **12C,** 275
Plateau, **5A,** 119
Plated contacts, **6B,** 52
Plating techniques, **6B,** 61
Platinum
black, **7A,** 23
boiling point, **14,** 351
hydrogen solubility, **14,** 354
melting point, **14,** 347, 351
physical properties, **14,** 373, 374
thermometer, **1,** 245ff, 256–257
vapor pressure, **14,** 347, 351
Young's modulus, **14,** 360
Platinum thermocouples, **11,** 359
Platinum,reflectance spectra, **31,** 212–213
PLD, *see* Pulsed laser deposition
Pleated sheet structure, of polypeptide chains, **16B,** 90–91
Pleochroism, **6B,** 250
PLL, *see* Phase-lock loop
Plot, parametric, **5B,** 135
Plume emission, **30,** 104–106, 154–166
from AlN, **30,** 249–252, 255–258
from MgO, **30,** 154–158
spectroscopy, **30,** 377–381
Plume expansion, **30,** 382–384
Plume formation, **30,** 152–168, 232–234, 376–386
Plume interferometry, **30,** 98–100
Plume, **30,** 225
Plunger, waveguide, **3,** 25
PLUTO reactor, Harwell, **23A,** 248
Plutonium critical assembly leakage neutron spectrum, **5A,** 477–478
PMA, *see* Parallel-plate electrostatic deflection analyzer
PMM/FM/AM, in telemetering, **2B,** 220
Pneumatic detectors, **13B,** 8, 51
PNPN diodes, **2A,** 92–94
PNPN structure, gate triggering in, **2A,** 97–99
PNPN switch, in integrated circuits, **2A,** 120
PNPN triodes, **2A,** 95–99
Po-α-Be neutron source, **5B,** 556–559
Pockels cell, **12A,** 378, 396; **18,** 717, 728–729
negative-electron-affinity semiconductor photocathode, **29A,** 22–23
Pockels electro-optic effect, **26,** 147
Pogo probe technique, **24B,** 239–240
Poincaré sphere, **12A,** 374; **6,** 112, 142; **29C,** 357–359
Poincaré relations, **1,** 216
Point contact, **27,** 16
diodes, **13B,** 111
squid, **11,** 285
Point contacts, **6B,** 89–91
defined, **11,** 258–260
in magnetometers, **11,** 281
quantum interference effects in, **11,** 263
Point estimation, **28,** 160–165, 213
desirable properties, **28,** 160–161
method of least squares, **28,** 162–164
method of maximum likelihodd, **28,** 164–165
method of moments, **28,** 161–162
statistic, concept of, **28,** 160
Point group, **23A,** 371; **23C,** 112
crystallographic, **16B,** 17
of the wave vector, **23C,** 113
Point groups, **3,** 273
Point kinematics, **1,** 74–81
Point light source, **18,** 356, 405, 710
Point process, *see also* Periodic signals, weak
Point processes, **28,** 240
Point projectile, in inner-shell vacancy production, **17,** 321–330
Point projection microscope, **6A,** 185–186
Point source, **22,** 368

Polar axis, **8,** 78Polarization studies, application, *see* Neutron polarization studies

Polarization techniques, **12A,** 361–414
Polarization variation, **23C,** 399
Polarization vector(s), **1,** 488ff; **18,** 347; **23A,** 373
Polarization, **1,** 376–385; **13A,** 73; **22,** 133, 143, 156, 159, 160, 171, 179, 180, 183, 360, 485
  applied electric field and, **15B,** 144
  atomic, **1,** 492; **3,** 641
  bremsstrahlung, **5A,** 60–61
  characteristics of toroidal grating monochromator (TGM), **31,** 135
  circular, **26,** 110, **31,** 7–10
  circular by reflection, **31,** 246–251
  circular by transmission, **31,** 252–253
  in crossed fields, **5B,** 227
  degree of, **12A,** 362, 684
  degree of incident γ-ray beam, **5B,** 141
  degree of polarization, equation, **29A,** 231
  determination, **26,** 159
  dielectric, **1,** 488, 491–493
  effect(s), **5A,** 241; **26,** 128
    in metals, **26,** 120
    in thin films, **26,** 125
  effects, **1,** 365, 378–379
  electric, of molecules, **3,** 637–645
  electromagnetic wave and, **15B,** 143
  electron
    from beta decay, **5A,** 64
    longitudinal, **5A,** 3
  electronic, **1,** 492; **3,** 641
  of electrons, **5B,** 214–274
  elliptic, **6B,** 258–262
  external field, **5B,** 169–176, 225
  elliptical, **26,** 113
  insertion devices, **31,** 20–21
  interfacial, **1,** 493
  Kerr component in, **11,** 721
  Kerr effects in, **11,** 705–706
  linear, **26,** 110, **31,** 228–229
  linear of γ-rays, **5B,** 138–142
  in longitudinal fields, **5B,** 225–226
  of magnetic hyperfine structure, **5B,** 176–183
  mathematical description, **26,** 159
  maximum, **31,** 236
  of media, **1,** 431
  of medium, **5A,** 231
  mirror/spectrometer combination, **31,** 132
  molar, **3,** 640–645
  negative-electron-affinity semiconductor photocathode
    circular polarization of incident radiation, **29A,** 22
    film growth, **29A,** 7, 9
    optimization, **29A,** 6–7, 9–10
    polarization formula, **29A,** 2–3, 6
  neutral beam, average, **5A,** 483–486
  neutron, **3,** 635–636
  nuclear, **5B,** 163, 174
    of assembly of nuclei, **5B,** 169
  optical, **26,** 107
  orientation, **1,** 492
  photon, **5B,** 196, 197, 214–274
  of scattered light, **3,** 122–123
  scramblers, **26,** 153
  secondary source standard, **31,** 128–129
  second-order dependence of on electrical field, **15B,** 143
  synchotron radiation, **31,** 123
  transition, **3,** 255
  in transverse electric field, **5B,** 226
  in transverse magnetic field, **5B,** 226–227
  types of, **1,** 376–377
  ultraviolet source, **31,** 141
  vacuum ultraviolet source, **31,** 141
Polarization-direction correlations, **5B,** 138
Polarized atoms by optical pumping, **4B,** 256
Polarized electrons, beams, **4B,** 256–257
Polarized fluorescent emission, **16A,** 412
  spectrum analysis, **16A,** 426–436
Polarized hydrogen atoms, **4B,** 253
Polarized ion source
  advantages, **4B,** 226
  applications, **4B,** 229–232
  production from polarized atomic beam, **4B,** 234
Polarized ions, acceleration of, **4B,** 251–253
  depolarization effects, **4B,** 253
Polarized light, **11,** 700
  photon detection efficiency, **7A,** 459
Polarized negative ions, source, **4B,** 255
Polarized neutrals, **4B,** 252–253
Polarized neutrons, **21,** 148; **23A,** 64, 216; **23C,** 93, 171
Polarized nucleons by scattering, **4B,** 232–234
Polarized targets and beams, **13A,** 139
Polarized x rays, **17,** 373
Polarized-photon photoemission, **22,** 156, 158
Polarizer/quarter-wave isolator, **29C,** 358–359
Polarizers, **13A,** 239
  Brewster angle, **10,** 104

dichroic, **31,** 232
four mirror, **31,** 239–242
grating, **26,** 157
infrared, **12A,** 364, 429
interference, **31,** 242–243
linear, **26,** 126, 130
made of four mirrors, **26,** 124
made of three mirrors, **26,** 123
metal grid, **26,** 157
multilayer, **31,** 244–245
for optical pumping, **7B,** 17
pile-of-plates, **31,** 229–230
pseudo-Brewster angle, **31,** 244–245
reflecting, **31,** 232–242
Rochon prism, **31,** 229
slit, **26,** 157
thin film, **31,** 230–231
three mirror, **31,** 239–242
transmitting, **31,** 229–232
waveguide, **10,** 104, 105
wire grid, **10,** 105–107
Wollaston prism, **31,** 229
Polarizing beam splitters, **12A,** 366–371
Polarizing field, **11,** 527
Polarizing filters, **23A,** 216
Polarizing microscope, **6B,** 265–269
observation of spherulites in, **16B,** 341
in thin film crystallization, **16B,** 361
Polarizing monochromators, **23A,** 216
Polarizing optics, **29C,** 353–355
Polarizing prisms, MacNeille, **26,** 130
Polarizing sheets, **29C,** 354
Polaroid, **1,** 379, 383
Polarons, **30,** 33, 34
Pole figure, **16C,** 157–161
Pole-dipole array, **24B,** 296, 303–305
Pole-zero cancellation, in pulse spectrometry, **2B,** 110–112
Poled ceramic pressure gage sensor, **18,** 542
Poling, **19,** 30, 52–54
Polish, chemical, **14,** 442, 444, 445
Polishing
guide to literature, **6A,** 152
mechanical, **14,** 443
Polishing methods, **6A,** 151–153
chemical, **6A,** 151
electrolytic, **6A,** 151
mechanical, **6A,** 151
Poly(2,6-dimethyl phenylene oxide)
alloy with polystyrene, **16C,** 308–309, 331
blend with polystyrene compatibility study by carbon–13 NMR, **16A,** 348
chain flexibility by NMR, **16A,** 305
glass transition, **16C,** 308
phenyl side chain motion, **16A,** 310
radicals generated during fracture, **16C,** 196, 197, 200–201
Poly(4-methyl pentene–1), chlorination of, **16B,** 248–249
Poly(α-methyl styrene)
chain flexibility by NMR, **16A,** 305
methyl and phenyl side group motion, **16A,** 310
radicals generated during fracture, **16C,** 198–199
syndiotactic, carbon–13 NMR spectrum, **16A,** 260
Poly(acrylic acid), replica for electron microscopy, **16C,** 242
Poly(acrylonitrile-*co*-butadience-*co*-styrene), (ABS)
density, **16C,** 72
elastic constants, **16C,** 81
ultrasonic properties, **16C,** 72, 74
Poly(carborance siloxane)
density, **16C,** 72
ultrasonic properties, **16C,** 72, 74
Poly(dimethyl siloxane)
as ultrasonic medium, **16C,** 64
density, **16C,** 72
ESR spectra, nitroxide-doped, **16A,** 464
glass-transition temperature, **16B,** 327
temperature of maximum hypersonic loss, **16A,** 195
ultrasonic properties, **16C,** 72
Poly(dimethylsiloxane) (PDMS), **30,** 124, 125
Poly(ε-caprolactum), alloy with PVC, **16C,** 290, 299, 300
Poly(ethylene glycol), dispersion curve and phonon frequency distribution, **16A,** 211
Poly(ethylene oxide)
bulk modulus, **16C,** 112
chain flexibility by NMR, **16A,** 305
density, **16C,** 72
glass-transition temperature, **16B,** 327
radicals generated during fracture, **16C,** 198–199
self-seeding crystallization, **16B,** 382
temperature of maximum hypersonic loss, **16A,** 195
ultrasonic properties, **16C,** 72, 73

Poly(ethylene terephthalate)
Brillouin splitting near T_g, **16A,** 202
crystallinity by X-ray diffraction, **16B,** 121
crystallization of, density balance method, **16B,** 373
degradation by aminolysis, **16B,** 276
DTA melting point, effect of heating rate, **16B,** 302
effect of crystallinity on gas diffusion, **16C,** 329
etching with argon and nitrogen plasma, **16B,** 244
intrinsic birefringence, **16C,** 162
oriented, **16C,** 145–147
(PeT), **30,** 285
Poisson's ratio, **16A,** 204
radicals generated during fracture, **16C,** 204, 207
selective degradation by hyrdolysis, **16B,** 274–275
small-angle X-ray pattern, **16C,** 182
work function, **16C,** 431
Poly(ethylene) (PE), **30,** 285
Poly(hexamethylene adipate)
density, **16C,** 72
elastic constants, **16C,** 81
ultrasonic properties, **16C,** 72
Poly(imide) (PI), **30,** 274–278, 285, 338, 360–361
Poly(methacrylic acid), methyl side group motion, **16A,** 310
Poly(methyl acrylate), radicals generated during fracture, **16C,** 198–199
Poly(methyl acrylate), temperature of maximum hypersonic loss, **16A,** 195
Poly(methyl methacrylate)
Brillouin splitting near T_g, **16A,** 200
Brillouin line width, **16A,** 203
carbon–13 NMR study
of solid state, **16A,** 362
of tacticity, **16A,** 289–294
chain flexibility by NMR, **16A,** 305, 307–308
craze opening displacement, **16C,** 248
creep, effect of moisture, **16C,** 55
density, **16C,** 72
dipole moment and dielectric loss, **16A,** 396
elastic constants, **16C,** 81
electrical treeing in, **16C,** 482
glass transition, **16A,** 362
methyl side group motion, **16A,** 310
(PMMA), **30,** 285
positron lifetime in, **16A,** 395
radicals generated during fracture, **16C,** 198–199
spin-lattice relaxation, **16A,** 322
tensile strength in various environments, **16C,** 259
ultrasonic properties, **16C,** 68, 71, 72, 73, 74
X-ray diffraction pattern (amorphous), **16B,** 54
Poly(ortho-methyl styrene)
crystal structure, **16B,** 88–89
density, **16B,** 88
repeat distance, **16B,** 88
unit cell dimensions, **16B,** 78
Poly(*p*-(2hydroxyethoxy)-benzoic acid), (PEOB), stable primary free radicals, **16C,** 204, 205, 207, 208, 209
Poly(phenyl quinoxaline)
density, **16C,** 72
elastic constants, **16C,** 81
ultrasonic properties, **16C,** 72, 73, 74
Poly(prolylene oxide)
chain flexibility by NMR, **16A,** 305
incoherent neutron scattering spectra, **16A,** 236–238
methyl side group motion, **16A,** 310
Poly(propylene glycol)
hypersonic loss data, **16A,** 195, 196
transition map, **16A,** 198
Poly(propylene oxide)
dielectric loss data, **16C,** 395
isotactic, melting behavior, **16B,** 304
radicals generated during fracture, **16C,** 198–199
spherulitic growth rate, **16B,** 363
Poly(styrene) (PS), **30,** 285
Poly(styrene-*b*-butadiene-*b*-styrene)
domain size *vs* molecular weight, **16C,** 296
glass transition, **16C,** 311, 312
spherulite patterns, **16C,** 299
Poly(styrene-*b*-ethylene oxide), morphology, **16C,** 298
Poly(styrene-*b*-isoprene)
glass transition, **16C,** 312
interphase thickness, **16C,** 304
small-angle X-ray pattern, **16C,** 302
Poly(styrene-*co*-buadiene), latex, **16A,** 474, 476
Poly(tetrafluoroethylene) (PTFE), **30,** 285

Poly(tetramethylene-*p*-silphenylene siloxane), spherulitic growth rate, **16B,** 363
Poly(vinyl acetate), temperature of maximum hypersonic loss, **16A,** 195
Poly(vinyl alcohol)
chain flexibility by NMR, **16A,** 305
glass-transition temperature, **16B,** 327
radicals generated during fracture, **16C,** 200–201
Poly(vinyl butyral)
density, **16C,** 72
ultrasonic properties, **16C,** 72
Poly(vinyl chloride)
alloy with poly(ethylene-*co*-vinyl acetate), **16C,** 293, 310
chain flexibility by NMR, **16A,** 305
dipole moment and dielectric loss, **16A,** 396
effect of plasticizer, **16A,** 473
glass-transition temperature, **16B,** 327
historic reference, **16A,** 1
positron lifetime in, **16A,** 395
radicals generated during fracture, **16C,** 204
work function, **16C,** 431
Poly(vinyl methyl ether), alloy with polystyrene, **16C,** 291, 309
Poly(vinylidene chloride)
dehydrohalogenation of, **16B,** 252–253
glass-transition temperature, **16B,** 327
historic reference, **16A,** 1
infrared bands, **16B,** 253
single crystal, fold surface, 253
Poly(vinylidene fluoride)
alloy with poly(methyl methacrylate), **16C,** 311
chain flexibility by NMR, **16A,** 305
density, **16C,** 72
elastic constants, **16C,** 81
glass-transition temperature, **16B,** 327
ultrasonic properties, **16C,** 72
Poly-L-glutamic acid, **19,** 203, 233
Poly-L-lysine, **19,** 203, 233
Poly-L-ornithine, **19,** 233
Polyacrylonitrile, chain flexibility by NMR, **16A,** 305
Polyacrylonitrile, glass-transition temperature, **16B,** 327
Polyamide, *see* Nylon
Polyatomic molecules, **3A,** 142–172, 292–301
Coriolis resonance, **3A,** 170
Fermi resonance, **3A,** 169
l-type doubling, **3A,** 171
Vibration-rotation, **3A,** 142–167
Polybutadience, radicals generated during fracture, **16C,** 200–201
Polybutadiene, positron lifetime in, **16A,** 395
Polybutadiene, staining for electron microscopy, **16B,** 216
Polycaprolactum, *see also* Nylon 6
density, **16C,** 72
elastic constants, **16C,** 81
ultrasonic properties, **16C,** 72
Polycarbonate
carbon–13 NMR spectrum, **16A,** 327, 333
ESR study, nitroxide-doped, **16A,** 465
Poisson's ratio, **16A,** 204
radicals generated during fracture, **16C,** 198–199
Rayleigh-Brillouin's spectrum, **16A,** 181, 182
temperature of maximum hypersonic loss, **16A,** 195
work function, **16C,** 431
yield stress and Young's modulus, **16C,** 131
Polychlorofluoroethylene
crazing-type creep, **16C,** 266, 268
crystallization behavior, **16B,** 264
spherulitic growth rate, **16B,** 363
stress-strain curves, **16C,** 130
yield stress and elastic modulus, **16C,** 131
Polychloroprene, glass-transition temperature, **16B,** 327
Polychlorotrifluoroethylene, carbon–13 NMR spectrum, **16A,** 360–361
Polycrystalline materials
as hosts, **15A,** 189
preferred orientation, **6A,** 240–241
Polycrystalline medium, description of, **19,** 237–238, 240
Polycrystalline specimen diffraction pattern, **6A,** 276–278
Polydimethylsiloxane, **23C,** 300, 308, 324, 327
Polydispersity, **16A,** 30
Polyelectrolytes
behavior in solution, **16A,** 22
fluorescent probes in, **16A,** 410
Polyenes, **20,** 156
Polyepoxide
density, **16C,** 72
elastic constants, **16C,** 81
secondary transition, **16C,** 105
Polyethylene, **10,** 96, 97, 99–101; **14,** 413–414; **23C,** 301

amorphous/crystalline contrast
enhancement, in electron microscopy, **16B,** 232
branching in, **16A,** 61, 285–288
chain conformation, **16B,** 20
conformation, infrared studies, **16A,** 144–146
crystal growth rate, **16B,** 365
crystal structure, **16C,** 151
crystallite size in highly-drawn sample, **16B,** 149
crystallization from dotriacontane, **16B,** 346
dispersion curve
crystalline, **16A,** 119
isolated chain, **16A,** 103
DTA melting curve, **16B,** 376
dynamic viscosity, **19,** 229–231
effect of crystallinity on gas diffusion, **16C,** 330
electric breakdown in, **16C,** 467–469, 472
by discharges, **16C,** 450
equilibrium melting point, **16B,** 379–381
etching with oxygen plasma, **16B,** 242–245
etching for electron microscopy, **16B,** 216, 240
fiber diffraction pattern, **16C,** 152, 158
gauche bonds, **16A,** 129, 142
glow-discharge polymerized, **16A,** 167–168
grating filters, **10,** 103, 104, 139
heat of fusion, **16B,** 325
high density
density, **16C,** 72
elastic constants, **16C,** 81
electric breakdown in single crystal, **16C,** 459
ultrasonic properties, **16C,** 72, 73, 74
homogeneous nucleation, droplet experiments, **16B,** 386–396
calculation of surface free energies, **16B,** 395
inelastic neutron spectra, **16A,** 221
infrared and Raman spectra, **16A,** 104, 125, 135
intercrystalline links in, **16B,** 231–233, 234
irradiation with X rays, **16B,** 281
infrared bands, **16B,** 251
lamellae, **16C,** 176, 177
low density
carbon–13 NMR spectrum, **16A,** 286
chain flexibility by NMR, **16A,** 305–306
density, **16C,** 72
nature of branches, **16A,** 285–288
ultrasonic properties, **16C,** 72
melt-crystallized, small-angle X-ray pattern, **16B,** 168
morphology, effect on melting point, **16B,** 298
normal vibration assignment in infrared and Raman, **16A,** 107
oxide, **19,** 232–233
perdeuterated
crystallization of, **16A,** 504
infrared bands, **16A,** 127–128, 131
SANS studies, **16A,** 504–506
photochlorination of, **16B,** 248, 249
photoelectric emission, **16C,** 435–436
polycrystalline and fiber-oriented, carbon–13 NMR spectra, **16A,** 352–356
positron lifetime in, **16A,** 379, 395
purification and thin film preparation, **16B,** 355
radicals generated during fracture, **16C,** 198–199, 204, 207
repeat distance, **16B,** 79
single crystals
annealed in suspension, **16B,** 231
branching location in, **16B,** 267
bright-field electron micrograph, **16B,** 190–191, 192
bromination of, **16B,** 250
core thickness by X-ray diffraction, **16B,** 152–153, 161
dark-field electron micrograph, **16B,** 195
DSC melting curve, **16B,** 295, 306, 309
oil suspended, **16B,** 311
effect of irradiation on melting, **16B,** 305
electron diffraction pattern, **16B,** 194
fold structure, **16B,** 229–230
from GPC of degraded samples, **16B,** 263–266
heat of fusion, **16B,** 250, 315
irradiation with γ rays, **16B,** 282–286
mats of, small-angle X-ray scattering pattern, **16B,** 162
mosaic blocks
in sedimented mats, **16B,** 152
in suspended samples, **16B,** 174
ozone degradation, **16B,** 260–261
preparation of *n*-paraffins from, **16B,** 268–269
self-seeding, **16B,** 260
twinned, **16B,** 229

small-angle X-ray pattern, **16C,** 181, 182
solvent-etching, **16B,** 240
space group and crystal structure, **16B,** 15, 16
spherulites in, **16B,** 341, 342
stepwise melting of, **16B,** 377
ultraoriented, **16C,** 145
unit cell dimensions, **16B,** 75
vibrational analysis by neutron scattering, **16A,** 220–235
X-ray fiber pattern, **16B,** 73–74
indexing of, **16B,** 75
Polygon scanners, **25,** 113
Polygonization, **1,** 425; **6A,** 339–340
Polyimide, work function, **16C,** 431
Polyisobutylene
chain flexibility by NMR, **16A,** 305
density, **16B,** 93
helical conformation and crystal structure, **16B,** 93
hypersonic loss data, **16A,** 194, 195
methyl side group motion, **16A,** 310
positron lifetime in, **16A,** 395
radicals generated during fracture, **16C,** 198–199
spin-lattice relaxation by NMR, **16A,** 304
transition map, **16A,** 197
Polyisoprene, hydrogenation, **16A,** 2
Polyisprene, radicals generated during fracture, **16C,** 200–201
Polymer analysis, **30,** 430, 431
Polymer films, **10,** 97, 98, 100
Polymer mixture, identification by thermal analysis, **16B,** 335–336
Polymer reactions, **16B,** 335
Polymer single crystals, **16A,** 8, 121, 475, 505–506; **16C,** 459
bromination of, **16B,** 247
chlorination of, **16B,** 247–248
crosslinking by electron beam, in electron microscope, **16B,** 190
gaseous halogenation of, **16B,** 248
halogenation in suspension, **16B,** 246–248
melting behavior
irradiated, **16B,** 315
in oil suspension, **16B,** 308–312
melting point, **16B,** 296
morphology, **16B,** 128, 174, 229–230, 237, 342, 343–344
precipitated mats
crystallinity profile, **16B,** 161
small-angle X-ray scattering in, **16B,** 162
self-seeding, **16B,** 382
"true" melting point, **16B,** 313–316
Polymer surfaces
chemical modification, **16B,** 245–255
degradation with chemicals, **16B,** 255–277
modification by plasma-etching, **16B,** 241–245
Polymerization, **16A**
ESR spin probe study, **16A,** 477
heat of, **1,** 262
kinetics, calorimetric (DSC) studies, **16B,** 331–334
or organic molecules in an electron microscope, **16B,** 208, 221
Polymers, **23C,** 287; **30,** 336
amorphous, **23C,** 299
blends, **23C,** 313
conformation, **23C,** 443
crystalline, **23C,** 301
diffusive motions, **23C,** 322
dynamics, **23C,** 314
in solution, **23C,** 302
mixtures, **23C,** 290
networks, **23C,** 306
phonons, **23C,** 321
rotations, **23C,** 318
solid-solvent interfaces, **23C,** 312
uniaxially deformed, **23C,** 309
vibrations, **23C,** 316
Polymethacrylonitrile, methyl side group motion, **16A,** 310
Polymethyl methacrylate, **23C,** 299, 319
Polynomial
fitting, **1,** 19
functions, **1,** 15
orthogonal, **1,** 18
Polynomials, *see* Legendre polynomials
Polyoxymethylene
chain flexibility by NMR, **16A,** 305
crystallization behavior, **16B,** 364
density, **16C,** 72
elastic constants, **16C,** 81
glass-transition temperature, **16B,** 327
helical-chain conformation, **16B,** 20–21
longitudinal acoustic mode, **16A,** 124
radicals generated during fracture, **16C,** 198–199
ultrasonic properties, **16C,** 72
unoriented, X-ray pattern, **16B,** 54, 59
X-ray fiber pattern, **16B,** 53, 55, 57

X-ray peaks, crystalline and amorphous, **16B,** 119

Polyphase rectifier, with induction load, **2A,** 202–207

Polypropylene, **14,** 413–414; **23C,** 316
- atactic, glass-transition temperature, effect of molecular weight, **16B,** 331
- density, **16C,** 72
- elastic constants, **16C,** 81, 82
- hard-elastic fiber, **16C,** 144
- intrinsic bifrefringence, **16C,** 162
- infrared spectra, crystalline and molten, **16A,** 141
- isotactic
  - crystallization behavior, **16B,** 364
  - DSC melting curve, **16B,** 305
  - effect of crystal form on melting, **16B,** 304
  - nitric acid degraded, molecular weight, **16B,** 263
  - photochlorination of, **16B,** 248, 249
  - repeat distance and chain conformation, **16B,** 80
  - solvent-etching of, **16B,** 240
  - typical crystallinities, **16B,** 125
  - unit cell and crystal structure, **16B,** 16–17
- methyl side group motion, **16A,** 310
- oxide, **23C,** 316
- positron lifetime in, **16A,** 395
- radicals generated during fracture, **16C,** 198–199
- small-angle X-ray pattern, **16C,** 182
- syndiotactic
  - repeat distance and chain conformation, **16B,** 81
  - single crystals, melting behavior, **16B,** 307
- tacticity by ESR spin probe, **16A,** 474
- temperature of maximum hypersonic loss, **16A,** 195
- ultrasonic properties, **16C,** 72, 74
- yield stress, **16C,** 131

Polystyrene, **23C,** 299, 308, 310
- atactic and isotactic, separation by ultracentrifugation, **16A,** 68
- Brillouin splitting near T_g, **16A,** 200
- bulk compressiblity, **16C,** 104, 105
- chain flexibility by NMR, **16A,** 305
- critical strain, in environmental crazing, **16C,** 257
- critical stress for crazing, **16C,** 251
- crosslinked gel, for GPC column, **16A,** 50
- density, **16C,** 72
- deuterated, conformation by SANS, **16A,** 499–501
- dipole moment, **16A,** 396
- elastic constants, **16C,** 81
- glass transition, in block copolymer, **16C,** 312
- glass-transition temperature, **16B,** 327
- high impact
  - crazing in, **16C,** 271–273
  - morphology, **16C,** 289, 292–293
- historic reference, **16A,** 1
- isotactic
  - crosslinked film, for electron microscope specimen support, **16B,** 213
  - crystallization behavior, **16B,** 364
  - DSC melting curve, **16B,** 304
  - single crystals, **16B,** 255
  - spherulitic growth rate, **16B,** 363
- labeled, fluorescent emission spectrum, **16A,** 426
- phenyl side group motion, **16A,** 310
- photochlorination of, **16B,** 248, 249
- positron lifetime in, **16A,** 395
- radicals generated during fracture, **16C,** 198–199
- random coil dimensions, **16A,** 4
- temperature of maximum hypersonic loss, **16A,** 195, 198, 199
- ultrasonic properties, **16C,** 72, 74
- unperturbed end-to-end distance, **16C,** 297
- work function, **16C,** 431

Polyterfluoroethylene
- carbon–13 NMR spectrum, **16A,** 360
- positron annhiliation in, **16A,** 379, 384, 395

Polytetrafluoroethylene
- chain conformation, **16B,** 20
- density, **16C,** 72
- phase transformations of the chain helix, **16B,** 114
- radicals generated during fracture, **16C,** 200–201
- thermally stimulated current spectrum, **16C,** 441
- ultrasonic properties, **16C,** 72
- work function, **16C,** 431
- X-ray fiber pattern and helical conformation, **16B,** 86–87

Polytetrahydrofuran, **23C,** 328

Polythioethylene, glass-transition temperature, **16B,** 327

Polyurethane
density, **16C,** 72
radicals generated during fracture, **16C,** 200–201
segmented block, **16C,** 299, 301, 302, 313
ultrasonic properties, **16C,** 72, 73
Polyvinylidene fluoride (PVF_2) piezoelectric sensor, **18,** 542
Polywater, **16A,** 11
Polzarization meters, **29C,** 357–358
Polzarization, electric, **16C,** 382
Ponderomotive effect, **16C,** 425
Pool-cathode tubes, **2A,** 180
Popcorn noise, **2B,** 467
POPOP, **5A,** 149
Population inversions, **29C,** 138
collisional scheme, **31,** 103
laser and, **15A,** 6–9, 20
Q-switching and, **15A,** 27
recombination scheme, **31,** 102–103
XUV lasers, **31,** 101–103
Population parameters, **28,** 158
Population-inversion mechanisms, **3A,** 334
Porcelain, *see* Ceramics
Pore pressure, **24B,** 450
Pore size, **14,** 237
Porod approximation, **23A,** 52; **23C,** 432
Porod characteristic function, **20,** 368
Porod's law, **23B,** 515
Poroelastic parameter, **24B,** 396
Porosity, **23C,** 441
metal, **14,** 579
Porphyrins, **20,** 150–154
Porro prisms, **25,** 64, 137
Porter-Thomas distribution, **5A,** 543–544
Posistors, **2A,** 49
Position angle, of polarization, **12A,** 363, 385, 411
Position measurements
lunar occultation in, **12C,** 99–100
VLBI techniques for, **12C,** 251, 261–276
Position-sensitive avalanche detector, **17,** 315–316
Position-sensitive pulse detector, **22,** 287
Position-sensitive solid-state detector, **17,** 86
Position-sensitive x-ray detectors, **20,** 350
Positioning, acoustic, **24B,** 110–120, *see also* Transponder; Pinger
Positive displacement flowmeter, **18,** 323–324
Positive ion emission, **30,** 142–149
Positive ion, electrostriction model of, **11,** 445–446
Positive logic, **2A,** 411
Positron annihilation-in-flight, **5B,** 238–239, 262, 485, 511–514
cross-section per Mev, **5B,** 512
with polarized electrons, **5B,** 262–264
Positron annihiliation; **21,** 6, 58
angular correlation analysis, **16A,** 392
data analysis, **16A,** 390–393
deconvolution of lifetime spectra, **16A,** 391
experimental, **16A,** 380–390
angular correlation measurements, **16A,** 383–385
specimen preparation, **16A,** 388–389
theoretical, **16A,** 377–380
Positron properties, table of, **4B,** 2
Positron-electron pair production, **5A,** 641–642
Positron-sensitive detector, for X rays, **16B,** 46, 155–156
Positronium, **4B,** 148–150; **21,** 78, 80
acceleration, **29A,** 61–62
annihilation by pick-off, **16A,** 376
lattice cell model, **16A,** 378
atom, **11,** 443–445
cooling methods, **29A,** 61
decay, **4A,** 389; **4B,** 148
fine structure, **4B,** 148–150
formation
activation energy, **29A,** 59, 61
electric field effect on, **16A,** 401–402
gas sources, **29A,** 56
Ore gap model, **16A,** 374
powder sources, **29A,** 57
probability as a function of temperature, **29A,** 60
spur model, **16A,** 376
temperature and mechanism, **29A,** 57–59
vacuum source, **29A,** 57–61
lifetime and collision, **29A,** 56–57
mean lifetime, **16A,** 376
modes of decay in polymers, **16A,** 398–401
ortho- and para-, **16A,** 374
para- and ortho-, **21,** 78
physical properties, **29A,** 55–56
pickoff, **21,** 81
work function, **29A,** 58
Zeeman effect, **4B,** 149–150

Positrons, *see also* Angular correlation; Bloch state of positron; Doppler broadening; Lifetime; Sources, positron; Trapping of positrons
β decay sources
debunching, **29A,** 55
emitter generation, **29A,** 40–41, 43–44
isotopes, **29A,** 40, 45
maximum energy, **29A,** 40, 42
mirror nuclei, **29A,** 40–41
principle, **29A,** 39–40
decay in magnetic fields, **5B,** 264
detection, **4A,** 389
electron momenta determined by, **21,** 79, 94
electron enhancement around the, **21,** 80
emission, **21,** 3
free lifetime in polymers, **16A,** 374
lifetime, sensitivity to morphology, **16A,** 397
lifetime measurements, **16A,** 380–383; **29A,** 44–45
spectrometer design, **16A,** 381
moderators
efficiency moderation, **29A,** 46–47
energy distribution from single-crystal negative-affinity moderator, **29A,** 47–48
geometry, **29A,** 49–50
negative affinity solids, **29A,** 47–49
physical properties, **29A,** 46
preparation, **29A,** 48–49
wide band gap solids, **29A,** 49
pair-production sources, **29A,** 45–46
penetration depth, **16A,** 373
physical properties, **29A,** 39
remoderation
large sources, **29A,** 52
microbeams, **29A,** 51–52
principle, **29A,** 51–52
ragging, **29A,** 55
range, **21,** 6
rate of energy loss, **16A,** 373
sources (emitters), **16A,** 388–389
sources, **4A,** 149–155
radioactive
backscattering effect of, **4A,** 154
Cu^{64}, **4A,** 152
effect of source thickness, **4A,** 152–153
fabrication of, **4A,** 152, 154–155
Na^{22}, **4A,** 152
reduction of positron energy, **4A,** 150
table of, **4A,** 151
use of electron accelerators, **4A,** 149
transport
electrostatic transport, **29A,** 51
imaging optics, **29A,** 50
magnetic transport, **29A,** 51
trapping
applications, **29A,** 53–54
gaseous moderators, **29A,** 54
harmonic potential bunching, **29A,** 55
radiofrequency bunching, **29A,** 54–55
resistive traps, **29A,** 54
Post ionization, **22,** 337, 399
Post-detector amplification, **10,** 225–228
Post-translation modifications, **30,** 435
Postacceleration detection of deep space plasma, **9B,** 72
Postacceleration stripping technique, **17,** 281–282
Postfoil beam velocity, **17,** 575–578, 593–595
Postfoil charge distribution, in beam-foil interactions, **17,** 551–552
Potash-soda-lead glass, **14,** 315
outgassing, **14,** 325
Potash-soda-varium, outgassing, **14,** 325
Potassium beam intensity distribution, **3,** 597, 598
Potassium chloride, **22,** 179, 180
Potassium chloride, F_2^+ centers in, **15B,** 21
Potassium chloride-lithium laser
construction and performance of, **15B,** 28–37
F_A (II) centers in, **15B,** 37
maximum efficiency of, **15B,** 34
Potassium dihydrogen phosphate (KDP), **12A,** 374
Potassium fluoride, F_2^+ center lasers in, **15B,** 45–48
Potassium iodide (Tl) scintillators, **5A,** 417
Potassium niobate ($KNbO_3$), as OPO material, **29C,** 257
Potential
additivity, **3,** 235
appearance, **3,** 526, 527–542
Buckingham, **23A,** 444
chemical, **1,** 319, 325–327
coefficient of, **1,** 435
distribution across cold-cathode arc, **2,** 88
distribution map, **1,** 445
electrostatic, **1,** 438–443
equalizers, **1,** 444
gravitational, **23A,** 328

interatomic, **23A,** 452
Lennard-Jones, **23A,** 444
measurement, **2,** 558–583, 649–650
neutron-nucleus scattering, **5A,** 318
nuclear at surface of nucleus, **5A,** 312–313
optical, **23A,** 263
model, **5A,** 310–313
proton-nucleus scattering, **5A,** 316
retarding, difference method, **3,** 533–535
of a sphere, **1,** 502
Potential energy
curve, **14,** 26
diagram, **14,** 26
of Ne-O system, **17,** 166
Potential function, double minimum, **3,** 69–71
Potential functions, least-squares fitting, **23A,** 497
Potential modulation, **22,** 203, 220
Potential probes, **1,** 443ff
measurements, **6B,** 34
Potentials, **13A,** 32
Potentiometer, **1,** 453ff
in analog computer, **2B,** 154, 161
dc voltage and, **2B,** 58–59
Diesselhorst, **1,** 246
measurements, **1,** 460
slide-back, **5B,** 486ff
Wenner, **1,** 512
White, **1,** 246
Potentiometric recording, **20,** 490–492
Potter-Bucky grid, **1,** 420
Potts model, **23B,** 34
Poulter window, **6A,** 431
Pound, **1,** 29
Pound weight, **1,** 30
Pound-Drever-Hall scheme, **29C,** 115, 116, 119, 124
Pound-Watkins oscillator, **3,** 513
Poundal, **1,** 30
Powder crystals, standard photographic methods, **6A,** 214–217
Powder diffraction, **23A,** 223, 474
Powder diffractometer, X-ray, **16B,** 61–63
Powder method, **3,** 309–311
X-ray, **1,** 423–424
Powder pattern (colloid) technique, in magnetic domain studies, **11,** 687–692, 695–697
Powder pattern, X-ray, **16B,** 51, 53
Powder spectra, EPR, **20,** 90–93
Powders, **23C,** 85
for phosphor preparation, **6B,** 296–297
resistivity measurements, **6B,** 65
Power, **1,** 93–97; *see also* Acoustooptic measurement of sound wave, power
acoustical, **1,** 202
angle dependence, **31,** 20
average, **1,** 467
density, spectral, **1,** 220
electrical, **1,** 231, 458ff
factor, **1,** 493
insertion device, **31,** 18
instantaneous, **1,** 467
integrated, of coherently scattered neutrons, **3,** 354
level(s), **1,** 202
high, dispersive mode at, **3,** 423
sound, **1,** 201
limitations, oscillator, **2,** 269
measurement, **1,** 458–459, 467–469, 486
output
microphone, **1,** 231
signal-to-noise, **3,** 440
photon beam, **5B,** 508
relationships, **2,** 21–22
resolving, **14,** 83
scattered sound, **1,** 212
synchrotron radiation, **29C,** 24–25
thermoelectric, **1,** 508, 515
total sound, **1,** 208
transformer, maximum, **2,** 28
undulator, **29C,** 29–30
Power absorption, **8,** 137
Power amplifier, **8,** 13, *see also* Voltage feedback power amplifier
Power amplifiers, **2A,** 320–324, *see also* Amplifiers
design considerations in, **2A,** 320–324
push-pull, **2A,** 321
SCR (silicon-controlled rectifier), **2A,** 323–324
Power broadening, **29B,** 154
Power capacity, *see* Peak power capacity
Power consumption, **8,** 186
Power Darlington amplifier transistors, **2A,** 79
Power density, **8,** 156
Power failure, **14,** 456
Power flow angle, **19,** 503, 505, 508
Power line frequency, **8,** 189
Power measurements
microwave, **2,** 737–740
in microwave circuits, **2B,** 293–296
Power radiancy, *see* Hemispherical power radiancy
Power reactor systems, **8,** 233

Power reflectivity, **10,** 86, 87, 95, 105–107, 112, 253–257
Power relationships, in network, **2A,** 7
Power response, *see* Flat radiated power response
Power spectral density, **29C,** 104–106
calculation, time series with missing data, **28,** 305–309
function, *see* Spectral density function
Power spectrum, **8,** 5
Power spectrum, computation of, **12B,** 284–285
Power spectrum measurement, fast Fourier transform, **29C,** 330
Power supplies
Dandl high voltage, **2,** 360, 361
regulated, **2A,** 242–243
Power throughput, **25,** 195
Power transfer, **2A,** 11
Power transformations, **28,** 268
Power transformers, in electronic circuits, **2A,** 39
Power transistors, **2A,** 78–79, *see also* Transistor(s)
Poynting vector, **1,** 476
Poynting's theorem, **2B,** 287–288
Pr, **23C,** 201
Practical significance, versus statistical significance, **28,** 181–182
Prandtl number, **1,** 137, 160, 300
definition, **18,** 459, 830
wind tunnel, **18,** 777, 844
Preamplifier(s), **12A,** 299; **12A,** 432
charge-sensitive, **2B,** 101
electronics, **27,** 60
noise sources in, **2B,** 103–106
pulse amplitude measurement and, **2B,** 101–106
in pulse spectrometer, **2B,** 101–103
Prebreakdown current, **22,** 369
Precession
of atomic beam, **5B,** 70–71
camera, **19,** 248–251
X-ray, **16B,** 60, 69
combined forced and free, **3,** 399–400
fields, **23A,** 217
method, **3,** 317–319ff
proton, **1,** 44, 45
VLBI measurement of, **12C,** 276
Precipitate aging, **23C,** 442
Precipitate phases, **23C,** 447
Precipitates and grain anisotropy, **19,** 253–256
Precipitation, **6A,** 22; **23B,** 111
electricity, **1,** 505
Precise optical wavelength measurement, **4B,** 96–99
Benoit's method, **4B,** 97–98
Precision, **1,** 2, 3, 8; **2,** 2; **3,** 710; **6A,** 2
vs accuracy, **5B,** 779–780
grinder, **6A,** 368–369
of measurement, **5B,** 808
Precombustor, for supersonic chemical lasers, **15B,** 102–103
Prediction intervals, **28,** 171–173
Predictive distribution, **28,** 413
Predissociation, infrared radiation and, **15B,** 295
Preferential ablation, **30,** 398–400
Preferential sputtering, **22,** 324, 333
Preferred orientation
calculation of, **19,** 252
in pyrolytic graphite, **19,** 292–293
scattering, effect on, **19,** 251, 275–277
and texture, **19,** 251–253, 293–294
in tool steel, **19,** 293–294
and velocity, **19,** 239, 268, 276–277, 293–294
in zinc bar, **19,** 275–277
Preionized discharge lasers, ultraviolet, **15B,** 90–93, *see also* Rare gas halide lasers
Preionizer, in rare gas halide lasers, **15B,** 135–136
Preparation of laboratory glass samples, **6A,** 139–147
Press-Ewing long-period seismometer, **24B,** 56–58
Pressure, **14,** 9, 39
blank-off, **1,** 172ff
cone, **1,** 153
critical, *see* Pressure, breakdown
diffusion, **1,** 141
in gas flow, **1,** 164
impact, **1,** 153
liquid, **1,** 117–118
normal atmospheric, **1,** 38
pore, **24B,** 396
relation to stress tensor, **18,** 501
shut-in, **24B,** 395
sound, **1,** 200ff, 208, 225
standard, **1,** 119
static, **1,** 131, 153
units, **18,** 508
vapor, **1,** 179
Pressure amplitude, **19,** 195

Pressure bar gage, **18,** 593
Pressure broadening, **10,** 345
  of mass peaks, **14,** 100
  of spectral lines, **3,** 677–678
Pressure concept
  extension by thermodynamics, **18,** 503
  kinetic theory, **18,** 502
  mechanical, **18,** 500, 801
Pressure contacts, **6B,** 50–51
Pressure distribution
  atmospheric, **12B,** 138–139
  ionospheric, **12B,** 138–139
Pressure estimates, **8,** 161
Pressure fluctuations in vacuum systems, **7A,** 74, 188–189, 274
Pressure gage
  bar gage, **18,** 593, 602
  Bourdon tube type, **18,** 511
  calibration, dynamic, **18,** 555
  calibration at gigapascal range, **18,** 592, 606
  calibration at kilopascal and megapascal range, **18,** 507, 512, 557
  calibration at 100 gigapascal, **18,** 610
  calibration below 10 pascals, **18,** 509, 515
  calibration standards, **18,** 504, 514
  capsule type, **18,** 511
  characterization, **18,** 527
  deformation type, **18,** 510
  diaphragm, **18,** 559–576
  diaphragm below 10 pascals, **18,** 515
  diaphragm types, **18,** 570
  dilatational gage, **18,** 604
  dynamic calibration, **18,** 555
  fast response, **18,** 576
  free surface sensor, **18,** 606
  frequency response function, **18,** 527
  hold time, **18,** 527, 531, 557
  holographic method of recording many diaphragm gages, **18,** 575
  Hopkinson bar, **18,** 596
  inductance sensor, **18,** 548
  McLeod, **18,** 509
  meaning, **18,** 504
  miniature bar gage, **18,** 602
  miniature capacitance, **18,** 571, 573, 575
  miniature probe, **18,** 592
  miniature stub, **18,** 588
  optical sensor, **18,** 549
  peak pressure, **18,** 535
  piezoelectric sensor
    on bar gage, **18,** 593, 603
    on stub gage, **18,** 588
  piston and cylinder, **18,** 512
  probe, **18,** 591
  range, **18,** 531, 557
  recording methods, **18,** 552–555
  reluctance sensor, **18,** 548, 571
  resonant period, **18,** 527, 531
  response, to step function, **18,** 527
  response characteristics of diaphragm gage, **18,** 567–570
  response time, **18,** 527, 531, 557
  secondary, **6A,** 415–417
  sensitivity, **18,** 531, 557
    diaphragm type, **18,** 561, 562, 566
  sensors, **18,** 534–552
  slab type, **18,** 588
  standards, **18,** 504
  static calibration, **18,** 504, 514
  steady or slowly varying pressure, **18,** 505–515
  stub type, **18,** 588
  theory
    of diaphragm gage, **18,** 559–570
    of fast response gage, **18,** 579–588
  thin polymer piezoelectric, **18,** 573
  types, **18,** 505, 526, 577
  use in free-flight model, **18,** 781
  use in non-Newtonian flow, **18,** 801
  U-tube manometer, **18,** 505
  wall taps, **18,** 516–518, 801
Pressure gauge, **14,** 40–75, *see also* Manometer; Thermal conductivity gauge; Ionization gauge; Pressure measurement
  location, **14,** 479
  using optical analysis, **14,** 510, 511
Pressure generation, **6A,** 407–414
Pressure gradient, **14,** 12, 13
  in turbo-pump, **14,** 250
Pressure in cloud chambers, **5A,** 197–198, 202
Pressure jump, **19,** 194
Pressure measurement, **6A,** 414–417; **7A,** 28, 51, 291, 390–396; **14,** 35–80
  above 100 kilopascals, **18,** 512
  below 10 pascals, **18,** 515
  errors, **4B,** 302–304; **14,** 46–47
  gauge calibration, **14,** 80
  gauge location, **14,** 54, 70, 75–77
  general, **18,** 499–610
  in high vacuum, **1,** 189–192
  in moving fluid, **18,** 515
  ranges, **14,** 37–38

Rayleigh scattering method, **14,** 37
static probe, **18,** 516, 518, 521
techniques, **14,** 37–38
Pressure measuring devices, in permeation and sorption studies, **16C,** 340–341
Pressure probe in moving fluid, **18,** 515
Pressure range, **14,** 3
diffusion pumps, **14,** 144, 149
of forevacuum pumps, **14,** 111–113
glass vacuum systems, **14,** 320
Pressure ratio, **14,** 110
diffusion pumps, **14,** 143, 144, 158, 159
Pressure recovery, wind tunnel, **18,** 758, 760
Pressure regime, **14,**12
Pressure requirement, 65
limitations, 65
Penning discharge, 73
Pressure scanning, **25,** 222
Pressure seals
Bridgeman-type cone seals, **24A,** 70
Porophyllite gaskets, **24A,** 248–249
Pressure standards
barium transitions, **24A,** 262–264
bismuth transitions, **24A,** 261–263
cesium transitions, **24A,** 262
fixed point, **24A,** 261–262
Hg, freezing point, **24A,** 261
manganin wire coil, **24A,** 74, 215–217, 263–264
ruby R-line fluorescence, **24A,** 262, 265
thallium transitions, **24A,** 262
Pressure system for 0–15000 bars, **6A,** 408–411
Pressure transducers, *see* Pressure gage, sensors
Pressure vessel, **24A,** 69–70
Pressure-time recording, **18,** 552
Pressure-time relation, **14,** 192
Pressurized tank accelerator, **17,** 7–8
Prevacancy effects, positrons, **21,** 84
Prewhitening, data, **28,** 305, 320, 325–328
Primary electron emission, **16C,** 437
Primary electrons, **22,** 132, 148, 174
Primary extinction, **21,** 147, 159
Primary Mahan cones, **22,** 163, 164, 166
Primary mechanism of ablation, **30,** 283
Primary nucleus, **16B,** 348
Primary photoelectrons, **22,** 177
Primary source standards, **31,** 122–125
Prime-focus feed, in paraboloidal reflectors, **12B,** 30–47
Primitive cell, **23A,** 370
Principal mode in waveguide, **10,** 113, 117, 118
Principal planes, **25,** 23, 25
Principal ray, definition, **31,** 151
Principal refracting plane, **25,** 25
Principal solution, **12C,** 2
Principle angle, **26,** 122
Principle of reciprocity, **9A,** 26
Principle surface, definition, **31,** 156
Print-out devices, **2,** 532–533
Printers, digital, **2,** 533–534
Priors
choice, consequences of, **28,** 407–408
conjugate, **28,** 410–411
improper, **28,** 409–410
maximum entropy, **28,** 411–413
noninformative, **28,** 408–409
Prism(s), **1,** 337–338; **13A,** 3, 7, 279; **13B,** 15; **25,** 64; **29B,** 370–372
astigmatism, **1,** 365
beam splitters, **26,** 138
constant deviation, **13A,** 282
Cornu, **1,** 365, 383; **13A,** 285
disperser radiometer, **26,** 260
Foucault, **1,** 380
goniometer, **25,** 188
instruments, **3,** 160–161
and lenses, **23A,** 265
liquid, **1,** 364
materials, **1,** 365, 366, 398
electronic spectra studies, **3,** 169–170
Nicol, **1,** 379, 380
polarizers, **26,** 139
resolving power, **1,** 363, 364
spectrographs, **13A,** 270
spectrometer(s), **5A,** 359–364; **13B,** 17; **25,** 206
Prismatic dislocation, **6A,** 328–329
Prismatoid theorem, **1,** 60
PrO_2, **23C,** 197
Probabilities, combined, **5B,** 780–781
Probability
additivity, mutually exclusive events, **28,** 10–11
of attachment per collision, measurement, **7B,** 56
definition, **28,** 4–7
classical, **28,** 6
frequency, **28,** 7
quantum, **28,** 5
experimental science, **28,** 1–4

function, **28,** 9
of internal conversion, **5A,** 583
of observed data, **28,** 213–214
of passage, **14,** 17–19, 21, *see also* Clausing factor
of sticking, **14,** 28, 29
subjective, **28,** 405
of transmission in turbo-pump, **14,** 250
Probability distribution, **28,** 492
functions, weighting schemes, **28,** 453
random variables, **28,** 13–18
Probability modeling, **28,** 1–33
modeling measurement, **28,** 27–33
random errors, **28,** 29–33
systematic errors, **28,** 28–29
Probability plot, **28,** 157–158
normal, **28,** 268–269
Probability theory, **28,** 4
elements, **28,** 8–27
central limit theorem, **28,** 26–27
compound events, **28,** 12–13
conditional probability and independence, **28,** 11–12
covariance, **28,** 22–25
expected values, **28,** 18–19
law of large numbers, **28,** 25–26
moment and mean and variance, **28,** 19–22
random variables and probability distributions, **28,** 13–18
sample space, **28,** 8–11
Probe(s)
deuterated polyethylene chain, **16A,** 504
electron beam gas density, **1,** 171
fluorescent, **16A,** 437
free molecule wire, **1,** 171
hot-wire electron emitting type, **9A,** 13
paramagnetic (spin), **16A,** 457–465
positron, **16A,** 372
Probe gauge for dynamic pressure measurement, **18,** 591
Probe hole, **22,** 359, 366, 414
Probe methods, **6B,** 54
for pressure measurement, **18,** 515–516, 518, 521, 591
for temperature measurement, **18,** 457–463
for velocity measurement, **18,** 240–341
Prober-layer technique, for x-ray yield measurements, **17,** 260–262
Process failure, **14,** 455
Processes, independent random, superposition of several, **5B,** 781–783
Processing of holograms, **26,** 204
Production of neutrons, **23A,** 104, 129
Production of ultrahigh vacuum, **14,**183
Profile likelihood, **28,** 224–225
Weibull distribution, **28,** 234
Profile likelihood functions, **28,** 178
limited failure population model, **28,** 236–238
Profile symmetry of x-ray reflection lines, **5A,** 604
Profile *t* function, **28,** 263
Projectile charge dependence, **17,** 297–302
target fluorescence yields, **17,** 290–297
target inner-shell cross sections, **17,** 284–289
Projectile ion, charge equilibrium in matter, **17,** 73–148
Projectile ionization, **17,** 113–114
Projectile velocity, and energy loss, **17,** 37–39
Projection lenses, **25,** 99
Projection machining, **30,** 192–195
Projection matrix, **28,** 255
Projection quantum number, **4B,** 110, 111
Projection system, in an electron microscope, **16B,** 211–212
Projection theorem, **19,** 566
Projectors, **25,** 143
Prony dynamometer, **1,** 94, 95
Propagating wave model, continuous wave, **19,** 112–116
attenuation, resonance width, **19,** 114
complex particle velocity, **19,** 113
phase velocity, **19,** 114
resonance condition, **19,** 114
wave superposition, **19,** 112
Propagation constant, **1,** 477
complex, of electromagnetic waves, **16C,** 409
Propagation of error, **28,** 228
Propagation of sound
adiabatic in non-viscous fluid, **3,** 682–685
non-adiabatic in viscous fluid, **3,** 685–688
Propagation of waves, *see* Wave propagation
Propagation, *see* Acoustooptic measurement of sound wave
Propeller anemometer, **18,** 254
Proper motion
VLBI measurement of, **12C,** 269
Properties of nitrogen, table of, **4B,** 261

Proportional baseband, in telemetering, **2B,** 238
Proportional counter, **17,** 86, 356–358, 374; **7A,** 104, 408, 468
  for X rays, **16B,** 44, 45–46, 155
Proportional counters, **5A,** 108–118, 471–473, 533–534, 573ff
Proportional flow counters, **21,** 34, 50
  alpha, beta particles, **21,** 60
  dead time, **21,** 68
Proportional limit, **1,** 107
Protective device, **14,** 449–456
  rotary oil sealed pumps, **14,** 128
Protein crystallography, **20,** 244–246; **23C,** 335
  data collection, **23C,** 339
  instrumentation, **23C,** 337
Protein Data Bank, **20,** 268
Protein ions, **30,** 417, 423–440
Protein multilayers, **22,** 417
Protein sequencing, **30,** 432
Protein solution, **19,** 233
Proteins, **23C,** 335
  dynamics, **23C,** 354
  fibrillous, **20,** 333–336
  solvents, **23C,** 351
Protheroe test, **28,** 358–359
Proton, **17,** 402–403
  antineutrino bombardment, **5A,** 685ff
  gap length distribution, **5A,** 256
  gyromagnetic ratio, **1,** 44
  gyromagnetic ration, **5B,** 122–123
  high energy, number of neutrons produced, **5A,** 527
  ionization loss in solids, **5A,** 47–48
  K-shell ionization, **17,** 229, 236–237
  linear accelerator, **5B,** 635–642
  L-subshell ionization, **17,** 242
  magnetic moment, **1,** 44–45
  from neutral-induced reactions, **5B,** 450–451
  passage through solids, **17,** 134–135
  precession, **1,** 44, 45
  range-energy relation, **5A,** 44, 45–51
    in aluminum, **5A,** 233–234
  recoil detectors, **5A,** 107–108
  resonance frequency, **1,** 523
  response
    of cesium iodide (Tl), **5A,** 415, 416
    of semiconductor detector, **5A,** 274–276
    of sodium iodide (Tl), **5A,** 412
  scattering, **5A,** 314–317
  synchrotron, **5B,** 656–663, 668–676
Proton charge, measurement, **4B,** 11
Proton exchange, **29C,** 386
Proton linac, **17,** 16–17
Proton stopping cross section, **17,** 28–30, 61
Proton whistler, **9B,** 120–131
Proton-nucleus scattering potential, **5A,** 316
Proton-transfer reaction, **19,** 193, 202, 222–229
Protoplasma, **30,** 157
Prototype spectrometer, **8,** 205–207
Proving ring, **1,** 67
PrSb, **23C,** 196
Ps, *see* Positronium
PSD, *see* Photon-stimulated desorption
Pseudo-Brewster angle polarizers, **31,** 244–245
Pseudo-equilibrium state, **1,** 305, 306
Pseudo-Rayleigh wave, **24B,** 586–588, *see also* Leaky mode; Stonely wave; Reverberant wave
Pseudogeometric factor, *see* Sonde, geometric factor
Pseudomagnetic moment, **23C,** 255
  table, **23C,** 266
Pseudomagnetic nuclear field, **23C,** 254
Pseudomagnetic precession, **23C,** 262
Pseudomagnetic resonance, **23C,** 257
Pseudonoise, as telemetering test, **2B,** 245
Pseudopotential theory of metals, **23A,** 381
Pseudoscopic configuration in holography, **26,** 187
Pseudosection, **24B,** 298, 309
Pseudostationary point, **24B,** 1
Pt(111)
  scanning tunneling microscopy of, **27,** 297
Pulfrich refractometer(s), **1,** 344–345; **25,** 188
Puli function, **26,** 68
Pulsar receivers, **12B,** 221; **12C,** 86
Pulsar signals, **12B,** 202
  dispersion measurements, **12C,** 87
  dispersion removal, **12C,** 84–87
  intensity variations of, **12C,** 80
  interferometer observations, **12C,** 90–91
  limitations to measurement of, **12C,** 83–84
  observing techniques for, **12C,** 78–91
  period measurement of, **12C,** 87–88
  periods of, **12C,** 78
  polarization measurements of, **12C,** 79, 88
  propagation effect with, **12C,** 80
  search techniques for, **12C,** 91
  spectral measurements of, **12C,** 89–90

spectral-line absorption of, **12C,** 90
spectra of, **12C,** 79
Pulse(s)
amplitude measurement, **2,** 583–624
multidimensional, **2,** 597–598
resolution, **2,** 599–601
baseline shift in, **2B,** 79
bipolar, **2B,** 98
burst of, **2B,** 77, 94
bursts of many, speed for, **2,** 602–604
characteristics of, **2B,** 78
circuits, thyratrons in, **2,** 305–307
data analysis, **2,** 605–606
defined, **2B,** 77
droop, **2,** 227
fast recovery in, **2B,** 79
lengthener, **2,** 347–351, 623
measurement
multichannel, **2,** 595–597
single channel, **2,** 593–595
single discriminator, **2,** 593
modulation, **2,** 265
random, recording speed for, **2,** 601–602
short-term recovery in, **2B,** 79
spaced in time by predetermined interval, **2,** 544–545
spectrometer
direct counting, **2,** 607–611
raw-data recording, **2,** 599–606
spectrum distortion, **2,** 598–599
straightforward curve plotting, **2,** 611–615
using computer techniques, **2,** 615–624
spectroscopy
basic problems, **2,** 584–592
counting losses, **2,** 588–590
data recording, **2,** 588
output, **2,** 591–592
temporary storage, **2,** 590–591
unipolar, **2B,** 99
Pulse 1315 nm iodine lasers, **15B,** 111
Pulse amplitude
amplitude channels in, **2B,** 81
comparison of, **2B,** 126
input variables in, **2B,** 80–81
performance limits in, **2B,** 81
variables in, **2B,** 80
Pulse amplitude measurements, **2B,** 77–132, *see also* Pulse amplitude, Pulse spectrometer
bipolar pulses in, **2B,** 98
buffer capacity losses in, **2B,** 94
bursts in, **2B,** 94
cathode ray tube in, **2B,** 119
centering error in, **2B,** 86
channel position errors in, **2B,** 101
channel width in, **2B,** 85–86, 90, 99–101
comparison of methods in, **2B,** 83–90
counting losses in, **2B,** 91
deadtime in, **2B,** 90
derandomizers in, **2B,** 95
differential discriminator in, **2B,** 85
digital data acquisition in, **2B,** 121–122
digital pulse analyzer techniques in, **2B,** 119–132
digital raw-data recording in, **2B,** 123
digital tape recording systems in, **2B,** 123
gating circuitry in, **2B,** 112–113
integral nonlinearity error in, **2B,** 101
jitter in, **2B,** 112
leading-edge timing in, **2B,** 112
limited buffer capacity in, **2B,** 94
linear gates in, **2B,** 113
main pulse amplifier and, **2B,** 106–107
multichannel measurements in, **2B,** 89–90
multichannel method in, **2B,** 85–99
noise filtering in, **2B,** 79
one-pulse buffer circuit in, **2B,** 93
pile-up in, **2B,** 90, 95–97
preamplifier configuration in, **2B,** 101–103
pulse spectrometer analog circuits in, **2B,** 101–114
single channel in, **2B,** 85–86
single discriminator in, **2B,** 83–84
sliding baseline in, **2B,** 120
sliding pulsers in, **2B,** 100
spectrum distribution due to peak pile-up in, **2B,** 95–97
spectrum shape in, **2B,** 90
statistical baseline fluctuations in, **2B,** 97–98
steady data flow in, **2B,** 94
systematic errors in, **2B,** 90–101
tail pile-up, **2B,** 97
temporary storage in, **2B,** 93–95
time interval inspector circuit in, **2B,** 97
unipolar pulses in, **2B,** 99
variable blocking time in, **2B,** 91
zero-crossover timing technique in, **2B,** 112
Pulse amplitude spectra, analog records and, **2B,** 81
Pulse amplitude time distribution in, **2B,** 80

Pulse amplitude time-division multiplex telemetering, **2B,** 220
Pulse analyzers, **5A,** 559–566
Pulse code modulation (PCM), **2B,** 220–221
carrier modulation for, **2B,** 231–233
pulse waveforms, **2B,** 228
PCM/FM radio telemetering, **2B,** 228, 233
PCM/FSM telemetering, **2B,** 232
PCM/PM telemetering, **2B,** 220, 233
block diagram for, **2B,** 232
carrier modulation for, **2B,** 231–232
receiver, block diagram for, **2B,** 225
telemetering system, **2B,** 222–225
block diagram of, **2B,** 223
Pulse counting systems, **21,** 55–59, *see also* Angular correlation, electronics; Doppler broadening; Lifetime, electronics
amplifiers, preamplifiers, **21,** 55
analog to digital converters, **21,** 55, 56
discriminators, scalers, times, **21,** 55
drift, **21,** 66
high voltage supply, **21,** 55
liquid scintillators, **21,** 52
multichannel analyzers, **21,** 56–57, 62
pulse height analyzers, **21,** 52, 55
single-channel analyzers, **21,** 55, 57
Pulse delay method, **6B,** 109
Pulse delivery, from stored-energy source, **2A,** 238–240
Pulse discharges, **3A,** 236
Pulse duration, **22,** 405
of rf accelerator, **5B,** 329
Pulse excitation in lifetime measurement, **6B,** 101–104
Pulse formation, **5A,** 95–99
Pulse generation techniques, in picosecond spectroscopy, **15B,** 198–201
Pulse generator, **3,** 435
Pulse height information, **8,** 259
Pulse height, **5A,** 102
analyzer, **17,** 181
distribution, **5A,** 658–659
gamma rays in NaI (Tl) crystals, **5A,** 623ff
helium–3 spectrometer, **5A,** 575
maximum, **5A,** 129
from organic scintillators, **5A,** 421–422, 423
semiconductor detector, **5A,** 268, 269
spectrum, **17,** 181–182
Pulse interferometer methods, **19,** 99–102
Pulse ion lasers, operating pressure for, **15A,** 89, *see also* Laser(s)
Pulse ionization chamber, **5A,** 89–91, 107–109
Pulse light, very short time, **5A,** 192
Pulse measurements of magnetic losses, **6B,** 223
Pulse methods in resistance measurement, **6B,** 46, 55–56
Pulse methods in superfluid helium, **19,** 447–450
comparator techniques, **19,** 449
Pulse pile up, **21,** 225
Pulse pileup, **8,** 215
Pulse reactors, **23A,** 121, 124
Pulse recording devices, **2B,** 114–115
Pulse repetition rate, **8,** 123
Pulse reverse method of lifetime measurement, **6B,** 107–109
Pulse shape
epithermal, **23A,** 159
functions, **23A,** 178
low-energy, **23A,** 164
Pulse shaping
optimum, **2B,** 108
terminology of, **2B,** 78–83
Pulse spectrometer, *see also* Pulse amplitude measurement, Pulse spectrometry
analog circuits in, **2B,** 101
analog-digital converter in, **2B,** 120–122
analog recording type, **2B,** 114–119
analog storage in, **2B,** 115, 118–119
associative memories in, **2B,** 127–128
block diagram of, **2B,** 83, 145
circulating-type dynamic memories in, **2B,** 125–127
cathode-ray tube displays in, **2B,** 131–132
as computer, **2B,** 128
conversion capacitance in, **2B,** 102
delay-line clipper in, **2B,** 109
delay-lines in, **2B,** 126–127
discrete channel count storage in, **2B,** 115
electronic wedge in, **2B,** 115–118
electrostatic storage type, **2B,** 124–125
fast clipping circuits in, **2B,** 107
field-effect transistors in, **2B,** 105
gating and, **2B,** 112–113
gray-wedge technique in, **2B,** 115–118
individual-pulse recording in, **2B,** 114–115
instrumentation tape decks in, **2B,** 123
magnetic cores in, **2B,** 124
magnetic film in, **2B,** 124
magnetic tape in, **2B,** 114

main pulse amplifier and, **2B,** 106–107
mechanical curve plotting in, **2B,** 115
MOS devices in, **2B,** 125
noise in preamplifiers of, **2B,** 103–106
nuclear magnetic resonance and, **2B,** 145
on-line computer and, **2B,** 128–129
optical filter in, **2B,** 117
output risetime in, **2B,** 103
passive memory as, **2B,** 122
pole-zero cancellation in, **2B,** 110–112
preamplifier configuration in, **2B,** 101–103
pulse shaping in, **2B,** 107–109
pulse stretchers in, **2B,** 113
random access in, **2B,** 124
recovery time in, **2B,** 102
Sabatier effect in, **2B,** 116
static memory array in, **2B,** 123–25
time-of-flight in, **2B,** 126
timing and gating circuitry in, **2B,** 112–114
tube amplifiers in, **2B,** 105
two-dimensional analog storage in, **2B,** 118–119
Pulse spectrometry, **2B,** 81–83, *see also* Pulse amplitude measurement, Pulse spectrometer
basic problems of, **2B,** 78–83
data display in, **2B,** 129–132
Pulse stretchers, **2A,** 389–395; **2B,** 113–114
feedback and, **2A,** 393
Pulse superposition method
cycle-for-cycle matching of pulses, **19,** 95–97
in ultrasonic velocity measurements, **19,** 94–97
Pulse systems, **19,** 85–111
echo-overlap method, **19,** 97–99
gated double pulse superposition method, **19,** 90–94
long pulse buffer rod method, **19,** 102–105
pulse-echo method, **19,** 85–89, 134
pulse interferometer methods, **19,** 99–102
pulse superposition method, **19,** 94–97
sing-around method, **19,** 89–90
spectrum analysis, **19,** 109–111
Pulse tails, **2B,** 79
Pulse time, Čerenkov radiation, **5A,** 192, 454
Pulse transformers, **2A,** 39
transistor drive for, **2A,** 404–405
Pulse variations, within train, **15B,** 187
Pulse-counting mode, **22,** 474
Pulse-echo method, basic
in ultrasonic attenuation measurements, **19,** 107–109
in ultrasonic velocity measurements, **19,** 85–89
Pulse-echo response, **19,** 31–32
Pulse-echo techniques,m in ultrasonic studies, **11,** 371–372, 399–400
Pulse-generator circuit, basic, **5B,** 498
Pulse-height analysis, **7A,** 105
Pulse-height analyzer(s), **1,** 474; **22,** 477, 478
Pulse-height distributions, **8,** 188
Pulse-height spectrum, **8,** 188
Pulse-pair resolution, **2B,** 1
Pulse-shaping network, **2B,** 107
Pulsed accelerator, **8,** 260
Pulsed Doppler ultrasonic velocity meter, **18,** 317–318
Pulsed electrical field method, **19,** 194, 229
Pulsed extraction ion source, **30,** 430
Pulsed field method
de Haas-Van Alphen effect in, **11,** 34
in magnetoresistance measurements, **11,** 25–26
Pulsed laser(s), **29C,** 137–138; **31,** 101
builder's guide, **29C,** 153–169
buyer's guide, **29C,** 150–152
cavity mode, **29C,** 144–146
dye lasers, **29C,** 146
error sources, **29C,** 198
gas lasers, **29C,** 147–148, 151
high-repetition-rate lasers, **29C,** 146–147, 150
injection seeding, **29C,** 144–146
Nd:YAG laser, **29C,** 143–144, 150, 151, 172
plasma production using, **29C,** 1–18
Q-switching, **29C,** 139, 141–143, 146
ruby laser, **29C,** 139, 140
safety, **29C,** 157–158
semiconductor lasers, **29C,** 149–150
theory, **29C,** 138–139
tunable solid-state lasers, **29C,** 148–149
Pulsed laser deposition (PLD), **30,** 3, 189, 399
apparatus, **30,** 477–478
characteristics, **30,** 478–480
from liquid target, **30,** 511–512
optimum conditions for, **30,** 554–556
scaleup in, **30,** 514–517
stoichiometry in, **30,** 518
target fragmentation in, **30,** 503–504
target preparation for, **30,** 506–508, 518

Pulsed magnets, in magnetoresistance measurements, **11,** 20
Pulsed microwave power, measurement, **8,** 123–127
  exercises, **8,** 126–127
Pulsed NMR, *see* Nuclear magnetic resolution spectroscopy
Pulsed noble-gas ion lasers, *see also* Noble-gas ion lasers
  characteristics of, **15A,** 74–76
  discovery and disappearance of, **15A,** 95
  longitudinal dc electric field in, **15A,** 80–81
  oscillation of, **15A,** 75
Pulsed systems, metal atoms or free radical production in, **15B,** 135–136
Pulsed-laser atom-probe, **22,** 406
Pulsers, precision, **2,** 655–658
Pulsewidth
  autocorrelator, **29C,** 187–188
  fiber laser, **29C,** 184, 187–188
Pump(s)
  booster, **1,** 176
  diffusion, **1,** 176–180
  evapor-ion, **1,** 182
  fore-vacuum, **1,** 174; **14,** 111–140
  fractionating, **1,** 178
  getter, **1,** 182
  ion, **1,** 180–182
  mechanical vacuum, **1,** 172–176
  molecular, **1,** 172, 176
  oil jet ejector, **1,** 168, 181
  pressure ranges, **14,** 108–109, 111–114
  pressure ratio, **14,** 110
  Roots, **1,** 172, 175–176
  rotary-vane oil-sealed, **1,** 173–174
  roughing, **1,** 174
  steam jet ejector, **1,** 176, 181
  vapor, **1,** 176–180
Pump capacity, **14,** 246
Pump laser, **29C,** 151, 259–260
Pumpdown curves, **14,** 506, 508–509
Pumpdown time, **14,** 104–107
  with condensable species, **14,** 107
  constant speed, **14,** 105
  constant throughput, **14,** 106
  with leak, **14,** 106
  with outgassing, **14,** 106
Pumpdown, **14,** 192, 240
Pumping
  differential, **1,** 189
  optical, **3,** 625
  Q-switching and, **15A,** 27
Pumping equation, **14,** 505
Pumping speed, **4B,** 265; **14,** 12–14, 102–107, 190, 232, 276, *see also* Molecular transport
  diffusing pump, **14,** 145–151, 173–174
  effect of traps, **14,** 145, 167, 168
  efficiency, **14,** 145, 147, 150, 173, 174
  orifice geometry, **14,** 103–104
  pipe flow geometry, **14,** 103–104
  of rotary oil sealed pumps, **14,** 117–118
  standards, **14,** 104, 145, 147
  in turbo-pump, **14,** 253–256
Pumping systems, **31,** 65
Pumpout, **14,** 284
Punches
  card, **2,** 534
    paper tape, **2,** 534
Pupil, **1,** 333. 334
  consideration in combining lenses, **29A,** 202–203
  definition of ion beam, **29A,** 192
Pupillary plane, **8,** 41
Pupils, **25,** 37; **31,** 150–151
Purcell absorption method, **3,** 390–394
Purely chemical electron transition laser, **15B,** 111–115
Purely chemical HCl and HCl-CO_2 lasers, **15B,** 114–115
Purely chemical laser
  new, **15B,** 111–116
  at visible wavelengths, **15B,** 115–116
Purely cw electronic transition chemical laser, **15B,** 113
Purging, nitrogen, **14,** 422
Purification
  by freezing methods, **6A,** 48–55
  general methods, **6A,** 21
Purification of materials, case histories
    compounds, **6A,** 28
  metals, **6A,** 27
  semiconductors, **6A,** 28
Purity of scintillator gas, **5A,** 432–433
Push-pull power amplifier, **2A,** 321–322
PUT (programmable unijunction transistor), **2A,** 353
PWBA, *see* Plane-wave Born approximation
Pyknometer, **1,** 118–119
Pyrex®, **29C,** 343–345
Pyroelectric detectors, **26,** 299; *see also* Detectors

Pyroelectric effects, lithium niobate, **29C,** 391–392
Pyroelectric materials, **26,** 301
Pyroelectric temperature sensor, **18,** 685
Pyroheliometers, **26,** 246
Pyrolysis of hydrocarbons, **3,** 571–572
Pyrolysis, **30,** 351–352
Pyrometers, **1,** 252–255
  Féry, **1,** 253
  optical, **1,** 254, 257
  radiation, **1,** 252, 254
Pyrometry, **30,** 206
PYS, *see* Partial-yield spectroscopy
PZR, *see* Piezoelectric transducer
PZT, *see* Lead zirconate-titanate

Q

Q branch of vibration-rotation spectrum, **10,** 327, 328
Q factor, **3,** 468, 473
Q machines, **9A,** 345
  in measurement of acoustic waves, **9A,** 15–16
Q meter, **1,** 465–466, 498
Q Regime, for small-angle neutron scattering, **16A,** 486
Q value(s), **2,** 242ff
  coupling networks and, **11,** 582–583
  for mesons and hyperons, **5B,** 846
  negatron decay, **5B,** 351
  nuclear reaction, **5B,** 9ff, 339–352, 354–357
  positron decay, **5B,** 351–352
  for real resonator, **11,**b 549
  for superconducting microwave resonators, **11,** 542–543
  of tune power amplifier, **2,** 258–259
Q values, *see* Nuclear Q values
Q, **19,** 21, 31–34, 48–51, 53–65
  of microwave cavity, determination of, **7B,** 94
  of a resonator, **10,** 85–87
  of tuned LC circuit, **3,** 391
Q-equation, **5B,** 341, 343
  inelastic scattering, **5B,** 350
  relativistic, **5B,** 344
Q-meter method of resistance determination, **6B,** 47–48
Q-resolution, **8,** 265
Q-surface representation of Thompson, **5B,** 346
Q-switched laser, **18,** 717–718
Q-switching, **3A,** 336; **29C,** 139, 141–143, 146
Q-switching, for laser pulses, **15A,** 27–28
Q-value reactions, kinematics, **8,** 267
QED, *see* Quantum-electrodynamical effects
QMS, *see* Quadrupole mass spectrometer
QPM, *see* Quasi-phase-matching
Quadrant detectors, **12A,** 215
Quadrant diode, as short-pulse detector, **29C,** 199
Quadrants, of X-ray fiber pattern, **16B,** 55
Quadratic approximation, **28,** 228
Quadratic chirp, **29C,** 195
Quadrature spectrum, **28,** 341
Quadrispherical curved-plate analyzer, **9B,** 64
Quadrupolar interactions, **20,** 22, 23
Quadrupolar relaxation, **21,** 449
Quadrupole, **14,** 86
Quadrupole doublet lens, **17,** 464–465
Quadrupole interaction, **21,** 429, 432
Quadrupole mass analyzer, **22,** 474
Quadrupole mass filter, **8,** 169; **22,** 328, 340; *see also* Mass spectrometer
Quadrupole mass spectrometer, **22,** 444, 487, 493
Quadrupole moment, **21,** 395
Quadrupole moment of deuteron, **3B,** 794
Quadrupole moment of Na, **4B,** 180–182
Quadrupoles, *see* Multipoles
Quality factor, **1,** 465; **8,** 2; *see also Q*
Quanta, **8,** 191, 204, *see also* Annihilation quanta; Back-scattered quanta; Compton scattered quanta; Gamma quanta; Incident quanta
Quanta, equivalent, **5B,** 508
Quantameter, **5B,** 544–546
Quantile-quantile plots, **28,** 192–200, 510–511
  completely specified distribution assessing fit, **28,** 192–195
  mechanical device failure times, **28,** 192–195
  parametric families, assessing fit, **28,** 196–197
  for randomly censored data, **28,** 205–206
  with simultaneous confidence bands, **28,** 197–200
  two-sample, **28,** 207–208
Quantitation, **30,** 397–400
Quantitative Auger analysis, **22,** 231

Quantitative energy surfaces, **15B,** 273
Quantity of information, *see* Information quantity
Quantity of information, *see* Information quantity
Quantity of matter, **1,** 325–327
Quantization of lattice vibrations, **10,** 367
Quantized vortex lines
  capture of ions by, **11,** 470–474
  ion escape from, **11,** 474–476
Quantum beat phenomena, **17,** 590–595
  beam-foil research, **17,** 530, 537
Quantum beats, **13B,** 244, 245, 249, 252
  classical picture, **29B,** 325–326
  in lifetime measurements, **29B,** 231–232, 234–235
  quantum description, **29B,** 327–328
  measurement, **29B,** 328–330
  for wave packets, **29B,** 330–332
Quantum conditions, **13A,** 65
Quantum crystals, **23A,** 397
Quantum detector, **26,** 303
Quantum detectors, **13B,** 6
Quantum efficiency, **5A,** 125ff, 134; **6B,** 306–307; **22,** 209
  of infrared detectors, **12A,** 418–419, 425
Quantum electronics, nonlinear optics in, **15B,** 249
Quantum interference
  acceleration induced, **23A,** 293
  gravitation induced, **23A,** 289
Quantum liquids, ions in study of, **11,** 443–484
Quantum number
  m, magnetic, **5B,** 147
  vibrational, **3,** 43–44
  "*z* component," **5B,** 353
Quantum probability, **28,** 5
Quantum theory of e-m fields, **13A,** 64
Quantum yield, **1,** 397
  absolute, **11,** 81
  in photoemission measurements, **11,** 81–84
  in photon emission, **16A,** 407
Quantum-electrodynamical effects, **17,** 587–588
Quarter wave matching, *see* Impedance matching
Quarter wave transformer, **8,** 120
Quarter wave transformer, **8,** 120
Quarter-square multiplier, **2B,** 158–159
Quarter-wave isolator, **29C,** 358–359
Quarter-wave plate, **1,** 380ff; **12A,** 373
  use with horn feeds, **12B,** 67
Quarter-wave stack, **26,** 35
Quarter-wavelength line, **8,** 119
Quarter-wavelength line, **8,** 119
Quarter-waveplate, **29C,** 354
Quartz, **10,** 94, 96, 99; **12A,** 374, 379; **14,** 386, 387; **19,** 30–32, 51–60; **25,** 36, *see also* Glass; Silica
Quartz crystal microbalance, **14,** 542–547
  errors, **14,** 544
Quartz fiber microbalance, **14,** 542, 543
Quartz piezoelectric pressure gage sensor, **18,** 542
Quartz spring, in sorption measurements, **16C,** 351
Quartz transducers, **11,** 392; *see also* Transducers, quartz
Quartz, transmission, 310–312
Quartz-crystal clocks, **1,** 70
Quartz-crystal oscillator,in thin-film measurements, **11,** 654–655
Quasi-adiabatic collision, **17,** 151–152
Quasi-atomic behavior, **22,** 231
Quasi-continuum, in multiphoton dissociation of molecules, **15B,** 267
Quasi-elastic approximation, **23C,** 15
Quasi-elastic scattering, **23A,** 42, 511
  coherent, **23B,** 151, 224
  instruments, **23A,** 236
Quasi-electric scattering, **23B,** 120, 132, 223
  incoherent, **23B,** 229
Quasi-harmonic model, **23A,** 473
Quasi-hydrostatic arrangement for studying resistance, **6A,** 427
Quasi-ideal detector, **2B,** 490
Quasi-molecular treatment, *see* Molecular orbital excitation
Quasi-optical components, **10,** 112, 113
Quasi-particle steps, Josephson effect and, **11,** 222–225
Quasi-periodic oscillator, **28,** 368
Quasi-phase-matching (QPM), **29C,** 233, 234, 243–247, 390–391
Quasi-resonance neutralization, **22,** 304
Quasi-viscous flow, **1,** 109, 110
Quasielastic scattering, **21,** 201, 210, 219
  time-of-flight measurements, **21,** 203, 213–214
Quasimolecular transitions in PIXE background, **21,** 256
Quasimolecular x- rays, **13A,** 182

Quasistatic methods for anelasticity, **6A,** 304
Quaternions, **23A,** 310, 504
Quenched surface states, **22,** 154
Quenching
  external, **3,** 515
  gas pressure, in rare gas halide lasers, **15B,** 87
  insensitivity to Raman scattering, **18,** 413
  of molecular excited states, **16A,** 408
  role in electron beam fluorescence, **18,** 438, 447–448
  in superregenerative oscillators, **2B,** 140
Quenching of metastable atoms
  electrostatic, **4B,** 134, 136–137, 144
  magnetic, **4B,** 136, 165–166
  by motional electric field, **4B,** 137
  radio-frequency, **4B,** 133–134, 137–138, 145, 167
Quinary divider, **2B,** 20
Quinquefilar pendulum magnetometer, **6B,** 204

R

R branch of vibration-rotation spectrum, **10,** 327, 328
R branch, **3,** 144
R factor in XRDA, **20,** 264
R6G laser, *see* Rhodamine laser
Ra-α-Be neutron source, **5B,** 558–560
Rabi oscillations, **29B,** 132, 156
Rad Lab coupling (fitting), **14,** 295, 296
Radar, **1,** 55; **2B,** 251
  beacon formula, **2,** 702
  cathode ray tube, **2,** 769
  cathode-ray tube fore, **2B,** 343–344
  CW, **1,** 77
  technology, **11,** 371–372
  traffic control, **1,** 77
Radechon, **2,** 783
Radechon storage tube, **2B,** 355–356
Radial angular aberration, **8,** 184
Radial distribution function, **23B,** 256
Radial field components, **8,** 74
Radial parameter, **8,** 90
Radial shear interferometer, **26,** 18
Radial temperature distribution, calculation, **8,** 247–248
Radian, **1,** 56
Radiance temperature, **26,** 241
Radiance, **12A,** 193
Radiancy expressions, **8,** 30
Radiant energy measurement, **8,** 27
Radiant energy of light sources, **18,** 689, 706–707, 718–719
Radiant power, **12A,** 193
Radiation, **1,** 134; *see* Synchrotron radiation. *See also* Blackbody radiation
  attenuation, **1,** 316
  beta, angular distribution, **5A,** 60
  Čerenkov, energy loss due to, **5A,** 8–10
  curve, slope, **8,** 33
  electric dipole, **4B,** 215
  by electrons, **5A,** 56–61
  field energy density **3,** 453
  impedance, **1,** 203ff
  incident,
    density of, **3,** 242
    fraction radiated, **3,** 243
  infrared, **1,** 398–402
  intensity, Čerenkov counter, **5A,** 163ff
  length, **5A,** 59
  pattern, **8,** 118
  pyrometers, **1,** 253
  resonance, **1,** 396
  scattered, **1,** 393
  scattering per second, **3,** 117
  of sound, **1,** 202–208
  temperature, **1,** 255; **26,** 239
  thermometer, **1,** 252–255
  ultraviolet, **1,** 402–405
Radiation assay technique for diffusion study, **6A,** 369–371
Radiation boundary condition, **18,** 665, 667, 669
Radiation chopping, **13B,** 7
Radiation constants, blackbody, **18,** 466, 468
Radiation conversion and detection by photovoltaic effect, **6B,** 372–373
Radiation damage, **6A,** 357–363; **22,** 177, 333; **23A,** 340
  parameters of, **6A,** 358
Radiation damage to dielectric material, 314–317

Radiation damage to dielectric material, 314–317

Radiation detectors, **18,** 407
Radiation force, **19,** 303–312
  and acoustic drying, **19,** 345
  and aggregation of particles (cells), **19,** 341–342
  Langevin, **19,** 303–306, 310
  Rayleigh, **19,** 303–305
Radiation hazard, **14,** 454

Radiation hazards, **21,** 27–33
  dose rate effect, **21,** 31
  dose, equivalent, **21,** 28
  dose, response, threshold, **21,** 28
  exposure, **21,** 28, 31
  lethal dose, **21,** 29, 30
  maximum body burden, **21,** 32
Radiation noise, *see* Noise, radiation
Radiation Pattern, paraboloidal reflector, **12B,** 34–38
Radiation pattern. *See* Dipole pattern
Radiation patterns, of semiconductor diode lasers, **15A,** 237–242
Radiation pressure, **19,** 363, 380–381, 405
  and bioeffects, **19,** 408
Radiation pressure technique, **19,** 133
  attenuation measurement by, **19,** 133
Radiation protection, **21,** 33–39
  fume hoods and glove boxes, **21,** 33
  personnel monitoring, **21,** 35
  radioactive waste, decontamination, **21,** 36
  shielding, **21,** 36
  spills, **21,** 38
  standards, licensing requirements, **21,** 29, 40
  survey instrumentation, **21,** 33
Radiation protection, beamline design, **31,** 36
Radiation source, *see* Light source
Radiation standards, **4A,** 295–297, 312–314
Radiation theory, **13A,** 31
Radiation tolerance of transistors, **2,** 159
Radiation trapping, **29B,** 161, 230
  in noble-gas ion lasers, **15A,** 87–90
Radiation unit in pure emulsion stack, **5A,** 680
Radiation units
  activity (becquerel, curie), **21,** 9
  dose (rad, gray), **21,** 29
  dose equivalent (rem, sievert), **21,** 28
  exposure (roentgen), **21,** 28
Radiation-induced cavitation, **19,** 366–368
Radiation-induced chemical changes in polymers, **16B,** 278–280
  location of, **16B,** 281–286
Radiative attachment, **7A,** 205–208
Radiative Auger effect, **17,** 401
Radiative beam-foil measurements, **17,** 532
Radiative capture, **23A,** 532
  table, **23A,** 548
Radiative decay, **22,** 495
Radiative electron capture, **13A,** 186; **17,** 94, 136, 144–148
  cross section, **17,** 146–148
Radiative electron rearrangement, **17,** 400–401
Radiative heating, study in hypersonic tunnel, shock tube, **18,** 679, 683–685, 762, 791–792
Radiative lifetimes, measurement techniques
  beam-foil spectroscopy, **29B,** 232–235
  fast beam methods, **29B,** 235–239, 312
  general approach, **29B,** 219–220
  ion traps, **29B,** 239–243
  laser-induced fluorescence, **29B,** 220–232
Radiative loss, temperature sensor, **18,** 461
Radiative recombination, *see* Radiative electron capture
Radiative thermal conductivity
  calculation, **24A,** 297–299
  measurement, **24A,** 293–297
  method of J. F. Schatz, **24A,** 299–301
Radiative transfer, **9A,** 143, 171–172; **26,** 223
  equation for, **12B,** 142–145
  in transparent media, **26,** 231
  in a vacuum, **26,** 228
Radiative transition(s), **13A,** 31
  allowed, **17,** 579–583
  in doubly-excited state, **17,** 561–564
  forbidden, **17,** 583–586
  measurement of excited states, **17,** 88–89
  probabilities, **29B,** 218
Radiator(s), **1,** 483
  Čerenkov counter, **5A,** 168–170
  ionization loss in, **5A,** 178
Radical
  generation, effect of temperature, **16C,** 210–212
  observation by ESR, **16C,** 191–194
  production from mechanical fracture, **16C,** 189
  reaction of, **16C,** 197, 202
  reactivity, **16C,** 205
  species, various polymers, **16C,** 195–201
  stable, **16C,** 188
Radical concentration
  determined by ESR, **16C,** 187–188
  of various polymers, **16C,** 206–207
Radicals, *see* Free radicals
  in flames, **7B,** 196–198
Radio frequency spectroscopy, **29B,** 309–312
Radio propagation equation, **2,** 702

Radio telescopes, comparison with optical, **12B,** 8–9
Radio wave scattering, in ionosphere, **9B,** 139–186
Radio-frequency power, **8,** 137
Radio-frequency spectroscopy
apparatus, **4B,** 127–131, 183–184
"cross beam" detection, **4B,** 180, 183–185
lock-in amplifier, detector, **4B,** 129–131, 141–142, 144–145
magnets, **4B,** 128–129
history, **4B,** 106–107
Radio-frequency transition, **4B,** 245
Radioactive atomic beam, **4B,** 152, 169–170
Radioactive decay, **21,** 1–10
chains, **21,** 9
constants, **21,** 9, 10, 68, 71
Radioactive decay rate, **8,** 210
Radioactive detection of atomic beams, **4A,** 410–412
Radioactive ionization gauge, *see* Alphatron
Radioactive isotope, **8,** 236
Radioactive isotopes, production of, *see also* Atomic beams of radioisotopes
by neutron bombardment, **4A,** 199–202
by proton, deuteron, and alpha particle bombardment, **4A,** 202–206
range-energy curves, **4A,** 206, 256
sample chemistry cave, **4A,** 215
sample holders, **4A,** 209–215
Radioactive source strength, **8,** 207
Radioactive sources, *see also* Sources, positron
diffusion, thin layer, **21,** 40
for Mössbauer experiments, **11,** 350–351
packaging, **21,** 38
preparation, **21,** 40, 41–43
weightless, **21,** 3, 60
Radioactive substances, **8,** 177
Radioactivity, alpha particle, **5A,** 313–314
Radioactivity, measurement, **24B,** 213
Radiochemical separations, **8,** 207
Radiofrequency discharge ion source
discharge types, **29A,** 116–119
lifetime, **29A,** 117
mechanism of discharge, **29A,** 117–118
types
capacitatively coupled source, **29A,** 119–120
high-frequency source, **29A,** 119–121
multicusp-field source, **29A,** 121–122
proton beam source, **29A,** 120–121
waveguide sources, **29A,** 123
Radiofrequency linear accelerator
applications, **29A,** 314–315
principle, **29A,** 314
Radiofrequency methods of resistance determination, **6B,** 47–49
Radiofrequency region, **13A,** 12; **13B,** 134
astronomy, **13B,** 132
capacitance cells, **13B,** 136
time-domain techniques, **13B,** 163
Radiofrequency spectrometers, **3B,** 511
Radiography, **1,** 409, 419–421; **18,** 407; **23C,** 421
dynamic, **23C,** 421
resonance, **23C,** 424
stroboscopic, **23C,** 422
subtraction, **23C,** 421
transmission, **23C,** 421
Radioisotopes
carrier free, **21,** 23
commercially available, **21,** 11, 12–14, 26
impurities, **21,** 11
production, **21,** 11–26
as thermometers, **11,** 525
Radiometer(s), **12B,** 201–224; **26,** 246; *see also specific types*
astronomical, **26,** 246
designs, considerations in, **12B,** 223
filter, **26,** 257
imaging, **26,** 252
interferometric, **26,** 261
prism disperser, **26,** 260, 261
scanning, **26,** 252
space defense, **26,** 250
standard, **26,** 246
Radiometric instruments, **26,** 246
Radiometric measurements, **12B,** 1–6; **26,** 263
Radiometric output, stabilization, **31,** 48–49
Radiometric properties, determining, **31,** 119
Radiometric standards, **26,** 264
Radiometric temperature measurements, **26,** 239
Radiometry, **26,** 213
of visible light, **26,** 284
Radionuclei, short-lived, investigation by atomic beams, **5B,** 58–85
Radiorefractivity, atmospheric, **12B,** 186–190
Radiosonde data, **12B,** 141

Radius
collision, **5A,** 290
of confusion, of a point image, from spherical and diffraction errors, **16B,** 201
curvature, of, **14,** 87
of curvature
field emitter tip, **22,** 358
of tracks, errors, **5A,** 381ff
of gyration, **1,** 84; **16A,** 498; **16C,** 291, 305; **20,** 364, 365, 378–380; **23B,** 516, 556; **23C,** 304, 307, 381
nuclear, **5A,** 309, 314
Shelton triode, **22,** 11
standard surface, **22,** 22
Rain
attenuation coefficient of, **12B,** 179
measure attenuation by, **12B,** 184–185
radar measurements of, **12B,** 185
scattering by, **12B,** 177–178
Rainbow
cylindrical hologram, **26,** 189
holograms, **26,** 184
scattering, **7A,** 316–320
Raman active vibrators, **10,** 422, 423
Raman effect, **3,** 111–115; **10,** 420–431
apparatus for investigations, **3,** 125–141
selection rule, **3,** 118
theory, **3,** 115–125
Raman laser, **10,** 70–74; **18,** 722–724
spectrometer, **10,** 430
Raman scattering, **1,** 395; **3A,** 421–429; **15B,** 260
advantages over other density measurement techniques, **18,** 414, 418–421
basic features, **18,** 412–414
calibration, **18,** 429
density, temperature, composition diagnosis, **18,** 408–455, 643–645
energy and momentum transfer diagram, **16A,** 207
in gases, **3A,** 427
light sources, **18,** 419
line intensity use for density, concentration measurement, **18,** 428, 431
meaning, **18,** 411–414
molecular rotation, **18,** 411
molecular vibration, **18,** 411
nitrogen vibrational line contour, **18,** 421–428
pulsed laser illumination, **18,** 419–421, 425
removal from Rayleigh-Brillouin spectrum, **16A,** 187
rotational line contribution, **18,** 425–430, 488
scattering amplitudes, **18,** 413
stimulated, **3A,** 446–450
Stokes and anti-Stokes line, meaning, **18,** 422
temperature effects on density measurement, **18,** 421–425
Raman shift, **18,** 413, *see also* Raman scattering
Raman shifters, **29C,** 233, 266–268
Raman spectroscopy, **20,** 123–161
anti-Stokes spectrum, **20,** 124, 125
distinction from laser-induced fluoresence, **29B,** 400–403
experimental considerations
collection optics, **29B,** 410–413
detectors, **29B,** 417–420
laser Raman spectrometer, **29B,** 406–408
laser sources, **29B,** 408–410
spectrometers, **29B,** 413–417
fundamentals, **16A,** 77–79
general, **29B,** 394–396
instrumentation, **16A,** 84–87
molecular dynamics and wavepacket propagation, **29B,** 396–400
pump-probe methods, **29B,** 404–406
resonance, **20,** 125, 128–130
sample preparation, **16A,** 88
Stokes spectrum, **20,** 124, 125
Raman techniques, **3A,** 423, 447
Raman tube, mirror, **3,** 132–133
Raman-Nath regime, **26,** 151
for light scattering, **19,** 76–81, 461, 470
Ramp function, unit, **2,** 465
Ramsauer apparatus, **7A,** 3
Ramsauer-Townsend effect, **7B,** 48
Ramsay fringes, **29B,** 260–265, 311–312, 319, 351–353
Ramsden circle, **25,** 131
Ramsden eyepiece(s), **1,** 339; **25,** 62, 132
Ramsey
diamagnetic term, **3,** 367
separated field technique, **4B,** 123–124, 155
separated oscillating field method, **3,** 622–624
Random component of beam, **21,** 295, 309, 327
Random error, **28,** 29–33
instrumental, **28,** 30–31

statistical, **28,** 31–33
Random fields, **23C,** 47, 61
Random number generator, **28,** 126
Random phase distribution, **5A,** 543–544
Random powder diffraction patterns, **6A,** 267–277
Random process, Markoffian, **2,** 797
Random reentry model of chain folding
infrared studies, **16A,** 127
SANS studies, **16A,** 505–506
Random selection, **1,** 3
Random solid, general theory for, **7B,** 232–236
Random spectrum in channeling, **21,** 301, 309
Random strain, **22,** 247
Random time series, in Monte Carlo simulation, **28,** 462–463
Random truncation, **28,** 239
Random variables, probability distribution, **28,** 13–18
Random-access semiconductor memories, **2A,** 140
Randomes, **12B,** 16
Range, **28,** 157
beta rays in matter, **5A,** 71–72
electron mean, **5A,** 62
energy relation, **5A,** 2, 44–55, 437–438
heavy ions in emulsion, **5A,** 238
multiply charged particles, **5A,** 237–240
in nuclear emulsions, **5A,** 45, 229–236
protons, **5A,** 44, 45–51, 233–234
extension, **5A,** 237
germanium detectors, **3,** 83
in heavy materials, **5A,** 436–437
lead sulfide detectors, **3,** 81
mean, **5A,** 44
measured with cloud and bubble chambers, **5A,** 436–438
particles in emulsions, **5A,** 226–240
practical, **5A,** 71
straggling, **5A,** 51–55, 236
Range-energy relation of particle in emulsion, **5B,** 37
Ranger seismometer, **24B,** 61
Rankine-Hugonoit relation, **1,** 194
Raoult's law, **14,** 346
Rare earth ions, **15A,** 182–184
in solid state lasers, **15A,** 173–176
Rare earth lasers, optical pumping efficiency and output power of, **15A,** 193–194, *see also* Laser(s)
Rare earths, *see* Lanthanides
Rare gas
high order harmonic, **31,** 109
spectral distribution, **31,** 43
Rare gas continuum, **13A,** 206
Rare gas halide excimer lasers, **15B,** 141
Rare gas halide lasers, **15B,** 86–93
apparatus and power supply in, **15B,** 88–90
discharge electrodes in, **15B,** 88
e beam in, **15B,** 91–93
fast-discharge devices for, **15B,** 91
glow-to-arc transition in, **15B,** 88
long laser pulses in, **15B,** 91
mechanism of, **15B,** 87–88
results in, **15B,** 90–93
schematic of, **15B,** 89
tuning of, **15B,** 93
Rare gas monohalide excimer lasers, **15B,** 125–126
Rare gases, adsorbed on graphite, **23B,** 28, 41, 57
Rare gases, photoionization, **7A,** 157–161
Rare gases, resonances in, **7A,** 42
Rare-earth–transition-metal compounds, **23C,** 214
Rare-gas ion-atom collision, **17,** 155–156
Rarefied gas flow
density measurement by EBF, **18,** 434–455
visualization, **18,** 398
wind tunnel, **18,** 762–763, 784–785
Rate coefficients, for atomic processes, **9A,** 120, 126
Rate constant, in unimolecular decay, **15B,** 279
Rate constants, from flame studies, **7B,** 230
Rate equation description, for short-pulse dye lasers, **15A,** 269–271
Rate equation models, **30,** 337–338
of free-carrier heating, **30,** 177–120
of ion emission, **30,** 150–152
of lattice heating, **30,** 117–120
of photoemission, **30,** 164–166
of positive-ion emission, **30,** 150–152
Rate formula, optimum, Shannon's, **2,** 695–696
Rate growing, **6A,** 67
with melt back, **6A,** 67–68
Rate law, **19,** 207
Rate of chemical reaction, **19,** 184

Rate, pumping, **14,** 183
Ratio of amorphous to crystalline material, **6A,** 245–246
Ratio temperature, **26,** 241
  difference, **26,** 243
Ray aberrations, **31,** 151–152
Ray projection, **19,** 565
Ray sum, *see* Ray projection
Ray tracing, **19,** 577; **25,** 10, 53, **31,** 32–35
  aberration reduction, **31,** 172–173
  through a grating, **31,** 358
Rayl, **1,** 201
Rayleigh
  calibration, **9A,** 107
  criterion, **1,** 349; **10,** 132, 133; **13A,** 22, 276; **25,** 166, 303
  criterion for a prism, **25,** 207
  disk, **1,** 217
  formula, **1,** 130, 153
  fourth power law, **1,** 212
  intensities, ratio of Raman to, **3,** 124
  interferometer, **26,** 22
  length, **30,** 191
  limit, **12B,** 8
  linewidth, **16A,** 181
  loops, **6B,** 218–220
  piston, **1,** 204, 206, 211, **8,** 12
  power, distribution, **28,** 369–370
  ratio, **16A,** 27
  refractometer, **1,** 360; **25,** 190
Rayleigh scattering, (Rayleigh-Tyndall scattering), **1,** 394, 395; **3,** 112, 117; **5A,** 673; **7A,** 174, 178; **9A,** 106; **10,** 420, 421; **16A,** 85, 170; **19,** 241–247, 267, 272–274, 280–290; **20,** 126; **26,** 99
  advantages over emission and absorption spectroscopy, **18,** 414
  basic features, **18,** 412–414
  calibration for, **9A,** 111–112
  cross section, **7A,** 178
  line intensities used for diagnostics, **18,** 417–418
  line shape used for diagnostics, **18,** 414–417
  meaning, **18,** 411–414
  spontaneous, **3A,** 395–397, 453
  stimulated, **3A,** 453, 460
Rayleigh test, **28,** 355–358
  power of, **28,** 360
Rayleigh wave(s), **1,** 113; **19,** 496–498, 500
Rayleigh wing, **3A,** 453
Rayleigh-Brillouin scattering
  energy and momentum transfer diagram, **16A,** 207
  experimental, **16A,** 181–188
  in low-viscosity fluids, **16A,** 188–192
  sample preparation, **16A,** 183
  theory, **16A,** 171–181
  in viscoelastic fluids, **16A,** 192–200
Rayleigh-Brillouin spectrum, **16A,** 170
  depolarized, **16A,** 180
Rayleigh-Jeans law, **10,** 13, 288
Rayleigh-Taylor instability, **9A,** 429
Rays
  paraxial, **1,** 330
  residual, **1,** 399
Rays, a and b, **25,** 22
Rb_2CoF_4, **23C,** 22
Rb_2CrCl_4, **23C,** 21, 26
$RbCoF_3$, **23C,** 20
$RbMnF_3$, **23C,** 20, 29
RC active networks, **2A,** 333
RC filter, **8,** 7
RC network, in electronic feedback systems, **2B,** 452
RC networks, **2A,** 28
RC transients, **2A,** 15
RC-type oscillators, **2B,** 180
RCE, *see* Resonance coherent excitation
Re(001)
  scanning tunneling microscopy of, **27,** 297
Reaction
  chemical, by molecular beam methods, **3,** 600–602
  coordinate, **30,** 40, 59
  cycles, **5B,** 14–16
  endoergic, **5B,** 340
  exoergic, **5B,** 340
  nuclear, reaction energy of, **5B,** 339–352
  zones in horizontal flame, **3,** 569–570
Reaction kinetics in uniform supersonic flow technique
  gas cooling, **29A,** 293–295
  ion sampling, **29A,** 294
  operation, **29A,** 294
Reaction mechanism, **19,** 207
Reaction product
  energy states, disequilibrium in, **15B,** 97–102
  masses, **5B,** 429–430
Reaction rate, in afterglow studies, **7B,** 81

Reaction surfaces, *see* Isochronous reaction surfaces
Reaction systems, in new chemical lasers, **15B,** 126–135
Reaction zone, **8,** 30
Reactions
  antineutrino, **5A,** 682
  neutrino, **5A,** 682–683
Reactive power, **8,** 137
Reactive scattering, **7A,** 320–325
Reactivity oscillator, **8,** 233–237
  construction details, **8,** 235–236
  exercises, **8,** 236–237
  results, **8,** 236
Reactivity, **8,** 249, 250
  temperature dependence, **8,** 237
  void dependence, **8,** 237
Reactivity, reactor, **5B,** 596ff
Reactor core position, **8,** 235
Reactor dynamics, **8,** 251
Reactor fuel research, mass spectrometer usage, **8,** 177–180
  exercises, **8,** 180
  experimental methods, **8,** 178
  sample preparation, **8,** 178–179
  solid sample material, ion source, **8,** 179–180
Reactor fuel, **8,** 245
Reactor(s), **23A,** 112, 119, 120
  Brookhaven cold neutron studies, **5A,** 510–511
  coolants, **23A,** 113, 136, 141
  engineering, **5B,** 601–605
  fuel handling, **5B,** 605
  graphite moderated, **5B,** 611–614
  heavy water
    enriched uranium, **5B,** 618–619
    natural uranium, **5B,** 619
  high flux
    beam, **5B,** 620–622
    isotope, **5B,** 622
  kinetic behavior, **5B,** 595ff
  light water moderated pool, **5B,** 614–618
  neutron flux, **5A,** 498
  neutron source, **5B,** 590–622
  nuclear, **5A,** 498
  pulsed, **5A,** 499
  saturable, **2,** 793–794
  tank-type light water, **5B,** 618
  thermal column, **5B,** 609–610
  water boiler, **5B,** 611
  zirconium hydride enriched uranium, **5B,** 619–620
Reactor, regulation/control, **8,** 249–252
  exercises, **8,** 252
Read-only semiconductor memories, **2A,** 140
Ready-made optics, **25,** 3
Real gas effects, in hypersonic apparatus, **18,** 782
Real materials, ultrasonic study of, **11,** 421–433
Real solution, **19,** 201
Real-time holographic interferometry, **26,** 191
Real-time oscilloscope, diagnostics with, **29C,** 189–190
Reaumur scale, **1,** 32
REC, *see* Radiative electron capture
Receiver, telemetering, **2,** 681–683
Receivers, *see also specific types*; Spectrometers
  basic types of, **12B,** 210–212
  for electron plasma waves, **9A,** 20–21
  for ion waves, **9A,** 13–14
  special purpose, **12B,** 219–223
Receiving aperture size, effect on LDV performance, **18,** 138, *see also* Antenna theorem
Recessed specimens, for electric breakdown, **16C,** 452
Reciprocal Bragg spacing, **6A,** 250
Reciprocal dispersion, **7A,** 54
Reciprocal lattice, **1,** 422; **6A,** 191–194, 251–254; **16B,** 22–25, 193; **16C,** 152; **20,** 237–238; **22,** 241, 243, 244, 245, 247
  antiphase island, **22,** 254, 256
  in cylindrical coordinates, for fiber pattern indexing, **16B,** 71–73
  of defects, **22,** 247
  p(1 × 1) structure, **22,** 257
  relationship to real lattice vectors, **16B,** 24
  superlattice, **22,** 253, 254
  terrace, **22,** 252
Reciprocal lattice layers, **3,** 311
Reciprocal peak fraction, **8,** 225
Reciprocal-lattice rods, **22,** 244, 247, 250, 254
  broadening, **22,** 247
Reciprocity
  calibrations, in superfluid helium, **19,** 451, 453
  law, photographic, **1,** 411
  in network theory, **2A,** 10
  references for, **12B,** 63
Recognition of objects, **25,** 124

Recoil
counting rate, **5A,** 462–463
effect, for heavy ions on soil targets, **17,** 222–224, 227
energy, neutron interaction, **5A,** 463–464
energy spectrum, **17,** 152, 154, 167, 180
experiments, **5A,** 692–698; **7A,** 89–95
scatterer measurements, angular distribution, **5A,** 481–482
techniques, **5A,** 461–494
Recoil(s)
methods, nuclear, for lifetime measurement, **5B,** 333–338
proton, in nuclear emulsion, **5B,** 455
Recoil-free gamma-ray resonance absorption, **11,** 307, *see also* Mössbauer spectroscopy
Recoil-particle method, of inelastic energy-loss determination, **17,** 159–161, 163, 173–174, 177, 185–187, 189
broadening effects, **17,** 170–173
Recombination, **5A,** 100–101, 109–110; **30,** 35–36
centers, *see* Trapping centers
coefficient, **5A,** 94
of electrons and holes, **6B,** 95–99
non-radiative, **30,** 51–56
parameters, **6B,** 99
radiative, **30,** 50
sample effective lifetime, **6B,** 96–98
separation of bulk and surface parameters, **6B,** 99
Recombination lines
examples of, **12C,** 20
frequencies of, **12C,** 23–24
Recombination noise, *see* Noise, recombination
Recombination phonons, **11,** 150–151
Recombination probabilities, **6B,** 353
Recombination radiation
arrangements used, **6B,** 322
by carrier injection, **6B,** 322–323
by pair creation, **6B,** 322
production and detection, **6B,** 321–324
Recondensation, **30,** 278–281
Reconstruction, **19,** 565
of scalars, **19,** 568
of temperature, **19,** 575
of vectors, **19,** 573
Recorders, sound level, **1,** 219–220
Recording
binaural, **1,** 229
of displays, **2B,** 8–9, *see also* Tape recording systems
sound, **1,** 221
Recording methods
infrared, **18,** 750–753
light, **18,** 725–753
pressure gage, **18,** 552–555
wind tunnel, **18,** 768–769
Recording of holograms, **26,** 204
recording, **2,** 783
Recovery factor, **1,** 139, 154
Recovery factor, temperature
Couette flow, **18,** 459
definition, **18,** 459
flat plate boundary layer, **18,** 459
wind tunnel, **18,** 777–778
Recovery pressure, wind tunnel, **18,** 776
Recrystallization (solution), **6A,** 23
Rectangular aperture, **26,** 65
Rectangular cross sectional area, solenoid field
magnetic field calculation, **8,** 81, 85
Rectification, **2,** 344–356
dc restoration, **2,** 437, 438
Rectified diffusion, *see* Bubble motion
Rectifier(s), **2,** 177–181
in ac measurement, **2B,** 68–69
application regions in, **2A,** 191
bridge, **2A,** 234–235
cartridges and stacks, **2,** 132–133
cascade, **5B,** 581–584
cells, **2,** 127–132
circuits, **2,** 177–211, 299–300
comparison, **2,** 204–209
components, ratings of, **2,** 209–211
controlled, **2,** 208–209; **2A,** 220
avalanche, **2A,** 217
crystal, **3,** 28–29
full-wave, *see* Full-wave rectifier
bridge, **2,** 189–191
with capacitive load, **2,** 186–191
center-tapped, **2,** 186–189
with inductive load, **2,** 191–195
gas-filled, **2,** 205–206; **2A,** 219–220
general, **2A,** 189–193
germanium, **2A,** 220
grid-controlled, **2,** 99–101
half-wave, **2,** 182–186; *see also* Half-wave rectifier
with capacitive load, **2,** 182–185

with resistive of inductive load, **2,** 185–186
high current, resistance to thermal fatigue, **2,** 131
hot-cathode gas, **2,** 98–99
ideal, **2,** 53–54
polyphase with inductive load, **2,** 191–195
selection of, **2A,** 216–220
selenium, **2,** 206–207; **2A,** 220
silicon, **2,** 207–208; **2A,** 216–218
-controlled, **2,** 164
single-junction, **2,** 127–141
transformers, **2,** 209–210
vacuum diode and, **2A,** 218
voltage multiplying, **2,** 195–200; **2A,** 207–211
Greinacher cascade, **2,** 198–200
voltage regulation, **2,** 135–136
Rectifier block diagrams, **2A,** 189
Rectifier cartridges and stacks, **2A,** 60
Rectifier circuits, **2A,** 189–223
controlled, **2A,** 224–241
Rectifier components, ratings of, **2A,** 220–223
Rectifier output, filtering of, **2A,** 214–216
Rectifier system, design of, **2A,** 190
Rectifier transformers, rating of, **2A,** 221–222
Recurrence formula, *see* Legendre polynomials
Red noise spectrum, generation, **28,** 468–469
Red shift, *see* Gravitational red shift measurement, Gravitational red shift
optical spectra, **12C,** 279
radio spectra, **12C,** 279–280
Reduced mobility, **7A,** 377–378
Reduced variable, *see* Time-temperature superposition
Reduction rule, **1,** 316
Reference beam in holography, **26,** 170
Reference beam interferometer, **18,** 375–383
Reference distribution, **1,** 3
Reference energy, **22,** 149
Reference junction, **1,** 251
Reference signals, for interferometers, **12A,** 518
Reference surface, **22,** 18
Reference vacuum, **14,** 40
Reference wavelengths, **29C,** 279, *see also* Optical wavelength standards
Referencing, resonance position, **3,** 408–412
Refinement of structure from XRDA, **20,** 263–268
Reflectance, **1,** 335, 351ff
diffuse, **31,** 200–203
directional measurement of, **26,** 274
specular, **31,** 184
Reflectance spectra, **31,** 205–206
Reflected microwave signal measurements, **7B,** 93
Reflected-electron microscope, **22,** 22
contact potential, **22,** 22
patches, **22,** 22
scanning, **22,** 22
Reflecting filters on substrates, **31,** 329–338
Reflecting interference filter, **31,** 340
Reflecting polarizer, **31,** 232–242
Reflecting power, **6B,** 250
Reflection, **19,** 9–20, *see also* Acoustooptic measurement of sound wave, reflection
angle of, **1,** 329, 330
in buffer-specimen system, **19,** 262–264
cicular polarizer, **26,** 124
coefficient, **1,** 479; **19,** 9–18
cavity, **3,** 490–491
from crystals, **10,** 363–365
diffraction gratings, **26,** 78
in grain boundaries for very short wavelengths, **19,** 246–247
grating, **1,** 368
holograms, **26,** 172
integrated, **3,** 324–325
of molecules, **1,** 161
polarization, **1,** 378
dependence on, **19,** 19–20
polarizers, **26,** 122
selective, **1,** 421
at sidewalls of specimens, **19,** 294–296
of sound, **1,** 212–213
specular, **14,** 15
sphere, **3,** 280
from superconductors, **10,** 436–438
X-ray total, **1,** 429
Reflection coefficient, **31,** 265
Reflection filters, *see* Filters
Reflection grating, efficiency, **31,** 370
Reflection high-energy electron diffraction (RHEED), **11,** 635–636; **22,** 241, 264, 283; **30,** 294–297
detector, **22,** 283, 285, 288
measurement of angular profile, **22,** 295
Reflection infrared spectroscopy, **22,** 28, 29, 75

Reflection law, **31,** 147
Reflection measurements, **6B,** 255–262
Reflection microscope, **6A,** 183
Reflection minimum, **10,** 255–257
Reflection of radiation from crystallographic planes, **6A,** 195–197
Reflection spectra, **13B,** 47, 323a
Reflective coatings, **13A,** 232
Reflective devices, **25,** 296
Reflective optics, **25,** 169
Reflectivities, for polarized light, **6B,** 256
Reflectivity, **10,** 95
  definition, **6B,** 250
  measurement, **6B,** 256–257
  measurement of, **26,** 272
  metals, **10,** 112
  reststrahlen crystals, **10,** 107–109, 363–365
  vacuum ultraviolet, **1,** 405
Reflectivity measurement, **8,** 23–25
Reflectivity, multilayer coatings, **31,** 271–272, 276
  definition, **31,** 183
  design, **31,** 187–200
  different angle of incidence, **31,** 192–193
  fixed angle of incidence, **31,** 184–185
  high vacuum, **31,** 193–195
  instrumentation, **31,** 191–192
  mechanical deviation error, **31,** 188–190
  mechanical displacement error, **31,** 187–188
  oblique angle of incidence, **31,** 185–187
  systematic errors, **31,** 187–191
  ultrahigh vacuum, **31,** 195–200
Reflectometer, Bennet-Koehler, **26,** 274
Reflectometer, in microwave measurements, **2B,** 291–292
Reflectometry facilities, **31,** 203
Reflectors for scintillation counters, **5A,** 136–137
  dual, **12B,** 48–51
  dual, shaped, **12B,** 55–55
  mechanical measurements of, **12B,** 94–96
  paraboloidal, **12B,** 29–63
  surface measurements of, **12B,** 94–95
Reflectors, *see also* Antennas
  pulsed source, **23A,** 182
  reactor, **23A,** 115
  ultraviolet, **1,** 403
Reflectron, **30,** 430, 432
Reflex feed system, **12B,** 58
Reflex klystron(s), **8,** 147; **29B,** 276–277
  in microwave circuits, **2B,** 267–268
Refraction(s), **19,** 9–20
  angle of, **1,** 329, 330
  double, **1,** 379, 388
  effect on diffraction pattern, **6A,** 256–257
  index, **3,** 645, 647–648
  ionospheric, **12B,** 128–130
  of light, **3,** 645–648
  lunar ionosphere as cause of, **12C,** 116–117
  magnetic, **1,** 387
  molar, **3,** 645–648
  of molecular beams, **3,** 609–611
  neutral atmospheric, **12B,** 186–200
  polarization, dependence on **19,** 19–20
  solar corona as cause of, **12C,** 116–117
  sound, **1,** 212–213
Refractive behavior of fluids, **18,** 346, 347, 411
Refractive index, **1,** 329, 330, 331, 338, 344, 359, 379, 409; **5A,** 180; **6B,** 250–251, 262–263, 279–281; **13B,** 46, 135; **16C,** 162; **23A,** 259
  carbon dioxide, **5A,** 176
  crystal quartz, **10,** 96
  density dependence, **18,** 348
  determination, **3A,** 344–352
  diamond, **10,** 96
  flow tracing particle materials, **18,** 58–59
  gas mixture, **18,** 349
  germanium, **10,** 96
  irtran, **10,** 96
  of opaque metallic films, **6B,** 280–281
  polyethylene, **10,** 96
  sapphire, **10,** 96
  of semitransparent absorbing films, **6B,** 281
  silicon, **10,** 96
  TPX, **10,** 96
Refractometer(s), **1,** 344–346, 359–361; **3,** 647–648; **25,** 183; **26,** 39
  gravity, **23A,** 269
Refractory materials, *specific materials*
Refrigerators, **12A,** 429; **14,** 235, 234
Regenerative detector, **11,** 302
Regenerative frequency dividers, **2B,** 18–19
Regenerative oscillators, short-pulse dye lasers and, **15A,** 287–293
Regenerative-oscillator detectors, **3B,** 549
Regenerators, pulse, **2,** 437, 438
Registered phases, **23B,** 17
Registers
  mechanical, **2,** 526–527
  print-out, **2,** 532–533

Regression, method of maximum likelihood, **28,** 220–221
Regressor
errors, **28,** 277–280
random, **28,** 278
versus residual, **28,** 269
variances, **28,** 279
Regulation, **2,** 453–518
characteristics, **2,** 489–491
current, **2,** 489–503
electronic, **2,** 488–510
specifications, **2,** 489–491
voltage, **2,** 489–503
Regulators
closed-loop, **2A,** 244, *see also* Voltage regulators
current, **2B,** 458–460
defined, **2B,** 376
nondissipative, **2B,** 456–458
voltage, **2B,** 452–458
Reinhold distribution, of sizes in semicrystalline polymers, **16B,** 168–169
Reinserter, dc, **2A,** 385
Rejection filters, **25,** 292
Rejection ratio improvement, **5A,** 460
Rejection region, **28,** 180, 181–182
Relative entropy, **28,** 412
Relative photometry, **12A,** 170–178
Relative quantum counter, for efficiency determination, **6B,** 307–308
Relativistic effects, **1,** 53
in beam-foil experiments, **17,** 598
Relativistic electron wavefunctions, in target ionization, **17,** 209–211
Relativistic electrons, **10,** 42–45
Relativity measurements, **26,** 38
Relativistic projectiles, in inner-shell ionization, **17,** 211–212
Relativistic rise in ionization, **5A,** 444
Relativistic semiclassical Coulomb approximation, in K-shell ionization, **17,** 324–327
Relativistic time, **1,** 90
Relaxation
activation energy, **13B,** 187
chemical, **19,** 184, 186
constraint, **1,** 322
dielectric, **13B,** 168, 175
in dielectrics, **1,** 493–495
dipolar electron, **3,** 428
electric quadrupole, **3,** 384–386
of electronic excitation, **15B,** 284–286
flexible molecules, **13B,** 196
in free liquid systems, **3,** 426–428
magnetic dipole, **3,** 375–384
magnetic resonance, **21,** 364–367
Bloch equations, **21,** 367–371
spin-lattice relaxation, **21,** 366, 367
spin-spin relaxation, **21,** 364, 366
measurement, **3,** 434–438
mechanisms, **3,** 377–379
molecular, **1,** 208, 213
in molecular adsorption, **3,** 429–430
multiphase, **3,** 431–433
nuclear in metals, **3,** 430
paramagnetic, aqueous systems, **3,** 428–429
quadrupole, deuterium, **3,** 428
rotational, **15B,** 287–288
in solid state, **3,** 430–431
solids, **13B,** 199
spectra, **13B,** 168, 198
of spin systems, **3,** 456–464
spin-spin, **3,** 456–459
thermal, of sound waves in gas, **3,** 689–693
total translation, **3,** 427
vibrational, **15B,** 286–287
Relaxation amplitude, **19,** 188, 200
Relaxation effects, **6A,** 295–296
measurement by echoes, **9A,** 32–53
Relaxation energy, **22,** 224, 229
Relaxation frequency, **3A,** 412; **19,** 23, 182
Relaxation map, dielectric, **16C,** 394
Relaxation oscillators, **2A,** 348–359
capacitor or inductor storage elements in, **2A,** 348
hook circuit of, **2A,** 356
negative resistance devices for, **2A,** 351–355
voltage-stable negative resistance and, **2A,** 350
Relaxation process, **15B,** 283–287; **19,** 21–27
in bimolecular encounters, **15B,** 283
electronic excitation in, **15B,** 284–286
structural, **19,** 25
thermal, **19,** 25
viscous, **19,** 22–24
Relaxation processes, atomic forces and, **15A,** 23
Relaxation spectroscopy, **19,** 181–235
Relaxation strength, dielectric, **16C,** 388
Relaxation theory, of Brillouin scattering, **16A,** 178

Relaxation time(s), **1,** 433, 493–494, 495; **3,** 425ff; **3A,** 118; **13B,** 173, 181, 194; **19,** 23–27, 185, 199
  benzene, **3,** 427
  in creep equation, **16C,** 16
  for different density matrix elements, **15A,** 24
  dielectric
    distribution in polymers, **16C,** 403
    model for, **16C,** 387–388
    spectrum, **16C,** 389
    temperature dependence, **16C,** 393
  effects of structure and motion, **3,** 379–384
  in liquids, **10,** 348
  Maxwell, **19,** 138
  nmr, **19,** 233
  of polymer-penetrant system, **16C,** 336
  for vibrational modes, **16A,** 191
  rotational, **16A,** 432–433
  spin-lattice, **3,** 375–377, 382, 385, 386, 456
  structural, at the glass transition, **16A,** 201
  thermal, **3,** 376
  and ultrasonic measurements, **16C,** 88
  water protons, **3,** 427
Relaxed excited state, in laser color centers, **15B,** 4
Relaxing compressibility, **19,** 183
Relaxing heat capacity, **19,** 199, 200
Relay
  control systems, **2,** 515–517
  photoelectric, **2,** 304–305
  unit, **25,** 137
Relays, electromechanical, **2A,** 378
Reliability, **1,** 1, 7
  factor, **22,** 271
  statement of, **6A,** 1
  transistor, **2,** 159–160
Remanence, **1,** 557
Remanent magnetization, natural
  mechanisms, **24A,** 330
  Nèel model, **24A,** 328–330
  sediments, **24A,** 328
  thermoremanent magnetization, **24A,** 328–331
Remelting with additions, **6A,** 63–64
Remote sensing, in regulator design, **2B,** 456
Remote sensing, lithium niobate, **29C,** 392
Removal efficiency, **8,** 219
Reneutralization, **22,** 510
Renner effect, **3,** 198–200
  in ammonia, **3,** 199, 201
Renner-Teller effect, **3A,** 262
Renormalization group, **23C,** 5, 47
Reorganizational processes, during melting, **16B,** 313
Reorientation, molecular, **23A,** 48
  rates, **23A,** 511
Repeat distance, of a polymer chain, **16B,** 19
  determination by X-ray diffraction, **16B,** 79
Repeat unit, **16B,** 19
  chemical, **16B,** 18
  of a crystal, **16B,** 3
Repeller modulation, **8,** 135
Replica gratings, **13B,** 16, 21; *see* Grating
Replica method, for electron microscopy, **11,** 692–693
Replica technique, **6A,** 177–181
Replication, of specimens for electron microscopy, **16B,** 217
Representation analysis method, **23C,** 111
RER, *see* Radiative electron rearrangement
Reradiation after interaction with medium, **18,** 407, 411–412
RES, *see* Relaxed excited state
Reservoir, **1,** 322
Residence time, **14,** 232
Residual color, **25,** 34
Residual evaluation, **3,** 295–296
Residual gas analysis, *see* Partial pressure measurement
Residual gas composition in turbo-pump, **14,** 257
Residual nucleus, **8,** 268, 269
Residual plus component plot, **28,** 270
Residual stresses, **23C,** 458
  experimental techniques, **23C,** 461
Residual sum of square function, weighted, **28,** 391–392
Residual sum of squares, **28,** 257
Residual surface resistance, **11,** 553–560
Residual, **28,** 247
  versus regressor, **28,** 269
Resistance
  bleeder, **2,** 193
  bridge, for resistance thermometers, **11,** 499–501
  dynamic screen, **2,** 224
  negative, **1,** 466
  saturation, transistor, **2,** 146–147
  sheet, **2A,** 118
  temperature coefficient, **18,** 262, 264

temperature scale, **1,** 246ff
wave, **1,** 201
Resistance measurements, **1,** 456–458
A. C. bridge methods, **6B,** 59–60
A. C. current methods, **6B,** 56
ammeter-voltmeter methods, **6B,** 58
bridge measurements, **6B,** 54–55
condenser discharge methods, **6B,** 45–46
D. C. bridge methods, **6B,** 58–59
eddy current methods, **6B,** 56–57
on electrolytes, **6B,** 67–70
electrometer methods, **6B,** 38–40
on films, **6B,** 63–65
four-point probe methods, **6B,** 36–38
high frequency methods, **6B,** 47–49, 60
on liquid metals and semiconductors, **6B,** 66–67
by microwave absorption, **6B,** 48
ohmmeter method, **6B,** 33, 57
on powders, **6B,** 65
probe method, **6B,** 54
problems at high impedance level, **6B,** 38–40
pulse methods, **6B,** 46, 55–56
R. F. methods, **6B,** 57
special methods for semiconductors, **6B,** 65–66
spreading, **6B,** 35–36
Resistance per unit length, resistive anode, **29A,** 263, 265
Resistance thermometers, **1,** 245–249, 256–257; **11,** 489–501; **18,** 313–314, 461
interference from, **11,** 495–496
resistance bridge in, **11,** 499–500
shielding of, **11,** 494–498
Resistance-reactance networks, **2B,** 40
Resisting vane anemometer, **18,** 254, 258
Resistive ladder D/A converter, **2A,** 421
Resistive summing network, **2A,** 420
Resistive-anode encoder, **22,** 287
Resistivity, *see* Electrical conductivity
electrical, **1,** 245
hot-wire material, **18,** 262, 264
of n-Ge, **2,** 115ff
strontium titanate, **2,** 117
temperature coefficient of, **1,** 245; **2,** 122
Resistor transistor logic (RTL), **2A,** 138
Resistor(s), **2,** 41–42
aging, **1,** 458
diffused, **2A,** 118
in electronic circuits, **2A,** 34–336
in integrated circuits, **2A,** 117–118
noise, **1,** 458
parasitic, **1,** 458
temperature coefficient, **1,** 457–458
thermometry, **11,** 492–494
wirewound, **2A,** 35
Resistor-capacitor coupling network, **2A,** 283
Resolution, **14,** 82; **21,** 94–97, 217–219; **23C,** 12
angular, in neutron scattering, **21,** 178–179
angular correlation, *see* Angular correlation (of annihilation radiation)
of appearance-potential spectroscopy, **22,** 205
Abbe limit for, **27,** 77
bandwidth, **28,** 323
cyclotron resonance spectrometer, **10,** 193, 194
Čerenkov counter, **5A,** 168–170, 456
correction, finite instrumental, **5A,** 428–429
corrections, **23C,** 16
of cylindrical microscope, **22,** 354
doppler vacancy, *see* Doppler broadening (of annihilation radiation)
of electron microscope image, **16B,** 185, 220–221
Fabry-Perot spectrometer, **10,** 156
of field-ion microscope, **22,** 384, 385
film, **5A,** 448
finite angular, correction for, **5B,** 419–420
finite energy, correction for, **5B,** 247–248
function, **23B,** 199
gas scintillation spectrometer, **5A,** 435
grating spectrometer, **10,** 133, 139
helium–3 spectrometer, **5A,** 573
high, with ion-collection chambers, **5A,** 410
ideal velocity spectrometer, **5A,** 518–519
lifetime, *see* Lifetime (of positron)
magnetic diffraction, **23C,** 87
magnetic field, **3,** 412
mass, **22,** 303, 344, 405
of mass spectroscope, **5B,** 4
microwave spectrometer, **10,** 153
multiplex spectrometer, **10,** 144
in neutron scattering, **21,** 151
nuclear spectroscopy, **5A,** 344
of point-projection microscope, **22,** 358, 359
resolution time, *see* Counting
of scanning-electron microscope image, **16B,** 225

semiconductor detector, **5A,** 268, 275, 276
time-of-flight system, **5A,** 517
wavelength, crystal monochromator, **5A,** 569
Resolution, **31,** 282–283
Resolution, depth
in prompt reaction analysis, **21,** 247, 248
in RBS, **21,** 228–229, 232–233
in resonant reaction analysis, **21,** 250
in sputtering, **21,** 264
Resolution, energy
Doppler broadening, **21,** 99, 100
of Ge(Li) system, **21,** 53, 62
of NaI(Tl) system, **21,** 51
relationship with gamma ray energy, **5A,** 634ff
time-of-flight spectrometer, **5A,** 545
Resolution, *see* Compton spectrometer
correction, **8,** 199
peaks, **8,** 197
Resolution, spectral, alkai halide prism, **3,** 28
Resolution, spectral, of a fluctuation process, **2,** 796–798
correlation method, **2,** 796–797
short time average, **2,** 797–798
Resolved shear stress of single crystal, **6A,** 315
Resolving instruments, *see* Gratings, Interferometers, and Prisms
Resolving limit, **1,** 341
Resolving power
grating, **3,** 157
spectrometer, **3,** 103
in gratings, **26,** 72
of optical instruments, **26,** 79
Resolving power in electron diffraction, **6A,** 271–272
Resolving power of interferometer, **13B,** 62
Resolving power of magnetic spectrometers, **5A,** 343ff
Resolving power of multiple-beam interferometer, **18,** 196
Resolving power, **1,** 348; **4B,** 80–82; **6B,** 286–288; **13A,** 21, 276; **22,** 258, 263, 286, 295; **25,** 196; **29B,** 364–365; **31,** 368–370; *see also* Resolution
echelon, **1,** 373
grating, **1,** 367
interferometer, **1,** 372, 373
in LEED, **22,** 264
prism, **1,** 363, 364
in RHEED, **22,** 264
theoretical, of diffraction grating, **5B,** 53
Resonance, **1,** 90; **3,** 65–71; **8,** 136, 137; **22,** 382, *see also* Series resonance
angular frequency, **8,** 130
anomalie, **26,** 157
antiferromagnetic, *see* Antiferromagnetic resonance
bandwidth, **8,** 9
bridge, **1,** 462–463
curve saturation, **5B,** 109
curves, universal, **2,** 23–24
cyclotron, *see* Cyclotron resonance
devices, **1,** 469
double, **3,** 625–626
cavities, **3,** 479–480
electron-nuclear, **3,** 498–500
electric
apparatus with quadrupole focusing fields, **3,** 614
molecular beam methods, **3,** 615–626
spectra of molecules, **3,** 634–635
electron, spin, **3,** 441–500
electron-nuclear double, **3,** 498–500; **5B,** 127–129
electron spin, **2,** 633–636
frequency, **1,** 498, 523; **8,** 138
giant (ψ, *n*), **5A,** 527
isolated Breit-Wigner theory for, **5A,** 538–541
levels, slow neutron in U^{238}, **5A,** 540–541
line
second moment, **3,** 380–381
shapes, **3,** 459–464
width, **3,** 618–620, 624–625
magnetic, **2,** 624–636
atomic beam apparatus, **3,** 596
in atomic beams, **2,** 625–626
atomic hyperfine structure by, **3,** 626–630
in bulk matter, **2,** 626–628
corrections in, **5B,** 121–122
molecular beam methods, **3,** 591, 615–626
neutron beam, **3,** 635–636
quadrupole splitting of, **5B,** 119–120
spectra of molecules, **3,** 630–634
molecular beam electric, **5B,** 28–29
neutron, in compound nucleus Cd^{114}, **5A,** 538–539

nuclear magnetic, **1,** 523–524; **2,** 628–633; **3,** 359–440
elementary theory, **3,** 360–386
experimental methods, **3,** 386–404
quantitative analysis, **3,** 421–422
nuclear magnetic and quadrupole, **5B,** 104–129
experimental apparatus, **5B,** 111–115
nuclear quadrupole, **3,** 501–525
applications, **3,** 523–525
experimental methods, **3,** 507–523
position referencing, **3,** 408–412
quadrupole, **5B,** 119–121
radiation, **1,** 396
Raman effect, **3,** 125
scattering, **1,** 394; **9A,** 144; **18,** 413; **23C,** 372
states, particle, **5B,** 847–848
width, **2,** 627
Resonance at low temperature, **6B,** 240
Resonance coherent excitation, applications, **29A,** 303
Resonance detectors, **11,** 347–348
Resonance experiments, **3B,** 772–846; **28,** 349–350
Resonance fluorescence, **1,** 397, **7B,** 16; **13A,** 145
Resonance frequency, **19,** 374–380, 383–385, 399–402
of gas bubbles, **19,** 374, 399
of vapor bubbles, **19,** 383–385
Resonance ionization spectroscopy
introduction, **29B,** 171–173
ionization schemes, **29B,** 173–176
laser sources, **29B,** 176–178, 182
sensitivity, **29B,** 181, 185
single atom counting
general, **29B,** 178
cesium, **29B,** 178–179
noble gas atoms, **29B,** 179–182
using sputter initiation or laser desorption, **29B,** 182–187
with thermal atomization, **29B,** 187–189
Resonance ionization, at metal surface, **29B,** 194
Resonance line emission, **9A,** 128–129, 142
Resonance methods in rf region, **13B,** 147
Resonance methods, of shear impedance measurement, **19,** 166–170
Resonance modes, **23B,** 124
Resonance neutralization, **22,** 304
Resonance phenomena, **6B,** 227–243
properties observed, **6B,** 228
primary quantities measured, **6B,** 227–229
Resonance processes, **13B,** 189
Resonance radiation, **4B,** 175–180, 188–189
imprisonment of, **7A,** 48–50, 60, 132, 133
polarization of, **4B,** 177–179
Resonance reverberation method, continuous wave, **19,** 130–131
accuracy of, **19,** 131
calibration of, **19,** 131
wall losses, minimizing, **19,** 131
Resonance shunt ammeter, *see* Ammeter
Resonance spectrometers, **23A,** 243, 254
Resonances in electron scattering, **7A,** 40, 53, 88
Resonant capture, **13A,** 136
Resonant cavities, **10,** 124–126
Resonant cavity, **3B,** 556; **13B,** 55, 127
Resonant charge exchange, by merging beams, **7A,** 478
Resonant circuit
coupled, **2A,** 14
impedance of, **2A,** 13–14
pole and zero locations in, **2A,** 12–13
Resonant circuit, *see* Lumped-constants resonant circuit
Resonant circuits, for high-frequency dielectric measurements, **16C,** 404–408
Resonant circuits, in dielectric measurement, **6B,** 15–20
Resonant electrons, of large amplitude waves, **9A,** 5–6
Resonant elements, **3,** 473–480
Resonant enhancement, multiphoton processes and, **15B,** 211–216
Resonant frequency, **8,** 132, 135, 136, 148
Resonant mode absorption, **10,** 368, 371, 372
Resonant modes, **21,** 186–190
Resonant transfer and excitation, accelerator applications, **29A,** 304–305
Resonating gas column, **1,** 279
Resonator(s), **2,** 710–711; **19,** 31–34, 39–41, 43, 46–51, 55–65, 166–170, *see also* Transducers
astigmatic, **15A,** 343–344
for flashlamp-pumped dye lasers, **15A,** 321–324
high-Q, *see* High-Q superconducting resonators
in microwave circuits, **2B,** 256–257

modifications of characteristics in, **15A,** 26–30
Q of, **11,** 549, 565, 568–570, 577–579
superconducting microwave, *see* Superconducting microwave resonators
two-mirror, **15A,** 287–290
ubiquity of, **11,** 541
Resonator, *see* Box resonator; Microwave cavity resonator
Resonator methods, in superfluid helium, **19,** 435–447
annular, **19,** 441–447
cylindrical, **19,** 435–437
doppler shifts, **19,** 444–447
Helmholtz, **19,** 438–439
loss mechanisms, **19,** 437–439
mechanically driven, **19,** 439–440
Resonator stability criteria, for cw dye lasers, **15A,** 338–343
Response
forced, of a system, **2,** 469–472
frequency, **1,** 225, 230–231
loudspeaker, **1,** 231
phase, **1,** 231
semiconductor detector to radiation, **5A,** 273–279
sodium iodide (Tl) spectrometers, **5A,** 621–637
spectral
lead sulfide cells, **3,** 82
lead telluride detectors, **3,** 82–83
transient, **1,** 231–232
Response
function, *see* Photo-Peak spectrometer
nonlinearity, *See* Compton spectrometer
Response
function, *see* Photo-Peak spectrometer
nonlinearity, *See* Compton spectrometer
Response function, **5A,** 637
for coincidence scintillation spectrometer, **5B,** 529, 530
for single-crystal scintillation spectrometer, **5B,** 516–517
Response time (τ) of infrared detectors, **12A,** 418, 421–423
Response time, *see also* Frequency response
bolometer, **10,** 169–171
dilatational pressure gage, **18,** 604
diode detector, **10,** 165, 166
electron cyclotron detector, **10,** 193, 194
flow tracing particles, **18,** 26–32, 38
free surface motion Raman scattering, **18,** 609
free-carrier photoconductive detector, **10,** 187
germanium detector, **10,** 171
heat transfer gage, **18,** 668–669, 674–678
hot-wire anemometer, **18,** 293–295
Josephson junction, **10,** 208
liquid manometer, **18,** 508
Pitot probe, **18,** 252–253
pressure bar gage, **18,** 599
with end sensor, **18,** 603
pyroelectric detector, **10,** 173, 174
radiation scattering diagnostics, **18,** 409
superconducting bolometer, **10,** 170
tracer particle, **18,** 26–32, 38
Response variable, **28,** 245
Responsivity (S) of infrared detectors, **12A,** 421–423, 440
Responsivity of photon detectors, **4A,** 415–416
frequency dependence, **4A,** 416
Responsivity, **13B,** 12
Rest mass, **1,** 61
Restorer, dc, **2A,** 385
Restricted maximum likelihood, fitting semivariogram models, **28,** 392
Restricted motions, **20,** 25
Restricted system, **1,** 318
Restricted trace approximation, **23C,** 271
Reststrahlen, **13B,** 25
absorption, **10,** 94, 95
crystals, **10,** 94, 95, 99, 100, 103, 107, 109, 364, 365
in polyethylene, **10,** 102, 103
filters, **10,** 103, 107–109
reflection, **10,** 107–109, 363, 364
Retardance
units, **29C,** 353
optical, **12A,** 372
Retardation sheets, for optical pumping, **7B,** 17
Retardation time, **19,** 143
in creep equation, **16C,** 16
Retarders, optical, **12A,** 371–379
achromatic, **12A,** 373–378
variable, **12A,** 378
Retarding field analyzer (RFA), **11,** 84–89, 99; **30,** 319
electronics of, **11,** 89–92
Retarding field, defined, **11,** 85

Retarding lens system, of parallel-plate analyzer, **17,** 440–449
Retarding potential, **7A,** 8, 42, 43, 65–67
measurements, **6B,** 379–380
Retarding waveplates, **26,** 142
Retarding-potential analyzer, **22,** 227, 362
Retarding-potential method, **22,** 11, 20, 21
band structure, **22,** 20
emission current density, **22,** 11
energy distribution, **22,** 20
experimental arrangement, **22,** 20
field-emission retarding potential technique, **22,** 20
nonuniform surface, **22,** 22
reflection mechanism, **22,** 20
scanning arrangement, **22,** 21
scanning low-energy probe technique, **22,** 20, 21
Reticles, **25,** 131
Retinoscopic lamp, **25,** 244
Return beam vidicons, **2B,** 349
Reverberant wave, **24B,** 588–589, *see also* Leaky mode; Pseudo-Rayleigh wave; Stonely wave
Reverberation
chamber, **1,** 233
time, **1,** 229–230
Reversal shear interferometers, **26,** 19
Reversible process, **1,** 309–310
adiabatic, **1,** 310
infinitesimal, **1,** 311
Rewelding, **14,** 279
Reynolds number, **1,** 133, 143, 145, 148, 158, 159, 300, 301
components, electron spin resonance spectrograph, **3,** 469–480
definition, **18,** 830
hot-wire and hot-film convection, **18,** 269
low, apparatus, **18,** 796–801
nuclear signal receiver, **3,** 437
receiver stability, **3,** 417
systems, pulsed, **3,** 425–438
temperature sensor, **18,** 461
transmitter, **3,** 435–436
wind tunnel, **18,** 758–759, 773–774, 844–847
Rf field, superconducting surface impedance and, **11,** 560–566
rf period, **8,** 173
rf phase, **8,** 171
RF spectrum analyzer, **29C,** 188–189
rf voltage, **8,** 169–171
RFFT2, **12C,** 292, 294, 318
RG–99/U waveguide, **8,** 151
Rh(111)
scanning tunneling microscopy of, **27,** 297
RHEED, *see* Reflection high-energy electron diffraction
Rheinberg illumination, **25,** 173
Rhenium
boiling point, **14,** 351
energy gap anisotropy in, **11,** 143
filament, **14,** 70
melting point, **14,** 348, 351
physical properties, **14,** 372–373
vapor pressure, **14,** 348, 351
Rheological fluid, **18,** 801
Rheology, *see* under specific viscoelastic functions
Rheovibron instrument, **16C,** 35
Rhodamine 6G (R6G) laser, **29C,** 48, 56, 66–67, 68, 146, 155
Rhodamine G6 dye lasers, **15A,** 331, 357
characteristics of, **15A,** 256–261
laser threshold for, **15A,** 336
output of, **15A,** 336
Rhodium, reflectance spectra, **31,** 215–216
Rhombohedral crystal system
definition, **16B,** 4
interplanar spacing formulae, **16B,** 26
reciprocal lattice relationships, **16B,** 24
Rhombohedral lattice, definition, **16B,** 5
Ribbonlike lamellae in melt-crystallized polymers, **16B,** 345, 346
twisting in, **16B,** 347
Riboflavin binding protein, **19,** 202
Ribonuclease, **19,** 202; **23C,** 361
Ribonucleic acid, **23C,** 400
Ribosomes, **23C,** 377, 380, 398, 401
Richardson equation, **22,** 5, 11
Richardson line, **6B,** 130–131, 134
Richardson plot, **4A,** 1
Richardson's equation, **3,** 555
Richardson-Dushman equation, **2,** 47–48; **2A,** 147–148; **4A,** 1, 125; **6B,** 128–129
Richter magnitude, **24B,** 56
Riedel singularity, **11,** 225–229
Josephson frequency and, **11,** 242
Righi-Leduc effect, **6B,** 166, 169–170
Right truncation, **28,** 220
Right-censored observations, **28,** 219
Rigid rotor, **3A,** 12

spectral patterns, **3A,** 37
Rigid-body mechanics, **1,** 81–97
Rigid-molecule model, **23A,** 448
Rigidity, shear, **1,** 208
Ring dye laser, **29C,** 62–63
Ring geometry regenerative oscillator, **15A,** 291–292
Ring interferometer, **26,** 42
Ring puckering, **3A,** 77, 82
Ringing, **28,** 296
Ripple, **2,** 177
Ripple, spectrometer baselines with, **12C,** 33, 48
RIS, *see* Resonance ionization spectroscopy
Rise time, *see* Response time
RL transients, **2A,** 15
RLC bridges, in equipment testing, **2B,** 176–177
RLC circuit, admittance and impedance of, **2A,** 11–12
RLC filter, **2A,** 335
RLC network, **2A,** 1–3
transfer function of, **2A,** 28
voltage-current sources in, **2A,** 3
RLC transients, **2A,** 16–17
Rms divergence, calculation, 6–7
rms noise, **8,** 23
Rochelle salt pressure gage sensor, **18,** 547
Rochon prism polarizer, **31,** 229
Rochon prism, **12A,** 370; **26,** 138
Rock magnetism
ac loop plotter, **24A,** 433
magnetic characterization, **24A,** 425–427
magnetic phases, identification, **24A,** 418–422
magnetocrystalline anisotropy transitions, **24A,** 423–425
stress effects, **24A,** 452–456
susceptibility
anisotropy, **24A,** 442–452
initial, **24A,** 438–442
thermomagnetic analysis, **24A,** 422–423
Rock salt, *see* NaCl
Rocket photometry, bayesian methods, **28,** 426–428
Rocksalt, second-order spectrum, **3,** 154
Rod drop method, **8,** 249, 251–252
Rod worth determination
asymptotic period method usage, **8,** 249–251
measurements, **8,** 252
Roentgen, **1,** 415
below 3 Mev, **5B,** 539
Rogowski coils, **9B,** 272–276
Rogowski profile, in rare gas halide lasers, **15B,** 88
Rogowski-contour electrodes, **6B,** 27
Role of surface electronic structure, **27,** 11
Roll, **1,** 147
aerodynamic, **18,** 768–769
Ronchi ruling, **25,** 159
Ronchi schlieren, **18,** 370
Ronchi-test, **26,** 39
Roof Prism, **25,** 65
Root mean square velocity, **29B,** 9
Root-locus diagram, **2B,** 405–409
Root-locus techniques, in feedback control systems, **2B,** 405–410
Root-mean-square (rms) ac measurements, **2B,** 63
Roots pumps, **1,** 172, 175–176; *see* Lobe pump
Rotameter, **18,** 322, 331
Rotary joints, **1,** 482
Rotary oil-sealed pump, **14,** 112–128, *see also* Fore-vacuum pump
air stripping, **14,** 124
boil off, **14,** 124
comparison, **14,** 120–121
compression ratio, **14,** 114
decanting, **14,** 122–124
installation, **14,** 126–128
maintenance, **14,** 467–471
oil contamination, **14,** 124–126, 469
oil reclamation, **14,** 124
operation, **14,** 127–128
power requirement, **14,** 118–119
pumping speed, **14,** 117–118
single and compound stages, **14,** 119–120
ultimate pressure, **14,** 115–117, 122
water vapor pumping, **14,** 121–124
Rotary pump, *see* Rotary oil-sealed pump
Rotating and oscillating crystal methods, **6A,** 206–208
Rotating anode, in X-ray tube, **16B,** 41, 158
Rotating crystal method, X-ray, **1,** 422
Rotating flow apparatus
construction, **18,** 806–809
data transmission and photography, **18,** 817–819
examples of studies, **18,** 801–819
experimental configurations, **18,** 811–813

moving boundaries, **18,** 814–816
precision and control requirements, **18,** 809–812
pumping, **18,** 816–817
Rotating mirror camera, *see* Framing camera; Streak camera
Rotation, **1,** 86–87
lines, nuclear spin effect, **3,** 71–73
method, **3,** 311–313
of molecular groups, **3B,** 984
of nuclear magnetic resonance sample, **3,** 414
optical, **3,** 679–680
of polarization plane, **1,** 383
selection rules, **3,** 123–124
about top axis, rotational constant of, **3,** 146
Rotation axis, in crystal symmetry, **16B,** 8, 9
Rotation effect, **30,** 258–261
Rotation matrix, in deep space plasma measurements, **9B,** 79
Rotation oscillations, in de Haas-Van Alphen effect, **11,** 53
Rotation spectra, **13B,** 29
hindered, **13B,** 44
Rotation-vibration spectra, **13B,** 30
Rotational chemical lasers, **15B,** 116–120
Rotational correlations, **23B,** 483
Rotational disorder, **23B,** 108
Rotational hysteresis, **6B,** 223–224
Rotational isomerism, **3A,** 90
Rotational levels of HCl, **10,** 89, 311, 312
Rotational magnetic moments, **3B,** 787, 802
Rotational reference frame, **21,** 454–456, 465–466
Rotational relaxation, **15B,** 287–288
Rotational shear interferometers, **26,** 19
Rotational spectra of molecules, **10,** 306–321
Rotational structure of electronic spectra, **3A,** 271
Rotational temperature, **30,** 123–124, 598–600
EBF technique, **18,** 436, 493–497
measurement, **18,** 457, 464, 466, 480, 493–497
spectral emission technique, **18,** 480
Rotational-electronic interactions, **3A,** 276
Rotator
diatomic, **3,** 46, 47, 49
Faraday, **2,** 715
rigid, kinetic energy, **3,** 46
symmetric, **3,** 59–65
vibrating, **3,** 48–53
energy levels, **3,** 52
Rotometer, float, **1,** 155
Roton gas, liquid helium and, **11,** 464
Roton system, **11,** 465–466
Rotons in liquid helium–4, **23B,** 326
Rotor, **14,** 247, 249, 255
Rough polishing, *see* Lapping
Roughing pumps, *see* Fore-vacuum pumps
Roulette, nonalgebraic solutions, **28,** 441–443
Rounding off, **1,** 21
Rounding off of numbers, **2,** 14–15, 27; **3,** 722–723
Routh criterion, **2B,** 398–401
Row lines, **3,** 312
Row structure, **16C,** 143–145, 156
Rowland circle, **1,** 369; **13A,** 227, 292; **25,** 215; **31,** 28
definition, **31,** 366
monochromator, **31,** 131
non-symmetrical system, **31,** 164–166
Rowland ghosts, **26,** 76
Rowland grating mount, **1,** 369
Rowland principle, **1,** 369
Royal Radar Establishment
interferometer at, **12C,** 165
RSCA, *see* Relativistic semiclassical Coulomb approximation
RT (rotational-translation) event or collision, **15B,** 64, 79
RTEA, *see* Resonant transfer and excitation
RTL (resistor-transistor logic), **2A,** 138
Rubber, **14,** 407, 408–411, *see also* Elastomer
butyl, **14,** 481
Rubber-toughened plastics, **16C,** 271
Rubbers, *see* Elastomers
Rubbing contact, charge generation by, **16C,** 432–433
Rubidium, **12C,** 87
resonant frequency of, **12C,** 210
Rubidium clock, **12C,** 179
Rubidium isotopes, **5B,** 81–85
Rubidium liquid, **23B,** 458
Rubidium-gas-cell resonator, **12C,** 208
description of, **12C,** 221–223
Rubredoxin, **20,** 268–275
Ruby high pressure gage, **18,** 610
Ruby lasers, **3A,** 430; **9A,** 81, 103; **15A,** 202–203; **29C,** 139, 140
giant pulse, **18,**717–720
properties, **18,** 714–716, 725, 746

pump lamp, **18,** 693–694
Ruby maser, **3,** 154–155
Ruci laser, **25,** 265
Ruland's method, for X-ray crystallinity, **16B,** 122–127
Rule of greatest simplicity, **3,** 2
RULED, **12C,** 300, 327–330
Ruled-surface mapping
programs for, **12C,** 299–307
Run-away, thermal, **2,** 130
Russell-Saunders coupling, **3,** 206; **4B,** 109–111
pure L-S coupling, **4B,** 91, 110–111, 195
Russell-Saunders limit, **4B,** 111
Rutherford backscattering (RBS), **21,** 300–301; **24A,** 119–120; **30,** 291, 306, 309–310
application, **21,** 231–235
in channeling, **21,** 300–301
collisions, **21,** 276
cross-section, **21,** 223–225, 228–231
for defect distribution profiles, **21,** 311
detection sensitivity, **21,** 229
edge, spectrum, **21,** 226, 229, 233
equipment, **21,** 224
law, **21,** 303
limitations, **21,** 230
spectrum, **21,** 228, 230, 232, 300, 301
Rutherford formula, **5A,** 55
Rutherford scattering, **5A,** 289–293; **17,** 30–31
Rydberg atoms
characteristics, **29B,** 115
fine structure intervals
by rf spectroscopy, **29B,** 310–311
by quantum beats, **29B,** 329–330
lifetimes, **29B,** 123
production
charge exchange, **29B,** 119–121
collisional optical excitation, **29B,** 125–127
electron impact excitation, **29B,** 117–119
in fast beams, **29B,** 120–121, 126–127
optical excitation, **29B,** 121–125
overview, **29B,** 115–117
wave packet evolution, **29B,** 330–332
Rydberg constant, **1,** 42–43; **13B,** 330
Rydberg states, **3,** 189; **13B,** 231; *see also* Hydrogenic state
Rytov's approximation, **19,** 578

S

S (responsivity), **12A,** 421–423, 440
S branch, **3,** 144
S function, **28,** 504–509
least squares, **28,** 506–508
S on Cu(111)
chemisorption, scanning tunneling microscopy of, **27,** 297
S on Mo(001)
chemisorption, scanning tunneling microscopy of, **27,** 296
S on Re(001)
chemisorption, scanning tunneling microscopy of, **27,** 297
S system, **28,** 484
bootstrapping code for median, **28,** 494
code for bootstrapping nonlinear regression coefficients, **28,** 497
extending, **28,** 487–488
on Unix workstation, **28,** 484–488
S-bends, **29C,** 381
S-polarization, **26,** 114
S-test, **28,** 201
s-wave production, **8,** 267, 268
s.h.o. *see* Simple harmonic oscillator
S/B ratios, **8,** 193
S/N ratio, *see* Signal-noise-ratio
SA, *see* Stationary afterglow
Sabot, **18,** 780
Safety, **14,** 245
during chemical etch, **14,** 501, 502
device, **14,** 449
sputtering systems, **14,** 571–572
using acetone, **14,** 512
vacuum systems, **14,** 478, 541
Sagitta
constant, **5A,** 397, 403
formula, **6A,** 255
mean, **5A,** 402
method, **5A,** 376–377
Saggital
astigmatism, **25,** 42
coma, **25,** 42
Sagittal ray, definition, **31,** 159
Sagnac absorbers, modelocking with, **29C,** 180
Sagnac effect, **23A,** 295
Sagnac interferometer, **26,** 42; **29C,** 180
Saha equation, **31,** 50

Saha function, **9A,** 121, 362
Saha-Langmuir equation, **7A,** 270
Salts, paramagnetic, for nuclear orientation, **5B,** 181–182, 184–186
Saltus problem, **26,** 52
Sample(s)
cell, vacuum tight, **3,** 86–88
finite, **1,** 3
location, electron spin resonance, **3,** 476–478
mean, **5B,** 810
nuclear magnetic resonance, **3,** 413–415, 419–421
paramagnetic standard, **3,** 495–498
powder, techniques, **3,** 134–136
Raman spectra, **3,** 131–136
scattering, for transmission total cross-section, **5B,** 373–374
Sample and tip condition
experimental problems in scanning tunneling microscope operation, **27,** 68
Sample environment, **23C,** 414
Sample holders, in magnetoresistance measurements, **11,** 15–20
Sample mean, **1,** 4; **2,** 4; **28,** 156
Sample median, **28,** 156
Sample positioner mechanical structure, **27,** 57
Sample preparation, **22,** 289
contacts, **6B,** 50–52
evaluation of homogeneity, **6B,** 52–53
for irradiation, **21,** 19, 20, 24, 25
for resistance measurement, **6B,** 49–53
Sample space, **28,** 8–11
Sample variance, **3,** 712; **28,** 161
Sample-out measurement, **8,** 192
Sampled continuous wave, **19,** 57–60
Sampled technique, continuous wave, **19,** 123–124
benefits of, **19,** 124
frequency domain operation, **19,** 124
steplike response, **19,** 124
time domain decay operators, **19,** 124
Sampling, **28,** 20
autocorrelation spectrometer, **12B,** 281–284
correlated, **28,** 454
distribution, **28,** 492
error, **18,** 4–6
of interferograms, **12A,** 503–506
theorem, Shannon's, **2,** 690–692
Sampling interval, **8,** 198
Sampling theorem, **8,** 198
San Andreas fault, **24B,** 405
Sandwich heat transfer gage, **18,** 664
Sandwich holography, **26,** 194
Sandwich method, **5A,** 299
Sandwich-type light amplifier, **6B,** 356
Sanford-Bennett permeammeter, **1,** 559
Sapphire (Al_2O_3), *see* Alumina
Sapphire, **10,** 96, 99; **14,** 395, 403
SAS, **28,** 484
Satellite Auger decay, **17,** 469–474
Satellite lines, **13A,** 16; **17,** 564
Satellite orbits
VLBI measurement of, **12C,** 271–273
Satellite production, Coulomb ionization, **17,** 496–497, 499
Satellites, *see* Geodetic satellites
Satellites, whistler observation by, **9B,** 111–137
ducted whistlers, **9B,** 111–112
ion effects, **9B,** 112–120
Saturable reactors, **2B,** 369–370
Saturable-field excitation, of laser-cooled and trapped atoms, **29C,** 298–299
Saturated absorption spectroscopy, **15B,** 236
Saturated lasers, **31,** 104–106
Saturation broadening, **3,** 10–11
Saturation FM spectroscopy, **29C,** 120
Saturation magnetization, **6B,** 200–201
Saturation of spectral lines, **3A,** 113
Saturation spectroscopy, **15B,** 233–238
narrow resonances in, **15B,** 235
Saturation transfer, **20,** 33–36
Saturation, in excitation process, **15B,** 279
SAW, *see* Surface acoustic wave
SAXS, *see* Small-angle X-ray scattering
Sayre sign relation, **3,** 295
Sb on GaAs(110)
spatially resolved spectroscopy of, **27,** 266
surface structure of, **27,** 266
SCA theory, *see* Semiclassical approximation theory
Scalar potential, *see* Current loop
finding, **8,** 70–71
functions, **8,** 74
Scalar regularizing action, **5B,** 771
Scalar theory region, **26,** 194
Scale effect, **18,** 846
Scale of metal oxide, **14,** 442
Scalers, **2,** 384–393, 519–523
beam switching tube, **2,** 392–393

bistable circuits, **2,** 384–389
cascaded stages, **2,** 521–522
decade, **2,** 527–535
automatic recording, **2,** 532–535
visual representation, **2,** 529–532
distribution of random events, **2,** 520–521
glow transfer tube, **2,** 389–392
resetting and gating, **2,** 519–520
Scaling, **23C,** 6
function, **23C,** 6
Scanner mechanical strucure, **27,** 53
Scanners, **25,** 113, 114, 140
Scanning drives, **12A,** 516
Scanning electron microscopy, in magnetic domain studies, **11,** 734–735
Scanning electronics, **27,** 63
Scanning etalons, **25,** 222
Scanning force microscopy, **27,** 86
Scanning ion micrographs, **22,** 403
Scanning low-energy electron probe, **22,** 20, 21
Scanning Michelson interferometer
accuracy and resolution, **29C,** 315–316
alignment, **29C,** 322–325
data acquisition, **29C,** 326–331
principles, **29C,** 312–315
reference laser, **29C,** 325
structure, **29C,** 316–322
Scanning microprobe, **30,** 324
Scanning microscope, **6A,** 184; **25,** 174
Scanning multiple-beam interferometer, **18,** 201–220
Scanning noise microscopy, **27,** 79, 85
Scanning radiometers, **26,** 252
Scanning transmission electron microscope (STEM), **20,** 436–438, 440, 441; **30,** 292
Scanning tunneling microscope (STM) design, **27,** 31
common problems, **27,** 66
feedback oscillation, **27,** 73
noise, **27,** 71
sample and tip condition, **27,** 68
thermal drift and piezoelectric hysteresis, **27,** 73
computer interface and data acquisition, **27,** 64
control electronics, **27,** 60
feedback, **27,** 61
preamplifier, **27,** 60
scanning, **27,** 63
mechanical structure, **27,** 51
sample positioner, **27,** 57
scanner, **27,** 53
tip, **27,** 51
vibration isolation stage, **27,** 58
theory, **27,** 33
Feedback, **27,** 33
vibration isolation, **27,** 43
Scanning tunneling microscope (STM), **30,** 111–113, 291–292, 299–303
Scanning tunneling microscopy of
adsorbate covered semiconductor surfaces, **27,** 191
adsorbate covered Si, **27,** 191
gases, **27,** 191
H, **27,** 194
NH_3, **27,** 192
O, **27,** 194
H, wet chemical preparation, **27,** 197
metals, **27,** 199
Ag, **27,** 199
Al, **27,** 202
As, **27,** 205
Au, **27,** 200
B, **27,** 202
Cu, **27,** 206
Ga, **27,** 202
In, **27,** 202
Ni, **27,** 206
Pd, **27,** 201
adsorbate induced structure, **27,** 293
adsorbates on Ge(111), **27,** 244
O, **27,** 244
adsorbates on metal surfaces, **27,** 293
chemisorption, **27,** 297
chemisorption of
C_6H_6 on Rh(111), **27,** 297
I on Pt(111), **27,** 297
O on Cu(111), **27,** 303
O on Cu(110), **27,** 300
S on Cu(111), **27,** 297
S on Re(001), **27,** 303
metallic, **27,**3
band gap states, **27,** 255, 270, 273, 275
clean Ge surfaces and alloys, **27,** 227
clean metal surfaces
reconstruction
Au(111)
dynamics, **27,** 291
surface structures, **27,** 286
Au(110), **27,** 287
Au(111), **27,** 292
clean semiconductor surfaces, **27,** 151

clusters, **27,** 266, 271, 273
diffusion, **27,** 208
diffusion of Au on Au(111), **27,** 292
diffusion of O on Cu(110), **27,** 302
dimer structures, **27,** 176
dynamics of adsorbates on metal surfaces, **27,** 298
epitaxial growth of
Ag on Au(111), **27,** 296
Au on Ni(110), **27,** 295
Ni on Au(111), **27,** 296
epitaxy of
Ge on Si(001), **27,** 247
Si(111), **27,** 208
Si(001), **27,** 208
GaAs(100), **27,** 259
GaAs(110), **27,** 110, 255
surface buckling, **27,** 254, 258
GaAs(111), **27,** 261
GaAs surface reconstruction, **27,** 254
GaAs surfaces, **27,** 251, 255
Ge(001)–2 × 1, **27,** 235
Ge surfaces, **27,** 225
GeSi(111) 5 × 5, **27,** 239
kinks and step-kink interactions, **27,** 185
metal overlayers on Ge(111), **27,** 241
As, **27,** 243
As–1 × 1, **27,**241
Sn, **27,** 243
metal surfaces, **27,** 277
resolution, **27,** 288
phase transitions, **27,** 212
semiconductors, fermi level pinning, **27,** 265, 268, 273, 274
Si(110), **27,** 87
Si(111), **27,** 151
Si(111)–2 × 1, **27,** 65
Si(111)–7 × 7, **27,** 152
Si(111) 2 × 1, **27,** 106
Si(111) laser annealed surfaces, **27,** 218
Si surfaces, **27,** 149
surface density of states, **27,** 156
stepped structures, **27,** 180
at strained surfaces, **27,** 185
superconductors, **27,** 427
experimental techniques, **27,** 430
homogeneous phase, **27,** 427
inhomogeneous phase, **27,** 429
surface dimers, **27,** 260, 261
surface trimers, **27,** 261
vortices in superconductors, **27,** 429, 434
Scanning tunneling microscopy of charge density waves in
1T-TaS_2, **27,** 358
1T-$TaSe_2$, **27,** 358, 374
1T-$TiSe_2$, **27,** 381
1T-VSe_2, **27,** 359
2H-$NbSe_2$, **27,** 386
2H-TaS_2, **27,** 391
2H-$TaSe_2$, **27,** 388
4Hb-TaS_2, **27,** 393, 400
4Hb-$TaSe_2$, **27,** 393
Scanning tunneling potentiometry, **27,** 79, 81
Scanning, emulsion, photoelectric, **5A,** 264
Scanning-electron microscopy (SEM), **16B,** 224–227
Scanning-transmission electron microscopy, **16B,** 227
space charge effects in, **16B,** 205
Scatter matrix, **1,** 478
Scatter-plate interferometer, **26,** 20
Scattered light, characteristics of
harmonious fields, **19,** 467
standing sound waves, **19,** 466
unidirectional sound waves, **19,** 465
Scattered neutron, **8,** 254, 263
energy, **8,** 263
Scattered particles
energy analysis, **17,** 185–187, 189
energy spectra, **17,** 180
kinetic energy, **17,** 159–161
relative energy spread, **17,** 177
Scattered photons, **8,** 205
Scattered-particle method, of inelastic energy-loss determination, **17,** 159, 173–174, 177, 188
beam energy fluctuations, **17,** 187
broadening effects, **17,** 170–172
Scattering, **17,** 30–31, 306–310; **26,** 49, 62; **31,** 259–260; *see also* Coulomb scattering; specific type
alpha particle, **5A,** 309–313
amplitude
atomic, **3,** 336
coherent neutron, **3,** 35
complex coherent, **3,** 357
anomalous, **5A,** 290
antiproton, **5A,** 321
attenuation due to, **19,** 28, 237–257, 267–291
in bainitic steel, **19,** 285–287
boundary conditions at grains, **19,** 240

boundary roughness, **31,** 276–278
in bubble chamber, **5A,** 385–386
in carbon steel (low alloys), **19,** 284–290
Čerenkov photons, **5A,** 177
centers, density distribution, **3,** 286–287
changed by phase transformations,
    19, 279–290
changed by precipitates within grains,
    19, 282–284
in cloud chamber, **5A,** 385–386
Compton, **1,** 415; **5A,** 631–632
constant K, **5A,** 394, 403–407
Coulomb
  deviation from, **5A,** 309
  probability of elastic, **5A,** 389
dependence on frequency, **19,** 242–247,
    272–274
dependence on grain size, **19,** 242–251
dependence on prior austenite grain size in
    steel, **19,** 285–290
dependence on velocity, **19,** 242–245
dependence on wavelength, **19,** 240
differences among grain sizes at constant
    microstructure, **19,** 288–290
differences among microstructure at
    constant grain size, **19,** 279–289
differential, linear momentum measurement
    of neutron, **5A,** 481–483
diffraction, of neutrons, **5B,** 381
electron
  by atoms, **3,** 335–338
  elastic, **5A,** 322–329
    from hydrogen beams, **3,** 600
  inelastic, **3,** 336–337
  by nuclei, **5A,** 63–71
electron-electron, **5A,** 63–71
of electrons, **16B,** 187–190
experiments, in polycrystalline media,
    19, 267–291
formulas for grain scattering, **19,** 241–244
function, Karle and Karle, **3,** 341–342
grain anisotropy in, **19,** 242–243, 267
by grains, **19,** 237–257, 267–291
heavy particles by atoms, **5A,** 55–56
incoherent, **1,** 394
intermediate region, **19,** 242–247, 267,
    272–274
in iron-nickel alloy (polycrystalline),
    19, 268–274
by large obejcts, **26,** 94
law, Rayleigh fourth power, **1,** 212
light, **1,** 393–396
of light, *see* Light scattering
low angle, **20,** 337–390
in martensitic steel, **19,** 279–281,
    285–290
mean deviation, **5A,** 400
measurement of trajectories of slow
    particles, **5A,** 403
mode conversion at grains, **19,** 240–241
molecular beams, **3,** 599–600
multiple, **5A,** 389–395
  in absorbers, charged particle loss due to,
    5B, 420–422
  events, **21,** 281, 304, 309
  measurement of electron pair energy,
    5A, 676–678
  Molière theory, **5A,** 392ff
  of reaction products in gas target, **5B,**
    422–423
  in spherical shell, **5B,** 400, 401–403
  transition to single scattering, **5A,** 392,
    394
  Williams theory, **5A,** 390ff
neutron, inelastic, **5A,** 317–319
  by liquids and solids, **5A,** 504–507
  spectrum in vanadium, **5A,** 512
  spinning sample method, **5A,** 514–515
neutrons
  by nucleus, **3,** 350–351
  by paramagnetic atoms, **3,** 352
  slow, by atoms, **3,** 349–352
in nickel (polycrystalline), **19,** 268–275
nucleus-nucleus, **5A,** 321–322
in pearlitic steel, **19,** 267, 285–287
photoemission and, **11,** 77–78
pion, **5A,** 319–321
plural, **5A,** 390
plural and multiple in Mott scattering
    target, **5B,** 256–257
power, **1,** 393ff
preferred orientation modifies, **19,** 242,
    251–253, 275–277
process, idealized, **5A,** 462
proton, **5A,** 314–317
of protons and deuterons from b^{10}, **5B,**
    355–356
Raman, **1,** 395
Rayleigh, **1,** 394, 395; **3,** 112, 117; **5A,** 673;
    see also Rayleigh scattering
remaining after tempering of martensite,
    19, 287–288

removed by tempering of martensite, **19,** 287–288
Rutherford, **5A,** 289–293
single events, **21,** 281, 309
by small particles, **26,** 95
sound, **1,** 212–213
spontaneous, **15B,** 259–260
stimulated, **15B,** 259–260
theory of, for polycrystalline media, **19,** 237–257
in titanium alloy, **19,** 281–284
Tyndall, **1,** 395–396; **3,** 112
of X-rays, **1,** 394, 415–417, 429; **3,** 306–308
in zinc bar, **19,** 275–277

Scattering amplitude, **23A,** 522; *see also* Scattering length
  of diffracted wave train, **6A,** 188–189

Scattering angle, **5A,** 289; **17,** 159–160, 162, 166–167
  deflection effects, **17,** 328
  and electron loss, **17,** 117
  mean absolute, **5A,** 394, 396, 398, 400–401
  mean relative, electrons in emulsion, **5A,** 677
  uncertainty of, **17,** 187

Scattering angle, **8,** 263

Scattering chamber, **17,** 175–176, 436–439
  for recoil energy spectrum studies, **17,** 154

Scattering chambers, **7A,** 28

Scattering cross section(s), **8,** 239; **17,** 164–165, 167–168, 174; **21,** 147; **22,** 303, 304; **23B,** 87, 194; **26,** 100
  coherent, **21,** 148–152, 195–198, 203, 207–211
  constant **Q** scans, **21,** 180–181, 186
  for defects, **21,** 152
  for disordered systems, **21,** 152–154
  effect of lattice dilatation, **21,** 154, 156, 169
  elastic, **21,** 198
    Born approximation, **21,** 149–151
    dipole force tensor P, **21,** 156, 158
  frequency dependence, **21,** 172
  incoherent, **21,** 148–149, 172, 181, 195–206, 210
  inelastic, **21,** 199
    coherent, **21,** 174–176
  theoretical limitations, defects, **21,** 182–183
  Mott, **5A,** 55
  neutron, **5A,** 466, 484
  neutron-He^4, **5A,** 493–494
  one-phonon coherent, **23B,** 155
  polarized neutron, **23B,** 89

Scattering function(s), **23B,** 86, 194, 306; **23C,** 161
  coherent, **23A,** 21
  incoherent, **23A,** 22
  intermediate, **23A,** 22; **23B,** 195
  magnetic system, **23C,** 161
  moments, **23A,** 36
  reality, **23A,** 35
  resonance, **23C,** 379
  spin-dependent, **23B,** 307

Scattering geometry, in Mössbauer spectroscopy, **11,** 309

Scattering length, **16A,** 208; **23A,** 5, 523; **23C,** 242
  bound, **23A,** 524
  coherent, **21,** 148–150, 152; **23A,** 525
    of organic molecule, **16A,** 484
  densities, neutron and x-ray, **20,** 375
  density, **23A,** 51; **23C,** 374, 407
    table, **23C,** 410
  hydrogen, **23C,** 371
  incoherent, **23A,** 525
  magnetic, **23C,** 73
  nuclear, **23C,** 72
  operator, **23C,** 72, 370
  sign of imaginary part, **23A,** 528

Scattering lengths
  measurement of, **23A,** 527
  interferometric, **23A,** 286
  spin-dependent, **23A,** 64, 91; **23C,** 261, 371
    table, **23C,** 266
  table, **23A,** 533

Scattering probability, **22,** 49

Scattering signal, **20,** 341–343

Scattering theory, classical, **16B,** 27

Scattering vector **Q**, **21,** 149, 151, 152; **9A,** 62
  experimental values, **21,** 160, 163, 164

Scattering-range measurement of mass in cloud chamber, **5B,** 37

Scavenger methods, in flame combustion studies, **7B,** 223–224

Schawlow-Towned limit, **29C,** 104

Schelkunoff gain, **12B,** 85

Schering (transforming arm ratio) bridge, **16C,** 399–400

Schering bridge, **1,** 464, 498; **6B,** 9–12
  guard circuit, **6B,** 11–12

substitution method, **6B,** 11
Scherrer equation, **16B,** 130
Scherrer shape constant, **16B,** 130, 133–135
Schiebold-Sauter method, **6A,** 211
Schlieren electron optical, **1,** 172
Schlieren instruments, **25,** 175
Schlieren interferometer, **18,** 384
Schlieren method(s)
combined with holography, **18,** 368
combined with laser Doppler, **18,** 369
dephasing schlieren system, **18,** 373
effects of diffraction, **18,** 364
optical interferometry, **9B,** 20–22
sharp focusing, **18,** 368
Schlieren refractometer, **1,** 345
Schlieren studies of sound waves, **19,** 472–475
Schlieren systems, **1,** 154
color, **18,** 367
construction and principles, **18,** 361–374
double knife edge, **18,** 367
double pass, **18,** 367
infrared, **18,** 750
schlieren head, **18,** 361
spark light source, **18,** 695–703
Toepler, **18,** 361
Schlieren techniques, **26,** 39
in gas flow, **1,** 171
Schlumberger array, **24B,** 296–297, 303–305
Schmidt corrector, **1,** 340
Schmidt system, **1,** 340
Schmidt-camera spectrograph, **12A,** 480
Schmitt trigger, **2A,** 395–397; **29C,** 327
in counting, **2B,** 2
Schottky barrier diode(s), **2A,** 62; **10,** 165, 166
Schottky barrier gate FET, **2A,** 84–85
Schottky barrier rectifiers, **2A,** 59
Schottky barrier, **2B,** 275–276
Schottky clamp, **2A,** 117
Schottky diodes, **29B,** 282
Schottky effect, **2,** 47–48; **2A,** 147–148, 152–153; **4A,** 60; **22,** 6
Schottky hump, **22,** 397, 398
Schottky noise, **2,** 701
Schottky plot, **6B,** 129–130
patchy metal, **6B,** 133–134
Schottky's theorem, **2B,** 463–464
Schottky-barrier height, **22,** 155
Schottky-barrier junctions, **2B,** 470
Schottky-type potential barrier, **5A,** 270
Schriever inertial balance, **1,** 63
Schrödinger equation, **8,** 277; **10,** 3, 233, 323, 329, 339, 367; **23A,** 260
in electron capture problems, **17,** 93, 95
high-order harmonic generation, **31,** 112
Schumann-Runge band system, **15B,** 223
Schwarzchild coordinates, **12C,** 262
Schwinger scattering, **23A,** 332
Scientific inference, definition, **28,** 156
Scintillating crystal radiation detector, **18,** 407
Scintillation counting, **2B,** 316–317
Scintillation detector(s), **2B,** 80; **7A,** 406–409; **11,** 346–347
Scintillation, **12A,** 100–102, 387, 391
bandwidth effects in, **12C,** 135–137
characteristics, **5A,** 125–132
duration, **5B,** 307
efficiency, **5A,** 126–128
absolute, **5A,** 421
gas scintillators, **5A,** 433
vs particle mass, **5A,** 416–417
sodium iodide (Tl), **5A,** 413
energy response, **5A,** 128–129
interplanetary, **12C,** 118, 126
by interstellar medium, **12C,** 118
lifetimes, **5A,** 129–131
measurements, **12C,** 118–137
monitors, **5B,** 501
processes, **5A,** 121–125
properties of anthracene, **5A,** 421ff
in pulsar signals, **12C,** 81, 135–137
spectral distribution, **5A,** 131–132
spectrometer measurement of photon total beam energy, **5B,** 543–544
spectrometry of charged particles, **5A,** 411–435
spurious, **5A,** 396, 399, 401
Scintillation counter(s), **1,** 413; **5A,** 120–157; **8,** 188, 189
boron-loaded liquid, **5A,** 535–536
general applications, **5A,** 141–142
large, **5A,** 142–149
light pipes for, **5A,** 136–137
reflectors for, **5A,** 136–137
for X-rays, **16B,** 43–45, 155
Scintillators
absorption spectrum, **5A,** 126
cesium iodide (Tl), **5A,** 413ff, 618–619
emission spectrum, **5A,** 126, 131–132
gamma ray, **5A,** 617–619
inorganic, **5A,** 411–420
ionization loss in, **5A,** 37–38

large liquid, **5B,** 408–409
light output, **5A,** 147–148
liquid, **5A,** 132–134
energy loss in, **5A,** 143–146
lithium iodide, **5A,** 417
loaded, **5A,** 137
noble gas, **5A,** 137–141, 431–435
organic, **5A,** 420–431
electron backscattering, **5A,** 425–426
pulse heights from, **5A,** 421–422, 423
scintillation decay time, **5A,** 425
plastic, **5A,** 135–136
potassium iodide (Tl), **5A,** 417
sodium iodide (Tl), **5A,** 411ff, 617–618
transparency, **5A,** 148–149
Scintillators, **8,** 256, 258, 266
material, **8,** 255
Sco X–1, **12A,** 331
SCR (silicon-controlled rectifier), **2A,** 95, 224
SCR power amplifier, **2A,** 323
Screen wire apertures, **8,** 134
Screened hydrogenic wavefunctions, in ion-atom collision, **17,** 198–202
Screening, **17,** 163–166; **22,** 229, 231
coefficient, **5A,** 83
distance, **17,** 164–165
function, **17,** 163–165
length, **23C,** 304
dynamic, **23C,** 327
of nuclear charge by atomic electrons, **5A,** 63–71
parameter, **5A,** 56
and projectile velocities, **17,** 495–496
protons in solids, **17,** 134–135
radius, *see* Channeling, theory
Screw axis, in crystal symmetry, **16B,** 9, 10
conventional notation, **16B,** 21
Screw dislocations, **6A,** 323
SDT, *see* Static drift tube
Sea Beam echo sounder, **24B,** 81, 91, 97
Sea MARC, **24B,** 98, 103
Seal
bakeable, **14,** 290, 294, 296
ceramic-to-metal, **14,** 365, 367, 373
gland, **14,** 301
glass-to-metal, **14,** 373, 375
Kovar, **14,** 379
liquid, gallium, **14,** 375
metal-to-glass, **14,** 301
mercury, **14,** 382
Sealant, **14,** 414–417
epoxy, **14,** 519
Sealing force, **14,** 285, 287, 288
Seals, hermetic, **4B,** 275–277, 287
ceramic-to-metal, **4B,** 275–277
liquid metal, **4B,** 287
metal-to-glass, **4B,** 275
quartz-to-metal, **4B,** 277
sapphire-to-glass and metal, **4B,** 275–277
Searching techniques, for spectral lines, **12C,** 49
Searchlight mirrors, **12A,** 350
Second, **1,** 29
sidereal, **1,** 69
solar, **1,** 69
Second Brewster angle, **26,** 121
Second harmonic generation (SHG), **15B,** 148–150, 157; **29C,** 235–236, 240–242
in large-signal conversion, **15B,** 159
in mid- and far-infrared, **15B,** 181–183
phase-matched, **15B,** 162
in picosecond spectroscopy, **15B,** 199–200
recent work in, **15B,** 157
short-wavelength limit for, **15B,** 212
in VUV spectroscopy, **15B,** 211–212
Second order loop system, **8,** 79
Second sound, **19,** 417–453
Doppler shift of, **19,** 445, 446
properties of, **19,** 421
transducers for, **19,** 429–434
velocity of, **19,** 419, 420
Second-harmonic generation, **29C,** 233–237
external enhancement cavity, **29C,** 239–243
intracavity frequency doubling, **29C,** 237–239
quasi-phase-matching, **29C,** 233, 234, 243–247
synchronously pumped frequency doubling, **29C,** 243
Second-order autocorrelation, **29C,** 205, 206, 208
Second-order autocorrelators, **29C,** 209–213
Second-order desorption kinetics, **22,** 425, 457, 460
Second-order light, **22,** 481
Second-order parametric interaction, **29C,** 252
Second-order structure, estimation, **28,** 388–395
Secondary, *see* Secondary electron
Secondary color, **25,** 34
Secondary crystallization, **16B,** 303

Secondary differential coefficient, **8,** 94
Secondary electron, **22,** 131, 148, 150, 153, 174, 176, 178, 179
Secondary electron ejection coefficients, **29B,** 197–201
Secondary electron emission (SEE), **4A,** 69–84, 356; **16C,** 437; **21,** 257
  backscattered electrons, **4A,** 70–71, 73–75, 80, 84
  energy distribution measurement, **4A,** 78
  field enhanced, **4A,** 79–80
  high yield surfaces, **4A,** 78–79
  processes, **4A,** 69–70
Secondary electron imaging, in scanning-electron microscopy, **16B,** 225
Secondary electron processes in solids, **7B,** 301–304
Secondary electron production by electrons, **5A,** 63–71
Secondary electrons, from ions, **7A,** 432–433
Secondary emission, **12A,** 14–16
Secondary emission by ions, **4A,** 348–355
  Auger process, **4A,** 352–355
    resonance neutralization, **4A,** 352–354
  by Cs ions, **4A,** 351
  by hydrogen and nitrogen ions, **4A,** 349–352
  by noble gas ions, **4A,** 352–355
  from Ag-Mg and Cu-Be surfaces, **4A,** 351–352, 355–356
  kinetic process, **4A,** 348–352
  yield constant (factor), **4A,** 348–351
Secondary emission detection of metastable atoms, **4A,** 408–409
Secondary emission detector, **7A,** 400
Secondary emission of electrons, **7A,** 24
Secondary emission ratio δ
  dependence on angle of primary beam, **4A,** 72, 79
    on primary energy, **4A,** 71
  first and second crossover energies, **4A,** 71–72
  measurement of
    for conductors, **4A,** 72–75
    for insulators, **4A,** 75–78
  table of, **4A,** 81–82
  temperature dependence, **4A,** 75
Secondary excitation, **7A,** 457
Secondary ion mass spectrometry (SIMS), **30,** 292, 315–318, 415; **21,** 264–268
  instrumentation for SIMS, **21,** 265
Secondary ion reactions in flowing afterglows, **7B,** 136–137
Secondary Mahan cone, **22,** 163, 164
Secondary mechanism of ablation, **30,** 284
Secondary neutral mass spectrometry (SNMS), **22,** 329; **30,** 316–318
  secondary quantities measured, **6B,** 229
Secondary source standard, **31,** 125–127
  calibration, **31,** 127–129
Secondary transition, by high-pressure measurements, **16C,** 105
Secondary-electron emission, **22,** 193, 202, 331
Secondary-electron multiplier, **22,** 320
Secondary-electron yield, **22,** 210
Secondary-emission spectrum, **22,** 227
Secondary-ion mass spectroscopy, **22,** 339, 403
  static, **22,** 339
Sectioning method for diffusion study, **6A,** 366–368
Sector spectrometer, **5A,** 359–364
Sector, magnetic, **14,** 87
Sector-focusing cyclotron, *see* AVF cyclotron
Secular equation, **19,** 210–217
  symmetrization, **19,** 214, 215
Secular equilibrium, **21,** 10
Sedimentation, effect on flow tracing particles, **18,** 41
Sedimentation analysis, of polymer solutions
  in a density gradient, **16A,** 68
  by ultracentrifugation, **16A,** 65–66
Seebeck effect, **1,** 460, 506
  definition, **6B,** 115
  measurement, **6B,** 118–125
Seeing, **12A,** 99–100
  infrared, **12A,** 441, 447, 451
Seemann-Bohlin method, **6A,** 216
Segmental motion, of a polymer chain
  by carbon–13 NMR, **16A,** 303–309
  by fluorescent probe techniques, **16A,** 410
Segregation coefficient, *see* Distribution coefficient
Seismic array, **24B,** 36, *see also* Large aperture seismic array
Seismic background noise
  detection, **24B,** 38
  filtering, **24B,** 38
  power spectral density, **24B,** 12–14
  sources, **24B,** 12
Seismic data compression, *see* Seismic event detection

Seismic event detection, **24B,** 48
Seismic monitoring system, *see* Seismograph, system
Seismic network, **24B,** 7, 35–37, *see also* Seismograph, network; World-Wide Standardized Seismograph Network; Microearthquake network
Seismic recorder, *see also* Seismograph
  analog tape, **24B,** 4, 47–48
  digital, **24B,** 4
  photosensitive, **24B,** 4, 45, 47
  smoked paper, **24B,** 4, 45
Seismic Research Observatories (SROs), **29C,** 265
Seismic Research Observatory, **24B,** 7, 67
Seismic signal, telemetered
  conditioning, **24B,** 38
  modes of reception, **24B,** 41–43
  recording, **24B,** 43–48
  timing, **24B,** 49–50
  transmission, **24B,** 38–42, *see also* Telemetry
Seismic unit, **1,** 81, 91ff
Seismic waves, *see also* Elastic waves
  boreholes, **24B,** 581–590
  porous media
    Biot media, **24B,** 592–593
    Gassman model, **24B,** 591–592
  velocity of compressional waves in various materials, **24B,** 479
Seismogram, **24B,** 1
Seismograph damping, **24B,** 25–28
  dynamic range, **24B,** 12
  early devices, **24B,** 2–3
  electromagnetic, **24B,** 6, 51–65
  long-period, **24B,** 14, 56–65
  magnification, **24B,** 5–6
  network, **24B,** 1
  output, **24B,** 61–63
  short-period, **24B,** 14, 51–56, 61–63
  system, **24B,** 2–3
Seismology, **1,** 113–114
Seismometer, *see also* Seismograph
Seismoscope, **24B,** 3
Selected ion flow drift tube
  apparatus, **29A,** 293
  applications, **29A,** 293
  principle, **29A,** 292–293
Selected ion flow tube
  apparatus, **29A,** 286–287
  applications, **29A,** 289, 295
  operation, **29A,** 287, 289
  principle, **29A,** 287
Selected-area aperture, **16B,** 212
Selection rules, **3A,** 30, 159, 277–282; **10,** 89, 304, 309, 316, 323, 327, 328, 423, 424; **13A,** 97; **16A,** 95, 104–105; *see also* Optical selectrion rules
  for atomic hyperfine structure, **3,** 627
  in electron beam excitation, **18,** 438–440
  electron spin resonance, **3,** 447–448
  in IETS, **16A,** 152, 163–164
  modification by crystal field, **16A,** 117
  for neutron scattering, **23A,** 464
  Raman, **3,** 118
  rotation, **3,** 123–125
  symmetry, **3,** 248
  vibrational, **3,** 121–123
Selective chemistry, by vibrational excitation, **15B,** 294–297
Selective deuteration, **23B,** 502, 514
Selective ionization, **22,** 410
Selenium rectifiers, **2A,** 59
Selenium-ion laser, **15A,** 149–151
  operating parameters of, **15A,** 150
Self-absorption, **7A,** 48–50, 60, 132, 133; **8,** 162, 187–188
  coefficient, **8,** 205
  effect on intensity of emitted radiation, **18,** 640
Self-compressed discharges, **15A,** 91
Self-consistent phonon theory, **23B,** 312
Self-diffusion, *see also* Diffusion
  Co in $Co_{1-x}O$, **24A,** 108
  ^{28}Mg in MgO, **24A,** 107–108
  Na in minerals, **24A,** 108
  ^{18}O in Al_2O_3, **24A,** 96–99, 104, 117
  ^{18}O in anorthite, **24A,** 123
  ^{18}O in olivine, **24A,** 112, 117, 123
  ^{18}O in quartz, **24A,** 117
  ^{18}O in UO_2, **24A,** 117
  ^{18}O in ZRO_2, **24A,** 100
  oxygen, **24A,** 90
  silicon isotopes, **24A,** 118
Self-focusing, **30,** 37–38, 177
  focusing, as nonlinear optical effect, **15B,** 261–262
Self-reversal of atomic lines, **7B,** 16
Self-reversal of spectral lines, **4B,** 59, 96
Self-seeding technique, of crystallization, **16B,** 260, 381–383
Self-similarity, **18,** 828

Self-sustaining oscillation, in semiconductor diode lasers, **15A,** 246–247
Self-trapped exciton recombination luminescence, **30,** 117–120
Self-trapped excitons, **30,** 56–57
SELFOC® lenses, **29C,** 351–352
Selove receivers, **12B,** 213
SEM (scanning-electron microscopy), **16B,** 224–227
Semiclassical approximation theory, **17,** 110, 112, 202, 229, 237, 241, 321–330, 337–342, 383
  Coulomb ionization, **17,** 486
  low-velocity effect, **17,** 205
  relativistic electron wavefunctions, **17,** 211
Semiconduction in earth, **24B,** 289
Semiconductor(s), *see* Li-drifted semiconductors
  I-III-VI_2, **30,** 524–537
  II-VI, **30,** 520–528
  III-nitrides, **30,** 541–543
  ablation of, **30,** 86–92
  apparatus for electron irradiation of, **6A,** 359
  applications , **30,** 630–632
  bandstructure of, **30,** 69–70
  bulk measurements, **6B,** 57–62
  cesiated, **11,** 112
  CIGS, **30,** 544–550
  circuit elements, **2,** 113–176
  CIS, **30,** 544–550
  conduction in, **2,** 115–118
  current noise, **2,** 807–808
  definition, **6A,** 16
  degenerate, **11,** 179–1480
  desorption from, **30,** 67–71, 78
  detector, **8,** 204, 213, 216, 223
  doped epitaxial, by PLD, **30,** 517–519
  evaluation of homogeneity, **6B,** 60–62
  experimental photoelectromagnetic measurements, **6B,** 360–364
  general description, **2,** 113–114
  generation-recombination noise, **2,** 805–806
  high frequency effects, **6B,** 110–114
  intrinsic, **6A,** 17
  laser annealing of, **30,** 364
  magnetic susceptibility, **6B,** 177–178
  minority carrier lifetime, **6B,** 351
  neutron energy spectrometer, **5A,** 276–277
  organic, **2**, 120–121
  photoemission data, **6B,** 381–384
  properties, devices employing, **2,** 121–127
  scanning tunneling microscopy of, **27,** 149
    surface density of states, **27,** 156
  special methods for measuring conductivity, **6B,** 65–66
  surface measurements, **6B,** 62–63
  surface modification of, **30,** 362–364
  temperature coefficient, **1,** 458
  thermoelectric properties, **1,** 514–515
  thermometers, **1,** 247ff
  in tunnel junctions, **11,** 177–187
  vacuum cleavage of, **11,** 181
Semiconductor circuit elements, **2A,** 45–146
  rectifier cartridges and stacks in, **2A,** 60
  temperature range in, **2A,** 46
  thermal resistance in, **2A,** 46
  voltage regulation rectifiers and diodes in, **2A,** 60–61
Semiconductor Compton spectrometer, principle, **8,** 202–203
Semiconductor controlled rectifier, **2A,** 224
Semiconductor crystals, **23A,** 413
Semiconductor detector, **5A,** 265–280; **17,** 358–361; *see also* Germanium (lithium) detector; Silicon (lithium) detector
Semiconductor devices, **2B,** 273–282, *see also* Discrete semiconductor devices; FET, transistors
  discrete, 47–102
  nonvolatile memory in, **2A,** 126–127
  symbols for, **2A,** 99–100
Semiconductor diode lasers, **15A,** 209–249, *see also* Heterojunction lasers
  carrier confinement in, **15A,** 219–220
  defined, **15A,** 209–211
  degradation of, **15A,** 242–244
  distributed-feedback lasers and, **15A,** 247–249
  facet damage in, **15A,** 242
  gain coefficient and threshold condition in, **15A,** 222–224
  heterojunction, *see* Heterojunction lasers
  homojunction, **15A,** 214
  injection in, **15A,** 216–219
  internal defect formation process in, **15A,** 242
  laser topology for, **15A,** 211–213
  materials used in, **15A,** 226–230

modulation characteristics in, **15A,** 244–247
performance of, **15A,** 234–237
radiation patterns for, **15A,** 237–242
self-sustaining oscillation in, **15A,** 246–247
temperature dependence of threshold current density, **15A,** 224–226
threshold condition for, **15A,** 222–224
vertical geometry of, **15A,** 213–216
Semiconductor laser, **25,** 265, 267; *see also* Lasers, p-n junction
Semiconductor memories, **2A,** 140
Semiconductor phenomena, non-equilibrium, **6B,** 78–109
Semiconductor photodetectors
advantages, **4A,** 449–451
basic processes, **4A,** 449–450
infrared detection, **4A,** 456–463
junction type, **4A,** 453–457, 463–469
Ga-As, **4A,** 464–465
heterojunction, **4A,** 469
high speed, **4A,** 467–469
silicon, **4A,** 465–467
photoconductive mode, **4A,** 451–453
current flow, **4A,** 452
gain, **4A,** 452
secondary photoconductivity, **4A,** 452
photovoltaic mode, **4A,** 453–456
Semiconductor photodiode, as short-pulse detectors, **29C,** 199
Semiconductor physics, short optical pulses in, **15B,** 207–209
Semiconductor saturable absorbers, modelocking with, **29C,** 180
Semiempirical tables, **8,** 272
Semilogarithmic plot, **8,** 189
Semiquantitive energy surfaces, **15B,** 273
Semivariogram, **28,** 380–381
estimating, **28,** 388–389
model, fitting, **28,** 390–392
nonparametric estimates, **28,** 389–390, 393–395
Senarmont prism, **12A,** 370
Senarmount compensator, **26,** 144
Sending/receiving antenna, **8,** 152
Sensible heat, **30,** 182
Sensitivity enhancement factors, continuous wave, **19,** 114–116
attenuation sensitivity, maximum, **19,** 116
dispersion sensitivity, maximum, **19,** 116
Sensitivity, **14,** 83; **22,** 258, 286
of electron spin apparatus, **3,** 490–498
inorganic scintillators to gamma rays, **5A,** 411
photographic emulsions, **5A,** 210–216
spectral, **1,** 391
Sensitors, **2,** 122
Sensor
light, **18,** 689, 715, 726, 735
temperature probe, **18,** 460–463
pressure gage, **18,** 534–552
Sensor, aperture, **8,** 50
Separation, **14,** 89
Separator, *see* Isotope separator
SEPP, *see* Single escape pair peak
Sequential logic design, **2A,** 414–416
Sequential operation, of SEC vidicon, **12A,** 304
Serial autocorrelation, **28,** 253, 259
Series and parallel components, of two-terminal network, **2A,** 6
Series compensation, in feedback control systems, **2B,** 415–421
Series reactance, **8,** 111
Series resistance (voltage clamp), **20,** 463–465
Series resonance, **8,** 13
Servo control receivers, **12B,** 217
Servo control, electronic frequency, **29C,** 103–104, 106, 286–288
Servo frequency tracking, applied to LDV, **18,** 169–174
Servo multiple-beam interferometer, applied to LDV, **18,** 211–220
Servo systems, frequency lock in, **12C,** 224–225
Servo-mirror dynamics, **12A,** 209
Servomechanisms, **2,** 511–518
components of, **2B,** 437–442
defined, **2B,** 376, 436
linear, common types of, **2,** 511–512
motors for, **2B,** 437–440
relay type, **2B,** 444
Ward-Leonard, **2,** 513–514
Ward-Leonard system as, **2B,** 442
Servomotor, two-phase, **2,** 511–512; **2B,** 439–440
SETE (New York University), **2B,** 173
Settling plenum, wind tunnel, **18,** 758
SEXAFS, *see* Surface-extended x-ray-absorption fine structure

Sextant, **1,** 56
Seya monochromator, **7A,** 167–168
Seya-Namioka monochromator, **22,** 144; **29C,** 35
Sezawa wave, **19,** 501, 502
SFG, *see* Sum frequency generation; *see* Sum-frequency generation
SGM, *see* Spherical grating monochromator
SGMOS (silicon gate metal oxide semiconductor), **2A,** 86
Shadow casting, **6A,** 182–183
Shadow cone, **22,** 310
Shadow microscope, **6A,** 184
Shadow technique, **22,** 418
Shadow test, **25,** 179
Shadow-cone effect, **30,** 312
Shadowgrams, in Schlieren methods, **9B,** 25–27
Shadowgraph
  high speed frames, **18,** 738
  light source, **18,** 694, 695–703
  method, **18,** 355, 356–360, 407, 738
Shadowing
  coefficient, **22,** 304
  of specimens for electron microscope, **16B,** 191, 214–215
Shannon number, **28,** 336–337
Shannon's
  optimum rate formula, **2,** 695–696
  sampling theorem, **2,** 690–692
Shannon-Jeynes entropy, **28,** 411–412, 415
Shape determination, **23C,** 391
Shape factor, **3,** 8
  for plasma containers, **7B,** 89
Shaped dual reflectors, *see* Reflectors
Shaping circuit, in counting, **2B,** 1–2
Shapiro-Wilk test, **28,** 203–204
Sharp cut-off model, **5A,** 309
Sharp quantum number, **4B,** 110–112
Shear
  bands, **16C,** 233
  compliance
    complex, **16C,** 10
    measurement of, **16C,** 27
  constant, **30,** 89–90
  failure, **6A,** 311
  flow, and crazing, **16C,** 269
  fluctuations, **16A,** 180
  rigidity, **1,** 208
  stress, **1,** 100, 123
Shear horizontal (SH) wave, **19,** 501
Shear impedance measurements, of liquids, **19,** 148–174
  guided waves, **19,** 159–166
  high pressure methods, **19,** 170–172
  inclined incidence, **19,** 148–149, 153–156
  normal incidence
    1–200 MHz, **19,** 149–156
    40–510 MHz, **19,** 156–159
  plane waves, **19,** 159–161
  reflection of shear waves, **19,** 148–159
  resonance, **19,** 166–170
  torsional waves, **19,** 161–166
Shear modulus, **1,** 106ff, 110, 214; **6A,** 297, 299–301; **19,** 5, 23–24, *see also* Elastic Moduli
Shear thinning, **16C,** 47
Shear viscosity
  dynamic, complex, **16C,** 10
  hypersonic, **16A,** 192
  measurement, **16C,** 45
Shear wave(s), **16C,** 57
  propagation in liquids, **19,** 139–141
  propagation in solids, **19,** 241–245, 273
  velocities, **24A,** 3
Shearing apparatus, **6A,** 422
Shearing interferometer(s), **18,** 375, 383–389, 740; **26,** 17
Sheet resistance, **2A,** 118
Shell correction term, **5A,** 4, 19–20
Shell correction, in energy-loss calculations, **17,** 36–37
Shell effect, **17,** 102, 107
  on equilibrium charge-state distributions, **17,** 119, 121, 124, 128
Shell model of nucleus, **23C,** 268
Shell model, **23A,** 380, 458
Shellac, **14,** 417
Shelton method, **22,** 13, 14
  application, **22,** 14
  retarding-potential, **22,** 13
  thermionic emission, **22,** 13
SHG, *see* Second harmonic generation
Shielding (neutron scattering), **21,** 178, *see also* Radiation protection
Shielding effect
  diamagnetic, **3,** 366
  intermolecular, **3,** 409
  intramolecular, **3,** 366–369
Shielding, **12A,** 40, 60–62; **14,** 453; **23A,** 199
  in electronic circuits, **2A,** 40
  magnetic and electric, **2,** 45–46

Shields, graded, **5A,** 631
Shift factor, **16C,** 14
Shift mechanisms, **8,** 161
Shift phenomena, **3,** 410–412
Shift-register techniques, in telemetering, **2B,** 224
Shifted Maxwellian distribution, **30,** 241
Shifts
  antiferromagnetic in single crystals, **3,** 374–395
  chemical, **3,** 368–369, 404–407
  conduction electron in metals, **3,** 369–370
  Dickenson, **3,** 371
  dilution, in strong hydrogen bonding systems, **3,** 424–425
  ferromagnetic in single crystals, **3,** 374–395
  Knight, **3,** 369–370
  molecular interaction in liquids, **3,** 370–374
  paramagnetic in single crystals, **3,** 374–395
  Van der Waals spectral, **3,** 234–238
Ship flow dynamics, dimensional analysis, **18,** 847–848
Shish kebob crystals, **16B,** 347, 384; **16C,** 143
Shock propagation, **1,** 193
Shock strength, **1,** 194, 195; **18,** 787
Shock tube(s), **1,** 192–197; **7A,** 122–123, 156, 181, 201–205; **7B,** 141–187
  chemical kinetic studies, **18,** 792–795
  combustion driver, **18,** 788
  composition measurement actions, **18,** 632–634, 656–659
  description of, **7B,** 141
  description as research apparatus, **18,** 785–796
  electric driver, **18,** 790
  electromagentic type, **7B,** 148
  flows, deviations from ideal calculations, **7B,** 157–167
    measurement techniques in, **7B,** 169–171
  gas and sound measurement by ultrasound, **18,** 339
  gas temperature measurement methods, **18,** 467–487
  measurement of surface heat transfer, **18,** 667, 677–685
  modified as shock tunnel, **18,** 791–792
  optical methods in measurements, **7B,** 170
  pressure gage test and calibration, **18,** 557
  production of high temperature gases, **18,** 787–790
  reflected shock region, **18,** 787–788
  reflected shock nonuniformities, **7B,** 168–169
  shock strength, **7B,** 145
  shock wave production in, **7B,**141–149
  strong shock generation, **7B,** 146–148
  x-t diagram, **18,** 786
Shock tunnel, research apparatus, **18,** 462, 666, 791–792
Shock wave(s), **1,** 129ff, 192–193, 194; **9A,** 169; **30,** 335, 350–351, 390–391
  apparatus, **6A,** 435–436
  collisionless type in laboratory plasmas, **9A,** 213–274
  dynamic yielding, **24A,** 200–203
  effects, in short-pulse dye lasers, **15A,** 318–319
  electromagnetic shock tubes, **9A,** 173
  generating systems, **24A,** 186–190
  governing equations, **24A,** 190–193
  at high pressure, **6A,** 434–436
  Hugoniot
    elastic limit, **24A,** 202
    Rankine-Hugoniot equations, **24A,** 192–193
  ionizing type, **9A,** 170
  isentrope, release, **24A,** 209–222
  light gas guns, **24A,** 190–191
  multiple shock fronts, **24A,** 201
  particle velocity gauge, **24A,** 213–214
  phase transitions, **24A,** 200–203
  plane impact of flyer plate, **24A,** 194–197
  plasma heating by, **9A,** 169–212
  pressure driven shock tubes, **9A,** 173–174
    contact surface, **9A,** 173
  Rayleigh line, **24A,** 193, 195, 202
  reflected temperature of, **9A,** 173
  reflection
    free surface, **24A,** 197–199
    impedance boundary, **24A,** 199–200
  temperatures, shock and postshock, **24A,** 228–233
  theory of, **9A,** 169–173
  velocity measurement
    electrical, **24A,** 207–208
    optical, **24A,** 203–207
Shock wave recorded
  by interferometer, **18,** 741
  by schlieren method, **18,** 365
  by shadowgraph, **18,** 358, 738, 768
  by shearing interferometer, **18,** 386
Shock waves, boundary layer, **7B,** 164

density measurements of, **7B,** 171–177
optical methods, **7B,** 171–173
schlieren and shadowgraphs, **7B,** 173–176
dissociation relaxation studies, **7B,** 185
electron beam densitometry of, **7B,**177–178
equilibrium gas studies of, **7B,** 181–183
equilibrium studies, **7B,** 183–187
ionization rates, **7B,** 186–187
in measurement of basic physical properties, **7B,** 181
optical absorption methods
partial density of particular species, **7B,** 178
production in shock tubes, **7B,** 141–149
properties of, **7B,** 149–157
recombination of atoms in, **7B,** 186
shock curvature of, **7B,** 164
temperature measurement of, **7B,** 179–180
hot wires, **7B,** 180
optical methods, **7B,** 179
speed of acoustic waves, **7B,** 180
vibrational relaxation, **7B,** 184
visualization of, **7B,** 173–174
X-ray absorption studies, **7B,** 178
Shock, use in formation of negative ions, **7A,** 201–205
Shock-tube data, **15B,** 74
Shock-tube driven supersonic mixing flames, **15B,** 137
Shock-wave relations, **7B,** 149–157
Short duration light sources, **18,** 717–718
Short laser pulses, **29C,** 193–194
by difference frequency generation (DFG), **29C,** 250–251
detectors, **29C,** 198–199
autocorrelation, **29C,** 203–208
cross-correlation, **29C,** 198–203, 220–223
intensity autocorrelators, **29C,** 209–215
phase-sensitive autocorrelators, **29C,** 215–220
streak camera, **29C,** 199–203
error sources, **29C,** 198
laser-produced plasma, **29C,** 10, 15–17
spatial characterization and focusing, **29C,** 196–198
theory, **29C,** 194–196
VUV and x-ray regions
cross-correlation, **29C,** 224, 226–227
ultra short pulses, **29C,** 223
x-ray streak camera, **29C,** 223–224
Short optical pulses, *see also* Picosecond spectroscopy; Ultraviolet spectroscopy
applications of, **15B,** 185
in semiconductor physics, **15B,** 207–209
Short pulses, interactions with, **15B,** 164–165, *see also* Short optical pulses
Short-circuit devices, **2A,** 182–183
Short-pass filters, **25,** 282
Short-pulse amplifiers, **15A,** 281–287
Short-pulse dye lasers, **15A,** 269–293, *see also* Dye lasers; Laser(s)
amplified spontaneous emission from, **15A,** 271–276
coaxial flashlamps in, **15A,** 319–321
dye flow system in, **15A,** 324–325
excitation volume for, **15A,** 280
intraresonator and, **15A,** 290–292
laser threshold in, **15A,** 291
resonators and tuning elements in, **15A,** 321–324
ring oscillator parameters and, **15A,** 291–292
short-pulsed amplifiers and, **15A,** 281–287
thermal limitations of, **15A,** 293
two-mirror resonator and, **15A,** 287–290
wavelength-tunable oscillator and, **15A,** 276–278
Short-pulse OPOs, **29C,** 261–264
Short-pulse single-pass amplifier, output and point-by-point efficiency of, **15A,** 284
Short-pulse wavelength-tunable oscillator, **15A,** 276–278
Short-range order in alloys, **23B,** 99
Shorted air-dielectric transmission line, **8,** 115
Shorted line technique of measuring luminescence, **6B,** 313
Shot effect, **4B,** 4
Shot noise, **2B,** 463–464, 499; **7A,** 23, 73, 189–191, 274; **22,** 259; *see also* Noise
in amplifier, **2A,** 251
Shottky-barrier junctions, **11,** 172–174
Shoulder height, **25,** 165
Shower
absorption, total, Čerenkov counters, **5A,** 189–191
detectors, **5A,** 652–668
Cassels, **5A,** 664–665
Čerenkov, response to positrons, **5A,** 663
limitations, **5A,** 665–668
size, **5A,** 655–660

Shower method measurement of photon total beam energy, **5B,** 544–546
Shrinkage factor, **5A,** 216, 223–224, 226–227
Shroedinger equation, **3,** 335
Shrouded thermocouple stagnation temperature probe, **18,** 463
Shubnikov groups, **23A,** 463
Shubnikov-de Haas oscillations, **11,** 9, 23–24
Shunts, **1,** 449–450
Shut-in pressure, **24B,** 395
Shutter
  elecro-optical, **1,** 386
  electromagnetic modulated, **7A,** 295
  methods, electronic, for lifetime measurement, **5B,** 327–333
  microwave cavity, **5B,** 330–331
  radiofrequency deflecting, **5B,** 327–329
  for single exposure photography, **18,** 727–732
Si detector, **8,** 222
SI system of units (Système International), **18B,** 822
Si(-Li) detectors, **8,** 214
Si(001), scanning tunneling microscopy of, **27,** 168
Si(001)–2 × 1
  scanning tunneling microscopy of
    low temperature, **27,** 173
    room temperature, **27,** 170
Si(111), scanning tunneling microscopy of, **27,** 151
Si(111)–2 × 1
  scanning tunneling microscopy of, **27,** 106, 165
  spectroscopy of
    current imaging, **27,** 132
    I-V measurements, **27,** 112
    plane wave model of STM contours, **27,** 107
    surface density of states, **27,** 112
    voltage dependent imaging of, **27,** 106
Si(111)–7 × 1, scanning tunneling microscopy of, **27,** 152
Si(111): As–1 × 1, scanning tunneling microscopy of, **27,** 205
Si(111): B, scanning tunneling microscopy of, **27,** 204
Si(111): Cu, scanning tunneling microscopy of, **27,** 207
Si(111): H, scanning tunneling microscopy of, **27,** 197
Si(111): Ni, scanning tunneling microscopy of, **27,** 206
Si, *see* Silicon
Side groups
  methyl and phenyl, reorientational motion by NMR, **16A,** 309–312
  torsional dynamics, **16A,** 235–236
Side-looking sonar, *see also* Sonar, bottom imaging
  method of imaging, **24B,** 99
  nature, **24B,** 98
  naval systems, **24B,** 104
  on-line processing, **24B,** 105
  patch width, **24B,** 103
  recording
    digitizing, **24B,** 106
    fascimile, **24B,** 93
    time varying gain (TVG), **24B,** 98
  slant-range transformation, **24B,** 105–106
  types, **24B,** 103–104
Side-pumped pulsed amplifiers, **29C,** 162
Sidereal day, **1,** 69
Sidereal second, **1,** 69
Sidereal time standards, **1,** 29
Sidewall fracturing, borehole, *see* Borehole, breakouts
Siegbahn X-ray scale, **1,** 40
Siemens process, for nobium, **11,** 571–573
SIFDT, *see* Selected ion flow tube
SIFT, *see* Selected ion flow tube
Sign convention, **25,** 14
  in polarization, **5B,** 222–223
Signal analysis
  classification, **18,** 161
  definition, **18,** 154
Signal averagers, **9A,** 15
  in ionic drift velocity measurements, **11,** 458–461
Signal averaging, **7A,** 64, 77, 277, 292–299
  in ac measurements, **2B,** 75
Signal conditioning, **18,** 162–165
Signal deflection, **2,** 662–663
Signal detectability, theoretical limit, **8,** 1–10
Signal dropout in LDV, **18,** 157–164, 172–173, 175
Signal filtering, **18,** 163
Signal gating, **22,** 313
Signal generator, **1,** 475
Signal generators, in equipment testing, **2B,** 180–183
Signal processing

classification, **18,** 161
definition, **18,** 154
effect of signal-to-noise ratio, **18,** 155
heat transfer gage, **18,** 666, 670
hot-wire anemometer linearizer, **18,** 302
Signal source, **8,** 137
Signal spectra, in LDV, **18,** 302
Signal strength in radioisotope atomic beams, **4A,** 206–208
Signal to noise ratio (S/N), **8,** 6, 8–10, 24
Signal to total noise ratio, **8,** 10
Signal, ideal minimum detectable, **3,** 490–491
Signal-converter storage tube, **2B,** 356
Signal-to-noise, **7A,** 72–73, 75, 187–191, 274–278
Signal-to-noise power output, **3,** 440
Signal-to-noise ratio, **4A,** 326, 330–335; **12A,** 74–94; **22,** 285, 292; **26,** 247; **28,** 74
in amplifier, **2A,** 77
amplifier selection and maximization, **29A,** 443–444
dependence on amplifier response, **4A,** 334
effect of refractive index variations, in LDV, **18,** 141
in electrographic recording, **12A,** 274
heterodyne configuration, **18,** 144
homodyne configuration, **18,** 145
improvement by pulsed measurement, **29A,** 437
in interferometers, **12A,** 516
multiple particle effects in LDV, **18,** 147–154
optimization, **4A,** 326, 332–335
Penning discharge source, **29A,** 370–371
photodetector, **18,** 140, 146
for quadrupole spectrometer, **3,** 508
requirements in LDV, **18,** 140–154, 234, 235
in telemetering, **2B,** 241, *see also* Noise
Signal-to-noise ratio for nuclear induction apparatus, **5B,** 112–113
in photographic recording, **12A,** 223
in photomultipliers, **12A,** 74–94
in television systems, **12A,** 282, 291
Signal-to-shot-noise ratio, **22,** 259
Signals, *see also specific types*
measurement of in radio astronomy, **12B,** 202–210
types of in radio astronomy, **12B,** 201–202
Signatures, **13B,** 66
Significance of results, **3,** 716; **6A,** 7–8
Significance test, **28,** 179–181
Significant digits, **1,** 21
Silica glass(es), **23C,** 459
hydrogen, from ultraviolet radiation, **14,** 330–333
Silica optical fiber, **29C,** 373–375
Silica replica technique, **6A,** 179
Silica, thermal outgassing, **14,** 325
Silicate glasses, binary, **6A,** 142
Silicon, **10,** 95, 96, 98, 99, 100; **22,** 154, 157, 175, 181, 182
diodes, **10,** 20, 21, 163–165
in MOS and other transistors, **2A,** 78–79, 85–86
rubber, **14,** 407, 410–411, *see also* Elastomer
Si(111), **22,** 154, 157, 169, 182
Si(111)-Al, **22,** 155
Si(111)-Cl, **22,** 157, 158, 169, 181
Si(111) (2 × 1) surface, **22,** 153, 175
Si(111) (7 × 7) surface, **22,** 153, 155
Silicon (lithium) detector, **17,** 88, 214, 218, 313, 358–361, 374–375
Silicon (Si), **30,** 91
Silicon annular transistor, **2A,** 69
Silicon barrier semiconductor detector, **7A,** 407–409
Silicon carbide, reflectance spectra, **31,** 224
Silicon detector, with thin layer, energy loss measurements on, **17,** 55–57
Silicon dioxide (SiO_2) , **30,** 152
Silicon dioxide, reflectance spectra, **31,** 222
Silicon gate metal oxide semiconductor (SGMOS), **2A,** 86
Silicon hot carrier rectifiers, **2A,** 59
Silicon junction diodes, **2A,** 365
Silicon mesa transistor, **2A,** 69
Silicon P-N junction switching diodes, **2A,** 61
Silicon photodiode arrays, **29B,** 418–419
Silicon planar transistor, **2A,** 69
Silicon solar cell, **2B,** 327
Silicon transistor chips, group of, **2A,** 102
Silicon transistor wafer, **2A,** 101
Silicon transistor, geometries of, **2A,** 101
Silicon unijunction transistor (UJT), **2A,** 90–91, 351–352, 355–356, 361
Silicon, reflectance spectra, **31,** 207–208
Silicon-controlled rectifier (SCR), **2A,** 95, 224
Silicon-on-insulating substrate (SOIS) structures, **2A,** 83–85

Silicon-on-sapphire (SOS) structures, **2A,** 125–126
Silly putty, **16A,** 5
Silver
  alloys, **14,** 375
  chemical properties
    boiling point, **14,** 350
    melting point, **14,** 347, 350
    vapor pressure, **14,** 347, 350
  chloride, **14,** 416–417
  electrical conductivity, **14,** 363
  hydrogen solubility, **14,** 354
  lubricant, **14,** 416
  as mirror coating, **29C,** 347
  oxygen filter, **14,** 422
  oxygen permeability, **14,** 375
  physical properties, **14,** 375, 376
  Young's modulus, **14,** 360
Silver
  reflectance spectra, **31,** 210–211
  transition radiation, 93–94
Silver ion conductors, **23B,** 203, 228, 232
Silver ion laser, **15A,** 162–164
Silver-oxygen-cesium cathode, **6B,** 389–390
Similarity laws, **2,** 78–79
Similarity rule, **5A,** 660–661
Similarity solution, examples, **18,** 832, 834, 835–842
Similarity, **18,** 828–829
  of gas flow, **1,** 142–143
SIMION
  electrostatic lens system design, **29A,** 205–206
  operation, **29A,** 205
Simple freezing-in, **6A,** 44
Simple harmonic oscillator, selection rules for, **15B,** 60
Simple linear regression, **28,** 162–163
Simpson's rule, **1,** 58
SIMS, *see* Secondary-ion mass spectroscopy
Simulation, *see also* Monte Carlo simulation
  $K^0{}_1 \rightarrow \pi^+\pi^-$ decay, **28,** 447–450
    motivation for, **28,** 435–436
    particle detector, **28,** 446–447
    roulette, **28,** 441–443
    sampling any distribution, **28,** 437–440
    within simulations, **28,** 454–455
    traveling salesman problem, **28,** 441
    validity testing, **28,** 450–451
    variance reduction, **28,** 439
    weighting schemes, **28,** 453
  of physical systems, **28,** 433–455
    analysis of experiments, **28,** 451–453
    annealing, **28,** 440–441
    available codes, **28,** 450
    background rejection, **28,** 441
    correlated sampling, **28,** 454
    examples, simple simulations, **28,** 436–437
    games, **28,** 443–444
    high-energy particle interactions, **28,** 444–445
    history, **28,** 434–435
Simultaneous segregation of two impurities, **6A,** 61
SiN glasses, **23B,** 270
Sinapic acid, **30,** 420
Sine-screw linkage system, **5A,** 603
Sine-wave-modulated magnetic structures, **23C,** 131
Sing-around methods, **19,** 89–90
Sing-around type flowmeter, **18,** 339
Single alignment, channeling, **21,** 290
Single antenna, spectral-line measurements with, **12C,** 20
Single binary collision peaks (SBC), **21,** 237
Single cavity maser amplifiers, **12B,** 250
Single collision, **17,** 306–310
  charge-changing probabilities, **17,** 79
  electron capture in, **17,** 89–94, 351
    multiple, **17,** 104–107
    single, **17,** 105
  heavy ions in gases, **17,** 272
  hypersatellite production, **17,** 408–410
  inelastic energy loss in, **17,** 149–192
  multiple-electron loss, **17,** 116–118
  multiple L-shell ionization, **17,** 401–408
  neon, **17,** 410–416
Single crystal diffraction patterns, **6A,** 273–276
Single crystal diffractometers, **6A,** 218
Single crystal growth, **6A,** 86–121
Single crystal instruments for X-ray scattering, **6A,** 218–219
Single crystal orientation, **6A,** 238–240
Single crystal technique for structure determination, **6A,** 229–231
Single crystal(s), **23C,** 83
  polymer, *see* Polymer single crystal
  tunneling into, **11,** 140–141
Single cylindrical curved-crystal spectrometer, **17,** 365–369

Single discriminator, in pulse amplitude measurements, **2B,** 83–84
Single electron capture cross section, **17,** 98, 103
Single electron loss cross section, **17,** 98, 106
Single escape pair peak (SEPP) intrinsic efficiencies, **8,** 215
Single frequency laser, **25,** 260
Single layer films, **26,** 125
Single longitudinal mode tunable dye laser, **29C,** 167–168
Single mode propagation, **8,** 150
Single modulator
  energy gain, **8,** 170–173
  phase condition, **8,** 170–173
Single molecule, **22,** 370
Single resonant oscillator, **15B,** 161
  conversion efficiency for, **15B,** 171
  dichroic beam splitter for, **15B,** 173
  risetime for, **15B,** 169
  threshold power density for, **15B,** 169
Single-channel tunable spectrometer, **12B,** 220
Single-crystal orientor, X-ray goniometer, **16B,** 63
Single-crystal X-ray monochromator, **16B,** 48
Single-electron collision process, **15A,** 109
Single-escape pair-peaks, **8,** 219, 220
  fractions, **8,** 223
Single-layer diffraction, **26,** 53
Single-plane crystal spectrometer, **17,** 361–365
Single-shot autocorrelator, **29C,** 208
Single-sideband receivers, **12B,** 212
Single-stage transistor amplifier, limitations in, **2A,** 272–281, *see also* Amplifier, Transistor amplifier
Singly ionized krypton, energy level of, **15A,** 68–69
Singly-excited state, beam-foil interactions, **17,** 551, 557–560
Singularity functions, **2B,** 389
SINQ, **23A,** 132, 137
Sinusoidal component, **8,** 24
SiO, transition frequencies of, **12C,** 27
SiO_2, **23B,** 274, correlated motions, **23B,** 296
SIRIS, *see* Sputter-initiated resonance ionization spectroscopy
SISAM, **13B,** 85
SIT (silicon intensifier-target) camera tube, **2B,** 347, 351
Six-pole magnets, for beam focusing and compression, **29B,** 109–110
Skimmer
  disturbance in nozzle beams, **7A,** 246–250
  for molecular beam, **18,** 654–656
Skimmers, **29B,** 30–31
Skin depth, **1,** 461; **2A,** 40; **3,** 479
  of metals, **30,** 21
Skin effect, **24B,** 523–528
Skin thickness, **5A,** 327–328
Skin-depth in plasma, **10,** 289
Sky subtraction, **12A,** 527
Sky-horn switching, **12C,** 33, 49
Slab pressure gage, **18,** 588
Slab target, in ion-atom collision, **17,** 371
SLEEP, *see* Scanning low-energy electron probe
Slewing rate, in equipment testing, **2B,** 208
Sliding pulsers, in pulse amplitude measurements, **2B,** 100, 120–121
Sliding seal, **7A,** 32
Sliding spark, **13A,** 273
  source, **7A,** 153
Slip flow, **18,** 763
Slip, isothermal, **1,** 160
Slit(s), **8,** 39–41, 43–46, 49, 170, *see also* Ion extraction slits; Monochromator; Object slit; Source
  diffracting, **26,** 57, 65
  source, **8,** 41, 42, 53, *see also* Finite width a slit source; Infinitely narrow slit source
  spectrograph, **3,** 86
  widths, **8,** 32, 171, 172, *see also* Collimator
Slit illumination, **1,** 364
Slit polarizers, **26,** 157
Slit smearing, **20,** 345
Slit-edge penetration by charged particles, **5B,** 423–424
Slotted line, **8,** 147
Slotted tube, **7A,** 34
Slow decay of luminescence, **6B,** 309–311
Slow ion-atom collision, **17,** 484–485
  charge-changing cross sections, **17,** 80
  electron capture, **17,** 102, 104
  excitation cross sections, **17,** 168
  multiple capture cross section, **17,** 107
Slow passage in NMR, **3B,** 515
Slowing down, time-dependent, **23A,** 160
Sludge or gel formation in turbo-pump, **14,** 263

Slug, **1,** 30
calorimeter, **18,** 676
SM system, *see* Center-of-mass system
Sm^{142} fission product, **5B,** 599
Smakula's equation, **6B,** 326
Small angle scattering of neutrons (SANS), **21,** 159
apparatus, **21,** 159–160
results, Al-Zn, **21,** 167
Small angle scattering of X-rays, **6A,** 244–245
Small-amplitude wave propagation, **11,** 384–387
Small-angle compromise, **20,** 344, 345
Small-angle electron diffraction, **16B,** 222
Small-angle light scattering, in polymeric alloys, **16C,** 299–300
Small-angle neutron scattering, in polymeric alloys, **16C,** 305
Small-angle scattering, **23A,** 49; **23B,** 116, 499; **23C,** 288, 368
cross section, **23B,** 507
experimental technique, **23C,** 293, 438
instruments, **23A,** 220
magnetic, **23A,** 74
technique, **23B,** 503
theory, **23C,** 289
two-phase systems, **23C,** 428
Small-angle x-ray scattering, **23B,** 499; *see also* X-ray diffraction
collimation, **16B,** 156–159
focusing methods, **16B,** 157, 158
Small-area single crystal photoconductor, **6B,** 343–345
Small-diameter sources
observation of, **12C,** 1–18
Small-single neutron scattering (SANS), *see* Neutron scattering
$SmCo_5$, **23C,** 106
Semiconductor lasers, **29C,** 77–78, 80, 149–150
diode lasers, **29C,** 77–100, 232
Semivariogram model, spherical, **28,** 394–395
SmIG ferrimagnetic exchange resonance, **10,** 419
Smith chart, **8,** 109
Smooth model, **5A,** 248ff
Smoothing
spectral data with, **12C,** 47
statistical computing, **28,** 498–500
window, leakage, **28,** 330
Sn adsorbed Ge(111)
scanning tunneling microscopy of, **27,** 243
Snell's law, **1,** 213, 223–224, 329, 379
in whistler ducting, **9B,** 95; **25,** 11; **31,** 273
SNMS, *see* Secondary neutral mass spectroscopy
Snowballing, **1,** 98
SNQ, **23A,** 132, 135
SNS junctions, **11,** 253–255, 268–273
SNS, *see* ISIS
SO_2 plasma, **7A,** 122
SO_4^{2-} ion, **19,** 202
Soda-lime glass outgassing, **14,** 325
thermal, **14,** 323
water, **14,** 326
Sodium chloride (NaCl), **30,** 586, 595–598; *see also* NaCl
Sodium continuum as a measure of relative [OH], **7B,** 212–214
Sodium fluoride, emission and absorption bands of infrared transition in, **15B,** 10–11
Sodium iodide (Tl) scintillators, **5A,** 411ff, 617–618
gamma ray absorption coefficient, **5A,** 621–623
preparation, **5A,** 619
Sodium iodide pair spectrometer, **8,** 228
Sodium iodide x-ray detector, **17,** 313
Sodium nitrate ($NaNO_3$), **30,** 142–143, 146–147, 150–152
Sodium salicylate, **7A,** 145, 458
films, in reference phototube measurements, **11,** 82–83
Sodium-chlorine method for absolute [H], **7B,** 212
Sodium-lithium mehtod for absolute [H], **7B,** 211–212
Sofar system, **1,** 213
Soft mode model, **23C,** 8, 57
Soft modes, **23C,** 2
of vibration, **23A,** 427, 500
Soft x-ray appearance-potential spectroscopy, **22,** 202, 207
Soft x-ray monochromator, **29C,** 34, 36–38
Soft x-ray, source comparison, **31,** 129–133
Software
method of maximum likelihood, **28,** 240
statistical, **28,** 483–484
SOIS (silicon-on-insulating substrate) structures, **2A,** 83
Solar blind photomultiplier, in VUV spectroscopy, **15B,** 222
Solar cell, silicon, **2B,** 327

Solar day, **1,** 69
mean, **1,** 29
Solar flux, correlation with ozone, **28,** 475–477
Solar radio-astronomy receivers, **12B,** 221
Solar second, mean, **1,** 69
Solar studies, beam-foil measurements, **17,** 600–605
Solar Ultraviolet Measurements of Emitted Radiation (SUMER), 72
calibration, **31,** 135, 138
Solar wind, plasma parameters of, **9B,** 51
Solar x rays, **13A,**193
Solder, **14,** 578
Soldered contacts, **6B,** 51
Soldering
flux, **14,** 426
soft, **14,** 425
technique, **14,** 426
Soleil compensator, **1,** 382; **6B,** 269; **26,** 143
Solenoids, **1,** 525–528
Solenoid field, *see* Finite cross-sectional area; Rectangular cross sectional area
Solenoid pair, field, **8,** 93–106
exercises, **8,** 97–106
references, **8,** 106
Solenoid volume, **8,** 89
Solenoidal spectrometers, **5A,** 365–370
Solid angle, **1,** 57ff
Solid electrolytes, **23B,** 190
Solid samples, *see* Reactor fuel research
Solid state cloud chamber, **5A,** 280
Solid state counter, **8,** 273
Solid state detector(s), **13A,** 153
noise level, **5A,** 271–273
Solid state detectors, **8,** 257, 266, 273
Solid state lasers, **15A,** 167–207; **25,** 265
characteristics of, **15A,** 169–170
crystals for, **15A,** 188–191
energy levels in, **15A,** 171–176
excitation of, **15A,** 169
fluorescence sensitization in, **15A,** 193–194
glasses for, **15A,** 191–193
hazards associated with, **15A,** 205–207
history of, **15A,** 167–168
host materials in, **15A,** 169, 187–193
ion-ion interactions in, **15A,** 179–181
laser materials in, **15A,** 181–194
nonradiative transitions in, **15A,** 177–179
Nd:glass, **15A,** 199–202
Nd:YAG, **15A,** 195–198
optical radiation hazards from, **15A,** 206
physical processes in, **15A,** 170–181
properties and comparison of, **15A,** 195–205
radiative transitions in, **15A,** 176–177
size of, **15A,** 169
stoichiometric, **15A,** 203–204
structure of, **15A,**168–169
transition probabilities for, **15A,** 176–179
upconversion in, **15A,** 194
Solid state relaxation, **3,** 430–431
Solid state studies under high pressure, **6A,** 407–437
Solid target, **17,** 191, 371–374, 379
angle resolution effects, **17,** 310
atomic transitions and charge equilibrium in, **17,** 139–144
average equilibrium charge, **17,** 127–128, 130–131
backscattering energy loss, **17,** 58–65
beam-foil experiments, **17,** 532
charge-state fractions for heavy ions in, **17,** 272–278
collisional quenching, **17,** 426–428
density effects in, **17,** 134–139
ionization and x-ray productions, **17,** 193–278
K-shell vacancy production, **17,** 335, 337
mean charge state of ions emerging from, **17,** 519–524
multiple scattering, **17,** 309
spectra, **17,** 418–424
Solid-state detector, **17,** 54–57, 86–87; **22,** 210; *see also* Germanium (lithium) detector; Silicon (lithium) detector
Solid-state detectors, in light-scattering optical systems, **9A,** 105–106
Solid-state effect
on stopping cross section, **17,** 69–70
in target x-ray production, **17,** 257
Solid-state extrusion, **16C,** 140–141
Solid-state lasers, **29C,** 151, 232
Solid-state microwave power sources, **2B,** 273–282
Solid-state phase transition, **23A,** 503
Solid-state photocells, **2B,** 323–334
Solid-state polymerization, positron annihilation study, **16A,** 402
Solids
absorption of sound in, **1,** 223–224

attenuation and dispersion of waves in, **11,** 389–391
atomistic description of, **7B,** 289–304
elastic, anelastic, and plastic properties, **6A,** 291–321
finite-amplitude waves in, **11,** 387–389
finitely strained, **11,** 381–389
high precision density determination, **6A,** 283–291
load-deformation-time tests, **6A,** 313–314
mechanics of, **1,** 53–114
plasticity and fracture, **6A,** 309–321
properties of, **11,** 371–441
scattering of electrons in solids, **7B,** 290–291
small-amplitude waves in, **11,** 384–387
strain in single crystals, **6A,** 314–317
stress-train relationships in, **11,** 381
tension test and other load-deformation tests, **6A,** 310–313
ultrasonic studies of, **11,** 371–441
ultrasonic waves in, **11,** 374–391
velocity of sound in, **1,** 222
wave equation for, **11,** 382–383
Solitary laser, **29C,** 83
Soliton energy, **29C,** 178–179
Soliton modelocking, **29C,** 180
Soliton supporting fiber lasers, **29C,** 184–187
Solitons, **23C,** 140
excitations, **23C,** 227
quantum effects, **23C,** 231
Soller, slit collimator, **21,** 176
Solubility coefficient
definition, **16C,** 318
of gases in polymers, empirical relations, **16C,** 327
Solubility, **14,** 29
Solutes, secondary, **5A,** 134
Solution mixing, of high polymers, **16C,** 56
Solution scattering, **20,** 337–390
Solution viscosity, **16A,** 16–17, 19
Solution, heat of, **1,** 262
Solvation equilibria, **19,** 89–90
Solvent effect, **3,** 639
Solvent, choice in crystal growing, **6A,** 99
Solvent, cleaning, **14,** 499–500, *see also specific substances*
Solvent-etching, **16B,** 239–241
Solvent-solute
interactions
spectral shift, **3,** 236, 237
strong, **3,** 238–241
separation, **3,** 238
Solvents, secondary, **5A,** 134
Sommerfeld constant, **1,** 42
Sonar, **1,** 217; *see also* Echo sounder
background noise
ambient, **24B,** 78
bubble sweepdown problem, **24B,** 81
Knudsen curves, **24B,** 79–80
local, **24B,** 80
reduction, **24B,** 81–82
relation to operating frequency, **24B,** 87
sea-surface reverberation, **24B,** 82–83
sources of interference, **24B,** 88
use of directional transmitters, **24B,** 82
volume reverberations, **24B,** 82
beam, angular resolution, **24B,** 84
bottom imaging, **24B,** 97–110
multibeam swath mapping, **24B,** 90–93
phase difference bottom mapping, **24B,** 108–110
scanning, **24B,** 98, 107
side-looking, *see* Side-looking sonar
systems, **24B,** 77
Sonde, **24B,** 441
borehole compensated density, **24B,** 553–555
combination, **24B,** 456–458
focused, **24B,** 520–523
geometric factor, **24B,** 451–453, 516–520, 528–530, 542–545
multiple sensor, **24B,** 453–456
types, **24B,** 449–458
Sones, **1,** 227
Sonic anemometer, **18,** 315–318
Sonic flow, **18,** 760
Sonic measurements in borehole
compressional interval-transit-time, **24B,** 593–599
depth-derived borehole compensation, **24B,** 597–599
shear interval-transit-time, **24B,** 608–609
wave forms, **24B,** 600–608
Sonic modulus, **16C,** 167–172
measurement of, **16C,** 167–169
Sonic nozzle flowmeter, **18,** 330
Sonic velocity, *see* Sound, velocity
SOP, *see* State of polarization
Sorbant, **14,** 244
properties, **14,** 238

Sorption, **1,** 180–181; **14,** 198, 238
apparatus, **16C,** 349–356
data analysis, **16C,** 343–349
dual-mode, **16C,** 333, 347, 361
measurements, calculating the diffusion coefficient, **16C,** 345
types of methods, **16C,** 342
utility, **16C,** 339
Sorption pump, **14,** 112, 113, 138–140, *see also* Fore-vacuum pump
adsorption isotherms, **14,** 238
high vacuum, **14,** 181–182
operation, **14,** 140
preroughing, **14,** 140
ultimate pressure, **14,** 139
Sorption ratio, **14,** 196
SOS (silicon-on-sapphire) device, **2A,** 125
Sound, **1,** 199–234
absorption, **1,** 213–215, 232–234
absorption in sea water, **24B,** 78–79
field, **1,** 224–230
general concepts, **1,** 199–216
instruments, **1,** 216–221
medium, properties of, **1,** 221–224
radiation, **1,** 202–208
reflection, **1,** 212–213
reflectivity
frequency dependence, **24B,** 86
sea floor, **24B,** 85
refraction, **1,** 212–213
scattering, **1,** 212–213
sources, **1,** 216–217
taping, **1,** 228–229
techniques, **1,** 222–234
transmission, **1,** 208–211
velocity
borehole, **24B,** 476, *see also* Interval transit
time effect of porosity, **24B,** 477–478
in ocean, **24B,** 83
Matthew's tables, **24B,** 84
sea floor, **24B,** 85
shadow zone, **24B,** 84
sounding, **24B,** 83–84
velocity of, **1,** 129, 222, 259, 277ff
Sound pressure, **19,** 5–7
Sound speed, **18,** 246, 315–317; **19,** 4–7
dispersion of, **19,** 27
Sound velocity, **16C,** 172; *see also* Sound speed
in gases and vapors, **3,** 682–700
in liquids, **3,** 700–708
Sound waves, **23A,** 374
Source, **8,** 47–49, 147, 157, 179, *see also* Arc-discharge source; Broad source; Finite width a slit source; Huygens sources; Infinitely narrow slit source; Light source; Linear source; Matched source; Microwave source; Point source; Reactor fuel research
coordinate, **8,** 95, *see* Cylindrical source coordinates
impedance, **8,** 107
numerical application, **8,** 46–47, 49, 52–53
wide slit, **8,** 44–47
width, **8,** 40
Source blocks, **23A,** 199
Source contribution, positrons, **21,** 104, 134, 135
Source function, **9A,** 143
Source mapping, **12B,** 26
Source position, measurements of, **12C,** 7
Source shape, measurements of, **12C,** 4–7
Source size measurements, **12C,** 4–5
complex sources, **12C,** 6–7
Source standard, primary vs. secondary, **31,** 126–127
Source structure, scintillation effects of, **12C,** 126
Source survey, 4C, **12C,** 170
Source-to-fiber coupling, **29C,** 379–381
Sources, **13A,** 24, 204, 224, 245, 246, 259; *see also appropriate spectral region*
artificial particle, **5B,** 580–690
low energy, **5B,** 580–622
medium and high energy, **5B,** 623
coherent, **10,** 18–81
filament, **25,** 265
incoherent, **10,** 13–18
neutron
nuclear reactor, **5B,** 590–622
radioactive, **5B,** 555–567
photon beam, **5B,** 509–515
of radiation
CO_2 laser, **3A,** 365
infrared, **3A,** 188
for Raman studies, **3A,** 423
visible and ultraviolet, **3A,** 223–233
radioactive, **5B,** 555–580
α-particle, **5B,** 567–572
β-particle, **5B,** 572–580
γ-particle, **5B,** 461–462, 572–580

Sources, positron
angular correlation, *see* Angular correlation (of annihilation radiation)
Doppler broadening, *see* Doppler broadening (of annihilation radiation)
isotopes, **21,** 81
lifetime, *see* Lifetime (of positron)
slow beam, **21,** 82
source contribution, **21,** 104, 134, 135
Space charge, **14,** 82, 219, 220; **16C,** 448, 467–469
in afterglow studies, **7B,** 80
buildup, **22,** 6
density, **1,** 504
detector for atoms, **4A,** 401–402
limitation, proportional counters, **5A,** 111
neutralization in ion beams, **4A,** 281–283
waves in plasma, **10,** 278–279
Space clamp, **20,** 449–452, 453–456
Space defense radiometers, **26,** 250
Space group(s), **3,** 274; **20,** 239; **23A,** 371; **23C,** 111
crystallographic, **16B,** 11
Space lattice, **16B,** 3, 5
Space, collisionless shock waves in, **9A,** 218
Space, thermodynamic, **1,** 308, 313
Space-charge effects, in electron imaging, **16B,** 205
Space-charge-limited diode, **11,** 461–463
Spacecraft, plasma interaction problems of, **9B,** 51–52
photoelectrons from spacecraft, **9B,** 51
secondary electrons, **9B,** 51
Spallation, **23A,** 100, 105; **30,** 106
sources (for neutrons), **20,** 355
Spark chambers, **5A,** 281–288
Spark cutting, **11,** 11
Spark discharge, electrical and fluid dynamical parameters, **18,** 695–698, 752–753
Spark formation, mechanism, **18,** 695–698
Spark gap
ratings of, **2A,** 186–187
triggered, **2A,** 184
variable, **1,** 455
Spark sources, **1,** 375
Spark vacuum, **1,** 407
Spark, light sources, **4A,** 309–311, 316
Sparks, **13A,** 210, 220, 269, 273
Sparks shifts, **15B,** 242–243
Sparrow resolution criterion, **25,** 218
Spatial autocorrelation, **28,** 253, 259
Spatial coherence, *see* Light source
Spatial data, statistical analysis, **28,** 375–400
application to sulfate decomposition, **28,** 393–395
extensions, **28,** 398–400
geostatistical model, **28,** 378–381
median polish, **28,** 383–385, 387–389
second-order structure estimation, **28,** 389–390
spatial labels, **28,** 375
spatial prediction, **28,** 396–398
sulfate deposition data, **28,** 376–378
application, **28,** 386–389
trend surface analysis, **28,** 381–383, 386
Spatial definition, **23C,** 413
Spatial echoes, **9A,** 28
Spatial filtering of images, **26,** 91
Spatial frequencies, **12C,** 3; **26,** 84
Spatial harmonics, **10,** 35, 36, 38
Spatial interdependence, field statistics, **28,** 459
Spatial ionization, **22,** 383
Spatial multiplexing method for color holograms, **26,** 181
Spatial prediction, **28,** 396–398
Spatial profile, **30,** 174
Spatial quantization, **10,** 308
Spatial resolution, **22,** 287; *see also* Measuring volume dimensions
density measurement Rayleigh scattering, **18,** 418
in LDV, **18,** 134, 137–138, 235
pressure gages, **18,** 525, 573–576
in radiation scattering diagnostics, **18,** 409, 420
Spatial separation, **8,** 183
Spatially resolved spectroscopy, of superconductors, **27,** 436
Special functions, **28,** 491
Species concentration, *see also* Composition measurement
electron beam fluorescence technique, **18,** 434, 451
mass spectrometer, **18,** 645–661
Raman scattering diagnostics, **18,** 428, 643–645

Specific acoustic impedance, *see* Impedance
Specific activity, **21,** 9, 17
  method to measure half-life, **5B,** 279, 286–292
Specific detectivity D^*, *see* D^*
Specific heat, **1,** 261, 263–280; *see also* Heat capacity
  expansion method, **1,** 276
  gases, **1,** 272–280
  liquids, **1,** 269–272
  per mole, gas, at constant volume, **3,** 683, 684, 690
  ratio of C_p/C_v, **1,** 276
  relaxing, **3,** 689
  solids, **1,** 263–269
  spectroscopic method, **1,** 275
  of superconductors, **6B,** 74–75
  ultrasonic method, **1,** 279
Specific heats, **19,** 4–5, *see also* Heat capacity
Specific intensity, **12A,** 193
Specific volume, in crystallinity determination, **16B,** 120
Speckle, laser, **18,** 712–714; **30,** 341–342
Spectometry, in particle-solid studies, **7B,** 241–242
Spectra
  electronic, **3,** 155–264
    apparatus, **3,** 156–179
    external perturbations, **3,** 222–241
    simple free molecules, **3,** 179–222
    transition probabilities, **3,** 241–261
  forbidden polyatomic, **3,** 251ff
  infrared, **3,** 42–73
  measurement of characteristics, **3,** 138–141
  microwave, Stark effect in, **3,** 650–669
  molecular recombination, **1,** 407
  optical, hyperfine structure in, **5B,** 44ff
  Raman
    ammonia, **3,** 150
    applications, **3,** 141–154
    benzene, **3,** 152
    of CO_2 Fermi diad, **3,** 148
    of diacetylene, **3,** 149
    light sources, **3,** 126–131
    naphthalene, **3,** 152
    quartz, **3,** 152, 153
    samples and cells, **3,** 131–136
    second-order, **3,** 122
    solid hydrogen, **3,** 152–153
    of water, **3,** 151
  of some simple polyatomic molecules, **3,** 208–221
  sound, **1,** 220
  vibrational, **3,** 121–123
  X-ray continuous, **1,** 413–414
  X-ray line, **1,** 414–415
Spectral analysis, **28,** 297–305, 313–346
  aliasing, **28,** 309–310
  bandwidth, **28,** 299–300
  bivariate time series, **28,** 340–346
  data windowing, **28,** 297–302
  Hanning window, **28,** 300–301
  leakage, **28,** 300–301
  removal of background trends, **28,** 302–305
  smoothing, **28,** 301–302
  spectral power density, **28,** 303–304
  spectral window, **28,** 316–340
Spectral broadening, effect on LDV resolution, **18,** 133–140; *see also* Ambiguity noise
Spectral calibration standards, **4B,** 61, 93
Spectral cross relaxation, **15B,** 66
Spectral density
  calculation of, **2B,** 496–497
  defined, **2B,** 496
  function, **28,** 313–315
Spectral distribution
  flicker of frequency, **12C,** 204
  flicker of phase, **12C,** 205
  random walk of frequency, **12C,** 204
  white frequency noise, **12C,** 205
  white phase noise, **12C,** 206
Spectral distribution, **8,** 167
Spectral emissivity, **4A,** 291–292
Spectral estimator
  lag window, **28,** 329–333
  multitaper, **28,** 335–338
  WOSA, **28,** 333–335
Spectral fitting method
  capture-τ mode, **24B,** 573–577
  inelastic mode, **24B,** 578–580
Spectral lamp, **25,** 253
Spectral line broadening
  in atom interferometry, **29B,** 355–356
  in microwave spectroscopy, **29B,** 277–278, 291–292
Spectral line intensity, **31,** 52
Spectral line shape, *see* Line shapes
  distortion by Brillouin scattering, **18,** 415–417
  Raman and Rayleigh scattering, **18,** 412

use for temperature measurement, **18,** 414–417, 490
Spectral line(s), *see* Line; Lines
frequencies of, **12C,** 23–24
relative intensity of, **1,** 391
strengths of, **12C,** 24–29
Spectral line-shape determinations at low intensity, **7B,** 106–108
Spectral line-shape studies, in spectroscopy, **7B,** 102
Spectral linewidths, **17,** 524–527
Auger spectroscopy, **17,** 476, 497–498, 516–519
Doppler broadening, in beam-foil interactions, **17,** 556–557
measurement of, with parallel-plane analyzer, **17,** 452–454
Spectral measurements, of luminescence, **6B,** 304–306
Spectral power density
filter, **28,** 285
spectral analysis, **28,** 303–304
Spectral purity, **12A,** 464, 465; **25,** 196
Spectral radiance, **4A,** 284, 290, 292, 295; **18,** 690
Spectral resolution, **12A,** 464, 465, 471, 487, 500, 532; **25,** 196
Spectral response characteristics, of phototubes, **2B,** 304
Spectral response of Josephson junction, **10,** 209
Spectral response of the eye, **18,** 688–689, 692
Spectral response photocathodes, **26,** 307
Spectral responsivity of radiation detectors, **4A,** 416
Spectral sensitivity calibration, 71–72
Spectral unit, **25,** 190
Spectral units, radiometric and photometric, **4A,** 284–289
Spectral width of light, effect on LDV performance, **18,** 137
Spectral window, **6B,** 286–287; **28,** 321
periodogram, **28,** 317
Spectral-density models, **12C,** 202–207
Spectral-line measurements
equipment requirements for, **12C,** 20–23
fundamentals of, **12C,** 19–44
history of, **12C,** 19–20
instrumental correlation of, **12C,** 30
noise in, **12C,** 52
requirements of, **12C,** 19
use of, **12C,** 31
Spectral-line narrowing, **12C,** 30
Spectral-line observations
radial-velocity corrections to, **12C,** 277–283
Spectral-line receivers, **12B,** 219–221
Spectral-line signals, **12B,** 202
Spectrochemical analysis, X-ray, **1,** 417–419
Spectrograph(s), **5A,** 349; **25,** 191
in beam-foil experiments, **17,** 539
concave grating, **4B,** 65–69
Eagle mounting, **4B,** 67–68
Paschen-Runge mounting, **4B,** 67
Rowland circle, **4B,** 66
cross-dispersion, **4B,** 73–74
eschelle, **4B,** 73
electron spin resonance, **3,** 464–469
reflection type, **3,** 466
sample location, **3,** 476–478
superheterodyne, **3,** 467
transmission type, **3,** 465
grating, **1,** 367–370; **3A,** 174, 180, 207–213
grazing incidence, **1,** 406
infrared, **1,** 399–401
calibration, **3,** 90–91
light sources, **1,** 373–376
magnetic, **5B,** 433–443
mass, **1,** 64
microwave, **1,** 66; **4B,** 72–73
coma distortion, **4B,** 73
prism, **1,** 362–366; **4B,** 65
Littrow design, **4B,** 65
Raman, **3,** 136–138
for Raman studies, **3A,** 425
time-of-flight, **1,** 65
ultraviolet, **1,** 406
vacuum ultraviolet, **4B,** 74
grazing incidence, **4B,** 74
Spectrograph, **8,** 162
Spectrographic speed, **12A,** 465, 467–469, 480
Spectrographs and spectrometers
Czerny-Turner, **13B,** 23
Doppler-tuned, **13A,** 164
Ebert-Fastie, **13A,** 235; **13B,** 23
gamma ray, **13A,** 131
Girard grille, **13B,** 85
grating, **13B,** 20
Hadamard, **13B,** 87
optical region, **13A,** 274
prism, **13A,** 274; **13B,** 17
submillimeter wave, **13B,** 129
ultraviolet, **13A,** 234

Walsh double pass, **13B,** 19
x ray, **13A,** 149
Spectrometer, **10,** 130–140; **25,** 191, 200; *see also* Spectrographs; Spectrographs and Spectrometers; specific types
Spectrometer resolution function, 17, 516–517
Spectrometer slit function
effect on spectral line shape, **18,** 412
Raman scattering analysis, **18,** 427–428
Spectrometer transfer function, **25,** 196
angle-resolved, **22,** 140
anticoincidence scintillation, **5B,** 524–528
autocorrelation, **12B,** 220–221
basic qualities of, **12B,** 269
in beam-foil experiments, **17,** 539–544
beta ray, **5A,** 342–347
calibration of multichannel, **12B,** 277–278
Cauchois photographic x-ray, **5A,** 599–601, 605
coincidence scintillation, **5B,** 528–531
Compton, **5B,** 533–534
curved crystal, **5A,** 599–616
applications and scope, **5A,** 607–616
transmission type, **5A,** 599ff
double beta coincidence, **5A,** 363
double focusing, **5A,** 354–359
double-crystal, **1,** 426
double-perfect crystal, **23A,** 266
epithermal neutron time-of-flight, **5A,** 517–544
applications, **5A,** 538–544
ESR, **2B,** 148, *see also* Pulse spectrometer
fast neutron time-of-flight, **5A,** 544–554
applications, **5A,** 551–553
gas scintillation, resolution, **5A,** 435
gas maser, **5B,** 29
general considerations, **12A,** 463
grating, **3,** 97–103
helium–3 neutron, **5A,** 570–581
hollow-crystal scintillation, **5A,** 427–428
intermediate image, **5A,** 372–374
Kuyatt-Simpson electron, **22,** 35, 40
lens, **5A,** 365–375
magnetic, **5A,** 347–375; **5B,** 431–433, 531–537; **22,** 219
beta-ray, **5A,** 588–590
performance parameters, **5A,** 342–347
resolving power, **5A,** 343ff
mass
double focusing, **5B,** 2–3
Nier type, **3,** 531
rapid-scanning, **3,** 581–582
Mössbauer, *see* Mössbauer spectrometers
microwave, **3,** 31–36
cavity modulation, **3,** 35–36
double modulation, **3,** 32–33
simple sweep, **3,** 31–32
Stark modulation, **3,** 33–35
unmodulated, **3,** 31
millimeter wave, **5B,** 28
molecular beam
deflection systems, **3,** 613–615
magnet pole faces for, **3,** 609–611
neutron velocity, high-energy electron accelerator, **5A,** 527ff
nuclear induction, **5B,** 113–115
orange, **5A,** 361ff
pair, **5B,** 534–537
three-crystal, **5B,** 446, 447
Walker and McDaniel, **5B,** 443
photoelectric, speed of, **3,** 137
photoemission measurements and, **11,** 99
photon, **5B,** 515–537
prism, **3,** 91–97; **5A,** 359–364
double beam, **3,** 92–95
fast scanning, **3,** 95–96
ratio-recording, **3,** 94, 95
single beam, **3,**91
in quantum beat experiments, **17,** 594–595
rf, broad line continuous, **3,** 438–440
rotating crystal time-of-flight, **5A,** 514–515
sector field, **5A,** 359–364
semicircular focusing principle, **5A,** 348–353
single-crystal scintillation, **5B,** 516–524
sodium iodide (Tl), response, **5A,** 621–637
solenoidal with uniform magnetic field, **5A,** 365–370
spiral orbit, **5A,** 364–365
split-crystal scintillation, **5A,** 426
Stark-modulation, **5B,** 25–26
high temperature, **5B,** 27
third order focusing, **5A,** 353–354
total absorption, **5B,** 515–531
transmission, **5A,** 345–346
triple-axis, *see* Triple-axis spectrometers
two-crystal Compton, **5A,** 629
velocity resolution of ideal, **5A,** 518–519
velocity scanning, **11,** 319–320, 342, 454–456
Spectrometers, comparison, **8,** 213–226
background, **8,** 213–214

exercise, **8,** 226
Spectrometers, *see* Anti-Compton spectrometer; Beta-ray spectrometer; Electrostatic time-of-flight spectrometer; Glove-box-type spectrometer; Neutron time-of-flight spectrometer; Photo-Peak spectrometer; Prototype spectrometer; Time-of-flight mass spectrometer; Time-of-flight neutron spectrometer
comparison, **8,** 225–226
efficiency, **8,** 225
operation, **8,** 207
peak fractions, **8,** 225–226
principle, *see* Semiconductor Compton spectrometer
usage, *see* Upper atmosphere research
Spectrometry
charged particle scintillation, **5A,** 411–435
crystal diffraction, **5A,** 612–613
gamma ray scintillation, **5A,** 616–641
mass, **3,** 525–588
Spectrophone, **19,** 192
Spectrophotometer, **25,** 191
Spectroradiometer(s), **6B,** 304–305; **25,** 191; **26,** 246, 256
Spectroscope, **25,** 191
high resolution nuclear resonance, **3,** 404
hyperfine structure, **5B,** 53–55
microwave paramagnetic resonance, **5B,** 95–96, 100–103
spectroscope, **5B,** 53–55
Spectroscopic detectors, **4B,** 84–90
Spectroscopic instruments, **25,** 190
Spectroscopic methods of impurity detection, **6A,** 30
Spectroscopic splitting factor, **6B,** 228; *see also g*-factor
Spectroscopic standards, **3A,** 216–223
Spectroscopic system
calibration, **8,** 27–37
detector, photon flux, 31–35
experimental application, 35–36
Spectroscopic techniques, **3A,** 239–248
infrared, **3A,** 172–201
microwave, **3A,** 119–125
Spectroscopy of
1T-TaS_2, **27,** 363
1T-$TaSe_2$, **27,** 376
1T-TiS_2, **27,** 383
1T-$TiSe_2$, **27,** 381
1T-VSe_2, **27,** 380
2H-$NbSe_2$, **27,** 386
2H-TaS_2, **27,** 389
2H-$TaSe_2$, **27,** 389
4Hb-TaS_2, **27,** 400
4Hb-$TaSe_2$, **27,** 400
adsorbate covered surfaces, **27,** 110
Au(100)–5 × 20, **27,** 284
Au(111) , **27,** 285
band gap states, **27,** 270, 273
bulk band gap, **27,** 254, 255, 264, 265, 267
clusters, **27,** 266, 271, 273
evanescent surface states, **27,** 255, 272
GaAs(110)
Cs adsorbed, **27,** 123
Fe adsorbed, **27,** 127, 130
general I-V characteristics, **27,** 118
O adsorbed, **27,** 110
Sb adsorbed, **27,** 118, 266
GaAs/AlGaAs superlattices, **27,** 264
Ge(001), **27,** 237
Ge(111)-c2 × 8, **27,** 230
I-S, **27,** 280
for resonant transmission, **27,** 283
metal surfaces, **27,** 280
metal on GaAs(110), **27,** 65
charging effects, **27,** 265
$NbSe_3$, **27,** 418
ϕ -S, **27,** 281
Si(111) 2 × 1
I-V measurements, **27,** 112
surface density of states, **27,** 112
voltage dependent imaging of, **27,** 106
superconductor energy gap, **27,** 432
superconductors, **27,** 341
in zero field, **27,** 432
surface state density, **27,** 274
TaS_3, **27,** 418
transmission resonances, **27,** 282
Spectroscopy, **1,** 362–376, 422; **17,** 529; **26,** 43; **27,** 95, 351, 354, 421; *see also* Ion-induced Auger electron spectroscopy; Ion-induced x-ray spectroscopy
barrier resonances in, **27,** 103
coherent anti-Stokers Raman, **15B,** 258
coincident charge state analysis for spectral identification, **17,** 347–350
current imaging spectroscopy, **27,** 129
differential conductance, **27,** 157
differential conductivity, **27,** 282
electronic structure, **27,** 282

Doppler-free, *see* Doppler-free laser spectroscopy
experimental technique, **27,** 431
fixed separation I-V measurements, **27,** 98, 101, 112
heavy-ion-atom collisions, **17,** 151–152
inelastic energy-loss measurements, **17,** 191–192
infrared, **3,** 38–111
applications, **3,** 103–111
sampling techniques, **3,** 86–89
instrumentation, **27,** 96
mass, **5B,** 1–9
microwave, **3,** 7–38
microwave, **5B,** 23–29, 85–103
applications, **3,** 37–38
sample handling, **3,** 36–37
molecular, **3,** 7–264; **15B,** 58–63
neutron, time analyzers for, **5A,** 554–566
normalization methods for
acquisition methods, **27,** 135
average conductivity method, **27,** 138
band gap effects, **27,** 134
fixed separation, **27,** 112
general method, **27,** 141
optical, **5B,** 44–58
picosecond, *see* Picosecond spectroscopy
polarization, *see* Polarization spectroscopy
spatial semicondcutor characteristics, **27,** 123
spatially resolved, **27,** 123
surface band bending effects, **27,** 122, 126
surface density of states, **27,** 112
theory of, **27,** 22
tunneling transmission, **27,** 102
of tunneling transmission dependence on tunnel gap, **27,** 284
ultraviolet, **5B,** 44–58
vacuum ultraviolet, *see* Vacuum ultraviolet spectroscopy
variable separation measurements, **27,** 134
voltage dependent imaging, **27,** 99, 104, 106
Spectroscopy, of afterglows, **7B,** 102–110
in combustion studies, **7B,** 190
line-shape determinations at low intensity, **7B,** 106–108
low-intensity optical emission measurements, **7B,** 103–104
for measuring atomic and radical concentrations, **7B,** 209–210
photomultipliers for, **7B,** 108
time-sampling techniques, **7B,** 102–103
weak optical absorption measurements, **7B,** 104–106
Spectroscopy, *see* Fast neutron spectroscopy; Gamma spectroscopy
SPECTROSIL, **31,** 309, **31,** 326
Spectrum
absorption
scintillator, **5A,** 126
total large crystal, **5A,** 629
ammonia inversion, **3,** 15–16
antineutrino from fission fragments, **5A,** 686ff
distribution, scintillator, **5A,** 131–132
electromagnetic, **3,** 39–40
emission, scintillator, **5A,** 126, 131–132
first-order, **3,** 153
heavy charged particles, **5A,** 411
HF rotational, **3,** 48
leakage neutron of plutonium critical assembly, **5A,** 477–478
mercury lamp, **3,** 127ff
neutron
calculation from observed pulse distribution, **5A,** 577–579
from scattering on vanadium, **5A,** 512
nuclear γ and x-ray lines from neutron activated Ta, **5A,** 607–609
of relaxation times, **19,** 143–144
of retardation times, **19,** 143–144
time
delay, **5A,** 443
of fast neutron interactions, **5A,** 551, 552
vibration-rotation, **3,** 49, 50
isotope effect, **3,** 51, 52
Spectrum analysis, **19,** 265–267, 274–275
Bayesian, **28,** 417–422
ultrasonic pulse, **19,** 109–111
Spectrum analyzers, **20,** 314, 315
Spectrum scanning, applied to LDV, **18,** 154, 159–162, 165–174, 201–211
Spectrum unfolding in resonant nuclear reactions, **21,** 249
Spectrum-scanned LDV, requirements, **18,** 201–203
Specular reflectance, definition, **31,** 184
Speed
distribution function, **14,** 39
of light, **3A,** 338–343, 394
most probable, **14,** 8

photographic, **3,** 137; **12A,** 228
pumping, **1,** 172ff, 182–183, 187ff
of sound, **3A,** 412
spectrographic, **12A,** 465, 467–469, 480
volumetric, **14,** 183
Sphere of position, **16C,** 154
Sphere of reflection, **16B,** 34; **16C,** 154
Sphere transmission measurement of total nonelastic cross-section, **5B,** 397–405
Spherical aberration correction, **31,** 173–175
Spherical aberration, **1,** 332, 337; **25,** 42
Spherical aberration, **31,** 154
circular mirror, **31,** 170–171
term derivation, **31,** 166–170
Spherical aberration, in electron microscope, **16B,** 199–200
coefficient, **16B,** 200
Spherical analyzer, **7A,** 161
Spherical antennas, **12B,** 18–19
Spherical Bessel functions, **8,** 280, 281
Spherical electrostatic analyzer, **22,** 317
Spherical energy analyzer (ESA), **21,** 236
Spherical Fabry-Perot, **25,** 220
interferometer, **26,** 29
Spherical grating monochromator (SGM), **29C,** 37–38
Spherical grating monochromator (SGM), **31,** 28, 29
Spherical top, **10,** 304, 319–320
Spherical-capacitor analyzer, **22,** 223
Spherical-grid retarding potential analyzer, **22,** 217, 227
Spherical-sector analyzer, **17,** 458–462, 467; **22,** 139
Spherocylindrical surfaces, **25,** 56
Spherulite growth, nature of, **16B,** 345
Spherulites, **16B,** 341–346; **16C,** 471
in polyethylene, revealed by plasma-etching, **16B,** 243
in polymeric alloys, **16C,** 299
ringed, **16B,** 347
size from light scattering, **16B,** 367
Spin
cases, secular equations for, **3,** 505–506
dependent electron-electron scattering cross section, **5A,** 64ff
determination, **5B,** 44–214
indirect methods, **5B,** 129–147
microwave method, **5B,** 85–103
spectroscopic method, **5B,** 44–129
echoes, **3,** 395–399; **5B,** 123
flip, mutual, **3,** 378–379
flippers, **23A,** 217
flipping, **3,** 431–432
Hamiltonian, **3,** 446–447
isotopic, **5B,** 353
measurement in liquids, **5B,** 117–119
neutrino, **5A,** 698
nuclear, **3,** 503, 628
effect on rotation line, **3,** 71–73
orbitals, molecular, **3,** 179
orientation, neutron beam, **5A,** 465–466
systems, relaxation, **3,** 456–464
transformers, **5B,** 227–233
of U^{235}, **5B,** 55–56
Spin angular momentum, **4B,** 100–102; **20,** 2
Spin correlations, **23C,** 252
Spin decoupling, **16A,** 264; **20,** 33
Spin diffusion, **21,** 392
Spin echo technique, *see* Measuring spin-spin relaxation time T_2
Spin echo, **3B,** 540
Spin exchange in optical pumping, **7B,** 18–21
nuclear spins, **7B,** 21
spin temperature, **7B,** 21
Spin exchange, **20,** 102, 104–109
collision, **4B,** 186
Spin fluctuations, **23A,** 84
Spin glasses, **23B,** 107, 123; **23C,** 54
ferromagnetic, **23B,** 119
Spin moment anomaly of electron, **4B,** 195
Spin operators, expectation values, **4B,** 226–227
Spin precession of free electrons, **4B,** 29–31
Spin precession rate of proton, **4B,** 17–18
Spin resonance technique, *see* Electron spin resonance
Spin state of beam, **4B,** 227
Spin superposition, **23A,** 300
Spin temperature, **21,** 456–457
Spin waves, **10,** 401–406; **23C,** 175, 209
Spin-analysis in electron-atom scattering, **7A,** 94, 95
Spin-density mapping, **23A,** 365
Spin-diffuse neutron scattering, **3,** 350
Spin-down, **18,** 809, 814–815
Spin-down, **8,** 264
neutrons, **8,** 265
Spin-echo experiments, **23C,** 54
Spin-echo instruments, **23A,** 234
Spin-echo spectrometer, **23B,** 124, 344
Spin-exchange cross section, **7B,** 19

Spin-exchange process, **7B,** 3
Spin-filter, **7A,** 95
Spin-flip frequency, **10,** 71, 426
Spin-flip Raman laser, **10,** 70–73, 430, 431
Spin-flip Raman scattering, **10,** 70, 426–428, 430–431; **15B,** 260
Spin-flip scattering, **11,** 175
Spin-flop transition, **23C,** 283
Spin-forbidden transition, **17,** 584, 586
Spin-label spectroscopy, **20,** 53, 198
Spin-lattice relaxation time (T_1), **16A,** 446
  measurement by pulse-FT NMR, **16A,** 279–282
  Temperature dependence, **16A,** 320–322
  theory, **16A,** 245–253
Spin-lattice relaxation time, **3,** 456
Spin-lattice relaxation, **10,** 387; **20,** 13, 21; **21,** 366, 367, 447–449; **23C,** 248
Spin-orbit
  coupling, **3,** 227, 228, 445; **15B,** 271
  interactions, **3,** 205–221; **17,** 598
  perturbations, **3,** 255–256
  splitting, **22,** 216
Spin-rotation interactions, **3,** 605; **3B,** 789
Spin-spin coupling (*J* coupling), **16A,** 259–264
Spin-spin coupling, **3B,** 483; **20,** 14, 20–22, 39, 102–104
Spin-spin interactions, **3B,** 784
  electron-coupled, **3,** 407–408
Spin-spin ion source, **17,** 586
Spin-spin relaxation time (T_2)
  in the rotating frame, **16A,** 255
  measurement by pulse-FT NMR, **16A,** 283
  theory, **16A,** 253–255
Spin-spin relaxation, **10,** 387; **21,** 364, 366, 441–443
Spin-spin relaxation, **3,** 456–459
Spin-up, **8,** 264
Spin-wave dispersion relation, **23A,** 80
Spine-and-ribs cross plot, **24B,** 555–558
Spinodal, **16C,** 277
  decomposition, **23B,** 111
Spins
  absorbing, **3,** 454
  number of, **3,** 491, 495–496
Spiral orbit spectrometer, **5A,** 364–365
Spiral pit, *see* Hopper crystal face
Spiral tube gauges, **14,** 47–48, *see also* Manometer, mechanical
Splicing, **14,** 285; **29C,** 172
Split coil system, **8,** 102
Split-coil magnet, **11,** 362
Splitting
  electric hyperfine, **5B,** 178, 183
  factor, spectroscopic, **3,** 443
  Jahn-Teller, of nuclear potential function, **3,** 196
  by ligand field, **10,** 388–394
Spoiled *Q*, **19,** 32–33, 61
Spontaneous emission, **13A,** 87
  requirements in EBF, **18,** 438, 441, 442–447
  temperature measurement, **18,** 465–482
Spontaneous magnetization, **6B,** 194
  compensation point, **6B,** 196
Spontaneous potential, **24B,** 449
Spot diagram formulas, **31,** 363
Spot welding, **14,** 307
  in magnetoresistance measurements, **11,** 13
Spot-shape effects, **30,** 259–261
Spot-welded contacts, **6B,** 51
Spreading resistance, **6B,** 35–36
Springback, elastic, **14,** 289, 290
Sputter cleaning, **22,** 487
Sputter etching, **22,** 290
Sputter-ion pump, **14,** 193
Sputtered contacts, **6B,** 51
Sputtered glasses, **23B,** 274
Sputtered metal vapor ion lasers, **15A,** 157–166
Sputtering, **2,** 95–97; **14,** 217, 222, 228, 550–575; **21,** 257–264; **22,** 321, 328, 366
  advantages, **14,** 573–575
  alloys, **14,** 558
  aluminum, **14,** 364
  angular distribution, **14,** 552
  bias, **14,** 566, 567
  cathode sputtering, *see* Diode sputtering
  comparison, vacuum evaporation, **14,** 572–575
  diode sputtering, **14,** 564–566
  disadvantages, **14,** 573–575
  dynamic systems, **14,** 556
  electronic, **30,** 176, 232, 282
  energy distribution, **14,** 552, 555
  exfoliational, *see* Exfoliation
  getter, **14,** 570, 571
  hydrodynamic, *see* Hydrodynamic sputtering
  ion production, glow discharge, **14,** 558–563
  ion source, **17,** 84
  ion-beam, **14,** 570–572
  in LEIS, **21,** 244

radio-frequency, **14,** 566, 568
rate, **22,** 367; *see also* Sputtering, yield
ratio, approximation, **29A,** 99
reactive, **14,** 568–570
safety considerations, **14,** 571
Sigmund theory, **21,** 257–259
static systems, **14,** 556
substrate potential and temperature, **14,** 563–564
surface topography, **21,** 263
system design, **14,** 459
target materials, **14,** 556–558
thermal-spike, *see* Thermal-spike sputtering
in thin flim preparation, **11,** 637–645
triode, **11,** 642–643; **14,** 566, 567
vacuum systems, **14,** 552, 556
yield, **2,** 95–97; **22,** 321, 330, 334, 367
yield, variation with atomic number, **14,** 551–555
Square dislocation loop, **6A,** 327–329
Square law device, **1,** 468
Square loops, **6B,** 220–221
Square root circuit, **2B,** 67
Square wave test
hot-wire and hot-film probe, **18,** 279, 285, 297–299
pressure probe, **18,** 563–570
system frequency response, **18,** 528–531
whirling vane anemometer, **18,** 259
Square wave train generator, in equipment testing, **2B,** 183
Square-loop core driver, **2A,** 405–408
Squid, *see* Cryogenic magnetometer; Superconducting quantum interference device
SRO, *see* Seismic Research Observatory
SRS, *see* Stimulated Raman scattering
$SrTiO_3$, **10,** 373, 380–384; **23C,** 57
SSHG, *see* Surface second-harmonic generation
Stability
chart, **14,** 97
condition, general, **1,** 323
criterion
Routh, **2,** 473–474
Nyquist, **2,** 474–476
degree of, second-order servomechanism, **2,** 481–483
in feedback control systems, **2B,** 398–410
feedback system, **2,** 455
linear, **2,** 472–476
Nyquist stability criterion in, **2B,** 401–405
Routh test in, **2B,** 398–401
theory, **1,** 321–324
Stability (voltage clamp), **20,** 467–470
Stabilization
of amplifier gain and zero, **2,** 503–506
amplitude of feedback oscillators, **2,** 508–509
bias, **2,** 313–316
factor, voltage regulator, **2,** 489–490
feedback system, **2,** 476–488
magnetic field, **3,** 415–416, 485–486
networks, **2,** 483–488
transconductance, current regulator, **2,** 491
Stabilizer
frequency
klystron, **3,** 470–471
Pound, **3,** 470
magnetic field, proton controlled, **3,** 486
Stage microscope, **25,** 159
Staging ratio, **14,** 118
Stagnation
condition, **1,** 129
detector, **7B,** 273
point, **1,** 153
Stagnation enthalpy, **18,** 245, 458, 788
Stagnation pressure
meaning, **18,** 244, 246, 503, 515
produced for shock tunnel, **18,** 791
ratio across shock, **18,** 247
wind tunnel, **18,** 776
Stagnation temperature, **1,** 138, 146
hypersonic apparatus, **18,** 782
meaning, **18,** 245, 457, 665
measurement, **18,** 460–463
produced for shock tunnel, **18,** 791
relation to gas speed and temperature, **18,** 458
in shock tube flow, **18,** 788
Staining specimens, for electron microscopy, **16B,** 216
Stainless steel
physical properties, **14,** 354–356
sulfur containing, **14,** 480
thermal conductivity
type 304, **14,** 356
type 304L, **14,** 480, 496
Standard
atomic wavelength, **1,** 55
end, of length, **1,** 54
frequency, **1,** 73

length, **1,** 54
line, of length, **1,** 54
radiometer, **26,** 246
time, **1,** 69
Standard deviation, **1,** 4, 7; **2,** 4; **6A,** 3–4; **28,** 20, 157
Standard error, **1,** 4; **28,** 166, 257; *see also* Standard deviation
Standard irradiance, **25,** 247
Standard lamps, **12A,** 112–116; **26,** 6
Standard molecular beam, **7A,** 233
Standard sources, **13A,** 224
Standard stars, **12A,** 117, 162, 451–463, 456–459
Standard waveguide, **10,** 113–121
Standards, pumping speed measurement, **14,** 104
Standards, radiometric and photometric, **26,** 264
Standing wave
measurements, **8,** 107
pattern, **8,** 138
ratio, **8,** 110, 121, *see also* Voltage standing wave ratio
Standing wave indicators, **6B,** 20–22
Standing wave measurement, in microwave circuits, **2B,** 289–290
Standing wave method
for dielectric measurements, using transmission lines, **16C,** 409–416
for sonic modulus determination, **16C,** 168
Standing wave ratio, **19,** 9–14, *see also* Acoustooptic measurement of sound wave, standing wave ratio
Standing wave techniques, **13B,** 158
Standing waves, **1,** 278–279; **19,** 31–32, 57–60
Standing-wave resonances, in tunneling, **11,** 152
Stanford University High Energy Physics Laboratory, **11,** 590
Stanski-Krastanov growth, **30,** 308
Star diameter measurements, **26,** 39
Star of wave vectors, **23C,** 80
Star-shaped block polymers, **16C,** 298
Stark beats, **13B,** 245
Stark effect, **1,** 386; **3,** 16–18, 606–607; **3A,** 98–105, 724, 837; **4B,** 108, 134, 144, 223–224; **7A,** 140, 210–211, 263–267; **9A,** 163
broadening, **5B,** 51
with hyperfine structure, **3,** 657–662
modulation, **5B,** 90; **13B,** 115
with no hyperfine structure, **3,** 653–657
in plasmas, **7B,** 349
Stark measurement, **3,** 634–635
Stark modulation, **3A,** 120; **29B,** 277–278
spectrometer, **3,** 33–35
Stark perturbation energies, **3,** 654, 655, 656
Stark spectra, **13B,** 340
State of polarization (SOP), **29C,** 173
State selection, **3B,** 723, 732
State selection, atoms
by magnetic field gradient, **4B,** 129, 154, 235–245
by radio-frequency field, **4B,** 133, 166–167, 245–248
State selector, **7A,** 259–267, 330–334
State vector, **13A,** 733
States
electron polarization, **5B,** 216–217
mixed, **5B,** 218
isobaric, **5B,** 363–366
nuclear, formed after radioactive β-decay, **5B,** 304
Static approximation, **23A,** 36; **23B,** 246
Static capacitor, **22,** 19
Static drift tube, electron attachment rate determination, **29A,** 290–291
Static electric field, phase shifts generated by, **29C,** 302–303
Static electricity
generated by contact, **16C,** 428–432
generated by rubbing, **16C,** 432–433
role in adhesion, **16C,** 434
Static graphical methods, **28,** 510, 514
Static high pressure, **6A,** 407–414
generation at low temperatures, **6A,** 413–414
range, 0–2000 bars, **6A,** 407–408
range, 2000–12000 bars, **6A,** 408–411
range, 12,000–30,000 bars, **6A,** 411–412
range, over 30,000 bars, **6A,** 412–413
Static memory arrays, **2B,** 123–125
Static pressure probe
in steady flow, **18,** 243, 518
in unsteady flow, **18,** 521
Static pressure, meaning, **18,** 247, 516
Static temperature, meaning, **18,** 457
Static vents on airplane, **18,** 520
Stationary afterglow
apparatus, **29A,** 276

applications, **29A,** 276–278
Cavalleri technique, **29A,** 280
ion monitoring, **29A,** 279
negative ion reactions, **29A,** 278–279
plasma formation, **29A,** 276
pulsed radiolysis technique, **29A,** 280
Stationary, intrinsic, **28,** 379
Stationary-phase evaluation, use in radiation integrals, **12B,** 62
Statistic, concept, **28,** 160
Statistical computing, **28,** 481–517, *see also* Graphics
bootstrap methods, **28,** 491–498
applications, **28,**495–498
implementation, **28,** 495–498
methodology, **28,** 492–494
motivation, **28,** 491–492
nonparametric bootstrap, **28,** 493–494
parametric bootstrap, **28,** 493–494
computer-intensive methods, **28,** 491–500
cumulative distribution functions, **28,** 491
difference between numerical methods and numerical mathematics, **28,** 489
differential equation models, **28,** 501–509
kinetics examples, **28,** 502–509
problem description, **28,** 501–502
environments, **28,** 482–488
hardware platforms and operating systems, **28,** 482–483
statistical software, **28,** 483–484
using S on Unix workstations, **28,** 484–488
important numerical mathematics areas, **28,** 490–491
literature, **28,** 489–490
methods, **28,** 488–491
smoothing, **28,** 498–500
types, **28,** 488
Statistical decision, **28,** 179
Statistical errors, **7A,** 72–73, 75, 187; **28,** 31–33
Statistical hypothesis, **28,** 179
Statistical inference, **28,** 155–185
basic ideas, **28,** 212–213
Bayesian, **28,** 403–407
descriptive statistics, **28,** 156–158
interval estimation, *see* Interval estimation
least squares, **28,** 261–265
likelihood function, **28,** 213–214
linear models, **28,** 184
parameter values, **28,** 264
point estimation, **28,** 160–165
population parameters, **28,** 158
robust estimation, **28,** 184
sequential methods, **28,** 184–185
statistical model, **28,** 158–159, 185
statistical tests, **28,** 179–183
Statistical methods, **1,** 3–6; **3,** 711–713; **6A,** 2–5
Statistical model, **28,** 157–159, *see also* Method of maximum likelihood
diagnostics, **28,** 266–277
electromigration, **28,** 180
fitting, **28,** 240–241
functional, **28,** 278
multivariate nonlinear, **28,** 248
nonparametric, **28,** 185
nonregular, **28,** 241–242
parametric, **28,** 185
statistical inference, **28,** 158–159, 185
structural, **28,** 278–279
theoretical and empirical, **28,** 278
variability, parametric distribution, **28,** 212
Statistical modeling
basic ideas, **28,** 213–214
least squares, **28,** 245–252
terminology, **28,** 245–246
Statistical significance, versus practical significance, **28,** 181–182
Statistical tensor, **4B,** 229
Statistical tests, **28,** 179–183
choosing test statistic and rejection region, **28,** 180–181
Neymann-Pearson hypothesis testing, **28,** 182–183
p-value, **28,** 181–182
relation to confidence interval, **28,** 181
significance test, **28,** 179–181
statistical versus practical significance, **28,** 181–182
Statistical thermodynamics of desorption, **22,** 457
Statistical uncertainties, estimation, **28,** 452
Statistical visualization, **28,** 512
Statistical weight(s), **1,** 9; **3A,** 135
nuclear spin, **3A,** 36
Statistics, **2,** 3–5
methods, **5B,** 809–811
Poisson, **5B,** 792–802
useful inefficient, **5B,** 803–806
Stator, **14,** 247, 249
Steady heat flow, **6A,** 402–404

Steady model, **28,** 74
Steady state amplitude, **8,** 2
Steady state induction method, **6B,** 203
Steady state measurement of lifetime, **6B,** 100–101, 104–107
Steady state, defined, **2B,** 186
Steady-state conditions
in equipment testing, **2B,** 186–191
interpretation, **2,** 667–668
testing, **2,** 674–651
Steady-state lasers, **15A,** 292–325
flashlamps and, **15A,** 297–307
general description of, **15A,** 293–294
vortex stabilized lamps and, **15A,** 307–310
Steady-state measurements in nuclear magnetic resonance, **2B,** 137–144
Steady-state methods, in electron spin resonance, **2B,** 147–150
Steady-state power balance, **15A,** 294–298
Steady-state response, **2A,** 4
in rheology, **16C,** 44–46
Steam calorimeter, **1,** 274
Steam ejector pump, *see* Ejector pump
Steam point, **1,** 237
Steatite, *see* Ceramics
Steel, *see also* Stainless steel
attenuation in, **19,** 284–290
microstructure of, effect on scattering, **19,** 267, 284–290
physical properties, **14,** 356–357
Stefan problem, **30,** 180
Stefan-Boltzmann constant, **18,** 690
Stefan-Boltzmann law, **1,** 253, 302, 375; **4A,** 290; **22,** 76; **29C,** 7
Stellarator, **9A,** 329
STEM (scanning-transmission electron microscopy), **16B,** 227
Stem correction, **1,** 244
Step function
hot-wire and hot-film response testing, **18,** 279, 285
loading
bar pressure gage, **18,** 598, 603
diaphragm pressure gage, **18,** 563
dilatational pressure gage, **18,** 605
response function, **18,** 527
unit, **2,** 465
Step height, **22,** 251
Step recovery diode, **2A,** 62–63
Stepped structures
scanning tunneling microscopy of, **27,** 180
at strained surfaces, scanning tunneling microscopy of, **27,** 185
Stepwise excitation, in beam-foil interactions, **17,** 549
Steradian, **1,** 58
Stereochemistry of polymer chain, **16A,** 288–294
Stereograms, holographic, **26,** 184
Stereopair, **16B,** 222
Stern-Gerlach analyzer, **29B,** 163
Stern-Gerlach effect, in atom interferometry, **29B,** 354
Stern-Gerlach experiment, **23A,** 262
Stern-Gerlach magnet, **7A,** 80, 87, 94, 261
Stern-Gerlach velocity selector, **7A,** 94
Stern-Volmer equation, **16A,** 408
Stevenson's rule, **3,** 547
Stewartson layer, **18,** 806, 813
Sticking coefficient, **14,** 195, 196, 231
Sticking probability, **22,** 324
Sticking-coefficient determination, **22,** 442
Stimulated emission, **3A,** 302; **4B,** 119; **29B,** 149–150
cross section, for dye lasers, **15A,** 5–6, 263–264
devices, **3A,** 319–322
Stimulated light scattering, **15B,** 259–260
Stimulated Raman laser, **10,** 70–74, 428–431
Stimulated Raman scattering (SRS), **10,** 70, 428–431; **20,** 193, 194; **29C,** 266
Stimulated Raman transitions, in laser cooling, **29B,** 131
Sting, wind tunnel mount, **18,** 765
Stirring during crystal growth, **6A,** 95
Stirring modulation method, **6A,** 69
Stoichiometric lasers, **15A,** 203–204
Stoichiometric materials, as host materials, **15A,** 190–191
Stokes, **1,** 123
Stoke's integral, **24B,** 147–148
Stokes drag formula, **18,** 10–15, 759, 798, 799
Stokes equation, **1,** 148
Stokes law, **1,** 125
Stokes line, **3,** 112, 113, 124
Stokes number, **18,** 28
Stokes parameter for photon and electron beams, **5B,** 221–222
Stokes parameters, **12A,** 362, 372, 384, 399; **26,** 159; **29C,** 357–358
interferometer measurement of, **12C,** 152
measurement of, **12C,** 35

Stokes Q-branch, **18,** 421
Stokes Raman diagnostics for temperature measurement, **18,** 422–425
Stokes Raman line, **18,** 422, 723
Stokes scattering, **10,** 70–74
Stokes spectrum, **16A,** 78
Stokes vector, **25,** 181
Stokes-Kirchhoff analysis, **1,** 214
Stoneley wave, **19,** 501; **24B,** 589–590, *see also* Leaky mode; Pseudo-Rayleigh wave; Reverberant wave
Stop detector, **8,** 254
Stop, *see* Aperture stop; Circular stop
Stopping cross section, **17,** 27–30, 68–72; **21,** 229–230, 263
  backscattering method of determining, **17,** 58–65
  electronic, **17,** 42
  inverted Doppler-shift attenuation method of measuring, **17,** 66–68
  measurements on thin layer on solid-state detector, **17,** 56–57
  nuclear, **17,** 42
Stopping power, **24A,** 119
  at very low energies, **5A,** 42–43
  energy loss factor, **21,** 227, 229, 230, 231, 246
Stops, **1,** 333–334; **25,** 37
Stops, **31,** 150–151
Storage
  clean components, **14,** 500
  devices
    passive, **2,** 356–357
    temporary, **2,** 623–624
  pulse spectrometer
    electrostatic, **2,** 618–619
    magnetostatic, **2,** 617–618
Storage ring
  development, **31,** 2
  as source standard, **31,** 122–125
Storage ring(s), **22,** 141, 142; **29B,** 319
  dipole or bending magnet radiation, **29C,** 24–28
  electron cooling, **29A,** 318
  facilities, **29A,** 317–318
  light polarization, **29C,** 30–33
  magnetic, **5B,** 687–689
  schematic, **29A,** 315–316
  strength, **29A,** 317
Storage tubes, **2,** 782–784; **2B,** 355–358
  computer, **2,** 783–784
  display, **2,** 784
Stored-charge analysis, **2,** 152
Stored-energy source, pulse delivery from, **2A,** 238
Straggling, **5A,** 436
  effect, for heavy ions on solid targets, **17,** 222–224
  parameter, **5A,** 53
  range, **5A,** 51–55
Straight edge, diffraction, **26,** 57
Strain gage(s), **1,** 56, 67, 80; **24B,** 379–380, 383, 386–393, 404; *see also* Strain measurements
  circuit, **2A,** 248
  electric, **1,** 102–103
Strain gauges, **24A,** 146–147
  bonded resistance-type, **16C,** 120
  mechanical, **16C,** 119
  mechanical-optical, **16C,** 120
Strain measurements, **6A,** 291–294
  electrical methods, **6A,** 293–294
  mechanical methods, **6A,** 292
  optical methods, **6A,** 292–293
Strain pulse
  dispersion, **18,** 580
  reflection at end or interface, **18,** 585
  theory of one-dimensional wave, **18,** 581
Strain sensitivity of diaphragm, **18,** 561, 562, 566
Strain sensor, **18,** 536–540, 542–549, 572–574
Strain, **1,** 100, 102–106; **19,** 497; *see also* Crustal deformation
  in crystals, effect on X-ray line shape, **16B,** 144–149
  definition, **16C,** 118
  in flexural test, **16C,** 125
  measurement of, **16C,** 119–121
  principal, **24B,** 380
  rosette, **24B,** 379–383, 386–387, 390–391
  in single crystals, **6A,** 314–317
  triaxial, **24B,** 389
Strainmeter
  Benioff, **24B,** 429–430
  comparisons, **24B,** 434
  design, **24B,** 415, 429–433
  drift, **24B,** 413
  hydraulic, **24B,** 431–432
  installation, **24B,** 421–424
    borehle, **24B,** 423–424
    cavity effect, **24B,** 422

optical anchor, **24B,** 423
thermal stability, **24B,** 422, 427–428
underground, **24B,** 422–423
laser, **24B,** 432–433
materials, properties, **24B,** 416
mechanical, **24B,** 429–431
noise
power spectra, **24B,** 414
sources, **24B,** 412
quantity measure, **24B,** 410
seismic
Stratified fluid, optical visualization, **18,** 355
Straumanis method, **6A,** 215
Stray light, **12A,** 108–110
Stray light, reflectance, **31,** 191
Streak camera, **29C,** 193, 199–203, 223–224
image converted recording, **18,** 696
light source, **18,** 705, 715
use with interferometer, **18,** 795
using rotating mirror, **18,** 733–734
Streak length in RHEED, **22,** 276
Streak tube, **29C,** 201–202
Streaming, **19,** 319–328
and agglomeration, **19,** 342
and dispersal of cell aggregates, **19,** 331
and drying ceramics, **19,** 343–344
effect on rectified diffusion, **19,** 317
and emulsification, **19,** 330
Strength function, **5A,** 544
Stress, **1,** 100, 102–106
compressive, **1,** 100
in-situ, **24B,** 377
principal, **24B,** 377, 380, 382, 386, 388, 394–397
shear, **1,** 100, 123; **24B,** 405
state of, **24B,** 377–378, 404–405
tensile, **1,** 100
thermal, **14,** 279; **30,** 236
viscous, **1,** 132
Stress birefringence, **12A,** 373, 379, 409; **29C,** 378
Stress criteria, for crazing, **16C,** 249–252
Stress field, **6A,** 327
of a craze, **16C,** 245–248
Stress relaxation, **1,** 109; **19,** 139
in crazed polymers, **16C,** 268
in penetrant-swelling polymers, **16C,** 335, 337–338, 349
measurement of, **16C,** 32
modulus, **16C,** 11
Stress relief, **14,** 286
method, **24B,** 379, 383, *see also* Strain, rosette; Overcoring
Stress tensor, **18,** 500
Stress-freezing, **1,** 105
Stress-strain curve
effect of temperature on, **16C,** 129, 132
effect of strain rate on, **16C,** 132
of crazing polymers, **16C,** 263–265
Stress-strain relationship, **1,** 107, 108
Stretch, elastic, of bolts, **14,** 287
Stretched pulse modelocking, **29C,** 183–184
Stretched string, **1,** 209
String model, **22,** 309, 310
Stripping of replicas, *see* Replica techniques
Stroboscope, **1,** 87
rectifier controlled, **2A,** 240
Strong collision theory, **21,** 456–457, 461–466
Strong coupling regime, **27,** 16
Strong fission neutron field, **8,** 227
Strontium
autoionizing states in, **15B,** 230
VUV signal generated in, **15B,** 220
Strouhal number, definition, **18,** 830
Structural anisotropy in glasses, **23B,** 270
Structural material, **8,** 84
Structural model, **28,** 278
Structure
amplitude, relative, **3,** 283
atomic hyperfine,
by magnetic resonance, **3,** 626–630
separations, **3,** 628
determination procedure, **3,** 289–295
Structure effect, on stopping cross section, **17,** 70
Structure factor(s), **20,** 238–241; **21,** 210; **23A,** 32, 383; **23B,** 245, 408
for body-centered lattice, **16B,** 37
crystal, **3,** 278, 283–286
definition, **16B,** 36
elastic, **23A,** 25
experimental evaluation, **16B,** 65–68
of a helix, **16B,** 83–84
inelastic, **23A,** 394
interparticle, **23B,** 510
magnetic, **23A,** 63; **23C,** 78
magnitude, probable value of phase, **3,** 293–294
one-phonon, **23A,** 28
in terms of electron-density distribution, **16B,** 39
unit cell, **23C,** 78

of unit cell, **5A,** 567
unit-cell, **23A,** 26, 344
Structure refinement, **23A,** 347
difference Fourier, **23C,** 344
real-space, **23C,** 343
reciprocal-space, **23C,** 344
Structure solution, **23A,** 344
Structure, hyperfine, **3,** 18–20
effects of nuclear moments and spins on, **5B,** 46–49
electric, **5B,** 183–184
magnetic, **5B,** 176–183
in optical spectra, **5B,** 44ff
STS seismometer, **24B,** 67–71
Stub pressure gage, **18,** 588
Stubs, **3,** 26
Student t function, **2,** 6–7
Student's *t* distribution, **1,** 12, 17, 23
Student's *t* function, **1,** 8, 23; **3,** 715; **5B,** 812–813; **6A,** 6
Subgrain structure, **22,** 248
Subharmonic generation, **19,** 379, 382, 405
from periodic bubble motion, **19,** 378–380
from transient cavitation, **19,** 394–395
Subharmonic structure, **11,** 212
Sublattice polarization, proton, **23C,** 275
Sublimation, **1,** 262; **6A,** 22; **14,** 530
curve, **1,** 242
heat of, **1,** 281
Sublimation pump, in ultrahigh vacuum systems, **14,** 482
Sublimation pumping, **14,** 193
Submicrocracks, **16C,** 230
Submillimeter, **13A,** 10
Submillimeter lasers, **3A,** 305–338
mechanisms, **3A,** 307–319
miscellaneous, **3A,** 374
use in spectroscopy, **3A,** 343
Submillimeter waves, *see* far infrared and microwave region
Subpicosecond amplification, **15B,** 192–193
Subpicosecond pulsewidths, **15B,** 186
Subprotonospheric whistler, ion-effect observations on, **9B,** 112–115
Subreflectors
ellipsoidal, **12B,** 51
hyperboloidal, **12B,** 50–51
nonoptical, **12B,** 52
Subshell transfer cross section, **17,** 96–97
Subsonic gas flow, **1,** 155
Substage condenser diaphragm, **25,** 171
Substitution rule, **1,** 316
Substitutional defects, magnetic, **23B,** 99
Substrate, **14,** 195, 196
design, **14,** 208, 209
Subsurface heating, **30,** 238–239
Subsurface melting, **30,** 503–506
Successive approximation A/D converter, **2A,** 419
Successive field corrections, **8,** 79
Sucksmith ring balance, **1,** 541
Sucrose gap (voltage clamp), **20,** 500–504, 508–511
Sudden perturbation model, of ion laser excitation, **15A,** 91
Sugden burner, **7B,** 192, 193–194
Suhl effect, **6B,** 99
Sulfate
decomposition, statistical analysis application, **28,** 393–395
deposition
data, **28,** 376–378
kriging application, **28,** 397–398
statistical analysis application, **28,** 386–389
Sulfur dioxide
absorption coefficient of, **12B,** 172
atmospheric concentration, **12B,** 170
Sulfur point, **1,** 247
Sulfur-fluoride excitation and dissociation, **15B,** 296–297
Sulfur-iodine eutectic, for impregnation of crazes, **16C,** 243–244
Sulphur, **23A,** 459
hexafluoride, **23A,** 514
Sum of squares for error, minimization, **28,** 163
Sum rule(s), **4B,** 217–218; **23B,** 308
Sum-frequency generation
in mid- and far-infrared, **15B,** 181–183
upconversion and, **15B,** 183
Sum-frequency generation (SFG), **29C,** 246–247, 254
Sum-frequency mixing, second-order, **15B,** 258
Sum-of-reciprocals, **14,** 18, 276
Summing
coincedent, **5A,** 430–431
random, **5A,** 431
Sun, **25,** 239
Super-regenerative method, **6B,** 234

Super-regenerative oscillator detectors, **3B,** 551
Superconducting balloons, **11,** 616–617
Superconducting bladders, **11,** 615–616
Superconducting bolometer, **10,** 170, 171; **13B,** 8
Superconducting cyclotron, **17,** 14
Superconducting device technology, **11,** 595–618
Superconducting interferometers, **11,** 294–296
Superconducting junction detector, *see* Josephson junction detector
Superconducting junction radiation, *see* Josephson junction generator
Superconducting linac, **17,** 16–17
Superconducting magnets, **11,** 595–608
  construction of, **11,** 597
  costs for, **11,** 607–608
  ductile alloys for, **11,** 598
  field measurement in, **11,** 602–603
  field uniformity in, **11,** 600–602
  high-field or compound type, **11,** 598
  laboratory arrangement for, **11,** 596
  liquid helium consumption and, **11,** 605–606
  materials of, **11,** 598–600
  operating procedure for, **11,** 606–607
  operation of in swept field, **11,** 604
  "persistent mode" operation of, **11,** 597–598
  power supply for, **11,** 603–604
  reversed field in, **11,** 605
  stabilization of, **11,** 600
  testing of, **11,** 599–600
Superconducting microwave resonators, **11,** 541–594
  applications of, **11,** 592–594
  breakdown of, **11,** 565–566
  coupling networks for, **11,** 581–584
  design and fabrication of, **11,** 567–570
  experiments using, **11,** 584–594
  frequency stability in, **11,** 585–587
  requency standards referenced to, **11,** 588–590
  high-field applications of , **11,** 590–591
  high-*Q*, **11,** 567–584
  low loss in, **11,** 593
  material property studies of, **11,** 591–592
  oscillator and, **11,** 589
  Q values for, **11,** 589
  tuning of, **11,** 587–588
Superconducting quantum interference device (Squid), **11,** 281, 489, 532–535
Superconducting rings, **11,** 275–289
  interference effects in, **11,** 289–296
  magnetometer circuits and multipliers in, **11,** 279–285
  noise limit in, **11,** 283
  quantized states of, **11,** 286–289
  quantum interference experiments in, **11,** 291–294
  with resistor, **11,** 286–289
  with single weak link, **11,** 285–289
  symmetrically biased, **11,** 290
Superconducting state, current flow in, **11,** 126
Superconducting surface impedance, rf field level and, **11,** 560–562
Superconducting surface reactance, **11,** 552–553
Superconducting surface resistance, **11,** 550–552
Superconducting thin films, **30,** 401
Superconducting transition, effect of stress, **6B,** 74
Superconducting tunnel characteristics, **11,** 126–128
Superconducting tunneling measurements, junctions for, **11,** 127
Superconducting tunneling, in high magnetic fields, **11,** 153–156
Superconductivity, **6B,** 71–78; *see* Surface acoustic wave applications
  BCS theory, **24A,** 349–352
  detection of, **6B,** 71–72
  penetration depth, **6B,** 75–76
Superconductor energy gap, **10,** 431–441
Superconductor shielding, **11,** 609–618
  dc fields and, **11,** 612–617
  design of, **11,** 611–612
  time-varying fields and, **11,** 610–611
Superconductor tunneling, **27,** 427
Superconductor(s)
  coupling energy in, **11,** 203
  definition, **6B,** 71
  density of states, **27,** 428, 432, 436
  energy gap measurement in, **11,** 135–140
  flux lattice, **27,** 429, 435, 437, 447
  interphase surface energy, **6B,** 76–77
  Josephson effect in, **11,** 201–208
  microwave properties of, **11,** 544–566
  miscellaneous measurements, **6B,** 77–78
  quasiparticles, **27,** 428, 441, 446
  recombination lifetimes in, **11,** 150
  residual surface resistance in, **11,** 553–560

scanning tunneling microscopy of, **27,** 427
specific heat, **6B,** 74–75
surface impedance of, **11,** 544–550, 560–562
thermometry and, **11,** 536–537
time-varying fields and, **11,** 610–611
tunneling in, **11,** 126–156
tunneling between normal metal and, **11,** 128–131
two-fluid model of, **11,** 54–7–548
vortex evolution, **27,** 443
vortex-vortex interactions, **27,** 447
vortex wavefunction, **27,** 441
weakly linked, **11,** 199–305
Superconductors, type II, **8,** 91
Supercurrent steps, **11,** 217–222
self-induced, **11,** 229–233
Supercurrent wave forms, for equivalent circuit, **11,** 272
Supercurrent, oscillating, **11,** 233–238
Superelectrons, Josephson oscillator and, **11,** 265
Superfluid helium, *see also* First sound; Second sound; Fourth sound; Hybrid acoustic modes of superfluid helium
critical velocity for flow of, **19,** 416
equations of motion, **19,** 415–416
^{3}He, **19,** 417, 437
He II, **19,** 415
lambda temperature of, **19,** 414
Landau theory, **19,** 415
persistent current in, **19,** 417
Superfluid surfaces and films, studies of, **11,** 481–484
Superfluid, angular fluid motion in, **11,** 469–470
Superfluidity, **23B,** 326
Superheating, in crystalline polymers, **16B,** 295
Superheterodyne acoustooptic techniques, **19,** 487–492, *see also* Heterodyne acoustooptic techniques, superheterodyne systems
Bragg cell modulation, **19,** 487–488
electrooptic modulation, **19,** 488–490
Engan's method, **19,** 490–492
Superheterodyne detection, **10,** 163–165, 184
Superheterodyne receivers, **12B,** 212–213
Superhyperfine interaction, EPR, **21,** 406
Superimposed beams, **7A,** 477
Superionic conductors, *see* Fast ion conductors
Superlattice, **22,** 253, 254
Superleak, **19,** 434, 436
acoustic index of refraction in, **19,** 434–435
Superleak (helium), **4B,** 320
Supermendur, **5B,** 247
Supermirrors, **23A,** 216; **31,** 283–284
Supernova explosion, -rays from, **12A,** 333–336
Superparamagnetic condition, **24A,** 327
Superposition
integral, **2B,** 381
network response, **2,** 27
of profiles method, **5A,** 604, 609–611
principle, **2A,** 10; **24A,** 47
principle of, **1,** 350; **2B,** 379
Superradiant light sources, **18,** 720–721
Supersaturation
cloud chambers, **5A,** 194–196
discharge chambers, **5A,** 201
produced by changing properties of the solvent, **6A,** 109
Supersonic beam sources
clustering
choice of nozzle, **29B,** 31–33
ionic clusters, **29B,** 42–43
neutral clusters by laser ablation, **29B,** 41–42
pulsed sources
for low vapor pressure and refractory materials, **29B,** 293–295
pulse duration, **29B,** 36–37
valve design, **29B,** 37–40
spectroscopic applications
radicals, **29B,** 39–41, 295
clusters, **29B,** 41–43, 295–296
supersonic expansion from circular nozzles
departure from idealized behavior
perpendicular temperature, **29B,** 27
terminal Mach number, **29B,** 26
idealized continuum model
adiabatic, isentropic expansion, **29B,** 22–24
Mach number, **29B,** 23
translational temperatures, **29B,** 24–25
velocity distributions, **29B,** 25–26
interactions with background gas
barrel shock, **29B,** 28
Mach disc, **29B,** 28–29
seeded beams, **29B,** 29–30

skimmed beams, **29B,** 30–31
supersonic expansion from slit nozzles, idealized continuum model
Mach number, **29B,** 33
shock wave structure, **29B,** 35–36
transition to axisymmetric flow, **29B,** 34–35
use in metastable atom sources
coaxial excitation, **29B,** 99–100
DC discharge, **29B,** 104–105
transverse excitation, **29B,** 100–101
Supersonic chemical lasers, **15B,** 102–111
Supersonic flow, **18,** 330, 760
Supersonic nozzle source for molecular beam, **4A,** 190–192
Supersonic wind tunnel, **18,** 771–779
Support wires, **25,** 242
Supported film, energy-loss measurements, **17,** 53–54
Suppressed-carrier modulation, **2B,** 442
SUPRASIL, **31,** 309
Surface
aplanatic, **1,** 330
constant concentration, **6A,** 73–74
contact, **1,** 194
creation and nature, **6A,** 148–149
depth of damage, **6A,** 148
current, **1,** 576
equipotential, **1,** 433
focus of, **1,** 330
isothermal, **1,** 308
physics, **3,** 602
potential, **1,** 435
tension, **1,** 122–123
texture, **1,** 427
Surface absorption, **30,** 77–78, 590–592
Surface acoustic wave, **19,** 495–532
characteristics of, primary, **19,** 503
characteristics of, secondary, **19,** 507
comparison with bulk waves, **19,** 508
experimental problems, **19,** 507–508, 520
frequency, **19,** 505
light scattering by, **19,** 464, 471–473, 475, 480, 484–492
properties of a selected media, **19,** 504
Surface acoustic wave applications, *see also* Surface acoustic wave devices, **19,** 520–530
crack detection, **19,** 528–530
elastic properties of surfaces, **19,** 528–530
imaging, **19,** 517
nondestructive evaluation, **19,** 528–531
nonmeasurement applications, **19,** 521
semiconductor surface properties, **19,** 526–528
superconductivity, **19,** 524
thin-film properties, **19,** 523–525
Surface acoustic wave devices
delay line, **19,** 510, 521
filter, **19,** 510, 521
optical, **19,** 521–522
oscillator, **19,** 523–525
pressure sensor, **19,** 525
resonator, **19,** 523
Surface acoustic wave transducers, **19,** 508–520
direct piezoelectric pickup, **19,** 509
electromagnetic (EMAT), **19,** 69, 518–519
interdigital transducer (IDT), **19,** 509–514
losses, **19,** 511–514, 515, 516, 519
mode conversion, **19,** 509, 513
noncontacting, **19,** 509, 518–519
optical, **19,** 509, 512, 516–518
thermoelastic, **19,** 509
wedge, **19,** 509, 511, 515–516
wideband, **19,** 509
x-ray, **19,** 509
Surface adsorbates, **22,** 356
Surface analysis, **22,** 313
methods, **30,** 292, 327
Surface area, **14,** 186
contacts, **6B,** 88–89
Surface area-volume considerations, **14,** 481
Surface barrier, **22,** 162, 171
Surface binding states, **22,** 428
Surface breakdown, **30,** 157, 163
Surface capacity, **6B,** 140
Surface charge transfer devices, **2A,** 129
Surface charging, glass, **14,** 393
Surface chemical shift, **22,** 224
Surface cleaning, **22,** 324, 354
Surface collision rate, **14,** 6
Surface composition, **22,** 231, 333
analysis by neutral and ion impact radiation (SCANIIR), **21,** 268–274
Surface conditions, **14,** 185, 190
Surface conductivity, **6B,** 137–140
Surface contributions, **25,** 46
Surface coverage, **22,** 363
Surface creep, **14,** 299
Surface crystallography measurements, **22,** 267

Surface currents, **8,** 63
Surface damage,
  depth from abrasion, **6A,** 149
  factors, **6A,** 149
Surface defects, **22,** 245
Surface degradation, techniques, **16B,** 255–277
Surface diffusion, **14,** 29, *see also* Surface migration; **22,** 363, 392, 393, 395
  activation energy, **22,** 395
Surface dimers
  scanning tunneling microscopy of, **27,** 260, 261
Surface discharges, elimination of, **16C,** 455
Surface domain gratings, quasi-phase-matching in, **29C,** 245
Surface effects, **2,** 121
Surface elastic wave, *see* Surface acoustic wave
Surface finish, **14,** 281, 283, 285
Surface free energy
  of lamellar crystals, **16B,** 296
  in lamellar growth theory, **16B,** 349
  temperature dependence, **16B,** 350
Surface generation, *see* Generation of ultrasound
Surface impedance, of normal and superconducting metals, **11,** 544–550
Surface impurities, **21,** 227, 230, 240, 263
Surface ionization, **3B,** 711
  for atom detection, **4A,** 393–399
  detection of negative ions, **7A,** 271
  detector, **3,** 595; **7A,** 269–272; **22,** 497
  method, partial pressure in, **3,** 554–558
  reaction equilibrium constant, **3,** 554, 555
Surface ionization source
  capillary-type cesium ion source, **29A,** 156–157
  curvilinear-cesium-surface ion source, **29A,** 153, 156
  ion current density, **29A,** 148
  ionized materials, **29A,** 147
  porous tungsten source, **29A,** 148–149, 151
  probability of positive ion formation, **29A,** 147–148
  self-extraction spherical-geometry source, **29A,** 151–153
  theory of ionization, **29A,** 147–148
Surface measurements on semiconductors, **6B,** 62–63
Surface melting, **6A,** 63–64
Surface migration, **14,** 28, 278
  of pumping fluids, **14,** 167–169
Surface modification of polymers, **16B,** 245–255
Surface neutralization, **29B,** 58–59
Surface oxides, **30,** 177
Surface patterning, **30,** 189
Surface photovoltage, **6B,** 140–141
  method, **6B,** 109
Surface physics, **14,** 25
Surface plasmons, **22,** 196, 382
  ablation mediated by, **30,** 65–67
  desorption mediated by, **30,** 65–68, 212
  resonance spectroscopy, **30,** 107–109
Surface potential, **6B,** 137; **22,** 2
  correlation energy, **22,** 2
  dipole layer, **22,** 2
  ion-core potential, **22,** 2
Surface preparation of solids, **6A,** 147–176
  reasons for etching, **6A,** 148
  steps involved in preparing, **6A,** 148
Surface properties, miscellaneous measurements, **6B,** 142–143
Surface recombination velocity, **6B,** 95–96, 141, 359
Surface reflectivity, **30,** 178, 201–205
Surface resistance
  silicone coating, **14,** 292, 297
  of superconductors, **10,** 433–435, 439, 440
Surface roughness, **22,** 367; **30,** 83–86
Surface second-harmonic generation (SSHG), **29C,** 213
Surface segregation, **22,** 324
Surface self-diffusion, **22,** 367
Surface sensitivity, **22,** 131, 152
Surface sites, **14,** 29
Surface state(s), **6B,** 136–143; **16C,** 434; **22,** 152, 153
  density, **27,** 285
  desorption induced by, **30,** 57–58, 68–69
  studies in electron distribution mode, **22,** 152
Surface steps, **22,** 248
Surface structure, **22,** 328, 333
Surface studies with positrons, **21,** 89
Surface studies, channeling, **21,** 282, 345–357
  adsorbates, coverage determination, **21,** 351
  adsorbed atoms, effect of, **21,** 280, 350, 357
  adsorbed hydrogen, **21,** 353
  basic principles, **21,** 346–351
  blocking dip, **21,** 350, 357
  distance of closest approach, **21,** 346

enhanced vibrational amplitude, **21,** 347, 353, 354
impurity atom location, **21,** 356–358
lateral displacement, **21,** 280, 347, 355
LEED studies, **21,** 354
relaxation, platinum, **21,** 347, 351–354
shadow cone, **21,** 346, 347
surface Debye temperature, **21,** 352
surface peak yields, **21,** 301, 347–351, 354
surface relaxation, **21,** 281, 347, 352, 357
Surface temperature sensor
fluorescent paint, **18,** 672
infrared pyrometer, **18,** 672
light transmitting paint, **18,** 671
thermocouple-thermopile, **18,** 666, 672–673, 675
thin film resistance, **18,** 664, 666, 672–685
Surface temperature, **30,** 179
Surface treatment, **6B,** 138–139; **14,** 442
Surface trimers, scanning tunneling microscopy of, **27,** 261
Surface vaporization, **30,** 183–186
Surface-barrier detector, **22,** 210
Surface-catalyzed reactions, in chemical kinetics, **15B,** 281
Surface-extended-x-ray-absorption fine structure, **22,** 232, 233, 234, 235, 236
Surface-state quenching, **22,** 152
Surface-to-volume ratio, **14,** 457–458
Surfaces, **23B,** 1; **23C,** 436
metal, scanning tunneling microscopy theory, **27,** 8
semiconductor, scanning tunneling microscopy theory of, **27,** 11
Surge limiter or suppressor, **2A,** 61, 220
Surprisals, in chemical lasers, **15B,** 99–102
Susceptibility, **3,** 468; **23C,** 5; *see also* Magnetic susceptibility
apparent, **1,** 544, 545
component, minimum detectable, **3,** 473
volume, **3,** 423–424
dielectric, **1,** 490
electric, **16C,** 383
function, generalized, **23A,** 78
initial, **1,** 517
isothermal, **23A,** 79
longitudinal, **23C,** 30
magnetic, **1,** 258, 537–546
Curie expression, **2,** 626–627
thermometers, **11,** 502–533
wave-vector and frequency-dependent, **23C,** 7, 162, 164
wave-vector-dependent, **23C,** 5
Susceptometer, **24A,** 423
Suspension, center of, **1,** 84
Swarm techniques, *see also specific technique*
applications, **29A,** 273, 275–276, 295
comparison to beam experiments, **29A,** 273–274
drift tube operation, **29A,** 290
ideal experiments, **29A,** 274–275
nonideal experiments, **29A,** 275
rate coefficient determination, **29A,** 274–275
Swarms, **7B,** 29–139
atomic collisions of electrons, *see* Atomic collisions
electron type, *see* Electron swarms
Sweep circuits, **2,** 394–412; **2A,** 401–402
gated, **2,** 394–396
glow discharge tube, **2,** 401
oscilloscope, **2,** 398–401
purpose, **2,** 394–398
switch-off tube, **2,** 402–404
thyratron, **2,** 402
Sweep, exponential, **2,** 405–406, 410
Swept oscillator wave analyzer, LDV signal processing, **18,** 165
Swinging choke, **2A,** 204
Switchboard model, **16B,** 344
Switches, series, **5B,** 311–314
Switching
avalanche-mode, **2,** 153
symmetrical five-layer, **2,** 166–167
Switching circuits
tests of, **2B,** 209–211
transistor, **2,** 412–451
Switching diode(s), **2A,** 61–63
four-layer, **2,** 160–162
Switching schemes, **12C,** 45–51, *see also specific types*
Switching techniques, **12C,** 31–35
hydrogen cloud observations with, **12C,** 66
Switching triodes, **2,** 162–164
SWR, *see* Standing wave ratio
Sylphon, **14,** 300–306
Symbols, *see* lists in each Part
Symmetric collision
energy spread, **17,** 178
K-shell vacancy production, **17,** 332–334, 335
orbital transitions, **17,** 321

scattering, **17,** 306
target x-ray cross section, **17,** 288
Symmetric dip, **8,** 138–139
Symmetric rotor, **3A,** 18, 37, 57, 74
Symmetric top, **10,** 303, 315–319
Symmetrical eyepiece, **25,** 62
Symmetry
anisotropy factors for grains, **19,** 242–244
crystallographic, **16B,** 3, 8–18
of grains, **19,** 242–244, 276
groups, **3,** 268–301
macroscopic, of polycrystalline materials, **19,** 275–277, 292–294
molecular, **3,** 122
of a molecule, influence on vibrational spectrum, **16A,** 95
Symmetry axis, **8,** 61, 68, 71, 77–79, 82
perpendicular mirror plane, **8,** 78
Synchro pair, **2B,** 441
Synchrocyclotron, **5B,** 645–651
Synchrometer, mass, **5B,** 5
Synchronous counter, **2B,** 4–6
Synchronous detection, **26,** 279
Synchronous detector, **22,** 285
Synchronous pumping, **15B,** 192
Synchronously pumped frequency doubling, **29C,** 243
Synchrotron
accelerator, *see* Electrodynamic accelerator
alternating-gradient, **5B,** 663–678, 693–694
electron, **5B,** 676–678
proton, **5B,** 668–676
constant-gradient, **5B,** 651–663
electron, **5B,** 653–656
proton, **5B,** 656–663
"racetrack," **5B,** 656
Synchrotron radiation, **4A,** 313, 315–316; **8,** 166, 167; **10,** 295–299; **13A,** 210; **20,** 339; **22,** 127, 128, 141, 144, 147, 171, 183, 184, 217, 225, 480, 491; **23B,** 3; **29C,** 23–24; **31,** 2–3;
applications, **29C,** 40–41
characteristics, **29C,** 24–33
components, **31,** 123
definition, **31,** 1
emission, **31,** 2–3
history, **29C,** 23
light mochromatization, **29C,** 33–39
as primary source standard, **31,** 122–125
source, **22,** 156, 171; **31,** 1
UV source, **7A,** 152–153
Syndiotactic chain configuration, **16B,** 19, 340
Synthetic polymers, **30,** 438
System
function, **2,** 461
information efficiency, **2,** 696
magnification, **25,** 24
noise temperatures, **12B,** 204–207
System temperature, **12B,** 203
definition of, **12B,** 5
Systematic errors, **28,** 28–29
Systematic extinctions (absences), in X-ray diffraction, **16B,** 37, 69
criteria for various symmetries, **16B,** 77–78
Systematic uncertainties, estimation, **28,** 452–453
Systems for cw measurements, **19,** 118–133
acoustic streaming, **19,** 132–133
reflection oscillator ultrasonic spectrometer, **19,** 129–130
sampled continuous wave technique, **19,** 123–124
transmission oscillator ultrasonic spectrometer, **19,** 124–129
ultrasonic interferometer, **19,** 118–120
wall losses, **19,** 131
Systems, pump ranges, **14,** 108–109
Szerny-Turner spectrometer, **12A,** 476–479

T

τ (response time), **12A,** 418, 421–423
T-equivalent circuit, **2A,** 268
T-tube, electrically driven shock tube, **18,** 790
T_1 minima, **21,** 448–449, 481–482
T_1, **20,** 7, 9, 12, 13, 16, 19, 25, 26
T_2, **20,** 7, 9, 13, 14
TAC, *see* Time-to-amplitude converter
Si see Time-to-amplitude converter
Tachometers, **1,** 86ff
dc motor as, **2B,** 439
in feedback compensation, **2B,** 423
Tacticity, **16B,** 18, 340
definition, **16A,** 3
effect on chain flexibility in methacrylate polymers, **16A,** 308
NMR studies, **16A,** 288–294
Tail pile-up, in pulse amplitude measurement, **2B,** 90, 97
Takagi-Taugin equations, **23A,** 326
Tamperature-programmed desorption, **22,** 428, 431, 446
from single crystal, **22,** 432

Tanaka rare-gas light sources, **3,** 160
Tandem acceleretor, **17,** 8–12, 84, 465–467, 533–535
Tandem mass spectrometry, **30,** 432
Tangent approximation, in linearization, **2B,** 425
Tangent ray, definition, **31,** 159
Tangential meridional plane, definition, **31,** 159
Tantalum
  boiling point, **14,** 351
  hydrogen solubility, **14,** 354
  melting point, **14,** 348, 351
  neutron activated, **5A,** 607–609
  physical properties, **14,** 371–372
  tunneling measurements in, **11,** 142
  vapor pressure, **14,** 348, 351
Tantalum, reflectance spectra, **31,** 219–221
Tape recording systems, in pulse amplitude measurements, **2B,** 123
Tape, magnetic, **2,** 534
Target conditionging, **30,** 508
Target debris, laser plasma source, 87–88
Target mass per unit area, **17,** 47
Target preparation, for PLD, **30,** 506–507
Target pulsing, of vidicon tubes, **12A,** 306
Target temperature, **30,** 201–218
Target(s)
  -detector geometry in scattering and reaction experiments, **5B,** 412–413
  in elastic scattering experiments, **5B,** 479
  for neutron cross-section measurement, **5B,** 372–373
  proton synchrotron, **5B,** 661
Target, in ion-atom collisions, **17,** 371–374; *see also* Gas target; Solid target; Thick target; Thin target ionization
    due to light particles, **17,** 110–113
    and x-ray production, **17,** 193–278
  density, and charge-changing probabilities, **17,** 131–139
  medium, and energy loss, **17,** 37
  thickness of, **17,** 47, 51
    in beam-foil experiments, **17,** 532, 549
    charge state fraction measurements, **17,** 84–85, 87
    equilibrium charge state distribution, **17,** 81
    and x-ray productions, **17,** 255–257, 287
Target-detector arrangements, **17,** 316–317
Targetry, *see* Experimental targetry
TaS_2
  1T phase
    atomic force microscopy of, **27,** 358
    scanning tunneling microscopy of charge density wave in, **27,** 358
    spectroscopy of, **27,** 363
  2H phase
    scanning tunneling microscopy of charge density wave in, **27,** 391
    spectroscopy of, **27,** 363
  4Hb phase
    scanning tunneling microscopy of charge density wave in, **27,** 393, 400
    spectroscopy of, **27,** 401
TaS_3
  scanning tunneling microscopy of, **27,** 402, 405, 416
  spectroscopy of, **27,** 418
$TaSe_2$
  1T phase
    atomic force microscopy of, **27,** 358, 374
    scanning tunneling microscopy of charge density wave in, **27,** 358, 274
    spectroscopy of, **27,** 376–378
  2H phase
    scanning tunneling microscopy of charge density wave in, **27,** 388
    spectroscopy of, **27,** 389
  4Hb phase
    scanning tunneling microscopy of charge density wave in, **27,** 393
    spectroscopy of, **27,** 400
Tau Law, **25,** 61
Taylor cone, **22,** 403
Taylor's expansion, **8,** 70, 77, 87
Taylor's series, **8,** 72, 97
Taylor-Proudman theorem, **18,** 803, 805
Tb, **23C,** 212
Tb^{159} hyperfine structure, **5B,** 57–58
$Tb_3Fe_5O_{12}$, **23C,** 129
TBBA, **23B,** 546, 556, 564, 572
$TbMnO_3$, **23C,** 114
TE10 mode, **8,** 144
TE10-mode rectangular waveguide, **8,** 135
Technical noise, **29C,** 103
Tectonic deformation, **24B,** 412–413
Tees, microwave, **1,** 483–484
Teflon, **10,** 97, 98; **14,** 411–413; **30,** 361
  creep, **14,** 304
  helium permeation, **14,** 515–516, 518
  in ultrahigh vacuum systems, **14,** 481

Telecentricity, **25,** 37, 157
Telemetering, **2,** 674–686; **2B,** 220–249
  bit and frame synchronization in, **2B,** 229–231
  bit detection in, **2B,** 233–235
  carrier modulation in, **2B,** 248–249
  constant bandwidth in, **2B,** 240
  data quality assurance in, **2B,** 245–247
  digital *vs* analog types, **2B,** 247–248
  double sideband, **2B,** 240
  dynamic nonlinearity in, **2B,** 237
  error-detecting code in, **2B,** 225–227
  frequency-division multiplexed, **2B,** 235–242
  information, theoretical methods, **2,** 686–704
  magnetic tape recording and, **2B,** 242–245
  modulation and multiplexing in, **2B,** 247–249
  modulation noise in, **2B,** 242
  nonreturn-to-zero (NRZ) level in, **2B,** 228–229
  notch-noise method in, **2B,** 246
  PCM (pulse code modulation) type, **2B,** 222–235
  quantization error in, **2B,** 221
  requirements of, **2B,** 221
  Shannon limit in, **2B,** 225–227
  shift register generated pseudonoise in, **2B,** 245
  signal-to-noise ratio in, **2B,** 239–241
  single sideband suppressed carrier AM (SSB), **2B,** 241
  system errors, **2,** 683–684
  time base error spectrum in, **2B,** 244
  time division *vs* frequency division in, **2B,** 248
  word error probability in, **2B,** 226
Telemetry
  acoustic, **24B,** 237
  links, **24B,** 35
  modes, **24B,** 35, 38
  seismic, **24B,** 2, 5, 36
Telephoto lens, **25,** 86
Telescope(s), **1,** 338–340; **25,** 30, 59, 132, 135
  gamma ray, **5A,** 668–671
  magnification, **1,** 338
  objective(s), **1,** 339; **25,** 64
  refracting, **1,** 338
Telescope, **8,** 162
Telescopes, in infrared systems, **12A,** 434–444
  Cassegrain, **12A,** 434–443
  cooled, **12A,** 416, 426, 443, 447
  efficiency, **12A,** 439–441
  emission form, **12A,** 435–444, 449
  Herschelian (off-axis), **12A,** 443
  infrared-designed, **12A,** 437–447
  large optical, **12A,** 434–442
Televiewer, borehole sonic, **24B,** 399–401, 403
Television data processing, **12A,** 307
Television pickup camera, **6B,** 354–355
Television systems, **12A,** 277–313
  comparison of tube types, **12A,** 282
  data processing, **12A,** 307
  integrating systems, **12A,** 279–284
  photon-counting systems, **12A,** 284
Tellurium ion laser, **15A,** 152–153
TEM, *see* Transverse electromagnetic modes
Temperature, **1,** 31–33, 235ff
  absolute, **1,** 311
  behind shock wave, **1,** 194, 196
  brightness, **1,** 255
  coefficients, of resistance, **1,** 457–458
  color, **3,** 178–179
  control, nuclear magnetic resonance sample, **3,** 418–419
  Curie, **1,** 549
  effect on length, **1,** 53
  empirical, **1,** 308, 309
  free stream, **1,** 153ff
  in gas flow, **1,** 170
  glass transition, **19,** 146
  ideal gas, **1,** 238
  intermediate, law of, **1,** 511, 512
  magnetic, **1,** 258–259
  probe, total, **1,** 154
  radiation, **1,** 255
  recovery, **1,** 153
  requirements, magnet, **3,** 417–418
  scale fixed points, **1,** 236ff
  scales, **1,** 32, 238–246, 255–257
  sound velocity, **1,** 259
  stagnation, **1,** 138, 146
  thermodynamic terminology for, **11,** 486–489
  very low, measurement, **5B,** 189–190
  wave, **1,** 293
Temperature amplitude, **19,** 195
Temperature at equilibrium between solid and liquid, **6A,** 42

Temperature calibration, **24A,** 77
Temperature coefficient of delay, **19,** 507
Temperature control, **6B,** 50; **16C,** 341
  bubble chamber, **5A,** 205
  cloud chamber, **5A,** 200
Temperature creep, **14,** 279, 289
Temperature cycle, **14,** 440
Temperature distribution, atmospheric, **12B,** 136–138
Temperature distribution, calculation, *see* Radial temperature distribution
Temperature fluctuations, measurement by Rayleigh scattering, **18,** 417
Temperature gradient, **8,** 248
Temperature gradient measurement, Raman scattering, **18A,** 424
Temperature in moving fluid, **18,** 457–460
Temperature in P-N junction, **2A,** 46, 54
  rectifier, **2A,** 58–59
Temperature increment, **19,** 5–7, *see also* Temperature amplitude
Temperature jump, **19,** 194, 233
Temperature measurement, **1,** 235–240; **26,** 213; *see also* Thermistor; Thermocouple
  by analysis of emitted and absorbed radiation, **18,** 465–487, 698–699, 704
  behind detonation front, **18,** 470
  by Doppler broadened line shape, **18,** 481–482, 490
  during drilling, **24B,** 217–218, 241–245
  electron beam fluorescence, **18,** 489–497
  emittance on two paths, **18,** 475–478
  at high pressure, **6A,** 417
  at high temperature, **24B,** 197
  hot-wire probe method, **18,** 313, 461
  infrared pyrometer, **18,** 672
  line reversal methods, **18,** 466–470
  method of absorption in two spectral regions, **18,** 472–475
  molecular scattering of radiation, **18,** 409, 414–418, 421–428
  in moving fluid by probe, **18,** 457–463
  by paint transparency and phosphorescence, **18,** 671–672
  probe methods, **18,** 457–463
  radiation analysis methods, **18,** 463–497
  radiometric, **26,** 239
  Raman scattering, **18,** 487–489
  by Rayleigh scattered spectral lineshape, **18,** 412, 414–417, 482–485, 487–489
  relative intensities, **18,** 478–481
  semiconductor integrated circuit, **24B,** 197
  by simultaneous detection of radiation emission and absorption, **18,** 470–472
  simultaneous with velocity measurement, **18,** 313
  in sparks, **18,** 698–699, 704
  two path absorption in thin foils of x-rays, **18,** 485–486
  vibrational temperature by analysis of emitted radiation, **18,** 473–474, 481
Temperature noise, *see* Noise
Temperature recording, **24B,** 235–238
Temperature sensors
  probe type, **18,** 460–463
  resistance films, **18,** 460, 677–685
  thermocouple, **18,** 460–461, 671–676
Temperatures (sub 4.2 degrees K), **8,** 19–21
TEMPO, **20,** 55
Temporal coherence, **26,** 6; *see also* Coherence, temporal
Temporal echoes, **9A,** 27–28
Tensile deformation
  ASTM test method, **16C,** 123
  orientation by, **16C,** 138–140
  radical species, **16C,** 203–205
  studies by mass spectroscopy, **16C,** 228
  study of radical generation, **16C,** 191–194
Tensile strength, **1,** 67
  hydrofracturing, **24B,** 396–397
Tensile strength of liquids, **19,** 360–372
  by acoustic levitation technique, **19,** 361–364
  electric field effects, **19,** 370–371
  of ether and hexane, **19,** 363–364
  limit of superheat, **19,** 362, 364
  of liquid helium, **19,** 368–371
  influence of motes, **19,** 360, 362
  of water, **19,** 361
Tensile stress, **1,** 100
Tension tests on solids, **6A,** 310–313
Tension, surface, **1,** 122–123
Tensor components of polarization, **4B,** 227, 244
Tensor operators, irreducible, **4B,** 229
Tensor polarizability, **10,** 423
Tensors, statistical, **5B,** 194, 197
  expansion in powers of 1/T, **5B,** 210–214
TeO-BaO glasses, **23B,** 281
Terbium gallium garnet (TGG), **29C,** 360
Terephtal-dibutyl-aniline, *see* TBBA
Terminal voltage, **8,** 257

Terminations, microwave, **1,** 483
Terpolymer, **16A,** 3
Terrace
distribution of sizes, **22,** 252
reciprocal-lattice rod, **22,** 252
structure factor, **22,** 250
Terrestrial telescope, **25,** 36, 59, 136
Tertiary butyl cyanide, **23A,** 507
Tertiary reflector antennas, **12B,** 77, 79
Test pulser, general-purpose, **2B,** 183
Test section, wind tunnel, **18,** 758, 764
Test statistic, **28,** 180
Testing
dynamic, **2,** 651–661
interpretation, **2,** 668–669
equipment, **2,** 647–674
steady-state conditions, **2,** 647–651
interpretation, **2,** 667–668
Tests, classification of in equipment testing, **2B,** 170–176, *see also* Equipment testing
Tetracritical points, **23C,** 53
Tetracyanoethylene, **23A,** 488
Tetragonal crystal system
definition, **16B,** 4
interplanar spacing formulae, **16B,** 26
reciprocal lattice relationships, **16B,** 24
Tetragonal lattice, definition, **16B,** 5
Tetrodes, **2,** 62–65
Texas, University of, **11,** 592
Texture, **23C,** 450; *see also* Preferred orientation; Symmetry, macroscopic
measurement, **23A,** 232
methods for determination, **23C,** 452
TGG, *see* Terbium gallium garnet
TGM, *see* Toroidal grating monochromator
Thallium ion laser, **15A,** 155–156
Theoretical plate, **16A,** 47
Theoretical visibility, coefficient, **8,** 51
Theory
of e-m radiation, **13A,** 31
scanning tunneling microscopy imaging of adsorbates, **27,** 14
Theory of
apparent barrier height, **27,** 20, 26
apparent size of adatom, **27,** 16
atom superpostiion modeling, **27,** 8
forces, role of, i, **27,** 24
graphite, imaging, **27,** 14
imaging band edge states, **27,** 14
local density of states, **27,** 7
point contact, **27,** 16
role of surface electronic structure, **27,** 11
spectroscopy, **27,** 22
strong coupling regime, **27,** 16
surfaces, metal, **27,** 8
semiconductor, **27,** 11
tip, models for, **27,** 4, 6, 14
tip, tip-sample interactions, mechanical, **27,** 24
tunneling barrier, **27,** 20
tunneling Hamiltonian, **27,** 5
vacuum tunneling, **27,** 1
transmission probability, **27,** 2
voltage dependeng in STM imaging, **27,** 11, 22
work function, apparent, **27,** 20, 26
Thermal ablation, **30,** 175
Thermal accommodation, **22,** 383
Thermal activation, **22,** 354, 406
Thermal analysis, instrumentation, **16B,** 287–291
see also Differential thermal analysis, Differential scanning calorimetry
Thermal beam sources
practical sources
for dissociated atoms, **29B,** 16
heated filament, **29B,** 15
heated ovens
design considerations, **29B,** 10–15
dimer and polymer production, **29B,** 14
efficient improvement, **29B,** 18
for atomic clocks, **29B,** 19
temperature requirements, **29B,** 15
laser desorption, **29B,** 15–16
non-condensible gasses
apparatus, **29B,** 11
operating considerations, **29B,** 9–10
recirculation, **29B,** 17–19
theoretical considerations
beam intensity, **29B,** 4–5
beam shape
aperture geometry, **29B,** 6
collimator effects, **29B,** 5–8
effusion from long channels, **29B,** 3–4
effusion from thin-walled apertures, **29B,** 2–3
velocity distribution
average velocity, **29B,** 9
expressions for, **29B,** 8–9, 22
median velocity, **29B,** 9
most probable velocity, **29B,** 9
root mean square velocity, **29B,** 9

Thermal boundary layer, **1,** 137
Thermal breakdown, **6B,** 29–30; *see also* Electric breakdown
Thermal broadening, **17,** 156, 168–170, 516–517
Thermal broadening of microwave lines, **12B,** 146
Thermal capacity, **1,** 136
Thermal conductance (*G*), **12A,** 421–424
Thermal conductivity, **1,** 135ff, 287ff, 296ff; **6A,** 385–405; **8,** 247; **18,** 665–666; **30,** 344
  apparatus for measurement by direct static method, **6A,** 393
  direct static method, **6A,** 388–395
  drill cuttings, **24B,** 209
  dynamic method of determination, **6A,** 399–405
  effect of porosity, **24B,** 213–215
  electrical methods, **6A,** 397–399
  estimates from well logging, **24B,** 212, 254
  at high pressure, **6A,** 424
  high pressure, **24A,** 289–292
  high temperature, **24A,** 277, 280
  in-situ measurement, **24B,** 251–253
  lateral heat flow, **6A,** 395–396
  mean harmonic, **24B,** 216
  measurement
    Angtrom method, **24A,** 282–285, 291–292
    Bridgman method, **24A,** 289–291
    divided bar, **24B,** 207, 251
    divided bar method, **24A,** 271–277
    linear source methods, **24A,** 285–288
    needle probe, **24B,** 210, 248–251
    thermal stack method, **24A,** 277–282, 289
    transient techniques, **24B,** 210–212, 251
  methods of measurement, **6A,** 386–405
  porous rock, **24A,** 275–276
  relation to water content in sediments, **24B,** 254
  soft sediment, **24A,** 276
  solid rock, **24A,** 277
  standard substances, **24A,** 279–280
Thermal conductivity gauge, **14,** 51–60
  gas sensitivity, **14,** 59–60
  operating modes, **14,** 53, 55–56
  Pirani, **14,** 54–57
  thermistor, **14,** 56–59
  thermocouple, **14,** 58–59
Thermal creep, **1,** 160, 169
Thermal decomposition, **6A,** 25
  process, in thin films, **11,**646
Thermal desorption from polycrystalline substrates, **22,** 427
  surface binding sites, **22,** 428, 429
Thermal desorption single crystal, **22,** 431
  preparation and mounting of single crystal, **22,** 432
Thermal desorption, **22,** 354, 430, 464, 499, 503; *see also* Outgassing
  apparatus, **22,** 436
  data treatmetn, **22,** 447
  detector, **22,** 444
  surface binding site, **22,** 428
  theory, **22,** 454
Thermal detectors, **10,** 166–174; **13A,** 325; **13B,** 6; **26,** 293, 303
Thermal diffuse scattering, **23B,** 206
Thermal diffusion, **1,** 141; **4B,** 319
  length, **30,** 344
  penetration depth, **30,** 178
Thermal diffusivity, **1,** 136, 137, 288, 294; **18,** 665
Thermal dissociation, **4A,** 179, 221
  atomic source, **7A,** 80, 83, 88
Thermal Doppler width, **8,** 161
Thermal drift in piezoelectric hysteresis
  experimental problems in scanning tunneling microscope operation, **27,** 73
Thermal emf, **1,** 249ff, 506, 507, 511–513
Thermal energy collision at surfaces, **7B,** 268–275
Thermal equilibrium, **1,** 236
Thermal evaporation
  temperature and support materials in, **11,** 624–627
  of thin films, **11,** 622–637
Thermal expansion coefficient, **19,** 6–7
  frozen, **19,** 182
Thermal expansion, **1,** 325
Thermal history, of poly(ethylene terephthalate), effect on melting point, **16B,** 302
Thermal impulse breakdown strength, **16C,** 447
Thermal ion-neutral reactions, **7B,** 124
Thermal ionization source
  ionization
    efficiency, **29A,** 159–160
    equilibrium, **29A,** 159

theory, **29A,** 159–160
types, **29A,** 160–161
Thermal lensing, **30,** 179
Thermal light source, **1,** 373
Thermal measurement of dielectric loss, **6B,** 48, 49
Thermal measurements at high pressures, **6A,** 423–424
Thermal microscopy, **27,** 83
Thermal motion
effect on X-ray diffraction line shape, **16B,** 140–141
effect on X-ray scattering, **16B,** 38
lattice distortions from, **16B,** 101
Thermal neutron, *see* Nuclear cross section; Nuclear reactions
detection, **24B,** 541–542
Thermal neutrons, *see* Neutron scattering
Thermal neutrons, transmission, **8,** 242
Thermal noise, **2B,** 461–463
Thermal penetration depth, in superfluid helium, **19,** 431
Thermal plasmas, electron temperature measurements, **9A,** 108
Thermal population distribution, **10,** 345, 346
Thermal processes, **30,** 340–341
Thermal processing, **30,** 343–347
Thermal properties of solids, **6A,** 375–406
Thermal radiators, **4A,** 290–293
processes, **4A,** 290
surface, **4A,** 291–292
volume, **4A,** 292–293
Thermal relaxation time, **1,** 246; **6A,** 386
Thermal resistance, **24A,** 274–275
Thermal spikes, **22,** 330
Thermal sputtering, **30,** 176
Thermal time constant, **10,** 166
Thermal transfer in vacuum systems, **14,** 494
Thermal transpiration, **1,** 169
Thermal vapor deposition, *see* Vacuum evaporation
Thermal velocities in target, corrections for, **7A,** 283–287
Thermal vibrations, **22,** 247
effect on channeling, **21,** 287, 289, 348
Thermal wind relation, **18,** 804, 810
Thermal-desorption spectroscopy, **22,** 508
Thermal-diffuse scattering, **22,** 274
Thermal-neutron-energy selector, **8,** 239
Thermal-spike sputtering, **30,** 238, 281
Thermalization of positrons, **21,** 82
Thermally stimulated currents, **16C,** 439–440
Thermionic cathode, electron injection from, **11,** 451–452
Thermionic constants, for a patchy metal, **6B,** 134
for semiconductors, **6B,** 134–135
Thermionic emission, **2A,** 147–152; **30,** 186; **6B,** 128–136; **22,** 5, 20, 21, 205
current density of electrons, **22,** 5
from the perveance line, **6B,** 136
polycrystalline metals, **6B,** 133–134
Schottky plot, **6B,** 129–130
single crystal metals, **6B,** 129–133
Thermionic energy conversion, **4A,** 28–29, 36
Thermionic method, **22,** 6
Thermionic work function, **6B,** 130–133
Thermisters, **13B,** 8
Thermistor(s), **1,** 167, 168, 248, 458; **2,** 121–122, 679, 729; **2A,** 48–49; **3,** 84; **30,** 205, 206; *see also* Temperature measurement
gauge, **14,** 56–59, *see also* Thermal conductivity gauge
marine heat flow, **24B,** 230–231
in microwave circuits, **2B,** 284
negative temperature coefficient type, **2A,** 48
positive temperature coefficient type, **2A,** 48–49
precision, **24B,** 200–206
use in borehole temperature measurement, **24B,** 195–196
Thermochemistry of substances at high temperatures, **3,** 558–565
Thermocouple, **1,** 249–252, 392; **14,** 278, 307; **18,** 460–461, 671, 672, 675–676; **24B,** 197, 229–230
beam detector, **7A,** 350–352, 398–399
calorimeters, **26,** 293
gage, **1,** 167
gauge, **14,** 58–59, *see also* Thermal conductivity gauge
meters, **1,** 460
Pt/Pt-Rh, **24A,** 81
systems, **1,** 513–514
Thermocoupler converter, **2B,** 66
Thermocouples for photon detection
comparison with bolometer, **4A,** 442–443
metal film, **4A,** 440
optimal conditions, **4A,** 438–439
pin-type, **4A,** 440
thermopiles, **4A,** 441–442

wire, **4A,** 440
Thermocouples, **3,** 83–84; **11,** 534–535; **13B,** 7
in thermoelectric measurements, **6B,** 122
Thermodynamic critical termperature, **30,** 226
Thermodynamic properties of fluids, **3,** 681–708
Thermodynamic properties of superfluid helium, determined acoustically, **19,** 437
Thermodynamic temperature, terminology for, **11,** 486–489
Thermodynamics, **1,** 303–328
First Law, **1,** 260ff, 285, 307–309
of polymer mixtures, **16C,** 276–278
Second Law, **1,** 309–324
temperature scale, **1,** 32
Third Law, **1,** 324–328
Zeroth Law, **1,** 309
Thermoelasticity theory, **11,** 433–438
Thermoelectric effect(s), **1,** 392; **6B,** 114–128
Thermoelectric emf, **1,** 506, 507, 511
Thermoelectric generator, **6B,** 126
Thermoelectric laws, **1,** 510–512
Thermoelectric power, **1,** 508, 515
alloys, **6B,** 118
measurement, **6B,** 118–125
metals, **6B,** 116–117
semiconductors, solid electrolytes and insulators, **6B,** 118
significance of, **6B,** 116–118
at temperature above 1008 K, **6B,** 121–122
below 1008 K, **6B,** 122–124
Thermoelectric properties of semiconductors, **1,** 514, 515
Thermoelectric refrigeration, **6B,** 126–127
Thermoelectric tests for junctions, **6B,** 61–62
Thermoelectric thermometer, **1,** 249–252, 512
Thermoelectric voltage determination, **6B,** 119–120
Thermoelements, in ac circuit, **2B,** 65
Thermography, medical, **26,** 243
Thermogravimetry, **16B,** 336
Thermoionic noise, in photoelectric detectors, **17,** 545
Thermoluminescence, **6B,** 295, 313–315; **16C,** 440
Thermomagnetic balance, **24A,** 421–422
Thermomagnetic effects, **6B,** 166–170
Thermometer(s)
Beckmann, **1,** 244
electrical, **1,** 245–252, 512
gas, **1,** 238, 239, 240–241
germanium resistors for, **11,** 491–492
high temperature gallium, **14,** 375
liquid expansion, **1,** 243–244
mercury, **1,** 243
miscellaneous types of, **11,** 533–538
nickel resistance, **14,** 358
nuclear, **5B,** 190
orientation and, **11,** 488
paramagnetic susceptibility and, **11,** 487
platinum, **1,** 245ff
radiation, **1,** 252–255
radioisotopes as, **11,** 525–526
resistance, **1,** 245–259
type, **11,** 489–501
semiconductor, **1,** 247ff
susceptibility type, **11,** 502–533
thermoelectric, **1,** 249–252, 512
vapor pressure, **1,** 241–243
Thermometry, **1,** 235–260
capacitors in, **11,** 538
cerous magnesium nitrate (CMN) in, **11,** 503–504
energy gap and, **11,** 537
high-temperature, **11,** 359–360
multichannel analyzer and, **11,** 522–523
nuclear magnetic resonance measurements and, **11,** 513–518
resonance, **23C,** 424
superconductors and, **11,** 536–537
thermocouples in, **11,** 535–535
at ultralow temperatures, **11,** 485–539
Thermonuclear plasmas, **7A,** 466
Thermopile, **1,** 252; **3,** 80–81
Theta solvent, **16A,** 5, 19, 497
Theta temperature, **16A,** 18–19
Thevenin's theorem, **2,** 27–28; **2A,** 10
THG, *see* Third harmonic generation
Thick target, in ion-atom collisions, **17,** 371–372
backscattering energy loss, **17,** 61–65
energy-loss measurements, **17,** 47
equilibrium charge state distribution, **17,** 81
heavy incident ions, **17,** 221–228
for inner-shell ionization, **17,** 218–219
Thick-absorber technique in particle-solid studies, **7B,** 241
Thickness measurement, **14,** 541–549
comparison of methods, **14,** 548, 549
galvanometer balance, **14,** 542, 543
quartz crystal microbalance, **14,** 542–547

in situ weighing, **14,** 542–547
source precalibration, **14,** 541–542
Thickness modes, **19,** 44–55
thickness expander, **19,** 51–54, 57–58
thickness shear, **19,** 51–54, 59–60
Thickness of films, **6B,** 279
Thickness uniformity in thin films, *see* Uniformity
Thickness, skin, **5A,** 327–328
Thimble trap, **14,** 278, 285
Thin crystal criterion, **3,** 355
Thin film, *see also* Vacuum evaporation; Sputtering; Thickness measurement
backscattering energy loss, **17,** 58–59
deposition parameters, **14,** 521–522
heat transfer gage, **18,** 664, 666–685
method of crazing study, **16C,** 244
properties, vacuum evaporation, **14,** 523
resistor, **2A,** 118
specimens, for crystallization, preparation of, **16B,** 354–356
thermometers, in shock-wave studies, **7B,** 170
vacuum deposition, **14,** 521–575
Thin film polarizer, **31,** 230–231
Thin films, **11,** 619–674; **25,** 274; **26,** 32, 125; **30,** 2
definition, **6A,** 122
for electron-solid interaction studies, **7B,** 284
anodization of, **11,** 649–650
chemical deposition of, **11,** 645–652
chemical transfer processes in, **11,** 647
electrical thickness measurement for, **11,** 658–659
electrodeposition of, **11,** 648–650
electron beam evaporation system in, **11,** 634–635
ellipsometry for, **11,** 671–674
epitaxial growth in, **11,** 621
Getter-sputtering system in, **11,** 641–642
glow-discharge sputtering in, **11,** 638–642
immersion spectrophotometry for, **11,** 670–671
interferometer measurements of, **11,** 657
internal stresses in, **11,** 660–663
Langmuir-Blodgett technique in, **11,** 650–651
LEED (low-energy electron diffraction) system in, **11,** 635–636
low-pressure sputtering in, **11,** 642–645
matrix, **26,** 129
measurements of, **11,** 652–674
mechanical techniques in, **11,** 660–665
Michelson interferometer and, **11,** 657
microbalance measurements in, **11,** 653–655
multilayer, **11,** 630–632
multiple-beam interferometric techniques in, **11,** 656
optical measurements, **6B,** 278–281, *see also* Films
optical measurements of, **11,** 655–657, 665–674
oxide layers and, **11,** 650
polarimetric methods in, **11,** 671–674
polarization effects, **26,** 125
preparation, **6A,** 122–134
evaporation in vacuum method, **6A,** 123–129
preparation of, **11,** 620–652
quartz-crystal oscillation measurements of, **11,** 654
reflection and transmission measurements of, **11,** 666–669
refractory materials in, **11,** 633–637
resistance measurements of, **11,** 658
resistive heating in, **11,** 623–634
RHEED (reflection high-energy electron diffraction) system in, **11,** 635–636
single-crystal, **11,** 622
spectrophotometric methods in, **11,** 669–671
sputtering in, **11,** 637–645
stresses in, **11,** 663
stylus measurements of surface in, **11,** 659
substrate for, **11,** 620
surface contamination in, **11,** 620
tensile properties of, **11,** 663–665
thermal evaporation of, **11,** 622–637
thickness measurements in, **11,** 653–659
triode sputtering in, **11,** 642–643
vacuum evaporation system in, **11,** 629–630
X-ray absorption and reflection techniques for, **11,** 659
Thin foil, energy-loss measurements, **17,** 47–51
Thin foils, **7B,** 238
Thin layer on solid-state detector, energy-loss measurements, **17,** 54–58
Thin lens
Gaussian formula, **31,** 149
sign convention, **31,** 149

Thin lenses, **25,** 28
Thin metal film filters, **31,** 320–324
Thin organic film filters, **31,** 324–326
Thin target, in ion-atom collision, **17,** 214, 371–373
  beam-foil experiments, **17,** 536–537
  charge-dependent cross section measurement, **17,** 283
  for inner-shell ionization, **17,** 218–221
  K-shell ionization cross sections, **17,** 250–252
  K-shell vacancy production, **17,** 335
  multiple scattering, **17,** 309
Thin-film bridges, **11,** 255–258
  Josephson oscillation in, **11,** 266
  in magnetometers, **11,** 281
  quantum interference effects in, **11,** 263
  superconducting rings and, **11,** 278–279
Thin-film coating, **30,** 607–616
Thin-film properties, *see* Surface acoustic wave applications
Thin-film squids, **11,** 285
Thiourea, **23A,** 498
Third order
  aberration coefficients, **25,** 46
  aberrations, **25,** 40
  spherical aberration, **25,** 42
Third order aberration theory, **31,** 361–364
Third sound, **19,** 421
Third-harmonic generation (THG), **29C,** 252–253
  in VUV generation, **15B,** 217
  in VUV spectroscopy, **15B,** 211–212
Third-order parametric processes, **29C,** 253
Thirteen moment equation, **1,** 160
Thomas precession, **4B,** 30–31, 35
Thomas-Fermi atomic model
  electronic energy loss, **17,** 40–41
  potential energy, **17,** 166
  scattering cross sections, **17,** 42, 164–165
Thomas-Fermi potential, **21,** 276; **22,** 311
Thompson scattering, **26,** 101
Thomson
  coefficient, **1,** 508, 510
  cross section, **9A,** 67
  double bridge, **1,** 457
  heat, **1,** 507–508
Thomson effect
  definition, **6B,** 115
  measurement, **6B,** 125
Thomson scattering, **9A,** 419
  cross section, **5A,** 77
  expression, **3,** 306–307
Thoriated tungsten, **14,** 368
Thorium, hydrogen solubility, **14,** 354
Three mirror circular polarizer, **31,** 251
Three mirror polarizer, **26,** 123
Three mirror polarizer, **31,** 239–242
  adjustable, **31,** 240–241
Three-body reactions, in afterglow studies, **7B,** 81
Three-component model, for target x-ray production, **17,** 257–272
Three-dimensional band structure mapping, **22,** 169
Three-electron ion, target x-ray cross section, **17,** 288
Three-electron line energies, **17,** 387–388
"three-five" photocathodes, **12A,** 25–27
Three-reference beam method for color holograms, **26,** 181
Three-reflector feed system, **12B,** 58
Three-step model, **22,** 127, 129, 130
Threshold
  laser, *see* Laser threshold
  neutron
    detector, **5B,** 404
    measurement, **5B,** 360–362
  for sputtering, **22,** 367
Threshold current density, temperature dependence of in semiconductor diode lasers, **15A,** 224–226
Threshold measurement, **22,** 479
Threshold measurements, *see* Neutron threshold measurements
Threshold parameters, **28,** 239
Threshold relationships, in dye lasers, **15A,** 258–263
Thresholds for cavitation, *see also* Tensile strength of liquids
  Blake threshold for bubble instability, **19,** 364–366
  gas saturation dependence, **19,** 371–372
  for rectified diffusion, **19,** 375
  for transient cavitation, **19,** 390–393
Through-focus series, of electron micrographs, **16B,** 209, 210
Throughput, **12A,** 193, 464, 465, 487; **14,** 11, 12, 101, 232; **29B,** 363–365, 414; *see also* Molecular transport
  diffusion pumps, **14,** 147–154

Throughput-resolution product, **12A,** 465, 466, 536
Thyratrons, **2,** 99–101; **2A,** 174–176
  basic properties of, **2A,** 225–228
  control circuits, **2,** 300–301
    dc signal, **2,** 303
    peaking-transformer, **2,** 303
    phase-shift, **2,** 301–303
  in controlled rectifier circuits, **2A,** 224–241
  hydrogen, **2,** 101–105
  vs thyristor, **2A,** 226–227
  typical circuits for, **2A,** 229
Thyristors, **2A,** 95–96, 100
  basic properties of, **2A,** 225–228
  bidirectional, **2A,** 96
  in controlled rectifier circuits, **2A,** 224–241
  firing signal requirements of, **2A,** 228–229
  gated turn-off, **2A,** 224
  optimum performance by, **2A,** 231
  switching action in, **2A,** 225
  temperature limitations in, **2A,** 226
  vs thyratron, **2A,** 226–227
  typical circuits for, **2A,** 229
Ti-sapphire laser, **30,** 335
Ti:Al_2O_3 laser, **29C,** 264
Ti:sapphire crystal, **29C,** 148, 151
Ti:sapphire laser, **29C,** 26, 260
Tie molecules, **16B,** 111, 345
  see also Intercrystalline links
Tie molecules, role of polymer fracture, **16C,** 219–224
TIG (tungsten-inert gas) welding, **14,** 429, 434, 435
Tilt, *see* Crustal deformation
Tiltmeter, *see also* Strainmeter
  bubble, **24B,** 425–427
  design, **24B,** 415, 424–429
  drift, **24B,** 413
  indicator equation, **24B,** 424
  installation, *see* Strainmeter, installation
  long baseline, **24B,** 427–429
  materials, properties, **24B,** 416
  noise
    power spectra, **24B,** 413
    sources, **24B,** 412
  pendulum, **24B,** 425–426
  quantity measured, **24B,** 410
  short baseline, **24B,** 424–427
Time, **1,** 29, 31, 69–73
  burst start, **5A,** 546–547
  coincidence resolving, **5B,** 325
  collision, **5A,** 7
  decay, spectrum of mu-mesons, **5A,** 156
  duration of a flash, **3,** 165
  effective blocking, per accepted event, **2,** 589
  electrical measurement of, **1,** 72–73
  expansion, of cloud chambers, **5A,** 200
  frequency response and, **2A,** 4–6
  gate, **5A,** 555
  lags, statistical and formative, **2,** 73–74
  mean, between collisions, **14,** 10
  moderation, for fast neutrons, **5A,** 519–521
  pump-down, **1,** 189
  recovery, **2,** 112
    plasma delay, **2,** 81–83
  relativistic, **1,** 70
  relaxation, **1,** 433
  resolving
    effects on coincidence and anticoincidence circuits, **5B,** 794–795
    effects on single-channel counting, **5B,** 792–794
    scaling circuit, **2,** 521
    transit, in scintillating counter, **5B,** 308
    two-source method, **5B,** 793–794
  of response, gage system, **1,** 169, 191–192
  reverberation, **1,** 229–230
  rise, semiconductor detector, **5A,** 272–273
  sensitive, of bubble chambers, **5A,** 207
  spectrum of fast neutron interaction
    with gold, **5A,** 551
    with U^{238}, **5A,** 552
  uncertainty of BF_3 counter, **5A,** 533
  variation of magnetization, **3,** 396, 400
Time analyzer(s), **2B,** 34–35
  for neutron spectrometers, **5A,** 554–566
Time averaged holographic interferometry, **26,** 194
Time behavior, **14,** 14
Time code broadcasting, **24B,** 49–50
Time constant
  hot-wire and hot-film anemometer, **18,** 293–295
  Raman scattering diagnostics, **18,** 420, 421–425
  V/S, **14,** 183
Time definition, **23C,** 415
Time delay
  ionospheric, **12C,** 263
  spectrum, **5A,** 443

Time dependence, *see* Zero power reactor
Time dependent response of pressure gage, **18,** 527
Time derivation methods, **2B,** 32–33
Time differentiation method, for energy distribution curves, **11,** 98
Time differentiation, **8,** 1
Time digitizer, **22,** 479
Time dilation, **8,** 163
Time domain of signal processing, LDV, **18,** 161
Time domain reflectometry, as oscilloscope technique, **2B,** 199–200
Time domain reflectometry, in high-frequency dielectric measurements, **16C,** 416–421
Time domain response, **19,** 31–34, 61–65
Time interval
- gated electronic counter, **2B,** 24–25
- inspector circuit, in pulse amplitude measurement, **2B,** 97
- measurements, **2B,** 33–36
- short, **14,** 85
- sorter, **5B,** 314

Time lag, in electric breakdown, **16C,** 470
Time measurement, **2,** 543–549; **2B,** 29–37
- analog processing in, **2B,** 34–36
- atomic clocks in, **2B,** 37
- coincidence detector, **2,** 547–548
- digital processing in, **2B,** 36
- errors in, **2B,** 29–32
- jitter in, **2B,** 30–32, 112
- linear sweep method, **2,** 545, 548–549
- long-time interval in, **2B,** 36–38
- scintillator-photomultiplier combinations in, **2B,** 31
- time derivation methods in, **2B,** 32–33
- using oscillator and electronic counter, **2,** 543–544
- walk in, **2B,** 29–30, 112

Time prediction
- in time standards, **12C,** 207–208

Time resolution, **8,** 256, 258
Time resolution. **5B,** 324ff; *see* Frequency response
- resolution,

Time response, *see* Hold time; Response time
- PbS detector, **3,** 81–82
- of proportional counters, **5A,** 113–115, 116

Time reversal, **23C,** 118
Time scales, **12C,** 182
Time series
- cross correlation between ozone and solar flux, **28,** 475–477
- data preprocessing, **28,** 467
- random generation methods, **28,** 468–470
- randomly generated reference point, **28,** 465
- univariate, *see* Univariate time series, **28,** 465

Time series analysis, **28,** 283–310
- aliasing, **28,** 309–310
- bivariate, **28,** 340–346
- data windowing, **28,** 297–302
- filtering, **28,** 283–297
- with missing data, power spectra calculation, **28,** 305–309
- spectral analysis, **28,** 297–305

Time standards, **1,** 69; **12C,** 198–227
- bibliography of, **12C,** 227

Time structure, **22,** 143
Time-coincident events, **8,** 205
Time-correlated coincidences, **8,** 223
Time-delay circuit, **2,** 303; **2A,** 240
Time-dependent Schr^dinger equation, in electron capture problems, **17,** 93, 95
Time-invariant system, **2B,** 380
Time-lag method, for permeability, **16C,** 357–362
- sources of error, **16C,** 374, 376

Time-of-flight, **1,** 55–56; **8,** 170, 197; **14,** 85, 89; **22,** 328, 404
- analyzers using digital storage techniques, **5A,** 559–566
- as background eliminator, **5A,** 549–551
- delay curve, antiproton, **5B,** 754
- distribution of Cl^{37} recoils from Ar^{37} decay, **5A,** 695
- in LEIS, **21,** 236
- method, **22,** 494
  - to measure charged particle velocity, **5A,** 435–443
- neutron spectrometer, **21,** 213–216
- spectra from proton diffusion, **21,** 203
- system, **22,** 319
- technique, **22,** 476, 491, 495
  - in neutron scattering, **21,** 161–163

Time-of-flight analyzers, **23A,** 208
Time-of-flight atom-probe, **22,** 404
Time-of-flight mass spectrometer, **22,** 404, 444; *see* Mass spectrometer
Time-of-flight mass spectrometer, **8,** 169
Time-of-flight mass spectroscopy, **22,** 480

Time-of-flight measurements, **17,** 180–183, 529, 532, 570
  forbidden Auger transition, **17,** 586
  of neutron interactions, **5A,** 495–566
  of neutron reaction cross-section, **5B,** 456–457
  postfoil beam velocity, **17,** 575
Time-of-flight neutron spectrometers, **8,** 197
Time-of-flight spectrometer, **1,** 65; **5A,** 514–554; **23C,** 167
Time-of-flight spectrometer, *see* Electrostatic time-of-flight spectrometer; Isochronous time-of-flight spectrometer; Neutron time-of-flight spectrometer
Time-of-flight spectrum, **22,** 478
Time-of-flight techniques, **8,** 253
Time-pressure superposition, **19,** 145–146
Time-resolved absorption measurements, **19,** 234
Time-resolved decay process, beam-foil studies, **17,** 529, 570
Time-resolved holography, **30,** 112
Time-resolved velocity measurements, **19,** 234
Time-temperature reduction, **16C,** 14
Time-temperature superposition, **19,** 144–145, 172, 175
Time-to-amplitude converter, **17,** 181; **22,** 477, 478
  dead time, **17,** 311
  spectrum, **17,** 311–313
Time-to-pulse-height converter, **5A,** 440
Timing diagrams, in ultrasonic studies, **11,** 403–404
Timing, electronic, of decay, **5B,** 306–333
Tin ion laser, **15A,** 154–155
Tin oxide (SnO_2), **30,** 552
Tin-plated cavity, rf critical field of, **11,** 563
Tip apex, **22,** 355
Tip mechanical structur, **27,** 51
Tip profile, **22,** 356
Tip, models for, **27,** 4, 6, 14
Tip, tip-sample interactions, mechanical theory of, **27,** 24
Tipper, **24B,** 323, 325–326, 332
TiS_2
  1T phase
    spectroscopy of, **27,** 383
$TiSe_2$
  1T phase
    scanning tunneling microscopy of charge density wave in, **27,** 381
    spectroscopy of, **27,** 383
Tissue
  ablation of soft, **30,** 449–452
  dielectric properties of, **30,** 452–454
  structural properties of, **30,** 451–452
Titanium
  beta alloy, attenuation in, **19,** 282–284
  boiling point, **14,** 351
  chemical reactions, **14,** 195
  efficient operation, **14,** 212
  film, **14,** 193
  flash deposition, **14,** 215
  heating, **14,** 203, 204, 206, 207
  hydrogen solubility, **14,** 354
  melting point, **14,** 345, 351
  outgassing, **14,** 203, 207
  periodic deposition, **14,** 214
  physical properties, **14,** 366–367
  precipitate changes scattering, **19,** 282–284
  pumping speed, **14,** 195, 198
  scattering by grains, **19,** 282–284
  sublimation pumping, **14,** 194, 196, 237
  sublimation rate, **14,** 202
  sublimators, characteristics and construction, **14,** 202
  vapor pressure, **14,** 348, 351
Titanium hydride, ceramic-to-metal seals, **14,** 367
Tl^{200} decay scheme, **5B,** 156
TMMC, **23C,** 223, 229
TO modes, *see* Transverse optical modes
Tobacco mosaic virus (structure), **20,** 288–293
Toepler refractometer, **1,** 345
Toepler schlieren system, **18,** 361
TOF measurements, *see* Time-of-flight measurements
TOF, *see* Time-of-flight
Tokamak, **9A,** 446
Tokamaks, **13A,** 191
Tolerance, **2,** 6
Tolerance interval(s), **1,** 7; **3,** 714; **28,** 184
Toluene mean free path
  gamma rays, **5A,** 144
  neutron, proton collision, **5A,** 145
Tomography, **19,** 565
Tones, variable, **1,** 230
Tonks-Dattner resonances, **9A,** 46
Tool steel, **14,** 357
Top
  prolate symmetric, **3,** 59

rigid symmetric, energy of, **3,** 59
Topographic effect
gravity exploration, **24B,** 151–152
heat flow, **24B,** 191
magnetotelluric survey, **24B,** 341
Topographical contrast imaging, in electron microscopy, **16B,** 222–224
Toric surface, **25,** 8
Toroid, **1,** 557
Toroidal grating monochromator (TGM), **29C,** 37; **31,** 28–29
polarization, **31,** 135
Toroidal mirror, astigmatism correction, **31,** 162–163
Toroidal optics, **31,** 173–175
Toroidal prism, **22,** 493
Toroidal-grating monochromator, **22,** 144
Torque, **1,** 84
converter, **1,** 95
on current carriers, **1,** 519
dynamic, **1,** 84–86
magnetic, **1,** 533, 547
Torr, **18,** 508
Torsion
angle detectors, **16C,** 29
balance, **1,** 84, 541
in ferromagnetic measurements, **6B,** 207–208
Couette geometry, **16C,** 31
pendulum, **1,** 82–83
susceptibility balance, **6B,** 180
torque production, **16C,** 30
Torsional braid instrument, **16C,** 41
Torsional pendulum, **6A,** 304–305; **16C,** 38–40
shear modulus measurement, **6A,** 299
Tortuosity factor, **16C,** 329
Total cross section, emission-yield data, **17,** 192
Total enthalpy, *see* Stagnation enthalpy
Total head (fluid), **18,** 333
Total internal reflection devices, **26,** 131
Total power receivers, **12B,** 212–216
stabilization of, **12B,** 78
Total pressure, *see* Stagnation enthalpy
Total reflection, **23B,** 67
Total scattering, **23A,** 10
spectrometer, **23B,** 247
Total temperature, *see also* Stagnation temperature
probe, **18,** 462–463
Total vaporization method, **6A,** 35
Total yield spectrum, **22,** 211
Totally isolated system, **1,** 305
Touchek effect, **29C,** 26–27
Towing tank dimensional analysis, **18,** 843, 847–848
Townsend discharge, **7B,** 63
electrons, energy distribution, **7B,** 64
Townsend ionization coefficients, **7B,** 64–66
first coefficient, **7B,** 64, 70–72
experiment evaluation of, **7B,** 72–73
second coefficient, **7B,** 65–66
experimental evaluation of, **7B,** 73–74
Townsend mechanism of spark formation, **18,** 695–696
Toxocity of halogens, **16B,** 246
TPF, *see* Two-photon fluorescence
TPX, **10,** 101, 102
TR, *see* Transmit-receive
Tracer particle tracking, *see* Flow tracing particles
Track
density
errors in measurement, **5A,** 451
photographic, **5A,** 449–451
distortion in cloud chambers, **5A,** 382–383
length, **5A,** 226
noise, **5A,** 381–382, 396, 398, 399
eliminator, **5A,** 400–401
radius of curvature errors, **5A,** 381ff
Track width
in cloud chambers, **5A,** 200
in emulsion, **5A,** 305–307
mean, **5B,** 42–43
Tracking bandpass filter, LDV signal processing, **18,** 169–174
Tracking detectors, simulation, **28,** 446–447
Tracking multiple-beam interferometer for LDV signal, **18,** 211–220
Trajectories, particle, in strong focusing magnet, **5B,** 694–696
Trajectory, **14,** 87
Transmission
bandwidth, **1,** 356
filters, **1,** 485
grating, **1,** 367
line, **1,** 477–478, 479, 481
loss, **1,** 230
of sound, **1,** 208–211
systems, **1,** 476–481, 499
Transmission-line bridge, **1,** 487

Trans conformation, in polyethylene, infrared bands, **16B,** 251
trans–1,4-Polychloroprene, infrared spectrum, **16A,** 132–133
Transconductance multipliers, **2B,** 159
Transducer correction, continuous wave, **19,** 116
  composite resonator, **19,** 116
Transducer(s), **1,** 75, 234; **2,** 678; *see also* Interferometer; Sensor
  action, linear, **1,** 216
  backing materials, **19,** 61–62
  barium titanate, **19,** 30, 54
  bonding of, **11,** 393–395
  broadband, **19,** 31–34, 47–51, 61–65
  capacitive, **24B,** 417, 420
  coefficient, **1,** 216
  commercial, on buffer rods, **19,** 265–266
  in counting, **2B,** 1
  design considerations, **19,** 51–65
  displacement, **24B,** 416–421
  electroacoustic, **1,** 216, 231
  electromagnetic, **11,** 331–337
  electron tube, **1,** 80
  in feedback control systems, **2B,** 440–441
  insertion loss, **19,** 61
  laser frequency stabilization, **29C,** 121–124
  lithium niobate, **19,** 150
  LVDT, *see* Linear variable differential transducer
  magnetostrictive, **19,** 164, 170
  optical lever, **24B,** 417
  piezoelectric, **19,** 29–65
  piston source necessary, **19,** 260–261, 265–266
  plates, for direct bonding to specimens, **19,** 257–258
  PZT, **19,** 30, 54
  quartz, **19,** 54, 150, 161, 166, 168
  radiation field from, **19,** 260–261
  seismic
    electromagnetic, **24B,** 29–32
    electromotive, **24B,** 29
    paramagnetic, **24B,** 29
    use for magnification, **24B,** 29
  for sound in liquid helium, *see also* First sound, Second sound, Fourth sound
    capacitive, **19,** 422–424, 433–434
    electrical impedance matching of, **19,** 424
    piezoelectric, **19,** 424–429
    poled ferroelectrics, **19,** 429
    properties of transducer materials, at low temperatures, **19,** 425
    resistive films, **19,** 431
    superconducting edge, bolometers, **19,** 431–433
    thin piezoelectric films, **19,** 425
  strain/tilt
    design criteria, **24B,** 411
    economic criteria, **24B,** 415
  in ultrasonic wave generation, **11,** 391–393
Transduction, **19,** 29–65, *see also* Surface acoustic wave transducers
Transfer
  efficiency, **5B,** 234
  polarization, **5B,** 233–239
  radiative, **5A,** 124
Transfer admittance, **2A,** 9
Transfer efficiency, in integrated cirucits, **2A,** 131–132
Transfer energy, **5A,** 133
Transfer function, **2,** 461, 477; **8,** 233, 234, *see also* Linear-dependent transfer function
  of lenses, **26,** 40
  optical, **26,** 82, 84
Transfer impedance, **2A,** 9
Transfer lock, **14,** 498
Transfer oscillator, **8,** 147
Transferred electron devices (TEDs), **2A,** 51
Transformation, first order, **1,** 262
Transformation, Lorentz, from laboratory to c.m. system, **5B,** 821–824
Transformer
  coupling, *vs* capacitor coupling, **2A,** 282–283
  differential, **1,** 74
  wide band astatic, **1,** 461
Transformer (ideal), **19,** 44, 47
Transformers, **2,** 45
  in electronic circuits, **2A,** 38
  filament, **2,** 210; **2A,** 222–223
  ideal, **2,** 31, 33; **2A,** 21
  inductance and, **2A,** 2–3
  in network structure, **2A,** 21–22
  polarization, **5B,** 227–233
  rectifier, **2,** 209–210; **2A,** 221–222
  spin, **5B,** 227–233
Transient cavitation, **19,** 358, 381, 385–403
  collapse, *see* Thresholds for cavitation
  inertial effects, **19,** 390
  laser induced, **19,** 386–387

radius versus time curves, **19,** 386, 388–389
threshold for, **19,** 391–393
viscosity, effects of, **19,** 387, 390
Transient cavity collapse, *see* Collapse of transient cavity
Transient currents, **16C,** 494
Transient digitizer(s), **29B,** 225–226
noise reduction, **29A,** 448
speed, **29A,** 447
Transient flow, **1,** 109
Transient laser lines, between lowest neutral noble-gas energy levels, **15A,** 51
Transient measurement of lifetime, **6B,** 101–104, 107–109
Transient methods
in dielectric measurements
high frequency, **16C,** 416–421
low frequency, **16C,** 401–404
in high-field conduction, **16C,** 492–495
for measurement of relaxation times, **7B,** 22–24
Transient noble-gas lasers, **15A,** 53–55, *see also* Noble-gas ion laser(s)
Transient phenomena, forced, **3,** 399
Transient response, **1,** 231–232
in feedback system response, **2B,** 385–388
of filter, **2A,** 27–28
of a system, **2,** 465–466
Transinet method of measurement of thermal conductivity, **6A,** 404–405
Transistor amplifier, **2A,** 255–256, *see also* Amplifier
feedback stabilization in, **2A,** 280–281
manufacturer's power specification for, **2A,** 258
single-stage, **2A,** 272–281
stage, limitations on, **2A,** 272–281
Transistor bias circuits, **2A,** 256–261
Transistor choppers, **2A,** 303
Transistor circuits
equivalent, **2,** 307–313
linear, **2,** 307–326
positive feedback in, **2A,** 355
Transistor core drivers, **2A,** 404–408
Transistor curve tracer, in equipment testing, **2B,** 177
Transistor gating circuit, **2A,** 378–379
Transistor noise
analysis, **2A,** 326–327
level in, **2A,** 77–78
Transistor(s), **2A,** 67–92, *see also* Semiconductor circuit elements
admittance and transconductance values for, **2A,** 271
beam-led air-isolated, **2A,** 113–114
behavior under transient conditions, **2,** 416–420
bipolar, **2,** 142–160; **2A,** 67–80, 94
operation and characteristics, **2,** 144–150
transport factor, **2,** 144
common-emitter configuration, **2A,** 73–74
common-emitter switch for, **2A,** 77
field-effect, **2,** 168–170
gain bandwidth product in, **2A,** 279
germanium, **2A,** 68, 78–79
high-frequency
amplifying, **2,** 153–156
high-power, **2,** 157
hook collector, **2A,** 98–99
Hook-collector, **2,** 164–165
invention of, **2A,** 45
insulated-gate, thin-film, **2,** 170–171
junction, *see* Junction transistor, P-N junction
large-signal behavior, **2,** 415–420
mesa, **2A,** 69, 71
multiregion devices, **2,** 160–167
parameter values for, **2A,** 270
performance of, **2A,** 79
piezo-, **2,** 157
planar, **2A,** 70
p-n-p, in common emitter configuration, **2,** 413–414
power, **2,** 150–151; **2A,** 78–79
punch-through voltage in, **2A,** 82
ruggedness of, **2A,** 80
saturation prevention, **2,** 420–422
as semiconductor gain element, **2A,** 67
silicon, **2A,** 78, 101–102
small-signal (common base) equivalent circuit for, **2A,** 76
specification sheets for, **2A,** 78
switching, **2,** 151–153, 202–203
circuits, **2,** 412–451
terminal voltages and currents for, **2A,** 270
tetrode, **2,** 156–157
as 2-port black box, **2A,** 79
unijunction, **2,** 167–168
unipolar, **2,** 168–170; **2A,** 80–84
vs vacuum tubes, **2A,** 45
vertical anisotropic etch, **2A,** 112–113

voltage discriminators, **2,** 443–446
wafer of, **2A,** 101
y parameters for, **2A,** 274–275
Transistor, noise in, **2B,** 461, 466, 474–476, *see also* FET, JFET, Solid-state microwave power sources, Solid-state photocells
Transistor-transistor logic (TTL), **2A,** 138
Transit angle, **8,** 130, *see also* Electron transit angle
Transit time, **4A,** 322–323
effects, in vacuum tubes, **2A,** 154
Transit time, **8,** 261
Transition
density, **22,** 201
energy, **17,** 386–392, 470–474
fine structure, **3,** 16
frequency
diatomic molecule, **3,** 12
electron spin resonance, **3,** 465
quadrupole, temperature dependance, **3,** 521, 522
symmetric top molecule, **3,** 13
metastable ion, **3,** 587–588
multiple-quantum, **3,** 620
multiplicity forbidden, **3,** 251, 255–256
n-π*, **3,** 250–251
pi, **3,** 627
Q-N, **3,** 250
radiative
induced, **4B,** 216
spontaneous, **4B,** 215–216
via double quantum emission, **4B,** 225
Rydberg, **3,** 250
shift, **3,** 17
sigma, **3,** 627
symmetry allowed, **3,** 247
vibrationally induced, **3,** 251–255
V-N, **3,** 250
Transition metal chalogenides, scanning tunneling microscopy of, **27,** 349–355
Transition metal dichalogenides, scanning tunneling microscopy of, **27,** 352, 357
Transition metal trichalogenides, scanning tunneling microscopy of, **27,** 352, 358, 402–422
Transition metals, **23C,** 175
Transition moment
electronic, **3,** 245
of infrared absorption, **16C,** 173
Transition noise, **2B,** 461
Transition probabilities, **3,** 453–455; **4B,** 119, 121, 123, 125; **7A,** 44, 45, 117–121, 124, 139; **9A,** 166; **13A,** 80, 100, 116, 141; **13B,** 234; **22,** 146
in electron beam fluorescence, **18,** 438–441
experimental, **3,** 256–261
molecular, **3,** 617–620
in molecular electronic spectra, **3,** 241–261
radiative, mass effect, **3,** 253–254
Transition radiation, 93–99
features, 97–98
maximum, 96
Transition temperature, **6B,** 73–74
Transitions
electron-capture, **5A,** 296
enforced dipole, **13B,** 41
forbidden, **3,** 251–256; **13B,** 260
"forbidden," microwave magnetic field induced, **5B,** 124–125
Transitron, **3,** 402
Translational absorption, **13B,** 42
hindered, **13B,** 44
Translational spectroscopy, **3B,** 769
Translational temperature, **18,** 457, 464, 490; **30,** 83–85, 106–107, 158–160, 598–600
Transmission
angle in turbo-pump, **14,** 251
atmospheric, **10,**91–93
diamond, **10,** 96
diffraction gratings, **26,** 78
fractional, of homogenous γ-ray beams, **5A,** 673–674
γ-ray, law of, **5B,** 766
germanium, **10,** 96
glass, **10,** 94
holograms, **26,** 172
irtran, **10,** 95, 96
link, telemetering, **2,** 680–681
polarizers, **26,** 122
polyethylene, **10,** 96, 99, 101
polymer films, **10,** 98, 100
quartz, **10,** 96, 99
and reflection of particles at solids, **7B,** 251–252
sapphire, **10,** 96, 99
silicon, **10,** 96, 98, 99, 100
spectrometers, **5A,** 345–346, 599ff
superconducting films, **10,** 437
TPX, **10,** 96, 102
volume color holograms, **26,** 183
Transmission cavity, **8,** 141

Transmission coefficient, **19,** 9–18
maximizing, *see* Impedance matching
Transmission coefficient, **31,** 265
Transmission electron microscopy and channeling studies of defects, **21,** 314, 315, 320
Transmission factor, x-ray, **3,** 302, 304
Transmission function, in deep space measurements, **9B,** 79
Transmission geometry, in gamma ray emission, **11,** 308–309
Transmission grating, **31,** 353–355
efficiency, **31,** 373–375
Transmission line(s), **2,** 37–40, 706–710; **2A,** 29–32; **8,** 111, 113, 115, 122, 138, *see also* Shorted air-dielectric transmission line
as circuit elements, **2,** 39; **2A,** 30–31
constants, **2,** 39
delay lines as, **2A,** 31–32
in high-frequency dielectric measurements
standing waves, **16C,** 408–416
transient methods, **16C,** 416–421
line constants for, **2A,** 30
lossless, **2A,** 31
low-loss, **2,** 38–39; **2A,** 30
in microwaves, **2B,** 252–260
notation for, **2A,** 29
system in electric breakdown studies, **16C,** 470–471
termination of, **2A,** 31
theory, **19,** 41–51
measurements, **8,** 109
Transmission measurements, **8,** 228, *see also* Neutron transmission measurements
Transmission probability, **14,** 18; **27,** 284
in turbo-pump, **14,** 252
Transmission resonances, **27,** 282
Transmission technique, continuous wave, **19,** 120–123
attenuation measurement, **19,** 120
electromagnetic rf leakage problem, **19,** 123
frequency tracking technique, automatic, **19,** 122
velocity measurement, **19,** 159–166
Transmission-electron microscopy, *see* Electron microscopy
Transmissivity, 30, 31, *see also* Optical system
Transmit-receive (TR) tubes, **2B,** 285–286
Transmittance, atomic beam, **4B,** 240–241
Transmittance, of birefringent crystals, **12A,** 367
Transmitted radiation, measurements on, **6B,** 262–263
Transmitter
oscillator stability, **3,** 416–417
rf, **3,** 435–436
telemetering, **2,** 674–780
Transmitting antenna, **8,** 158
Transmutation reaction, **21,** 20, 22, 23
Transonic wind tunnels, **18,** 771–779
Transparent-conducting oxides (TCO), **30,** 543–554
Transponder
accuracy of systems for positioning, **24B,** 112–114, 117
arrays, **24B,** 117–120
determining relative coordinates, **24B,** 118–120
interrogation, **24B,** 114, 116
long baseline system, **24B,** 111–112
nature, **24B,** 111
navigation configuration, **24B,** 115–116
navigation system, **24B,** 122
obstacle avoidance, **24B,** 122
short baseline systems, **24B,** 111–112
up-looking sounder, **24B,** 122
Transport noise, **2B,** 461
Transport, molecular, **14,** 11, 13, 17
Transport, of gases and vapors, *see also* Permeability, Diffusion, Sorption
anomalous, **16C,** 334–337, 348
basic theory, **16C,** 316–319
effect of polymer morphology, **16C,** 328–331
historic experiments, **16C,** 319–324
Transresistance ammeter, *see* Ammeter
Transsonic flow, **1,** 150
Transverse acoustic modes, **10,** 362
Transverse axial chromatic aberration, **25,** 32
Transverse charge, **30,** 22, 30
Transverse electric field, **2B,** 253
Transverse electric mode, **25,** 261, 266
in waveguides, **10,** 114, 116–121
Transverse electromagnetic modes (TEM), **2B,** 253–254; **30,** 173
in microwave measurements, **2B,** 287
Transverse instabilities, **9A,** 299–304
Transverse magnetic modes, **2B,** 253
in waveguides, **10,** 114, 117–119
Transverse magnification, **25,b 133**

Transverse optical modes, **10,** 362, 363, 378–380
Transverse rays plots, **25,** 43
Transverse spatial coherence, **31,** 21–24
Transverse vibrations, 3
Trap barrier, creep proof, **14,** 299
Trap contamination, **14,** 300
Trap distribution, **6B,** 313–315
Trap pumping speed, **14,** 300
Trap, **14,** 277, 278, 299
  cleaning, **14,** 503
  in diffusion pump systems, **14,** 158–159
  effect on pumping speed, **14,** 145
    on throughput, **14,** 150
  liquid nitrogen cooled, **14,** 300
  types, **14,** 482
TRAPATT, *see* Trapped Plasma Avalanche Triggered Transit
Trapezoidal monochromator bandpass, 36–37
Trapezoidal rule, **1,** 58
Trapped electron method, **7A,** 9
Trapped ions, **7A,** 372
Trapped Plasma Avalanche Triggered Transit (TRAPATT) mode, **2B,** 276, 279–280
Trapped-ion experiments, **7A,** 96, 113, 114
Trapping centers, **6B,** 340
Trapping effects, **6B,** 94
Trapping impurities, **23B,** 139
Trapping of positrons, **21,** 81, 83
  annealing studies, **21,** 93
  defect specificity, **21,** 93
  detrapping, **21,** 93
  in dislocations, **21,** 83, 84
  prevacancy effects, **21,** 84
  specific trapping rate, **21,** 90, 91
  specific trapping rate, temperature dependence, **21,** 88, 91
  transition or motion limited, **21,** 83
  trapping model, **21,** 85, 88
  at vacancies, **21,** 83, 89
  at vacancy-impurity sites, **21,** 92
  in voids, **21,** 89, 93, 94
Trapping, **14,** 231
Traps in vacuum systems, **4B,** 262, 277–278, 280–281
  refrigerated, **4B,** 262, 277–278, 280
  room temperature, **4B,** 263, 278–280
Travel, phase angle, **8,** 154
Traveling detector, **6B,** 21–22
Traveling salesman problem, **28,** 441
Traveling wave oscillator, **10,** 34–38
Traveling wave tube, **1,** 487
Traveling-wave dye laser (TWDL), **29C,** 251
Traveling-wave maser amplifies, **12B,** 251
  representative systems, **12B,** 262
Traveling-wave oscillator, **2B,** 271
Traveling-wave tubes, **2B,** 269–271
Tree, as subnetwork, **2A,** 7
Trees
  artificially produced, **16C,** 461, 485
  electrical, **16C,** 449, 471, 482–488
  electrochemical, **16C,** 449, 488
  water, **16C,** 449, 488
Trend surface analysis, **28,** 381–383, 386
Triac (bidirectional thyristor), **2A,** 96–98
Triad structures, in copolymers, **16A,** 294–295
Triangle generator, **11,** 340
Triangulation, **1,** 55; **23C,** 400
Triangulation procedure, **21,** 277
Triaxial strain cell, **24B,** 389–393
Triboelectric series, **16C,** 422–423
Triboluminescence, **6B,** 295
Trichloroethane, **19,** 202
Triclinic crystal system
  definition, **16B,** 4
  interplanar spacing formulas, **16B,** 26
  reciprocal lattice relationship, **16B,** 24
Triclinic lattice, definition, **16B,** 5
Tricritical points, **23C,** 37
Triethyl benzene purification, **5A,** 149
Trigger circuits, **2,** 371–382
  rectifiers in, **2A,** 239–240
  Schmidt, **2,** 377–380, 444–445
Triode amplifier, **2A,** 263
Triode ionization gauge, **14,** 62–63, *see also* Ionization gauge
Triode sputtering, **11,** 642–643
Triode structure of ion pump, **14,** 224
Triode(s), **2,** 57–62
  cathode feedback biasing of, **2A,** 266
  characteristics, **2,** 57–58; **2A,** 155
  class A operation, **2,** 59
  equivalent circuits, **2,** 59–62; **2A,** 157
  microwave, **2A,** 163–164
  plate resistance, **2,** 60
  semiconductor-metal emitter, hot-electron, **2,** 171
  switching, **2,** 162–164
  vacuum-tube, **2A,** 155–158
Triple axis neutron spectrometer, **21,** 176–182, 213
Triple correlations, **13A,** 139

Triple point, **1,** 32, 243
of water, **1,** 237, 239, 246
Triple-axis spectrometers, **23A,** 246, 401; **23C,** 166
focusing, **23A,** 408
resolution, **23A,** 407
Triplet lens, **25,** 79
Triplet quenching, CW dye lasers, **29C,** 60
TRIUMF, **23A,** 132
tRNA, **20,** 387
Trouble shooting procedures, **2,** 664–670
True secondary electrons, **22,** 194
Truncation error, **3,** 721
Truncation of Fourier transform, **23B,** 249
Truncation, errors of, **1,** 20
Trypsin, **23C,** 345, 358
inhibitor, **23C,** 361
TSC, *see* Thermally stimulated currents
Tsvankin distribution, of sizes in semicrystalline polymers, **16B,** 169
TTL, *see* Transistor-transistor logic
TTNF, **23A,** 132
Tube
cylindrical, **14,** 16, 19
photoemissive, **1,** 402
shock, **1,** 192–197
traveling wave, **1,** 487
X-ray, **1,** 409–411
Tube effect, nonadiabatic sound propagation, **3,** 686–688
Tube lens, **25,** 165
Tube wave, *see* Stonely wave
Tubes
ATR, **2,** 109
barrier grid, **2,** 619
camera, **2,** 773–782
cathode ray, **2,** 766–773
classes, **2,** 67
cooling, **2,** 67–68
dark-trace, **2,** 784
disk-seal, **2,** 69–70
electrometer, **2,** 570–573
electron, beam quantizing, **2,** 623–624
electron-ray indicator, **2,** 765–766
electrostatic memory, **2,** 624
flicker noise, **2,** 806–807
gas, **2,** 71–112
grid-control, at microwave frequencies, **2,** 716–718
lighthouse, **2,** 69
low-current, cold-cathode, **2,** 108–109
"magic eye," **2,** 765–766
microwave, **2,** 716–727; *see also* Microwave tubes; Vacuum tubes
gas, **2,** 109
multielement, **2,** 62–66
output stage, of tuned power amplifier, **2,** 256–258
pencil, **2,** 69, 70
physical characteristics, **2,** 68–70
ratings, **2,** 110–112
remote-cutoff, **2,** 65
selection, **2,** 66–70
storage, **2,** 782–784
TR, **2,** 109
traveling-wave, **2,** 724–726
vacuum, **2,** 47–70
voltage-reference, **2,** 109
voltage regulator, **2,** 493–495
Williams, **2,** 618–619, 783–784
Tuckerman strain gage, **6A,** 293
Tufts, for visualization, **18,** 241, 770
Tunable cyclotron resonance detector, **10,** 185–188, 188–201
Tunable dye lasers
optical frequency mixing with, **15B,** 174
optical parametric oscillator and, **15B,** 168
Tunable fiber coupler, **29C,** 377
Tunable frequencies, nonlinear optics and, **15B,** 144
Tunable laser spectroscopy, **13B,** 273
absorption, **13B,** 325
diode injection, **13B,** 29
dyes, **13B,** 305
millimeter region, **13B,** 54
molecular, **13B,** 280
Tunable lasers, **3A,** 368; **15A,** 205
continuous wave dye laser, **29C,** 45–73, 138
frequency stabilization, **29C,** 103–134
precise wavelength measurement, **29C,** 311–312, 338
Fizeau wavemeter, **29C,** 331–337
plane-parallel interferometer, **29C,** 337–338
scanning Michelson interferometer, **29C,** 312–331
solid state lasers, **29C,** 148–149
Tunable photoconductive detector, **10,** 185–188
Tunable short-pulse oscillator, **15A,** 276
Tunable shorts, **8,** 145, 147

Tunable vacuum ultraviolet generation, **15B,** 216
Tuned amplifier, **2A,** 315–320
 bias for, **2A,** 318
 design considerations in, **2A,** 317–319
Tuned radio-frequency receivers, **12B,** 213
Tuned-laser spectroscopy, **3A,** 363
Tuners, stub, **1,** 484
Tungsten
 boiling point, **14,** 351
 melting point, **14,** 348, 351
 physical properties, **14,** 367–379
 thoriated, **14,** 368
 vapor pressure, **14,** 348, 351
 W(001), **22,** 161
 W(100), **22,** 291
 W(110), **22,** 296
Tungsten carbide anvils, cemented, **24A,** 247
Tungsten evaporation, **25,** 244
Tungsten filament, **3,** 77
Tungsten halogen lamp, **25,** 246
Tungsten lamp, **25,** 239
Tungsten oxide, **14,** 368
 formation in ionization gauges, **14,** 80
Tungsten strip lamp, **7A,** 59–60, 124, 459
Tungsten, reflectance spectra, **31,** 216–217
Tuning diodes, **2A,** 63–64
Tuning plunger, **8,** 127
Tuning sensitivity of Josephson junction, **10,** 74
Tunnel current dependence on tunnel gap, **27,** 281
Tunnel current, self field of, **11,** 215
Tunnel diode, **2A,** 65–67, 355, 359, 367
 negative resistance of, **2A,** 368
 structure of, **2A,** 65–66
 voltage-current characteristic of, **2A,** 368
Tunnel diodes, injection of hot electrons by, **11,** 449–450
Tunnel junction, **16A,** 150
 preparation of, **16A,** 154–156
 resistance, **16A,** 156
Tunnel junctions, *see also* Weak links
 barriers in, **11,** 254
 bias voltage across, **11,** 161–162
 cavity modes and, **11,** 229–233
 configuration of, **11,** 252–253
 construction of, **11,** 252
 contacts for, **11,** 192
 defined, **11,** 209
 equivalent circuit model of, **11,** 268–273
 fabrication of, **11,** 124–125, 252
 I-V characteristics of, **11,** 124, 193–194
 Josephson radiation in, **11,** 231, 234
 lead-copper-lead type, **11,** 254
 lead-lead oxide-lead type, **11,** 213, 217
 molecular excitations in barriers of, **11,** 165–167
 oxide barriers of, **11,** 161–164
 phonon intensity at, **11,** 243–245
 phonons emitted from, **11,** 243–245
 plasma oscillations in, **11,** 238–239
 p-n type, **11,** 184–187
 power spectrum for, **11,** 245–246
 quality of, **11,** 253
 semiconductors in, **11,** 133, 177–187
 SNS type, **11,** 254–255
 temperature dependence in, **11,** 216
 testing of, **11,** 125–126, 192–193
 thin-film bridges and, **11,** 255–258
 as weak links, **11,** 251–254
Tunnel wall corrections, **18,** 773, 847
Tunneling, **22,** 382, 384; *see also* Tunnel junction
 aritificial barriers in, **11,** 191–192
 contacts in, **11,** 192
 and density of excited states, **11,** 144–149
 elastic and inelastic, of electrons, **16A,** 151, 400
 electron, *see* Electron tunneling
 energy gap determination in, **11,** 135–140
 field splitting and, **11,** 154–155
 Fowler-Nordheim equation for, **11,** 162
 geometrical resonances in, **11,** 152–153
 Hamiltonian, **27,** 5
 high-voltage, **11,** 161–163
 hydrogen in metals, **23B,** 172, 179
 inelastic, **11,** 163–164
 I-V characteristic curve in, **11,** 135–136
 junction fabrication in, **11,** 187–192
 junction preparation in, **11,** 187–188
 measurement circuits in, **11,** 193–197
 normal-metal, **11,** 157–177
 normal metal and superconductor, **11,** 128–131
 photon-assisted, **11,** 223
 phonon-assisted quasi-particle, **11,** 242
 point contacts in, **11,** 258–260
 as probe of electron-phonon interaction, **11,** 149–150
 quasi-particle, **11,** 223
 self-energy effects in, **11,** 183

in single crystals, **11,** 140–144
special topics on experimental techniques in, **11,** 187–197
spin-dependent, **11,** 154
superconducting energy gap and, **11,** 126
in superconductors, **11,** 126–156
surface plasmons in, **11,** 184
between two superconductors, **11,** 132–133
Tunneling barrier
preparation in, **11,** 188–189
theory of, **27,** 20
Tunneling behavior, deviation from ideal in, **11,** 133–135
Tunneling measurements, phonon spectra and, **11,** 144–149
Tunneling thermocouple, **27,** 85
Tunneling tip, resolution and corrugation, **27,** 278
Tunneling transmission probability, **27,** 314
Turbidimetric titration, **16A,** 34
Turbidity, **1,** 395
Turbine blades, **23C,** 443, 457
Turbomolecular pump, **14,** 247–273
backing with diffusion pump, **14,** 483
cleaning, **14,** 502
design, **14,** 260–266
rotor, **14,** 247, 248
in ultrahigh vacuum systems, **14,** 483, 488–489
Turbulence
application of LDV for measurement, **18,** 103–104
effect on LDV spectrum, **18,** 161, 166, 221
effect on photon counting correlation in LDV, **18,** 184
effect on pressure probe, **18,** 252–253
temperature fluctuations in flame, **18,** 417
visualize transition, **18,** 771
wind tunnel flow, **18,** 764
Turbulence level, wind tunnel, **18,** 764, 774
Turbulent boundary layer, **18,** 459, 767
heat transfer, **18,** 666
Turbulent flow, **1,** 133, 145
Turbulent line widths, **12C,** 29
Turnstile feed, **12B,** 79
TV picture tubes, **2B,** 342–343
TWDL, *see* Traveling-wave dye laser
Twentieth-century physics, **13A,** 17
Twiss parameters, **31,** 23
Twisted section, **1,** 482
Two dimensional array of holes, diffracting, **26,** 80
Two parallel dipole feed, **12B,** 79
Two-beam interferometers, **26,** 7, 22
Two-body rate coefficient, in afterglow studies, **7B,** 81
Two-component model for target x-ray production, **17,** 257–278
Two-dimensional angular profile, **22,** 292
Two-dimensional band structure mapping, **22,** 165, 166, 168, 181
Two-dimensional crystals, **22,** 165
Two-dimensional detection, absorption spectra, **20,** 180, 181
Two-dimensional layered compound, **22,** 169
Two-dimensional models, **23C,** 22
Two-dimensional ordered systems, **22,** 166
Two-dimensional structures, **23B,** 10
finite, **23B,** 15
line shape, **23B,** 13
Two-dimensional systems, **22,** 168
Two-electron ion
Lamb shift, **17,** 590
partial energy level diagram, **17,** 584–585
projectile cross section, **17,** 299
target x-ray cross section, **17,** 288
Two-electron line energies, **17,** 288
Two-electron transition, **17,** 398
Two-electron-one-photon transitions, **17,** 344, 400–401
Two-hole localization, **30,** 47–50
Two-phase clock, in integrated circuits, **2A,** 134–137
Two-phase model, of crystallinity in polymers, **16B,** 99, 116, 175, 373–374
size distribution, **16B,** 167–171
Two-photon absorption, **13B,** 332; **15B,** 258
first experiment in, **15B,** 238
Two-photon coloration process, dynamics of, **15B,** 18–20
Two-photon electric dipole decay process, **17,** 583–585
Two-photon emission, **9A,** 128
Two-photon exposure, induced absorption as function of, **15B,**20
Two-photon fluorescence
in picosecond spectroscopy, **15B,** 198–199
in relaxation of azulene, **15B,** 205
Two-photon fluorescence, in second-order autocorrelation, **29C,** 212

Two-photon ionization, in second-order autocorrelation, **29C,** 212–213
Two-photon lineshapes, generalized, **15B,** 246
Two-photon polarization rotations, **15B,** 241
Two-photon transitions
  experimental setup for, **15B,** 240–241
  monitoring of, **15B,** 241–242
Two-point X-ray pattern, **16C,** 179
Two-roton bound states in liquid helium-4, **23B,** 348
Two-step method for rainbow holograms, **26,** 184
Two-step photoionization, F_2^+ center production and, **15B,** 41–45
Two-terminal networks, noise in, **2B,** 498
Two-terminal-pair networks, **2A,** 17–23
  notation for, **2A,** 18
Two-top molecules, **3A,** 88
Twofold symmetric molecule, **22,** 372
Twyman-Green interferometer(s), **1,** 357ff; **25,** 223; **26,** 12
Tygon, **14,** 413
Tyndall scattering, **1,** 395–396; **3,** 112
Type II centers, tuning elements for, **15B,** 30
Types of optical systems, **25,** 7

U

U band, optical excitation of, **15B,** 21–22
U center
  in color center formation process, **15B,** 17–18
  F_A (II) centers created from, **15B,** 38
U-tube manometer, **14,** 40–43, *see also* Manometer
U^{235}
  nuclear properties, **5B,** 594
  spin, **5B,** 55–56
UAs-USe, **23C,** 145
UJT (silicon unijunction transistor), **2A,** 90–91, 351–352, 355–356, 361
Ultimate pressure, **14,** 107–108
  diffusion pumps, **14,** 158–165
  fore-vacuum pump, **14,** 122
  limits, ultrahigh vacuum system, **14,** 484–485
  lobe pump, **14,** 130
  rotary oil-sealed pump, **14,** 115–117, 122
  sorption pump, **14,** 139
Ultra centrifugation, **6A,** 26
Ultracentrifuge, **16A,** 65
Ultrahigh vacuum system, **14,** 477–489
  conductance considerations, **14,** 483–484, 487
  maintenance, **14,** 498–503
  materials, **14,** 477, 480–481, 486
  methods of construction, **14,** 480–481, 486
  operation, **14,** 491–503
  pump choice, **14,** 482–483, 486, 488–489
  residual gases, **14,** 485, 487
Ultrahigh vacuum, production of, **14,** 220
Ultralow temperatures, thermometry at, **11,** 485–539
Ultralow-loss optics, **29C,** 348
Ultramicroscopy, **25,** 173
Ultrapure gases, **4B,** 318
Ultrasonic absorption, *see* Absorption
Ultrasonic attentuation, *see* Attenuation
  in metals, **11,** 428–430
Ultrasonic cleaning, **19,** 405–406
Ultrasonic dispersion, of polymer droplets for nucleation studies, **16B,** 389, 393
Ultrasonic flowmeter, **18,** 337–338
Ultrasonic frequencies, **16C,** 60
Ultrasonic measurements, **11,** 371–441
  attenuation in, **11,** 373
  delay rod technique, **16C,** 75
  immersion technique, **16C,** 61–75
  magnetoacoustic measurement and, **11,** 430–431
  multiple echo technique, **16C,** 77
Ultrasonic method in anelasticity, **6A,** 306–307
Ultrasonic methods for determining elastic constants, **6A,** 301–302
Ultrasonic pulse, **16C,** 65
Ultrasonic range, **11,** 371–372
Ultrasonic resonator, *see* Resonators
Ultrasonic speed
  frequency dependence, **16C,** 74
  in various polymers, **16C,** 72
  measurement of, **16C,** 66
  temperature dependence, **16C,** 73
Ultrasonic studies
  continuous-wave oscillator systems and, **11,** 397–399
  dislocation effects in, **11,** 421–428
  dispersion effect in, **11,** 424
  electronic effects in, **11,** 428–431
  losses in, **11,** 431–433
  measurement systems in, **11,** 400–410

phonon viscosity and, **11,** 431–432
pulse superposition in, **11,** 406–408
sample preparation in, **11,** 396–397
timing diagrams in, **11,** 403–404
velocity and attenuation measurements in, **11,** 397–410
Ultrasonic velocity measurements, *see* Velocity measurements
Ultrasonic wave transmission
buffer rod, **24A,** 3, 5
echo methods
pulse superposition, **24A,** 7
pulse comparison, **24A,** 7
free vibrations, **24A,** 12–13
phase shift, transducer-bond, **24A,** 8–11
pulse transmission, **24A,** 2–6
transducers, **24A,** 3
Ultrasonic waves
attenuation and dispersion in, **11,** 389–391
direct rf generation of, **11,** 395–396
generation of by other means, **11,** 391–396
losses in, **11,** 431–433
physical property measurements of, **11,** 410–433
in solids, **11,** 374–391
thermoelastic losses in, **11,** 432–433
Ultrasound scattering, energy and momentum transfer diagram, **16A,** 207
Ultraviolet absorption spectroscopy, in flame kinetic studies, **7B,** 210–211
Ultraviolet ion lasers, **15A,** 93
Ultraviolet irradiation, charge production, **16C,** 436
Ultraviolet microscope, **1,** 342; **25,** 174
Ultraviolet monochromator, **1,** 365, 406
Ultraviolet noble-gas ion lasers, **15A,** 98–99, *see also* Noble-gas ion laser(s)
Ultraviolet photoelectron spectroscopy, **22,** 33, 484, 485, 503
Ultraviolet photoemission, **11,** 67–122, *see also* Photoemission
Ultraviolet preionized discharge lasers, performance of, **15B,** 90–93. *see also* Rare gas halide lasers
Ultraviolet radiation, **1,** 402–408
outgassing of glass, **14,** 329–333
sources, **7A,** 152
Ultraviolet region, **13A,** 13, 204
absorption spectra, **13A,** 248
continuum sources, **13A,** 206
detectors, **13A,** 241
emission spectra, **13A,** 247
gratings, **13A,** 226
near, *see* Optical region
sources, **13A,** 204, 215, 259
spectra of gases, **13A,** 247
spectra of solids, **13A,** 251
Ultraviolet source comparison with BESSY, **31,** 138–143
Ultraviolet sources, **1,** 376
Ultraviolet transmission, **1,** 402, 403; **14,** 394–396
Ultraviolet, **7A,** 145ff., 457–458
erythemal, **1,** 395
generation of frequencies in, **15B,** 174–181
vacuum, problems, **3,** 160
Unaided eye, **8,** 39, 43
Unbalance parameter, hot-wire and hot-film probe compensation, **18,** 294, 296
Unbiasedness, **28,** 161
Uncertainty
frequency, **29C,** 301–303
principle, **1,** 74
wavelength, **29C,** 288
Undulator(s), **10,** 44, 45; **22,** 184
synchrotron radiation generation, **29C,** 28–30, 32–33
Undulators, **31,** 2, 15–18
polarization, **31,** 20–21
Unfilled apertures, **12B,** 21–24
Uniaxial crystal classes, effective coefficients for, **15B,** 147
Uniaxial crystals, **26,** 135
Uniaxial orientation, **16A,** 429
Uniform impurity distribution, **6A,** 57
see also Zone leveling
Uniformity condition, **8,** 94
Uniformity, of film thickness in PLD, **30,** 514–516
Unimolecular decay, in chemical kinetics, **15B,** 279–280
Unit area function, **5A,** 521
Unit cell, **16B,** 3, 5; **16C,** 151
historic reference, **16A,** 2
internal structure, **6A,** 197–199
mass of, **6A,** 284
primitive, **3,** 268
structure factor, **5A,** 567
volume of, **6A,** 284
Unit impulse, in network function, **2A,** 5
Unit(s), **13A,** 2, 149
best values, **1,** 43–51

conversion, **18,** 823–824
definition, **1,** 25, 27–34
derived, **1,** 25–26
electrical, **1,** 33–34, 432
electromagnetic, **1,** 33, 34
English system, **1,** 26, 28
fundamental, **1,** 25–51
Gaussian, **1,** 431
gravitational, **1,** 30
high vacua, **1,** 173
metric system, **1,** 26, 27–28
mks, **1,** 431
photometric, **18,** 688–689
power, **1,** 93
pressure, **18,** 508
Roentgen, **1,** 415
SI (SystËme International), **18,** 822
work, **1,** 93
UNIVAC, **2B,** 166
Univariate time series, **28,** 316–340
data tapering, **28,** 320–325
direct spectral estimator, **28,** 321
discrete prolate spheroidal sequence tapers, **28,** 323
duration-bandwidth product, **28,** 323
periodic nonsinusoidal signal, **28,** 339
periodogram, **28,** 316–320
spectral peaks, evaluating significance, **28,** 338–340
variance reduction, **28,** 329–338
lag window spectral estimators, **28,** 329–333
multitaper spectral estimators, **28,** 335–338
WOSA spectral estimators, **28,** 333–335
Universal calibration, for GPC, **16A,** 55–59
Universal ionizer, **4B,** 168
Universal negative-ion source (UNIS), **17,** 533, 535
Universal testing machines, **6A,** 318
Universal time
definition of, **12C,** 182
variations in, **12C,** 274
Universality, **23C,** 5, 57
two-scale-factor, **23C,** 19
University of Oklahoma, **8,** 35
Univibrators, **2,** 368–369
Unix, workstation, using S, **28,** 484–488
Unsaturation in polymers
in irradiated polyethylene, **16B,** 286
location of, using degraded samples, **16B,** 266–267
Unsteady adiabatic expansion, **30,** 226–227
Unsupported area packing, **6A,** 408
UO_2 antiferromagnetic resonance, **10,** 411
UP, **23C,** 125
Up-conversion lasers, **29C,** 233, 268–270
Up-down multiscaling, **11,** 325, 330
Upconversion, concept of, **15A,** 194
Upconversion, of weak infrared signals, **15B,** 144, 183
UPd_3, **23C,** 218
Upper and lower critical solution temperatures, **16C,** 283–284
Upper atmosphere research, mass spectrometer usage, **8,** 169–175
calculations, **8,** 170–173
exercises, **8,** 175
Upper hybrid frequency, **10,** 276
UPS, *see* Ultraviolet photoelectron spectroscopy
Upscattering, in phonon-neutron collision, **16A,** 210
Uranium
compounds, **23C,** 218
heavy water moderated reactor, **5A,** 505
target neutron yield, **5A,** 527–528, 529
Urbach rule, **30,** 584
Urea, **19,** 232
USb, **23C,** 220
UT0
definition of, **12C,** 182
UT1
definition of, **12C,** 182
UT2
definition of, **12C,** 182
UTC
definition of, **12C,** 182
UTe, **23C,** 222
UV beam, in F_2^+ formation, **15B,** 22

V

Vacancy configuration effect, in target x-ray production, **17,** 257
Vacancy formation enthalpy, positrons, **21,** 89
Vacancy-impurity binding, positrons, **21,** 92
Vacancy-interstitial pair, **6A,** 357
Vacancy-sharing, **17,** 385, 485–486, 491, 493
Vacua, high, **1,** 172–192
pressure measurement, **1,** 189–192
Vacuseal, **14,** 415

Vacuum calorimeter, **1,** 266ff, 274
Vacuum chamber, **14,** 276, 278
Vacuum component, metal, **14,** 275
Vacuum cryostat, **8,** 214
Vacuum degreaser, 2-propanol, **14,** 500
Vacuum deposition
  by laser-produced plasmas, **29C,** 17
  materials, **14,** 525–530
  resistivity heated sources, **14,** 531–533
  substrate cleaning, **14,** 535–536
  techniques, **25,** 299
Vacuum diode, rectifier and, **2A,** 152–154, 218
Vacuum evaporation, **14,** 522–550
  advantages, **14,** 573, 573
  alloys, **14,** 525, 530, 534, 539
  comparison, sputtering, **14,** 572–575
  compounds, **14,** 525
  deposition procedures, **14,** 548, 550
  deposition rate, **14,** 523
  disadvantages, **14,** 572, 573
  effect of background gas, **14,** 524
  electron beam source, **14,** 533–534
  geometry, source, substrate, **14,** 524
  MgF_2, **14,** 525
  multiple sources, **14,** 530
  oxide films, **14,** 524, 525
  patterns, **14,** 539
  permalloy, **14,** 530
  reactive evaporation, **14,** 530
  shutters, **14,** 539–540
  SiO, **14,** 525
  sources
    electrical requirements, **14,** 540–541
    heaters, **14,** 530–535
    temperature, **14,** 523, 526–529
    stoichiometry, **14,** 525
  sublimation, **14,** 530
  substrates, **14,** 535–537
    temperature, **14,** 523
  support materials, **14,** 526–529
  vacuum systems, **14,** 537–541
Vacuum evaporation, in thin-film preparation, **7B,** 284–285
Vacuum gaps, triggered, **2A,** 184–185
Vacuum gauge, **14,** 278
  glass, **14,** 320
Vacuum grease, **14,** 320
Vacuum Hazards Manual, **14,** 449
Vacuum hose, **14,** 300
Vacuum joint, bakeable and coolable, **14,** 439
Vacuum level, **22,** 128
Vacuum photocells, for optical pumping, **7B,** 17
Vacuum photodiode, as short-pulse detector, **29C,** 199
Vacuum phototubes, *see also* Photomultiplier tubes
  cathode composition in, **2B,** 302–303
  characteristics of, **2B,** 301–307
  operating principle of, **2B,** 301
  spectral response characteristics of, **2B,** 304
Vacuum pumps, **1,** 172–182
  cryogenic, **4B,** 268–269
  cryosorption, **4B,** 260
  mechanical, oil sealed, **4B,** 260–261
  mercury diffusion, **4B,** 262–263
  molecular drag, **4B,** 264
  oil diffusion, **4B,** 262–264
  orbitron, **4B,** 267–268
  in photoemission studies, **11,** 114–116
  sputter-ion-getter, **4B,** 264–267
  titanium sublimation, **4B,** 267
Vacuum range, **14,** 1
  fore (or rough) vacuum, **14,** 1, 3, 4
  high vacuum, **14,** 1
  ultrahigh vacuum, **14,** 4
Vacuum seal, **14,** 290
Vacuum spark, **1,** 407
  method, **6A,** 33
Vacuum system(s), **1,** 186–189
  design, mechanical limits, **14,** 488
  for electron diffraction apparatus, **6A,** 269
  glass, **14,** 313–343
  maintenance, **14,** 503
  metal, **14,** 275, 278
  schematic, **14,** 3
  specifications, **14,** 478
  time constant of, **7A,** 74, 188
Vacuum tubes, **2A,** 147–164
  bias currents in, **2A,** 261–267
  cathode types and shapes in, **2A,** 148–152
  cost of, **2A,** 161
  as diodes, **2A,** 152–154, 218
  disk-seal type, **2A,** 163–164
  dispenser cathodes in, **2A,** 150–151
  electrometer type, **2A,** 299
  equivalent circuit for, **2A,** 269
  hexodes and heptodes, **2A,** 160
  low-noise, **2B,** 466
  in microwave circuits, **2B,** 264–273
  microwave triodes and tetrodes, **2A,** 163–164
  mixers and converters, **2A,** 160–161

monolayer emitters, **2A,** 149
multielement, **2A,** 158–161
nuvistors, **2A,** 164
oxide emitters, **2A,** 149–151
pencil tubes, **2A,** 164
pentodes, **2A,** 159–160
remote-cutoff, **2A,** 160
Schottky effect in, **2A,** 147–148
selection of, **2A,** 161–164
vs solid-state devices, **2A,** 161
special applications of, **2A,** 162–163
tetrodes, **2A,** 158–159, 163–164
vs transistors, **2A,** 45
transit-time effects in, **2A,** 154
triodes, **2A,** 155–158, 163–164
variable-mu, **2A,** 160
Vacuum tunneling, **27,** 100, 314, 320
theory of, **27,** 1
Vacuum tunneling, transmission probability theory of, **27,** 2
Vacuum ultraviolet generation, **15B,** 216–222
spectroscopic studies using, **15B,** 222–226
tuning ranges in, **15B,** 219
Vacuum ultraviolet monochromator, **15B,** 218
Vacuum ultraviolet radiation, tunable, **15B,** 221
Vacuum ultraviolet radiation, ultrashort pulses, **29C,** 224–227
Vacuum ultraviolet spectroscopy, **15B,** 210–231
future developments in, **15B,** 230–231
multiphoton ionization technique in, **15B,** 227–229
phase velocity matching in, **15B,** 214
Vacuum ultraviolet, **31,** 101–115
comparison with BESSY, **31,** 138–143
lasers, **31,** 101–115
mirror optics, **31,** 145
Vacuum ultraviolet, in picosecond spectroscopy, **15B,** 195
Vacuum valve, **14,** 296–298
Vacuum wax, **14,** 320
Vacuum, ultrahigh, **4B,** 311–316
all metal system, **4B,** 315–316
glass system, **4B,** 312–315
ultimate pressures, **4B,** 262–266, 277, 290, 314–316
Vacuum-arc ion source
electrode lifetime, **29A,** 131–132
low-voltage vacuum-arc discharge, **29A,** 129–130
principle of operation, **29A,** 129
triggered, high-voltage vacuum-arc discharge, **29A,** 130–133
Vacuum-isolated ions, **30,** 413, 414
Valance band, **12A,** 8
Validity testing, simulation, **28,** 450–451
Valine, **19,** 202
Valve actuator, **14,** 297, 298
Valve bakeout cycle, **14,** 298
Valve bakeout temperature, **14,** 298
Valve leak, **14,** 297
Valve, **14,** 278, 296–300
malfunction, **14,** 460
control, diffusion pump systems, **14,** 466–467
Valves, vacuum, **4B,** 280–283
VAMFO technique, in thin films, **11,** 656
Van Cittert-Zernicke theorem, **8,** 46
Van de Graaff accelerator, **17,** 5–12, 84, 312–313, 466, 533, 535; *see* Electrostatic accelerator
Van de Graaff generators, **5B,** 584–590
tandem, **5B,** 588–590
Van der Held dilution curves, **7B,** 216–217
Van der Merwe growth, **30,** 297, 308
Van der Pol equation, **2B,** 162
Van der Waals
clusters, **29B,** 92–93
fields, **3,** 229–238
gas, **1,** 122
interactions, **23B,** 495
spectral shift, **3,** 234–238
Van Hove critical points, **11,** 144
Van Hove singularities, **16A,** 212
Van Hove, L., **23A,** 30; **23B,** 245
Van Vleck paramagnetism, **23C,** 282
Van Vleck relation, **12C,** 183
Van Vleck-Wesskopf equation, **5B,** 87
Vanadium measurements, **23B,** 416, 442
Vanadium, hydrogen solubility, **14,** 354
Vane anemometer, principle, **18,** 255–257
Vapor bubbles, *see* Bubble motion, vapor filled
Vapor cells, *see* Heat pipes
Vapor condensation method, for magnetic films, **11,** 693–695
Vapor degreasing, **14,** 445
Vapor deposition, *see* Vacuum evaporation of films, **6A,** 132–133
Vapor ejector pump, *see* Ejector pump
Vapor furnace, **8,** 185, 186

Vapor phase diffusion, techniques, **6A,** 82
Vapor pressure(s), **1,** 179; **3,** 602
of elements, **14,** 346–352
of lubricants, **14,** 116
of oils, **14,** 415
of pump oils, **14,** 116
Raoult's law, **14,** 346
of sealants, **14,** 415–417
vs temperature, **14,** 186, 187
thermometer, **1,** 241–243
in vacuum evaporation, **14,** 522
of waxes, **14,** 415, 416
Vapor state of high melting compounds, composition, **3,** 563
Vaporization, **1,** 262; **30,** 234, 239
curve, **1,** 242
heat of, **1,** 281ff
Vapors
metal, composition, **3,** 564–565
velocity of sound in, **3,** 682–700
Varactor diodes, **2A,** 64–65
Varactors, **12B,** 230
Variable area flowmeter, **18,** 331
Variable attenuator, **8,** 124, 126, 143
Variable capacitor, **8,** 145
Variables
displacement, **1,** 312
electrical, **1,** 306
force, **1,** 312
mechanical, **1,** 306
Varian e-gun, **11,** 105
Variance, **1,** 4–5, 16ff; **2,** 3–4; **3,** 711–712; **5B,** 773–774, 809–810; **6A,** 3–4; **28,** 19
sample, **5B,** 774
Variance function, asymptotic, **28,** 228
variance, **1,** 4
Variance-covariance matrix, **28,** 256–257
Varicap, **2A,** 347
Varied-line-spacing (VLS) grating, **31,** 348–350
Varistors, **2,** 122–123
Vaseline, **14,** 416
VBLY systems, **12C,** 174–197
block diagrams of, **12C,** 174
Canadian, **12C,** 174, 187, 192–196
comparative sensitivity of, **12C,** 197
correlators for, **12C,** 184
data processor for, **12C,** 184
data recording for, **12C,** 183
digital recording in, **12C,** 186
frequency conversion in, **12C,** 177
frequency standards for, **12C,** 176
Mark I, **12C,** 183–184, 186–187
recorder in, **12C,** 174
Mark II, **12C,** 183–184, 187, 189–192
recorder in, **12C,** 174
playback system for, **12C,** 184
storage capacity of, **12C,** 187
synchronization of, **12C,** 179
tape recorders for, **12C,** 187
timing required for, **12C,** 178
VCO, *see* Voltage-controlled oscillator
Vector polarization, **4B,** 227, 231; **5B,** 218
Vector potential, **13A,** 73
Vector potential, 55
Vector-correlation function, **3,** 287–289
Vector-coupled angular momentum, **8,** 274
Vee-block refractometer, **25,** 185
Velocities, Maxwell-Boltzmann distribution of, **15B,**65
Velocity, **1,** 75–78; **19,** 3–4, 140, *see also* Acoustooptic measurement of sound wave, velocity
angular, **1,** 86
bar phase, **1,** 209
in bulk materials, **19,** 239, 268, 276–277, 291–294
charged particles, time-of-flight measurement, **5A,** 438–443
constants, chemical kinetic for ion-molecule reactions, **3,** 584–586
of flexural waves, **1,** 210
group, **1,** 210–211
interval selecting counter, **5A,** 183, 186
of light, **1,** 36–37; **3A,** 338–343, 394
long bar, **1,** 211
meters, **1,** 78, 93
particle, **1,** 200
projectile, **1,** 76
resolution, Čerenkov counter, **5A,** 168–170, 456
and rotational relaxation, **15B,** 66–68
in scattering formulas, **19,** 242–245
selector, Fermi-type neutron, **5A,** 502–503
of sound, **1,** 129, 222
in gases, **3,** 681, 682–700
in liquids, **3,** 700–708
spectrometer
pulsed accelerator, **5A,** 519, 520
resolution of ideal, **5A,** 518–519
stretched wire, **1,** 209
wave, **1,** 199, 208ff

Velocity components
  measurement by chronophotography, **18,** 67
  measurement by hot-wire probe, **18,** 306–311
  measurement by LDV, **18,** 190–195
Velocity dependence, of electron capture and loss cross sections, **17,** 100–101
Velocity dispersion, **19,** 41, 183, 187
Velocity distribution(s), **14,** 6; **30,** 427, 428
  atomic beam, **4B,** 241
  effusive thermal sources, **29B,** 8–9, 22
  in molecular beam, **3,** 593, 597–598
  supersonic beams, **29B,** 25–26
Velocity feedback control, in Mössbauer spectrometer, **11,** 326–331
Velocity filtering, in PLD, **30,** 512
Velocity gradient, measurement by hot-wire probe, **18,** 312
Velocity head, **18,** 333
Velocity interferometer, **24A,** 219–222
Velocity measurement; **5A,** 444–460
  in bubble chamber, **5A,** 451–453
  using Čerenkov counters, **5A,** 454–460
  by chronophotography, **18,** 64–93, 818–819
  direction by chronophotography, **18,** 67
  direction by hot-wire anemometer, **18,** 306–312
  direction by LDV, **18,** 190–195
  direction by Pitot probe, **18,** 254
  direction, **18,** 241
  by Doppler shift
    of emitted characteristic radiation, **18,** 341–345
    of scattered light, **18,** 93–240, 342
    of scattered sound from tracers, **18,** 317–318
  electromagnetic method, **18,** 318–321
  fluorescent radiation Doppler shift, **18,** 343
  by Hall voltage, **18,** 318
  by heat loss probe method, **18,** 259–314
  hot-wire and hot-film probes, **18,** 259–314
  laser Doppler from tracing particles, **18,** 96–240
  by laser Doppler velocimeter, **18,** 93–240
  LDV with direct spectrum analysis, **18,** 194–227
  Pitot probe, **18,** 242–254
  by pressure probe, **18,** 242–254
  probe methods, **18,** 240–341
  propeller anemometer, **18,** 254, 256
  resonant absorption of Doppler shifted radiation, **18,** 344
  rotating flow apparatus, **18,** 817–819
  sensitivity of measurement using trace methods, **18,** 36
  simultaneous with temperature measurement by hot-wire probe, **18,** 313
  by time sound pulses, **18,** 315–318
  tracer methods, **18,** 1–240
  tracer particle loading error, **18,** 38
  vane anemometer, **18,** 254–259
Velocity measurements
  error in
    from bonded transducers, **19,** 68–70, 89, 93–94, 95–97
    from diffraction, **19,** 71–76, 99
    from dispersion, **19,** 68, 101
    from phase cancellation, **19,** 70–71
  pulse methods
    basic pulse-echo methods, **19,** 85–89
    echo-overlap method, **19,** 97–99
    gated double-pulse superposition method, **19,** 90–94
    long pulse buffer rod method, **19,** 102–105
    pulse interferometer methods, **19,** 99–102
    pulse superposition methods, **19,** 94–97
    sing-around method, **19,** 89–90
Velocity potential, **19,** 5
Velocity scanning spectrometers, **11,** 319–320, 342–343
  in ionic drift velocity measurements, **11,** 454–456
Velocity selector(s), **7A,** 255–259; **23A,** 213
Velocity-measurement systems
  block diagram of, **11,** 401
  drawbacks of, **11,** 409
Velocity-selective coherent population trapping, **29B,** 131
Venetian blind collimeter, of modulated-potential Faraday cup, **9B,** 59
Vent groove, **14,** 289, 291
Vent hole, **14,** 285
Vent slot, **14,** 429
Ventilated wall, wind tunnel, **18,** 773, 775
Venting, **14,** 498, 502
Venturi flowmeter, **18,** 324, 331
Venturi meter, **1,** 118, 155, 156
Verdet constant, **6B,** 275
Verdet's constant, **1,** 387
Vernier activity, **25,** 124

Vernier chronatron, **2B,** 127
Vertical angular distribution, calculation, 6
Vertical anisotropic etch, **2A,** 112
Vertical seismic profile, **24B,** 477
Very long baseline interferometry, *see* VLBI
Vessicles, **23B,** 490
Vibrating reed, **1,** 223ff
  electrometer, **4A,** 343–345
Vibrating rotor, **3A,** 136
Vibration, **1,** 225–227
  isolation, forevacuum pump, **14,** 126–127
  lattice, **1,** 293
  measurement, **1,** 90–93
  modes, **21,** 172
  pickups, **1,** 217, 218–219
  Raman-active, **3,** 121–122
  with turbo-pump, **14,** 267
Vibration amplitude, **8,** 1
Vibration isolation stage mechanical structure, **27,** 58
Vibration isolation theory
  scanning tunneling microscope design, **27,** 43
Vibration-rotation bands, **10,** 327–328
Vibration-rotation interactions, **3A,** 51, 274; **15B,** 271
Vibration-rotation lines, Hcl, intensity relations, **3,** 47–48
Vibration-rotation spectra, **13B,** 30
  of acetylene, **3,** 145
  CH_3Cl, **3,** 62
  of HF, **3,** 49
  of NO, **3,** 50
Vibration-rotation states of HCN laser transitions, **10,** 53, 328
Vibrational absorption
  local modes, **10,** 368–370
  resonant modes, **10,** 368, 371, 372
  reststrahlen, **10,** 90, 94, 95, 100, 103, 108, 109, 363–365
Vibrational coupling, **22,** 62
  intermolecular, **22,** 64
  intramolecular, **22,** 62
Vibrational decay, **16A,** 405–406
Vibrational energy
  distribution of among product states, **15B,** 99–100
  in molecular spectroscopy, **15B,** 277–278
Vibrational energy dispersal, in laser action, **15B,** 100–101
Vibrational energy levels, **3A,** 260
Vibrational energy transfer
  carbon dioxide and, **15B,** 72
  research in, **15B,** 74
Vibrational equations, **16A,** 91–92
  of one-dimensional periodic chain, **16A,** 99
Vibrational excitation, selective chemistry of, **15B,** 294–297
Vibrational frequency, **20,** 124, 125
Vibrational from atom exchange reactions, **15B,** 98
Vibrational intensities
  infrared, **16A,** 95–96
  Raman, **16A,** 96–98
Vibrational modes, **16A,** 90–91
  in crystalline solids, **3B,** 986
  energy shifts, **16A,** 161
  in IETS, **16A,** 152, 154
  relaxation time, **16A,** 191
Vibrational photochemistry, direct excitation and dissociation in, **15B,** 295
Vibrational processes, **30,** 18, 103–104, 413
Vibrational relaxation, **15B,** 68–75, 286–287
Vibrational spectral line contour analysis, **18,** 425–428
Vibrational states, **23C,** 468
Vibrational temperature, **18,** 436, 457, 464, 466, 473–474, 481, 491–493
Vibrational theory, **16A,** 89–98
  density of states, **16A,** 100
  group coordinates (rocking, twisting, etc. modes), **16A,** 90–91
  of periodic chain molecule, **16A,** 98–116
Vibrational-electronic interactions, **3,** 190–200
Vibrations
  CO_2, **10,** 323, 324
  HCN, **10,** 325
  H_2O , **10,** 326
  linear triatomic molecules, **10,** 324, 325
Vibrators, **2,** 202
Vibrometer, **1,** 92
Vibronic coupling, **3A,** 270
Vibronic states, **3A,** 261
Vicinal pyramid, **6A,** 333
Vickers hardness number, **6A,** 307
Video circuit, **8,** 126
Video detection, **10,** 162, 163, 165, 184
Video detector, **8,** 137, *see also* Microwave crystal video detector
Vidicon camera, **6B,** 354–355; **22,** 285, 488
  tube, **2B,** 347–349

Vidicon, **2,** 778–781; **12A,** 280
Viewport, **14,** 278, 311
Vignetting, **25,** 37, 130
Vignetting, **31,** 151
Vineyard, G. H., **23B,** 202
Vineyard, G., **23A,** 45
Vinyl chloride laser emission, **10,** 50
Violin bow probe, **24B,** 233, 253
Virial coefficient, **1,** 314
Virial equation of state, **18,** 611–612
Virial fringes, in LDV, **18,** 118
Virtual leak, **14,** 460, 481, 487, 505, 516–518
Viruses, **23C,** 390
Viscoelastic behavior
  definition, **16C,** 1
  dielectric analogy, **16C,** 403
  effect of pressure on, **16C,** 51
  linear, **16C,** 3–44
  nonlinear, **16C,** 46–51
Viscoelastic liquid, creep in, **16C,** 3
Viscoelastic solid, creep in, **16C,** 6
Viscoelasticity, **1,** 110–112; **16A,** 5
Viscometry, **18,** 798–801
Viscosimeters, **1,** 124ff
Viscosity, **1,** 123–126, 213; **13B,** 194
  absolute, **1,** 131, 132
  effects of, **1,** 130–134
  kinematic, **1,** 131
  local (micro-), **16A,** 434
  measurements under pressure, **19,** 171–172
  of a particulate suspension, **18,** 38
  shear, **16C,** 4, 5; **19,** 138
  temperature dependence, **1,** 126
  variation with pressure, **19,** 145–146
  variation with temperature, **19,** 144, 172
Viscous absorption, **19,** 184
Viscous damping coefficient, **2B,** 439
Viscous drag pump, **14,** 247
Viscous drag tachometer, **1,** 86
Viscous flow, **1,** 161
Viscous fluid, **18,** 502, 797–800
Viscous penetration depth in superfluid helium, **19,** 438
Viscous relaxation frequency, **19,** 186, 187
Viscous stress, **1,** 132
Visibility
  coefficient, 50, 51, *see also* Experimental visibility; Theoretical visibility
  factor, **8,** 40, 44
Visible emission lasers, **15A,** 231–232
Visible light, radiometry of, **26,** 284
Visible region, *see* Optical region
Vision, **20,** 185–189
Visual acuity, **25,** 124
Visual displays, types of, **2B,** 6–8
Visual system, **25,** 120
Visualization of sound waves, **19,** 472–475, 484–487
Visualization, *see* Flow visualization
Viton, **14,** 407, 411, 481, 485, *see also* Elastomer
Vlasov equations, **9A,** 1, 277, 305
VLBI date reduction, **12C,** 228–260
  continuum fringe visibility, **12C,** 235
  delay tracking in, **12C,** 232
  fringe phase in, **12C,** 233
  fringe rate in, **12C,** 232
  incoherent averaging in, **12C,** 241–247
  misidentification of signal in, **12C,** 247–248
  shift correction in, **12C,** 232
VLBI measurements
  coordinate systems for, **12C,** 262
  dual antenna techniques for, **12C,** 269
  examples of, **12C,** 266–276
  observables in, **12C,** 261
  observational techniques in, **12C,** 266
  relativistic effects in, **12C,** 262
VLBI observations, **12C,** 228–260
  checklist for, **12C,** 258–260
  fringe amplitude estimation, **12C,** 228
  fringe identification from, **12C,** 228
  measurable qualities from, **12C,** 228
  measurement of fringe amplitude, **12C,** 228–238
  measurement of fringe phase, **12C,** 229–238
  operational problems, **12C,** 228–229
VLBI receivers, **12B,** 221
VLBI, definition of, **12B,** 26
Vodar monochromator, **13A,** 237
Voice coil, **8,** 14, 15
  excursion, **8,** 15
  resistance, **8,** 11
Voids, **23C,** 441
  positrons, **21,** 93, 94
Voight elastic coefficients, **11,** 434, 441
Voight profile, **29B,** 256
Voigt effect, **1,** 387; **6B,** 276
Voigt functions, **9A,** 163
Voigt notation, **19,** 34
Volmer-Weber growth, **30,** 308
Volt, **1,** 33, 34
  defined, **11,** 275

Voltage
common-mode, **2B,** 74
breakdown, **1,** 499
deflection, on condenser plates, **5B,** 228–229
dependence of gas multiplication, **5A,** 112ff
generators, Stark, **3,** 663–666
parasitic, **1,** 454–455
plateau, **5A,** 101
standing wave ratio, **1,** 481ff, 486; **3,** 25
Voltage clamp, **20,** 445–512
Voltage comparator, **2A,** 417
Voltage dependence in STM imaging theory, **27,** 11, 22
Voltage dependent imaging in scanning tunneling microscopy, **27,** 99, 104, 106
Voltage feedback power amplifier, **8,** 11
Voltage gain, **8,** 126
Voltage generator, *see* Constant voltage generator
Voltage measurement, **1,** 436, 452–455, 459–461; **2B,** 48
digital voltmeter in, **2B,** 56–57
dual-slope integration in, **2B,** 55
indirect, **2B,** 58
potentiometers in, **2B,** 58–59
voltmeters in, **2B,** 51–52
Voltage reference diodes, **2A,** 61, 171, 240
Voltage regulators, **2A,** 169–171, 242–244, 365–366
open-loop, **2A,** 243–244
Voltage regulators, **2B,** 452–458
Voltage standing wave ratio (VSWR), **2B,** 290–291
Voltage standing wave ratio (VSWR), **8,** 109, 119, 121, 122, 141
Voltage standing-wave ratio (VSWR) mismatch, **2A,** 80
Voltage variable capacitor, **2A,** 347
Voltage(s)
doubler, magnetic, **2,** 790–792
line, variation, **2,** 179–180
measurement, **2,** 558–583
alternating current, **2,** 576–583
direct current, **2,** 558–576, 649–650
potentiometer, **2,** 567–568
maximum peak, **2,** 110
normal fall, **2,** 85
regulation, **2,** 489–503
alternating current, **2,** 491–493
diodes, **2,** 135–136
direct current, **2,** 493–501
sources, **2,** 18–19
terminal, parameters relating to terminal currents, **2,** 28–29
Voltage-controlled oscillator, **2B,** 21; **24B,** 39–41
Voltage-current sources, in *RLC* network, **2A,** 3
Voltage-multiplying circuit, **5B,** 581–584
Voltage-to-frequency converter, **2A,** 354; **2B,** 53–54
Voltage-tunable oscillator, **11,** 579
Voltameter, silver, **1,** 41
Voltmeter
attracted disk, **1,** 439
in dc measurements, **2B,** 51–52
electrodynamic, **2B,** 64
electrostatic, **1,** 452, 453, 460
Kelvin absolute, **1,** 438, 439, 443
thermocouple, **1,** 460
Voltmeters, amplifying
differential-amplifier, **2,** 559–560
direct current, **2,** 559–561
drift and gain stabilized, **2,** 561–567
precision direct-reading, **2,** 569–570
Volume, **1,** 29, 59–60, 68
dead, **1,** 240
of fluids, **1,** 116–117
normal molar, **1,** 37–38
Volume changes, **19,** 202–203
Volume coefficient, **8,** 90
Volume dipoles, *see* Distributed volume dipoles; Magnetic volume dipoles
Volume flow, **14,** 101, 299, *see also* Molecular transport
Volume hologram, **26,** 172
techniques for color holograms, **26,** 183
Volume minimized-split solenoid system, **8,** 101
Volume susceptibility, **1,** 538
Volume viscosity, **19,** 230–231
measurement by Rayleigh-Brillouin scattering, **16A,** 190–191
Volume-reflection hologram, **26,** 174
Volumetric efficiency
liquid piston pump, **14,** 135
lobe pump, **14,** 128–129, 132
Volumetric technique
for permeability, **16C,** 363
for sorption, **16C,** 354

von Laue transmission pattern, **4A,** 147
Vortex flow, ion techniques in, **11,** 469–470
Vortex generator, **18,** 816
Vortex rings
  creation of with ions, **11,** 476–479
  drift velocity *vs* electric field in, **11,** 478
Vortex-antivortex pairs, **23C,** 153
Vortex-stabilized loops, in steady-state lasers, **15A,** 307–310
Vorticity meter, **18,** 241, 312
VR (vibrational-rotational) event, **15B,** 64
VSCPT, *see* velocity-selective coherent population trapping
VSe_2
  1T phase
    scanning tunneling microscopy of charge density wave in, **27,** 359
    spectroscopy of, **27,** 378
VSWR, *see* Voltage standing wave ratio
VSWR, *see* Voltage standing wave ratio
VT (vibrational-translational) collision, **15B,** 64, 79
VT deactivation, of upper laser level, **15B,** 68–69
VUV generation in, **15B,** 222–226
VUV, *see* Vacuum ultraviolet
VV (vibrational-vibrational) collisions or energy transfers, **15B,** 64, 69, 74, 76, 79

W

w-seal, **14,** 315
Wadsworth arrangement, **25,** 208, 216
Wadsworth grating mount, **1,** 370, 399, 400
Wadsworth monochromator, **29C,** 35, 36
Wadsworth mounting, **13A,** 294
Wald's method, **28,** 228
Walk
  in pulse amplitude measurement, **2B,** 112
  in time measurement, **2B,** 29–30
Wall effect, counter, **5A,** 573–574
Wall effects, **8,** 188
Wall stabilized arc, **31,** 45–52
  continuum emission, **31,** 121–122
Wall temperature discontinuity, effect on heat transfer measurement, **18,** 673
Warburg impedance, **24B,** 287–288
Ward-Leonard control system, **2B,** 442
Warren line shape, **23B,** 15, 22
Warren-Cowley short-range order parameter, **23B,** 105, 254
Warshawsky bridge, **6B,** 42–43
Water, **19,** 189, 190, 195
Water cleaning vacuum system, **14,** 500
Water exchange, **23B,** 481
Water extinction by condense, **12B,** 177–185
Water molecule spectrum, *see* H_2O spectrum
Water spectrum, **13B,** 44
Water trees, **16C,** 449, 488
Water vapor fundamental bands, **3,** 89
Water vapor laser, *see* H_2O laser
Water vapor masers
  VLBI measurement of, **12C,** 251, 267
Water vapor pumping
  forevacuum, **14,** 121–124
  speed, **14,** 234
  by traps, **14,** 159
Water vapor transport, **16C,** 331–332
  experimental precautions, **16C,** 375
Water vapor, **14,** 234
  absorption coefficient of, **12B,** 150–257
  distribution in atmosphere, **12B,** 139–140
  from glass, **14,** 323, 324
    ultraviolet irradiation, **14,** 329
  lower frequency spectral lines of, **12B,** 153
  outgassing from glass, **14,** 322
  physical properties, **14,** 421
  thermal emission by, **12B,** 172
  in vacuum systems, **14,** 477, 491
Water weakening, **24A,** 84
Water-cooled nuclear reactor, **8,** 227
Water-cooled-type reactors, **8,** 245
Watt-hour meter, **1,** 469
Wattmeter, **1,** 467
Wave
  admittance, **1,** 202
  dilational, **1,** 208
  equivoluminal, **1,** 209
  flexural, **1,** 210
  free space, **1,** 199
  impedance, **1,** 202, 477
  linear separation, **1,** 362
  longitudinal, **1,** 199, 208
  motion, mechanical, **1,** 200
  plane, **1,** 477ff
  plates
    achromatic, **26,** 144
    measuring, **26,** 145
  resistance, **1,** 201
  scattered, amplitude, **3,** 276ff
  scattered, measurements, **2,** 734–737
  shock, **1,** 129ff, 192–194
  standing, **1,** 278–279, 479–480

temperature, **1,** 293
transverse, **1,** 199, 209
velocity, **1,** 199, 208ff
vibrational, **1,** 208
Wave aberration polynomial, **25,** 42
Wave equation(s), **9B,** 3–9; **10,** 114; **26,** 108; *see also* Wave propagation
elastic coefficients in, **11,** 439–441
solution for collisions, **9B,** 6–7
solution for no collisions, **9B,** 5–6
solution with static applied magnetic field, **9B,** 7–9
Wave form generators, in Mössbauer spectrometer, **11,** 337–342
Wave length uncertainty, **29C,** 288
Wave machine, **18,** 796
Wave packet(s), **23A,** 305; **29B,** 330–332
Wave propagation, **1,** 112–114
in anisotropic media, **26,** 133
devices, **6B,** 20–26
effect of grains on, **19,** 239
equation, **19,** 4–5, 238
isotropic media, **26,** 113
longitudinal, **19,** 241–243
for metals, **26,** 120
modes, **1,** 476–477
parameters, **19,** 238–239
in periodic media, **23A,** 314
shear, **19,** 241–243
wave vector, **19,** 5
Wave propogation in a solid, **31,** 260–264
Wave reflection, **31,** 264–267
Wave transmission, **31,** 264–267
Wave vector **Q**, **21,** 151, 156, 167
Wave-front topography, **26,** 40
Wave-vector mismatch, **29C,** 234
Waveform(s)
analysis, **1,** 220–221
current, special, **2,** 409–412
generator, **1,** 475
measurement, **1,** 469–474
voltage, special, **2,** 404–409
Waveform processing
borehole sonic data, **24B,** 600–608
direct phase determination, **24B,** 602–604
fourfold correlation, **24B,** 600–602
semblance correlation, **24B,** 604–608
Wavefront aberrations, **31,** 151–152
Wavefront division, **1,** 349
Wavefunction
of rigid rotor, **10,** 307
vibronic, **3,** 191
Waveguide interferometer, **8,** 147–152
exercises, **8,** 151–152
Waveguide resonators, **6B,** 20–22
Waveguide techniques, **13B,** 54, 116, 127, 128, 150
Waveguide(s), **1,** 477, 481–482; **2,** 706–707; **3,** 24–26; **8,** 133, 134, 138, 147; **16C,** 410; **29C,** 369–371; *see also* Standard waveguide, Oversize waveguide, RG–99/U waveguide; X-band waveguide
channel waveguides, **29C,** 383–384, 386
characteristic impedance, **8,** 140
components, **10,** 126–128
coupling, *see* Microwave cavity/waveguide
current, **8,** 140
fabrication, **29C,** 384, 386
filters, **10,** 121
gas absorption cell, **5B,** 24–25
gas-filled, absorption, **5B,** 87–88
guide wavelength, **8,** 143
hybrid ring network, **3,** 472, 473
intersecting waveguides, **29C,** 382–383
iris-loaded, **5B,** 642–643
loop, **8,** 151
magic T, **3,** 472
polymer waveguides, **29C,** 383–384
quasi-phase-matching in, **29C,** 245–247
system, **8,** 141
Wavelength, **1,** 199; **13A,** 2, 246, 326, 327; *see* Half-guide wavelength
band, **8,** 36
blaze, **1,** 368
calibration, **8,** 162
conversion to energy, **5A,** 607
de Broglie, **1,** 47
definition of, **19,** 5
determination, error in, **5A,** 609
Ge I lines, **3,** 177–178
limit, X-ray, **1,** 46ff
resolution of crystal monochromator, **5A,** 569
to resolve neighboring atoms in a molecule, **3,** 265
scattering regions, **19,** 242–247, 267
shift in, **5A,** 149
in waveguides, **3,** 24
Wavelength measurements, **3,** 173–174; **3A,** 237

Wavelength standards, **3,** 174–178; **3A,** 218; **4A,** 297
Wavelength tuning, **29C,** 61–66
Wavelength-tunable oscillator, **15A,** 276
Wavelengths, **13B,** 26
Wavemeter
  calibration, **8,** 136
  dip, **8,** 147
Wavemeters, **2,** 535–537; **2B,** 15–17; **3,** 27
  constant-frequency sources of, **2B,** 17–22
  grid-dip, **2,** 536–537; **2B,** 16
  laboratory oscillator and, **2B,** 17–18
  LC, **2,** 535–536
  LC type, **2B,** 15–16
  transmission-line, **2B,** 16
Wavenumber, **4B,** 50; **10,** 1; **13A,** 2
  dispersion, **3,** 174
  shift, **3,** 138–139
Wax, **14,** 414–417
  vacuum, **14,** 320
Wax-lined cells, in optical pumping, **7B,** 14
Weak law of large numbers, **28,** 26
Weak links (weakly linked superconductors), **11,** 209, *see also* Tunnel junctions
  applications of, **11,** 273–305
  characteristics of, **11,** 251–273
  Clarke slug and, **11,** 260–261
  as computer elements, **11,** 305
  current-phase relations in, **11,** 260–268
  defined, **11,** 199
  detector applications for, **11,** 299–304
  digital devices for, **11,** 304–305
  e/*h* measurements for, **11,** 273–274
  harmonic generation and, **11,** 299–304
  I-V characteristic curves for, **11,** 209–212
  Josephson effects and, **11,** 201–251
  kinetic inductance and, **11,** 267
  and microwave-induced steps on *I-V* curve, **11,** 217–225
  mixing applications of, **11,** 302–303
  parametric amplification of, **11,** 299–304
  phenomenological descriptions of, **11,** 262–268
  plasma resonance and, **11,** 238–239
  point contacts and, **11,** 258–260
  quasi-particle steps in, **11,** 222–225
  response of to ac signals, **11,** 299–301
  rf voltage across, **11,** 21
  Riedel singularity and, **11,** 225–2229
  self-induced supercurrrent steps and, **11,** 229–233
  in superconducting rings, **11,** 275–296
  supercurrent steps and, **11,** 217–225
  tunnel junctions as, **11,** 251–254
  types of, **11,** 251
Weak signals, upconversion of, **15B,** 183
Weakly bound state, beam-foil experiments, **17,** 564
Weber law, **25,** 123
Wedge-and-strip anode
  configuration, **29A,** 268
  error sources, **29A,** 267–270
  resolution, **29A,** 268–269
Weibull, **28,** 230–234
  confidence intervals, **28,** 233–235
  confidence regions, **28,** 233–235
  cumulative distribution function, **28,** 231
  galaxy variability, **28,** 230
  likelihood function, **28,** 232–233
  limited failure population model, **28,** 235–238
Weibull distribution, three-parameter, **28,** 239
Weibull probability, **28,** 230–231
  plot, right-censored failure data, **28,** 235
Weighing densities of single crystals, **6A,** 291
Weighing density, methods, **6A,** 286–291
Weight
  atomic, **1,** 38
  equivalent, **1,** 41
  molecular, **14,** 10
Weighted mean, **1,** 10
Weighted measurement, **1,** 9–11
Weighting schemes, simulation, **28,** 453
Weir, **18,** 332, 334
Weir block, **18,** 334
Weiss field approximation, **23C,** 271
Weiss induction method, **6B,** 202–203
Weissenberg camera, **16B,** 60, 69
Weissenberg method, **3,** 313–317; **6A,** 208–210
  variation, **6A,** 210
Weissenberg rheogoniometer, **16C,** 35
Weiszacker semiempirical formula, **5A,** 335–336
Welch's overlapped segment averaging spectral estimators, **28,** 333–335
Weld materials, **23C,** 452
  residual stresses, **23C,** 463
Weld porosity, **14,** 431
Weld shrinkage, **14,** 430
Welding, **14,** 427–429; **19,** 348–350

design, **14,** 427, 429, 430, 432, 433, 435, 436
lip, **14,** 433
rod, flux-coated, **14,** 427
Weldment, **14,** 430
Well-thermalized neutrons, **8,** 242
Wenner array, **24B,** 296, 303–304
Wenner potentiometer, **1,** 512
Wet-gas meter, **18,** 323–324
Wetting transition, **23B,** 51
Wheatstone bridge, **1,** 246, 457, 462; **6B,** 40–41, 43; **16C,** 397
Whirling cup anemometer, **18,** 256–257
Whisker contact, **6B,** 35
Whiskers, **6A,** 335
Whistler instability, **9A,** 299–300
Whistlers, **9B,** 89–137
afternoon "bulge" and, **9B,** 110
antennas for, **9B,** 90–91
broadband installation for, **9B,** 91
ducting of, **9B,** 95–98
Eckersley law, **9B,**95
lifetimes of, **9B,** 95
satellite observation of, **9B,** 111–112
Snell's law, **9B,** 96
total internal reflections and, **9B,** 97
dynamic spectra of, **9B,** 91
echo trains of, **9B,** 92–93
causative atmosphere, **9B,** 92
electric field measurements of, **9B,** 108–110
electron density distribution along field lines, **9B,** 107–108
equatorial electron density of, **9B,** 98–103
diffusive equilibrium model of electron density, **9B,** 99
gyrofrequency model of electron density, **9B,** 99
R^{-4} model and, **9B,** 100
experimental methods for, **9B,** 90–91
flux-tube content of, **9B,** 98–103
fractional hop type, **9B,** 124–126
ground-based observations on, **9B,** 92–110
helium type, **9B,** 131–133
"impulse response" and, **9B,** 90
ion-cutoff type, **9B,** 134–135
knee type, **9B,** 103–107
lightning discharges and, **9B,** 89
lower hybrid resonance noise of, **9B,** 117–120
in nonducted mode, **9B,** 117
magnetospherically-reflected type, **9B,** 115–117
mode waves of, refractive index, **9B,** 96
multiple ion effects of, **9B,** 120–124
nose type, **9B,** 93–95
"one-hop" type, **9B,** 93
propagation delay of, **9B,** 89
proton type, **9B,** 120–124
magnetic field strength and, **9B,** 127–128
proton number density of, **9B,** 128–129
proton temeprature and, **9B,** 129–131
signal of, minimum detectable power of, **9B,** 91
subprotonospheric type, ion effects, **9B,** 112–115
"two-hop" type, **9B,** 93
Whistles, **19,** 329–330, 331–332, 335–336
White follower, **2,** 215
White noise, **1,** 220
weak periodic signals, **28,** 351–352
White potentiometer, **1,** 246
White-light compensation, **26,** 6
White-noise spectrum, 5
Whole number rule, **5B,** 1
Wick's limit, **5B,** 382
Wide slit, *see* Source
Wide-angle X-ray diffraction, *see* X-ray diffraction
Wide-strip laser, **29C,** 90
Width of energy distribution for field-ionized hydrogen, **22,** 381
Widths of spectral lines, **13A,** 23
Wiedemann-Franz ratio, **1,** 292
Wien bridge, **1,** 464
Wien displacement law, **4A,** 290; **18,** 690; **26,** 220
Wien effect, **19,** 229
Wien filter, **5B,** 231–232; *see also* Mass spectrometer
Wien radiation formula, **18,** 466
Wien's displacement law, **1,** 376
Wien's law, **8,** 29, 33
Wiener-Khintchine theorem, **2,** 797
Wiener-Khintchine theorem, **2B,** 497
Wiggle beats, **3,** 406, 408
Wigglers, **22,** 184
synchrotron radiation generation, **29C,** 28–30, 32–33
Wigglers, **31,** 2, 14
polarization, **31,** 20–21
Wiggles, **3B,** 516

phenomena of, **3,** 393–394
Wigner spin rule, **15A,** 104, 121
Wigner, **5B,** 365
Williams refractometer, **1,** 361
Williams theory of multiple scattering, **5A,** 390ff
Williams tube, **2,** 618–619
Wilson test plate, **1,** 502–503, 504
Wind tunnel, **1,** 145ff
blockage interference, **18,** 765
classification, **18,** 758–764
dimensional analysis applied to, **18,** 844–847
flow visualization, **18,** 769–771
heat transfer techniques, **18,** 670–672
lift interference, **18,** 765
low speed, **18,** 764–771
model testing principles, **18,** 843–849
open-jet testing principles, **18,** 764
research apparatus, **18,** 756–785
supersonic, **18,** 771–779
transonic, **18,** 771–779
turbulence, **18,** 764
wall corrections, **18,** 773, 847
Winding space, **8,** 93, 94
Window
consideration in combining lenses, **29A,** 202–203
definition of ion beam, **29A,** 191
microwave, **14,** 395–396
optical, **14,** 394–396
Window-attenuation factor, in beam-foil lifetime measurements, **17,** 573, 579
Windows, **12A,** 428, 440
atmospheric, **10,** 91–93
dielectric, **10,** 94–102
electronic spectra studies, **3,** 169–170
infrared, cleaning, **3,** 87
for ultraviolet, **13A,** 238
in vacuum systems, **4B,** 276–277
Windows, cleanliness, **31,** 307–308
Wings, rotational, **3,** 114
Wire grid polarizer, *see* Grating polarizer
Wire-grid analyzer, **12A,** 364
Wireline loging, **24B,** 192–200, 441–443
cable resistance, **24B,** 195
high-temperature insulation, **24B,** 195
Wirewound resistors, **2A,** 35–36
WKB approximation, **22,** 352, 379
WNR, **23A,** 132
Wobble stretching, **5B,** 32–33
Wobbler, **12A,** 432
focusing, in electron microscopy, **16B,** 218
Wollaston prism polarizer, **31,** 229
Wollaston prism, **4B,** 63; **12A,** 369–371, 381, 385, 388; **26,** 138
shearing interferometer, **18,** 387
Wolter optics, **31,** 173–175
Wood anomalies, **25,** 214; **26,** 157
Wood's discharge tube, **4A,** 181–182
Wood's metal, **22,** 403
Wood-Anderson torsion seismometer, **24B,** 54–56
Work, **1,** 93–97, 306
hardening, **1,** 107
Work function measurement, **22,** 15
capacitance method, **22,** 17
electron beam method, **22,** 14
emission method, **22,** 4
field-emission retarding-potential method, **22,** 15
retarding-potential method, **22,** 11
technique comparison, **22,** 19
thermionic method, **22,** 6
Work function, **2,** 47; **4A,** 1–3, 6, 14, 16–19, 21–24, 26–27, 31–34, 55–57, 60–61, 64; **12A,** 11; **22,** 2, 8, 14, 19, 21, 128, 137, 141, 145, 148, 155, 167, 206, 350, 356, 359, 375, 397
adsorbate, **22,** 9
adsorption, **22,** 3, 5, 9, 21
apparent
theory of, **27,** 20, 26
band structure effects, **22,** 11
crystallographic direction, **22,** 9
different planes, **22,** 21
dipole layer, **22,** 3
dipole moment discreteness, **22,** 10
geometric factor, **22,** 9
individual crystal planes, **22,** 9
ordering, **22,** 21
patch effects, **22,** 5
of semiconductors, **6B,** 378, 382–383
single crystal, **22,** 21
of single crystals, **4A,** 33
versus surface coverage, **22,** 439
tables of, **4A,** 16–18, 21–23, 31–34
temperature dependence of, **4A,** 1–2; **22,** 5
theory, **22,** 2
thermal desorption, **22,** 5
true, **22,** 10
variation, **22,** 9

with temperature, **22,** 14
various polymers, **16C,** 431
Work functions from photoelectric stopping potentials, **6B,** 135
Working hypothesis, **28,** 179
Working section, wind tunnel, **18,** 758
Workstation, Unix, using S, **28,** 484–488
World-Wide Standardized Seismograph Network, **24B,** 7, 21, 58
WOSA spectral estimators, **28,** 333–335
Wulf string electrometer, **1,** 439–440
WWSSN, *see* World-Wide Standardized Seismograph Network
WWV, **24B,** 49
WWVB, **24B,** 50

X

X unit, **1,** 27ff, 40
X-band waveguide, **8,** 144, 152
X-cut quartz, **19,** 52, 54
X-ogen
transition frequencies of, **12C,** 26
X-probe, hot-wire anemometer, **18,** 310
X-ray, **14,** 455
atomic scattering factor, **3,** 307–308
collimator baffle system, **5A,** 601–602
critical absorption edges, **5A,** 675–676
electron secondary energies, **5A,** 582
filters, **6A,** 205–206
fine structure measurements, **5A,** 334–335
interaction with matter, **3,** 302–308
level splitting, **5A,** 334
limit, *see* Ionization gauge
line width, **5A,** 611–612
quanta, energy of, **5A,** 294
from radioactive substances, **5A,** 296–298
sources, **3,** 308–309
spectra, characteristic, **5A,** 293–298
spectrometer
Cauchois photographic, **5A,** 599–601, 605
tubes, **6A,** 203–204
white and characteristic radiation, **6A,** 204–205
X-ray absorption, **3,** 302–306; **6A,** 36; **22,** 189
in shock-wave studies, **7B,** 178
X-ray astronomy, **12A,** 315
X-ray cameras, **16B,** 50–60
X-ray densities, **6A,** 285
X-ray detection systems, **16B,** 42–46
X-ray detector, **17,** 88, 213–218, 311–317, 356–371
X-ray determination of electronic band structure, **6B,** 281–292
X-ray diffraction analysis, **20,** 229–236
X-ray diffraction pattern recording, **3,** 309–324
X-ray diffraction patterns, applications, **6A,** 225–246
X-ray diffraction topography, **11,** 736–739
X-ray diffraction, **3,** 301–334; **6A,** 203, 246
background correction, **16B,** 64–65
collimators and beam stops, **6A,** 219–220
crystal structure determination, **16B,** 68–98
crystallinity determination, **16B,** 118–127
crystallite size determination, **16B,** 129–153
experimental details, **6A,** 219–223
fiber patterns, **16B,** 56–60
instrumentation, **16B,** 39–64
integrated intensity, **16B,** 65–68
intensity expressions, **3,** 327
powder (polycrystalline) patterns, **16B,** 51, 53, 54, 59
radial distribution, **6A,** 244
radiation choice of, **6A,** 220–221
small-angle, *see* Small-angle X-ray scattering
craze studies, **16C,** 245
of alloys and block copolymers, **16C,** 300–305
polymer fracture studies, **16C,** 230
specimen size, **6A,** 221
theory, **16B,** 25–39
trial structure technique, **16B,** 91–93
wide-angle, **16C,** 150–161
X-ray diffractometers, counter, **16B,** 60–64
X-ray emission, **22,** 189
spectroscopy, **6A,** 36
X-ray energy level diagram, **6B,** 283
X-ray excited Auger spectroscopy, **22,** 503
X-ray fluorescent spectroscopy, **6A,** 36
X-ray generation, by laser ablation, **30,** 629–631
X-ray intensity measurement, **6A,** 223–225
X-ray laser
applications, **31,** 107–109
efficiency, **31,** 105
vs. high-order harmonic generation, **31,** 115
X-ray line broadening
effect of crystal distortions, **16B,** 143–149

effect of crystallite shape, **16B,** 133–135
effect of crystallite size, **16B,** 129–135
instrumental broadening
types, **16B,** 135–143
typical corrections for, **16B,** 149–153
separation from crystal size effects, **16B,** 141–143
at small angles, **16B,** 171–172
X-ray method of impurity analysis, **6A,** 36–37
X-ray microscopy, **30,** 631
X-ray monochromatization, **16B,** 46–50
X-ray photoelectron spectroscopy (XPS), **11,** 68, 99; **16C,** 437; **22,** 189, 214, 503, 507; **30,** 318–322
X-ray production cross section, **17,** 285–288
for heavy incident ions, **17,** 221–228
inner-shell ionization and, **17,** 195–197
measurements, **17,** 228–278
and detector efficiency, **17,** 214–218
K-shell, **17,** 229–240
L-shell, **17,** 240–249
M-shell, **17,** 249–250
multiple-collisions effects, **17,** 225–227
thick target, **17,** 218–219
thin targets, **17,** 219–221
X-ray production, **17,** 140–144, 284–289
bombardment of solid targets by high-velocity ions, **17,** 193–278
coincidence measurements, **17,** 318–319
single-encounter experiment, **17,** 306–307
and target thickness, **17,** 255–257
two-component and three-component models, **17,** 257–272
X-ray radiation, **18,** 407–408, 705
X-ray regime, **22,** 150
X-ray region, **13A,** 14, 18
cross sections, **13A,** 200
detectors, **13A,** 149
focusing, **13A,** 158
muonic, **13A,** 189
plasmas, **13A,** 191
quasimolecular, **13A,** 182
solar, **13A,** 193
spectra, **13A,** 166
spectrometer, **13A,** 156, 160
X-ray scattering, **29C,** 40
amplitude and phase formula, **6A,** 189
by an array of atoms, **16B,** 30–34
by an atom, **16B,** 28–30
by complex crystal structures, **16B,** 35–38
by electrons, **16B,** 25–28
effect of thermal motion, **16B,** 38
energy and momentum transfer diagram, **16A,** 207
parasitic, removal of in diffraction work, **16B,** 159–160
small-angle, *see* Small-angle X-ray scattering
X-ray source (tube), **16B,** 39–42
for small-angle scattering, **16B,** 157–159
X-ray spectroscopy, **17,** 191–192; *see also* Ion-induced x-ray spectroscopy
advantages and disadvantages, **6B,** 283–285
alternate methods, **6B,** 290–292
corrections, **6B,** 288–290
for fluorescence yield calculations, **17,** 292–294
intensity measurement, **6B,** 285–286
monochromator, **6B,** 285
resolving power, **6B,** 286–288
two energy regions, **6B,** 285–286
X-ray spectroscopy, electron-beam ion trap
lifetime measurements, **29A,** 408, 410
transition energy measurements
above 2 keV, **29A,** 402, 404–405
below 2 keV, **29A,** 405–406
X-ray polarization, **29A,** 406, 408
X-ray spectrum, **17,** 375–379, 386–387, 396, 399–400
dependence of solid-target spectra on environment, **17,** 418–424
effects of outer-shell relaxation on, **17,** 416–418
neon, **17,** 413–414
spectral lines and transitions to L shell, **17,** 196
and vacancy production cross section, **17,** 380
X-ray streak camera, **29C,** 223–224
X-ray tube, anodes, tungsten, **14,** 368
X-ray window, aluminum, **14,** 364
X-ray-electron, special identification, **17,** 347–350
X-ray-ion coincidence measurements, **17,** 305–307
X-rays, **1,** 408–429
absorption, **1,** 155, 415–417
crystallography, **1,** 421–427
detected by ground-based observation of atmospheric fluorescence, **12A,** 315–343, *see also* Atmospheric fluorescence
detection, **1,** 411–413
diffraction, **1,** 105, 418, 421ff

interference, **1,** 421
microscopy, **1,** 428
scale, Siegbahn, **1,** 40
scattering, **1,** 394, 415–417
spectrochemical analysis, **1,** 417–419
spectroscopy, **1,** 422
X-rays in PIXE, **21,** 250
X-rays, ultrashort pulses, **29C,** 224–227
bremsstrahlung, **21,** 4
characteristic, **21,** 4
detectors, **21,** 50, 53
specific ionization, **21,** 48
x-t diagram, shock tube, **18,** 786
X-Unit, **4B,** 4, 13
XAES, *see* X-ray excited Auger spectroscopy
Xenon arc lamp, **7A,** 201
Xenon bromide laser, **15B,** 125
Xenon flash lamp, **18,** 693–694
Xenon flashlamps, Nd:glass lasers and, **15A,** 199
Xenon fluoride laser, **15B,** 86–88
Xenon gas, in coaxial flashlamps, **15A,** 321
Xenon ion lasers, **15A,** 100
Xenon laser emission, **10,** 49
Xenon ultraviolet radiation, generation of, **15B,** 221
Xenon vapor, tunable VUV radiation from, **15B,** 221
Xenon, **14,** 82, 419, 420
atmospheric concentration, **12B,** 170
Xerography, **16C,** 424
XPS, *see* X-ray photoelectron spectroscopy
XUV emission from laser plasma
applications, 84
XUV lasers, **31,** 101–109
XUV, *see* Xenon ultraviolet radiation

Y

Y-branches, **29C,** 382
Y-cut quartz, **19,** 54
YAG laser, **25,** 124
Yaw **1,** 147
aerodynamic moment, **18,** 768–769
card, **18,** 780
hot-wire probe correction, **18,** 306–308
meaning, **18,** 248
Pitot probe correction, **18,** 248–249
total temperature probe correction, **18,** 462
YbIG ferrimagnetic exchange resonance, **10,** 418, 419
Yield(s), **23A,** 104, 105, 128, 130, *see also* Specific activity
of E2 γ-rays, **5B,** 447
from F19 (π,α) γ reaction, **5B,** 443–444
fluorescence in PIXE, **21,** 252
flourescent, **1,** 398
LEIS, **21,** 238
of γ-rays per incident charged particle per unit solid angle, **5B,** 443
NRA, **21,** 245
nuclear reaction, **5B,** 362–363
photodisintegration, per roentgen, **5B,** 469–470
RBS, **21,** 226, 229
reactor, **5B,** 593
secondary ions, **21,** 267–268
sputtering, **21,** 257, 259, 266, 272
Yield point phenomenon, **6A,** 312
Yield strength, effect of pressure on, **16C,** 116
Yield stress (yield point)
definition, **16C,** 129
effect of pressure on, **16C,** 131
in crazing polymers, **16C,** 264
micro-, **16C,** 119
various polymers, **16C,** 131
YIG (yyttium iron garnet), **29C,** 360
YNU, method of paraxial ray tracing, **25,** 19
Young's double slit interferometer, for atoms, **29B,** 348–349
Young's double slit, **1,** 353, 354
Young's experiment, **26,** 1
measure spatial coherence, **18,** 710–711
Young's interference fringes, visibility, 39–53
Young's modulus, **1,** 100, 101, 106, 211, 214, 223; **6A,** 296–297, 299–300; **14,** 360; **28,** 502; *see also specific materials, various*
measurement by flexural oscillations of bar, **6A,** 299
measurement by use of longitudinal and torsional vibration of rods, **6A,** 299
from resonant frequency, **6A,** 300
Yrast level, **17,** 581
Yttrium x-ray laser, characteristics, **31,** 106

Z

Z-pinched discharges, **15A,** 91
z-system, **25,** 177
Z^2_m-test, **28,** 359–360
ZBLAN glass, **29C,** 375
Zeeman components, polarization and intensity, **4B,** 125–127
Zeeman effect, **1,** 385–386; **3B,** 726; **4B,** 108–110, 114–115, 119–120, 126, 133, 149–150, 153, 191–209; **13A,** 341

by atomic beam resonance, **4B,** 192–198
of excited atoms, **4B,** 203–205, 208
hydrogen cloud measurements of, **12C,** 76–77
linear effect, **4B,** 115, 152
magnetic sublevels, **4B,** 110, 114–118, 126
modulation, **13B,** 115
in molecular oxygen, **12B,** 162–163
quadratic effect, **4B,** 115, 211
quantum beats, **13B,** 249
of quenching spectrum, **4B,** 133
tune lasers, **13B,** 275
Zeeman Hamiltonian, **21,** 448
Zeeman levels, **21,** 448
Zeeman interaction, **23C,** 245
Zeeman level, **16A,** 443
Zeeman modulation, **4B,** 171–172, 200; **5B,** 90
Zeeman order, **23C,** 252
Zeeman shift, **29B,** 160
Zeeman slowing, **29B,** 133
Zeeman splitting, **10,** 386, 392, 418; **16A,** 243, 444, 449; **23A,** 332
in tunneling, **11,** 153–154
Zeeman study of chlorates, **3,** 524
Zener breakdown noise, **2B,** 470
Zener diode, **2A,** 60, 116, 217, 240, 243–244, 288, 361, 365–366; **2B,** 453–454
limiting with, **2A,** 373–375
Zener effect, **2A,** 366
Zeolite sorption pump, **14,** 227
Zeolite, **14,** 237, 244
Zeolites (molecular sieves), **4B,** 322
Zero power reactor, time dependence, **8,** 233–234
Zero-bias anomaly, **11,** 183
"giant," **11,** 176
in normal-metal tunneling, **11,** 170–177
in *p-n* tunnel junctions, **11,** 186
Zero-crossing detector, **11,** 339
Zero-crossover timing technique, **2B,** 112
Zero-field experiment, in quantum beam method, **17,** 593–595
Zero-field mobility, in liquid helium, **11,** 465–466
Zero-field quantum beats, **13B,** 252
Zero-length spring, **24B,** 6, 18–21, 136–137
Zero-order desorption kinetics, **22,** 425, 437, 438
Zero-order waveplates, **29C,** 353, 355
Zero-velocity point, for Mössbauer spectrometers, **11,** 344–345
Zerodur®, **29C,** 344, 345
Zeugmatography, *see* Imaging, NMR
Zimm analysis, **23C,** 299
Zimm plot, **16A,** 28, 449
Zimm-Rouse theory, **19,** 229
Zinc
attenuation in bar, **19,** 276–277
velocity in bar, **19,** 276–277
Zinc ion lasers, **15A,** 143–146
excitation mechanisms in, **15A,** 144–145
performance of, **15A,** 145–146
spectroscopy of, **15A,** 143–144
Zinc oxide (ZnO), **30,** 544–551
Zinc sulfide crystals, growth by sublimation, **6B,** 298
Zinc sulfide, reflectance spectra, **31,** 221
Zinc x-ray laser, characteristics, **31,** 106
Zinc-indium-oxide (ZnO_3), **30,** 553–554
Zinc-Phthalocyanine, **22,** 372
ZING-P, **23A,** 132
ZING-P′, **23A,** 132
Zirconium
boiling point, **14,** 351
hydrogen solubility, **14,** 354
melting point, **14,** 349, 351
physical properties, **14,** 364–365
vapor pressure, **14,** 349, 351
Zirconium arc, **3,** 77
Zirconium hydride, **14,** 365
$ZnGBeP_2$, as OPO material, **29C,** 257, 258
Zonal harmonic expansion techniques, **8,** 102
Zone apparatus, **7A,** 6
Zone leveling, **6A,** 59–61
Zone melting, **6A,** 25, 49–51, 69–70
Zone plate, **8,** 158
aberrations, **31,** 296–298
construction, **31,** 298–301
lens, *see* Fresnel zone plate lens
multilayer, **31,** 296
Zone plate geometry, **31,** 289–292
Zone plate, Fresnel, **26,** 59
Zone plates, **23A,** 273
Zoom eyepieces, **25,** 132
Zoom lenses, **25,** 95
ZrNiD glasses, **23B,** 285
$ZrSe_3$
scanning tunneling microscopy of, **27,** 403–404
Zwitterions, **19,** 202, 211, 225
Zytel V, **14,** 414

90065

9 780124 759817

ISBN 0-12-475981-5